T0210445

Classical Covariant Fields

This book discusses the classical foundations of field theory, using the language of variational methods and covariance. There is no other book which gives such a comprehensive overview of the subject, exploring the limits of what can be achieved with purely classical notions. These classical notions have a deep and important connection with the second quantized field theory, which is shown to follow on from the Schwinger Action Principle. The book takes a pragmatic view of field theory, focusing on issues which are usually omitted from quantum field theory texts. It uses a well documented set of conventions and catalogues results which are often hard to find in the literature. Care is taken to explain how results arise and how to interpret results physically, for graduate students starting out in the field. Many physical examples are provided, making the book an ideal supplementary text for courses on elementary field theory, group theory and dynamical systems. It will also be a valuable reference for researchers already working in these and related areas. This title, first published in 2005, has been reissued as an Open Access publication Cambridge Core.

MARK BURGESS obtained his PhD in theoretical physics from the University of Newcastle Upon Tyne in 1990. He held a Royal Society fellowship at the University of Oslo from 1991 to 1992, and then had a two-year postdoctoral fellowship from the Norwegian Research Council. Since 1994, he has been an associate professor at Oslo University College. Dr Burgess has been invited to lecture at universities and institutes throughout the world, and has published numerous articles, as well as five previous books.

CAMBRIDGE MONOGRAPHS ON
MATHEMATICAL PHYSICS

General editors: P. V. Landshoff, D. R. Nelson, S. Weinberg

† Issued as a paperback

Classical Covariant Fields

MARK BURGESS
Oslo University College
Norway

CAMBRIDGE
UNIVERSITY PRESS

Shaftesbury Road, Cambridge CB2 8EA, United Kingdom

One Liberty Plaza, 20th Floor, New York, NY 10006, USA

477 Williamstown Road, Port Melbourne, VIC 3207, Australia

314–321, 3rd Floor, Plot 3, Splendor Forum, Jasola District Centre, New Delhi – 110025, India

103 Penang Road, #05–06/07, Visioncrest Commercial, Singapore 238467

Cambridge University Press is part of Cambridge University Press & Assessment,
a department of the University of Cambridge.

We share the University's mission to contribute to society through the pursuit of
education, learning and research at the highest international levels of excellence.

www.cambridge.org
Information on this title: www.cambridge.org/9781009289900

DOI: 10.1017/9781009289887

First published 2005
Reissued as OA 2022

A catalogue record for this publication is available from the British Library.

ISBN 978-1-009-28990-0 Hardback
ISBN 978-1-009-28986-3 Paperback

Contents

Foreword

This book is a collection of notes and unpublished results which I have accumulated on the subject of classical field theory. In 1996, it occurred to me that it would be useful to collect these under a common umbrella of conventions, as a reference work for myself and perhaps other researchers and graduate students. I realize now that this project can never be finished to my satisfaction: the material here only diverges. I prefer to think of this not as a finished book, so much as some notes from a personal perspective.

In writing the book, I have not held history as an authority, nor based the approach on any particular authors; rather, I have tried to approach the subject rationally and systematically. I aimed for the kind of book which I would have appreciated myself as a graduate student: a book of general theory accompanied by specific examples, which separates logically independent ideas and uses a consistent notation; a book which does not skip details of derivation, and which answers practical questions. I like books with an attitude, which have a special angle on their material, and so I make no apologies for this book's idiosyncrasies.

Several physicists have influenced me over the years. I am especially grateful to David Toms, my graduate supervisor, for inspiring, impressing, even depressing but never repressing me, with his unstoppable 'Nike' philosophy: (shrug) 'just do it'. I am indebted to the late Peter Wood for kind encouragement, as a student, and for entrusting me with his copy of Schweber's now ex-masterpiece *Relativistic Quantum Field Theory*, one of my most prized possessions. My brief acquaintance with Julian Schwinger encouraged me to pay more attention to my instincts and less to conforming (though more to the conformal). I have appreciated the friendship of Gabor Kunstatter and Meg Carrington, my frequent collaborators, and have welcomed occasional encouraging communications from Roman Jackiw, one of the champions of classical and quantum field theory. I am, of course, indebted to my friends in Oslo. I blame Alan McLachlan for teaching me more than I wanted to know about group congruence classes.

Thanks finally to Tai Phan, of the Space Science Lab at Berkeley for providing some sources of information for the gallery data.

Like all software, this book will contain bugs; it is never really finished and trivial, even obvious errors creep in inexplicably. I hope that these do not distract from my perspective on one of the most beautiful ideas in modern physics: covariant field theory.

I called the original set of these notes: The X_μ Files: Covert Field Theory, as a joke to myself. The world of research has become a merciless battleground of competitive self-interest, a noise in which it is all but impossible to be heard. Without friendly encouragement, and a pinch of humour, the battle to publish would not be worth the effort.

Mark Burgess
Oslo University College

*"The Dutch astronomer De Sitter was able to show that
the velocity of propagation of light cannot depend on
the velocity of motion of the body emitting the light...
theoretical investigations of H.A. Lorentz...lead[s] conclusively
to a theory of electromagnetic phenomena, of which the
law of the constancy of the velocity of light in vacuo
is a necessary consequence."*

– Albert Einstein

"Energy of a type never before encountered."

– Spock, *Star Trek: The motion picture.*

Part 1
Fields

1
Introduction

In contemporary field theory, the word *classical* is reserved for an analytical framework in which the local equations of motion provide a complete description of the evolution of the fields. Classical field theory is a differential expression of change in functions of space and time, which summarizes the state of a physical system entirely in terms of smooth fields. The differential (holonomic) structure of field theory, derived from the action principle, implies that field theories are microscopically reversible by design: differential changes experience no significant obstacles in a system and may be trivially undone. Yet, when summed macroscopically, in the context of an environment, such individually reversible changes lead to the well known irreversible behaviours of thermodynamics: the reversal of paths through an environmental landscape would require the full history of the route taken. Classical field theory thus forms a basis for both the microscopic and the macroscopic.

When applied to quantum mechanics, the classical framework is sometimes called the *first quantization*. The first quantization may be considered the first stage of a more complete theory, which goes on to deal with the issues of many-particle symmetries and interacting fields. Quantum mechanics is classical field theory with additional assumptions about measurement. The term *quantum mechanics* is used as a name for the specific theory of the Schrödinger equation, which one learns about in undergraduate studies, but it is also sometimes used for any fundamental description of physics, which employs the measurement axioms of Schrödinger quantum mechanics, i.e. where change is expressed in terms of fields and groups. In that sense, this book is also about quantum mechanics, though it does not consider the problem of measurement, and all of its subtlety.

In the so-called *quantum field theory*, or *second quantization*, fields are promoted from *c*-number functions to operators, acting upon an additional set of states, called Fock space. Fock space supplants Slater determinant combinatorics in the classical theory, and adds a discrete aspect to smooth field theory. It quantizes the allowed amplitudes of the normal modes of the field and gives excitations the same denumerable property that ensembles of particles have; i.e. it adds *quanta* to the fields, or indistinguishable, countable excitations, with varying numbers. Some authors refer to these quanta simply as 'particles'; however, they are not particles in the classical sense of localizable, pointlike objects. Moreover, whereas particles are separate entities, quanta are excitations, spawned from a single entity: the quantum field. The second-quantized theory naturally incorporates the concept of a lowest possible energy state (the vacuum), which rescues the relativistic theory from negative energies and probabilities. Such an assumption must be added by hand in the classical theory. When one speaks about *quantum* field theory, one is therefore referring to this 'second quantization' in which the fields are dynamical operators, spawning indistinguishable quanta.

This book is not about quantum field theory, though one might occasionally imagine it is. It will mention the quantum theory of fields, only insofar as to hint at how it generalizes the classical theory of fields. It discusses statistical aspects of the classical field to the extent that classical Boltzmann statistical mechanics suffices to describe them, but does not delve into interactions or combinatorics. One should not be misled; books on quantum field theory generally begin with a dose of classical field theory, and many purely classical ideas have come to be confused with second-quantized ones. Only in the final chapter is the second-quantized framework outlined for comparison. This book is a summary of the core methodology, which underpins covariant field theory at the classical level. Rather than being a limitation, this avoidance of quantum field theory allows one to place a sharper focus on key issues of symmetry and causality which lie at the heart of all subsequent developments, and to dwell on the physical interpretation of formalism in a way which other treatments take for granted.

1.1 Fundamental and effective field theories

The main pursuit of theoretical physics, since quantum mechanics was first envisaged, has been to explore the maxim that the more microscopic a theory is, the more fundamental it is. In the 1960s and 1970s it became clear that this view was too simplistic. Physics is as much about *scale* as it is about constituent components. What is fundamental at one scale might be irrelevant to physics at another scale. For example, quark dynamics is not generally required to describe the motion of the planets. All one needs, in fact, is an effective theory of planets as point mass objects. their detailed structure is irrelevant to so many decimal places that it would be nonsense to attempt to include it in calculations. Planets are less elementary than quarks, but they are not less fundamental to the problem at hand.

The quantum theory of fields takes account of dynamical *correlations* between the field at different points in space and time. These correlations, called fluctuations or virtual processes, give rise to *quantum corrections* to the equations of motion for the fields. At first order, these can also be included in the classical theory. The corrections modify the form of the equations of motion and lead to *effective field equations* for the quantized system. At low energies, these look like classical field theories with renormalized coefficients. Indeed, this sometimes results in the confusion of statistical mechanics with the second quantization. Put another way, at a superficial level all field theories are approximately classical field theories, if one starts with the right coefficients. The reason for this is that all one needs to describe physical phenomena is a blend of two things: symmetry and causal time evolution. What troubles the second quantization is demonstrating the consistency of this point of view, given sometimes uncertain assumptions about space, time and the nature of fields.

This point has been made, for instance, by Wilson in the context of the renormalization group [139]; it was also made by Schwinger, in the early 1970s, who, disillusioned with the direction that field theory was taking, redefined his own interpretation of field theory called *source theory* [119], inspired by ideas from Shannon's mathematical theory of communication [123]. The thrust of source theory is the abstraction of irrelevant detail from calculations, and a reinforcement of the importance of causality and boundary conditions.

1.2 The continuum hypothesis

Even in classical field theory, there is a difference between particle and field descriptions of matter. This has nothing *a priori* to do with wave–particle duality in quantum mechanics. Rather, it is to do with scale.

In classical mechanics, individual pointlike particle trajectories are characterized in terms of 'canonical variables' $x(t)$ and $p(t)$, the position and momentum at time t. Underpinning this description is the assumption that matter can be described by particles whose important properties are localized at a special place at a special time. It is not even necessarily assumed that matter is made of particles, since the particle position might represent the centre of mass of an entire planet, for instance. The key point is that, in this case, the centre of mass is a localizable quantity, relevant to the dynamics.

In complex systems composed of many particles, it is impractical to take into account the behaviour of every single particle separately. Instead, one invokes the continuum hypothesis, which supposes that matter can be treated as a continuous substance with bulk properties at large enough scales. A system with a practically infinite number of point variables is thus reduced to the study of continuous functions or *effective fields*. Classically, continuum theory is a high-level or *long-wavelength* approximation to the particle theory, which blurs out the individual particles. Such a theory is called an *effective theory*.

In quantum mechanics, a continuous wavefunction determines the probability of measuring a discrete particle event. However, free elementary quantum particles cannot be localized to precise trajectories because of the uncertainty principle. This wavefunction-field is different from the continuum hypothesis of classical matter: it is a function which represents the state of the particle's quantum numbers, and the probability of its position. It is not just a smeared out approximation to a more detailed theory. The continuous, field nature is observed as the interference of matter waves in electron diffraction experiments, and single-particle events are measured by detectors. If the wavefunction is sharply localized in one place, the probability of measuring an event is very large, and one can argue that the particle has been identified as a bump in the field.

To summarize, a sufficient number of localizable particles can be viewed as an effective field, and conversely a particle can be viewed as a localized disturbance in an elementary field.

To envisage an elementary field as representing particles (not to be confused with quanta), one ends up with a picture of the particles as localized disturbances in the field. This picture is only completely tenable in the non-relativistic limit of the classical theory, however. At relativistic energies, the existence of particles, and their numbers, are fuzzy concepts which need to be given meaning by the quantum theory of fields.

1.3 Forces

In classical mechanics, forces act on particles to change their momentum. The mechanical force is defined by

$$\mathbf{F} = \frac{d\mathbf{p}}{dt}, \tag{1.1}$$

where $\mathbf{p}$ is the momentum. In field theory, the notion of a dynamical influence is more subtle and has much in common with the interference of waves. The idea of a force is of something which acts at a point of contact and creates an impulse. This is supplanted by the notion of fields, which act at a distance and interfere with one another, and currents, which can modify the field in more subtle ways. Effective mechanical force is associated with a quantity called the *energy–momentum* tensor $\theta_{\mu\nu}$ or $T_{\mu\nu}$.

1.4 Structural elements of a dynamical system

The shift of focus, in modern physics, from particle theories to field theories means that many intuitive ideas need to be re-formulated. The aim of this book is to give a substantive meaning to the physical attributes of fields, at the classical level, so that the fully quantized theory makes physical sense. This requires example.

A detailed description of dynamical systems touches on a wide variety of themes, drawing on ideas from both historical and mathematical sources. The simplicity of field theory, as a description of nature, is easily overwhelmed by these details. It is thus fitting to introduce the key players, and mention their significance, before the clear lines of physics become obscured by the topography of a mathematical landscape. There are two kinds of dynamical system, which may be called continuous and discrete, or holonomic and non-holonomic. In this book, only systems which are parametrized by continuous, spacetime parameters are dealt with. There are three major ingredients required in the formulation of such a dynamical system.

- **Assumptions**
  A model of nature embodies a body of *assumptions* and *approximations*. The assumptions define the ultimate extent to which the theory may be considered valid. The best that physics can do is to find an idealized description of isolated phenomena under special conditions. These conditions need to be borne clearly in mind to prevent the mathematical machinery from straying from the intended path.

- **Dynamical freedom**
  The capacity for a system to change is expressed by introducing *dynamical variables*. In this case, the dynamical variables are normally fields. The number of ways in which a physical system can change is called its number of *degrees of freedom*. Such freedom describes nothing unless one sculpts out a limited form from the amorphous realm of possibility. The structure of a dynamical system is a balance between freedom and constraint.

  The variables in a dynamical system are fields, potentials and sources. There is no substantive distinction between field, potential and source, these are all simply functions of space and time; however, the words *potential* or *source* are often reserved for functions which are either static or rigidly defined by boundary conditions, whereas *field* is reserved for functions which change dynamically according to an equation of motion.

- **Constraints**
  Constraints are restrictions which determine what makes one system with *n* variables different from another system with *n* variables. The constraints of a system are both dynamical and kinematical.

  - **Equations of motion**
    These are usually the most important constraints on a system. They tell us that the dynamical variables cannot take arbitrary values; they are dynamical constraints which express limitations on the way in which dynamical variables can change.

- **Sources: external influences**

  Physical models almost always describe systems which are isolated
  from external influences. Outside influences are modelled by introducing *sources and sinks*. These are perturbations to a closed system
  of dynamical variables whose value is specified by some external
  boundary conditions. Sources are sometimes called generalized
  forces. Normally, one assumes that a source is a kind of 'immovable
  object' or infinite bath of energy whose value cannot be changed
  by the system under consideration. Sources are used to examine
  what happens under controlled boundary conditions. Once sources
  are introduced, conservation laws may be disturbed, since a source
  effectively opens a system to an external agent.

- **Interactions**

  Interactions are couplings which relate changes in one dynamical
  variable to changes in another. This usually occurs through a
  coupling of the equations of motion. Interaction means simply that
  one dynamical variable changes another. Interactions can also be
  thought of as internal sources, internal influences.

- **Symmetries and conservation laws**

  If a physical system possesses a symmetry, it indicates that even
  though one might try to affect it in a specific way, nothing significant
  will happen. Symmetries exert passive restrictions on the behaviour
  of a system, i.e. kinematical constraints. The conservation of book-
  keeping parameters, such as energy and momentum, is related to
  symmetries, so geometry and conservation are, at some level, related
  topics.

The Lagrangian of a dynamical theory must contain time derivatives if it is to be
considered a dynamical theory. Clearly, if the rate of change of the dynamical
variables with time is zero, nothing ever happens in the system, and the most
one can do is to discuss steady state properties.

2

The electromagnetic field

Classical electrodynamics serves both as a point of reference and as the point of departure for the development of covariant field theories of matter and radiation. It was the observation that Maxwell's equations predict a universal speed of light *in vacuo* which led to the special theory of relativity, and this, in turn, led to the importance of *perspective* in identifying generally applicable physical laws. It was realized that the symmetries of special relativity meant that electromagnetism could be reformulated in a compact form, using a vector notation for spacetime unified into a single parameter space. The story of covariant fields therefore begins with Maxwell's four equations for the electromagnetic field in $3 + 1$ dimensions.

2.1 Maxwell's equations

In their familiar form, Maxwell's equations are written (in SI units)

$$\vec{\nabla} \cdot \mathbf{E} = \frac{\rho_e}{\epsilon_0} \tag{2.1a}$$

$$\vec{\nabla} \times \mathbf{E} = -\frac{\partial \mathbf{B}}{\partial t} \tag{2.1b}$$

$$\vec{\nabla} \cdot \mathbf{B} = 0 \tag{2.1c}$$

$$c^2(\vec{\nabla} \times \mathbf{B}) = \frac{\mathbf{J}}{\epsilon_0} + \frac{\partial \mathbf{E}}{\partial t}. \tag{2.1d}$$

ρ_e is the charge density, $\mathbf{J}$ is the electric current density and $c^2 = (\epsilon_0 \mu_0)^{-1}$ is the speed of light in a vacuum squared. These are valid, as they stand, in inertial frames in flat $(3+1)$ dimensional spacetimes. The study of covariant field theory begins by assuming that these equations are true, in the sense that any physical laws are 'true' – i.e. that they provide a suitably idealized description of the physics of electromagnetism. We shall not attempt to follow the path which

9

led to their discovery, nor explore their limitations. Rather, we are interested in summarizing their form and substance, and in identifying symmetries which allow them to be expressed in an optimally simple form. In this way, we hope to learn something deeper about their meaning, and facilitate their application.

2.1.1 Potentials

This chapter may be viewed as a demonstration of how applied covariance leads to a maximally simple formulation of Maxwell's equations. A more complete understanding of electromagnetic covariance is only possible after dealing with the intricacies of chapter 9, which discusses the symmetry of spacetime. Here, the aim is to build an algorithmic understanding, in order to gain a familiarity with key concepts for later clarification.

In texts on electromagnetism, Maxwell's equations are solved for a number of problems by introducing the idea of the vector and scalar potentials. The potentials play an important role in modern electrodynamics, and are a convenient starting point for introducing covariance.

The electromagnetic potentials are introduced by making use of two theorems, which allow Maxwell's equations to be re-written in a simplified form. In a covariant formulation, one starts with these and adds the idea of a unified *spacetime*. Spacetime is the description of space and time which treats the apparently different parameters x and t in a symmetrical way. It does not claim that they are equivalent, but only that they may be treated together, since both describe different aspects of the *extent* of a system. The procedure allows us to discover a simplicity in electromagnetism which is not obvious in eqns. (2.1).

The first theorem states that the vanishing divergence of a vector implies that it may be written as the curl of some other vector quantity **A**:

$$\vec{\nabla} \cdot \mathbf{v} = 0 \quad \Rightarrow \quad \mathbf{v} = \vec{\nabla} \times \mathbf{A}. \tag{2.2}$$

The second theorem asserts that the vanishing of the curl of a vector implies that it may be written as the gradient of some scalar ϕ:

$$\vec{\nabla} \times \mathbf{v} = 0 \quad \Rightarrow \quad \mathbf{v} = \vec{\nabla}\phi. \tag{2.3}$$

The deeper reason for both these theorems, which will manifest itself later, is that the curl has an *anti-symmetric* property. The theorems, as stated, are true in a homogeneous, isotropic, flat space, i.e. in a system which does not have irregularities, but they can be generalized to any kind of space. From these, one defines two *potentials*: a vector potential A_i and a scalar ϕ, which are auxiliary functions (fields) of space and time.

The physical electromagnetic field is the derivative of the potentials. From eqn. (2.1c), one defines

$$\mathbf{B} = \vec{\nabla} \times \mathbf{A}. \tag{2.4}$$

This form completely solves that equation. One equation has now been automatically and completely solved by re-parametrizing the problem in terms of a new variable. Eqn. (2.1c) tells us now that

$$\vec{\nabla} \times \mathbf{E} = -\frac{\partial}{\partial t}(\vec{\nabla} \times \mathbf{A})$$

$$\vec{\nabla} \times \left(\mathbf{E} + \frac{\partial \mathbf{A}}{\partial t}\right) = 0. \tag{2.5}$$

Consequently, according to the second theorem, one can write

$$\mathbf{E} + \frac{\partial \mathbf{A}}{\partial t} = -\vec{\nabla}\phi, \tag{2.6}$$

giving

$$\mathbf{E} = -\vec{\nabla}\phi - \frac{\partial \mathbf{A}}{\partial t}. \tag{2.7}$$

The minus sign on the right hand side of eqn. (2.6) is the convention which is used to make attractive forces positive and repulsive forces negative.

Introducing potentials in this way is not a necessity: many problems in electromagnetism can be treated by solving eqns. (2.1) directly, but the introduction often leads to significant simplifications when it is easier to solve for the potentials than it is to solve for the fields.

The potentials themselves are a mixed blessing: on the one hand, the re-parametrization leads to a number of helpful insights about Maxwell's equations. In particular, it reveals *symmetries*, such as the gauge symmetry, which we shall explore in detail later. It also allows us to write the matter–radiation interaction in a *local* form which would otherwise be impossible. The price one pays for these benefits is the extra conceptual layers associated with the potential and its gauge invariance. This confuses several issues and forces us to deal with constraints, or conditions, which uniquely define the potentials.

2.1.2 Gauge invariance

Gauge invariance is a symmetry which expresses the freedom to re-define the potentials arbitrarily without changing their physical significance. In view of the theorems above, the fields $\mathbf{E}$ and $\mathbf{B}$ are invariant under the re-definitions

$$\mathbf{A} \to \mathbf{A}' = \mathbf{A} + \vec{\nabla}s$$

$$\phi \to \phi' = \phi - \frac{\partial s}{\partial t}. \tag{2.8}$$

These re-definitions are called *gauge transformations*, and $s(x)$ is an arbitrary scalar function. The transformation means that, when the potentials are used as variables to solve Maxwell's equations, the parametrization of physics is not unique. Another way of saying this is that there is a freedom to choose between one of many different values of the potentials, each of which leads to the same values for the physical fields **E** and **B**. One may therefore choose whichever potential makes the solution easiest. This is a curious development. Why make a definite problem arbitrary? Indeed, this freedom can cause problems if one is not cautious. However, the arbitrariness is unavoidable: it is deeply connected with the symmetries of spacetime (the Lorentz group). Occasionally gauge invariance leads to helpful, if abstract, insights into the structure of the field theory. At other times, it is desirable to eliminate the fictitious freedom it confers by introducing an auxiliary condition which pins down a single ϕ, **A** pair for each value of **E**, **B**. As long as one uses a potential as a tool to solve Maxwell's equations, it is necessary to deal with gauge invariance and the multiplicity of equivalent solutions which it implies.

2.1.3 4-vectors and $(n+1)$-vectors

Putting the conceptual baggage of gauge invariance aside for a moment, one proceeds to make Maxwell's equations covariant by combining space and time in a unified vector formulation. This is easily done by looking at the equations of motion for the potentials. The equations of motion for the vector potentials are found as follows: first, substituting for the electric field in eqn. (2.1a) using eqn. (2.7), one has

$$-\nabla^2\phi - \frac{\partial}{\partial t}(\nabla \cdot \mathbf{A}) = \frac{\rho_e}{\epsilon_0}. \tag{2.9}$$

Similarly, using eqn. (2.4) in (2.1d), one obtains

$$c^2\vec{\nabla} \times (\vec{\nabla} \times \mathbf{A}) = \frac{\mathbf{J}}{\epsilon_0} + \frac{\partial}{\partial t}\left(-\vec{\nabla}\phi - \frac{\partial \mathbf{A}}{\partial t}\right). \tag{2.10}$$

Using the vector identity

$$\vec{\nabla} \times (\vec{\nabla} \times \mathbf{A}) = \vec{\nabla}(\vec{\nabla} \cdot \mathbf{A}) - \nabla^2\mathbf{A} \tag{2.11}$$

to simplify this, one obtains

$$c^2\left(\frac{1}{c^2}\frac{\partial^2}{\partial t^2} - \nabla^2\right)\mathbf{A} = \frac{\mathbf{j}}{\epsilon_0} - \vec{\nabla}\left(\frac{\partial\phi}{\partial t} + c^2(\vec{\nabla} \cdot \mathbf{A})\right). \tag{2.12}$$

It is already apparent from eqns. (2.8) that the potentials ϕ, **A** are not unique. This fact can now be used to tidy up eqn. (2.12), by making a choice for ϕ and **A**:

$$\vec{\nabla} \cdot \mathbf{A} + \frac{1}{c^2}\frac{\partial\phi}{\partial t} = 0. \tag{2.13}$$

The right hand side of eqn. (2.13) is chosen to be zero, but, of course, any constant would do. Making this arbitrary (but not random) choice, is called *choosing a gauge*. It partially fixes the freedom to choose the scalar field *s* in eqns. (2.8). Specifically, eqn. (2.13) is called the Lorentz gauge. This common choice is primarily used to tidy up the equations of motion, but, as noted above, at some point one has to make a choice anyway so that a single pair of vector potentials (scalar, vector) corresponds to only one pair of physical fields (**E**, **B**).

The freedom to choose the potentials is not entirely fixed by the adoption of the Lorentz condition, however, as we may see by substituting eqn. (2.8) into eqn. (2.13). Eqn. (2.13) is not completely satisfied; instead, one obtains a new condition

$$\left(\nabla^2 - \frac{\partial^2}{\partial t^2}\right) s = 0. \tag{2.14}$$

A second condition is required in general to eliminate all of the freedom in the vector potentials.

General covariance is now within reach. The symmetry with which space and time, and also ϕ and **A**, enter into these equations leads us to define spacetime vectors and derivatives:

$$\partial_\mu = \left(\frac{1}{c}\partial_t, \vec{\nabla}\right) \tag{2.15}$$

$$x^\mu = \begin{pmatrix} ct \\ \mathbf{x} \end{pmatrix}, \tag{2.16}$$

with Greek indices $\mu, \nu = 0, \ldots, n$ and $x^0 \equiv ct$. Repeated indices are summed according to the usual Einstein summation convention, and we define[1]

$$\Box = \partial^\mu \partial_\mu = -\frac{1}{c^2}\partial_t^2 + \nabla^2. \tag{2.17}$$

In *n* space dimensions and one time dimension ($n = 3$ normally), the $(n + 1)$ dimensional vector potential is defined by

$$A^\mu = \begin{pmatrix} \phi/c \\ \mathbf{A} \end{pmatrix}. \tag{2.18}$$

Using these $(n + 1)$ dimensional quantities, it is now possible to re-write eqn. (2.12) in an extremely beautiful and fully covariant form. First, one re-writes eqn. (2.10) as

$$-\Box \mathbf{A} = \frac{\mathbf{J}}{c^2\epsilon_0} - \vec{\nabla}\partial_\mu A^\mu. \tag{2.19}$$

---

[1] In some old texts, authors wrote $\Box^2$ for the same operator, since it is really a four-sided (four-dimensional) version of ∇^2.

Next, one substitutes the gauge condition eqn. (2.13) into eqn. (2.9), giving

$$-\Box\,\phi = \frac{\rho_e}{\epsilon_0}. \tag{2.20}$$

Finally, the $(n + 1)$ dimensional current is defined by

$$J^\mu = \begin{pmatrix} c\rho_e \\ \mathbf{J} \end{pmatrix}, \tag{2.21}$$

and we end up with the $(n + 1)$ dimensional field equation

$$-\Box\,A^\mu = \mu_0 J^\mu, \tag{2.22}$$

where $c^2 = (\mu_0\epsilon_0)^{-1}$ has been used. The potential is still subject to the constraint in eqn. (2.13), which now appears as

$$\partial_\mu A^\mu = 0. \tag{2.23}$$

2.1.4 The field strength

The new attention given to the potential A_μ should not distract from the main aim of electromagnetism: namely to solve Maxwell's equations for the electric and magnetic fields. These two physical components also have a covariant formulation; they are now elegantly unified as the components of a rank 2 *tensor* which is denoted $F_{\mu\nu}$ and is defined by

$$F_{\mu\nu} = \partial_\mu A_\nu - \partial_\nu A_\mu; \tag{2.24}$$

the tensor is anti-symmetric

$$F_{\mu\nu} = -F_{\nu\mu}. \tag{2.25}$$

This anti-symmetry, which was alluded to earlier, is the reason for the gauge invariance. The form of eqn. (2.24) is like a $(3 + 1)$ dimensional curl, expressed in index notation. The explicit components of this field tensor are the components of the electric and magnetic field components, in a Cartesian basis $\mathbf{E} = (E_1, E_2, E_3)$, etc.:

$$F_{\mu\nu} = \begin{pmatrix} 0 & -E_1/c & -E_2/c & -E_3/c \\ E_1/c & 0 & B_3 & -B_2 \\ E_2/c & -B_3 & 0 & B_1 \\ E_3/c & B_2 & -B_1 & 0 \end{pmatrix}. \tag{2.26}$$

In chapter 9, it will be possible to provide a complete understanding of how the symmetries of spacetime provide an explanation for why the electric and magnetic components of this field appear to be separate entities, in a fixed reference frame.

With the help of the potentials, *three* of Maxwell's equations (eqns. (2.1a,c,d)) are now expressed in covariant form. Eqn. (2.1c) is solved implicitly by the vector potential. The final equation (and also eqn. (2.1c), had one not used the vector potential) is an algebraic *identity*, called the Jacobi or Bianchi identity. Moreover, the fact that it is an identity is only clear when we write the equations in covariant form. The final equation can be written

$$\epsilon^{\mu\nu\lambda\rho}\partial_\mu F_{\nu\lambda} = 0, \tag{2.27}$$

where $\epsilon^{\mu\nu\lambda\rho}$ is the completely anti-symmetric tensor in four dimensions, defined by its components in a Cartesian basis:

$$\epsilon^{\mu\nu\lambda\rho} = \begin{cases} +1 & \mu\nu\lambda\rho = 0123 \text{ and even permutations} \\ -1 & \mu\nu\lambda\rho = 0132 \text{ and other odd permutations} \\ 0 & \text{otherwise.} \end{cases} \tag{2.28}$$

This equation is not a condition on $F_{\mu\nu}$, in spite of appearances. The anti-symmetry of both $\epsilon^{\mu\nu\lambda\rho}$ and $F_{\mu\nu}$ implies that the expansion of eqn. (2.27), in terms of components, includes many terms of the form $(\partial_\mu\partial_\nu - \partial_\nu\partial_\mu)A_\lambda$, the sum of which vanishes, provided A_λ contains no singularities. Since the vector potential is a continuous function in all physical systems,[2] the truth of the identity is not in question here.

The proof that this identity results in the two remaining Maxwell's equations applies only in $3+1$ dimensions. In other numbers of dimensions the equations must be modified. We shall not give it here, since it is easiest to derive using the index notation and we shall later re-derive our entire formalism consistently in that framework.

2.1.5 Covariant field equations using $F_{\mu\nu}$

The vector potential has been used thus far, because it was easier to identify the structure of the $(3+1)$ dimensional vectors than to guess the form of $F^{\mu\nu}$, but one can now go back and re-express the equations of motion in terms of the so-called physical fields, or field strength $F_{\mu\nu}$. The arbitrary choice of gauge in eqn. (2.22) is then eliminated.

Returning to eqn. (2.9) and adding and subtracting $\partial_0^2\phi$, one obtains

$$-\Box\phi - \partial_0(\partial_\nu A^\nu) = \frac{\rho_e}{\epsilon_0}. \tag{2.29}$$

---

[2] The field strength can never change by more than a step function, because of Gauss' law: the field is caused by charges, and a point charge (delta function) is the most singular charge that exists physically. This ensures the continuity of A_μ.

Adding this to eqn. (2.19) (without choosing a value for $\partial_\nu A^\nu$), one has

$$-\Box\, A^\mu = \frac{J^\mu}{c^2\epsilon_0} - \partial^\mu(\partial_\nu A^\nu). \tag{2.30}$$

Taking the last term on the right hand side over to the left and using eqn. (2.17) yields

$$\partial_\nu(\partial^\mu A^\nu - \partial^\nu A^\mu) = \frac{J^\mu}{c^2\epsilon_0}. \tag{2.31}$$

The parenthesis on the left hand side is now readily identified as

$$\partial_\nu F^{\mu\nu} = \mu_0 J^\mu. \tag{2.32}$$

This is the covariant form of the field equations for the physical fields. It incorporates two of the four Maxwell equations as before (eqn. (2.1c) is implicit in the structure we have set up). The final eqn. (2.27) is already expressed in terms of the physical field strength, so no more attention is required.

2.1.6 Two invariants

There are two invariant, scalar quantities (no free indices) which can be written down using the physical fields in $(3 + 1)$ dimensions. They are

$$\mathcal{F} = F^{\mu\nu} F_{\mu\nu} \tag{2.33}$$
$$\mathcal{G} = \epsilon^{\mu\nu\lambda\rho} F_{\mu\nu} F_{\lambda\rho}. \tag{2.34}$$

The first of these evaluates to

$$\mathcal{F} = 2\left(\mathbf{B}^2 - \frac{1}{c^2}\mathbf{E}^2\right). \tag{2.35}$$

In chapter 4 this quantity is used to construct the *action* of the system: a generating function the dynamical behaviour. The latter gives

$$\mathcal{G} = \mathbf{E}\cdot\mathbf{B}. \tag{2.36}$$

In four dimensions, this last quantity vanishes for a self-consistent field: the electric and magnetic components of a field (resulting from the same source) are always perpendicular. In other numbers of dimensions the analogue of this invariant does not necessarily vanish.   ·

The quantity $\mathcal{F}$ has a special significance. It turns out to be a Lagrangian, or generating functional, for the electromagnetic field. It is also related to the energy density of the field by a simple transformation.

2.1.7 Gauge invariance and physical momentum

As shown, Maxwell's equations and the physical field $F_{\mu\nu}$ are invariant under gauge transformations of the form

$$A_\mu \to A_\mu + (\partial_\mu s). \tag{2.37}$$

It turns out that, when considering the interaction of the electromagnetic field with matter, the dynamical variables for matter have to change under this gauge transformation in order to uphold the invariance of the field equations.

First, consider classical particles interacting with an electromagnetic field. The force experienced by classical particles with charge q and velocity $\mathbf{v}$ is the Lorentz force

$$F_{\mathrm{EM}} = q(\mathbf{E} + \mathbf{v} \times \mathbf{B}). \tag{2.38}$$

The total force for an electron in an external potential V and an electromagnetic field is therefore

$$\frac{\mathrm{d}p_i}{\mathrm{d}t} = -e(E_i + \epsilon_{ijk}v_j B_k) - \partial_i V. \tag{2.39}$$

Expressing E and B in terms of the vector potential, we have

$$\partial_t(p_i - eA_i) = -eF_{ij}\dot{x}_j - \partial_i(V + eA_t). \tag{2.40}$$

This indicates that, apart from a gauge-invariant Biot–Savart contribution in the first term on the right hand side of this equation, the electromagnetic interaction is achieved by replacing the momentum p_i and the energy E by

$$p_\mu \to (p_\mu - eA_\mu). \tag{2.41}$$

The Biot–Savart term can also be accounted for in this way if we go over to a relativistic, Lorentz-covariant form of the equations:

$$\frac{\mathrm{d}}{\mathrm{d}\tau}(p_\mu - eA_\mu) + F_{\mu\nu}I_l^\nu = 0, \tag{2.42}$$

where $I_l^\mu = -e\mathrm{d}x^\mu/\mathrm{d}\tau \sim I\mathrm{d}\mathbf{l}$ is the current in a length of wire $\mathrm{d}x$ (with dimensions current $\times$ length) and τ is the proper time. In terms of the more familiar current density, we have

$$\frac{\mathrm{d}}{\mathrm{d}\tau}(p_\mu - eA_\mu) + \int \mathrm{d}\sigma \, F_{\mu\nu}J^\nu = 0. \tag{2.43}$$

We can now investigate what happens under a gauge transformation. Clearly, these equations of motion can only be invariant if p_μ also transforms so as to cancel the term, $\partial_\mu s$, in eqn. (2.37). We must have in addition

$$p_\mu \to p_\mu + e\partial_\mu s. \tag{2.44}$$

Without a deeper appreciation of *symmetry*, this transformation is hard to understand. Arising here in a classical context, where symmetry is not emphasized,
it seems unfamiliar. What is remarkable, however, is that the group theoretical
notions of quantum theory of matter makes the transformation very clear. The
reason is that the state of a quantum mechanical system is formulated very
conveniently as a vector in a group theoretical vector space. Classically, positions and momenta are not given a state-space representation. In the quantum
formulation, gauge invariance is a simple consequence of the invariance of
the equations of motion under changes of the arbitrary complex phase of the
quantum state or wavefunction.

In covariant vector language, the field equations are invariant under a re-
definition of the vector potential by

$$A_\mu \to A_\mu + (\partial_\mu s), \tag{2.45}$$

where $s(x)$ is any scalar field. This symmetry is not only a mathematical
curiosity; it also has a physical significance, which has to do with conservation.

2.1.8 Wave solutions to Maxwell's equations

The equation for harmonic waves $W(x)$, travelling with speed v, is given by

$$\left(\nabla^2 - \frac{1}{v^2}\frac{\partial^2}{\partial t^2}\right) W(x) = 0. \tag{2.46}$$

If the speed of the waves is $v = c$, this may be written in the compact form

$$-\Box\, W(x) = 0. \tag{2.47}$$

It should already be apparent from eqn. (2.22) that Maxwell's equations have
wavelike solutions which travel at the speed of light. Writing eqn. (2.22) in
terms of the field strength tensor, we have

$$-\Box\, F_{\mu\nu} = \mu_0(\partial_\mu J_\nu - \partial_\nu J_\mu). \tag{2.48}$$

In the absence of electric charge $J_\mu = 0$, the solutions are free harmonic waves.
When $J_\mu \neq 0$, Maxwell's equations may be thought of as the equations of
forced oscillations, but this does not necessarily imply that all the solutions
of Maxwell's equations are wavelike. The Fourier theorem implies that any
function may be represented by a suitable linear super-position of waves. This is
understood by noting that the source in eqn. (2.48) is the spacetime 'curl' of the
current, which involves an extra derivative. Eqn. (2.32) is a more useful starting
point for solving the equations for many sources. The free wave solutions for
the field are linear combinations of plane waves with constant coefficients:

$$A_\mu(k) = C_k \exp(ik_\mu x^\mu). \tag{2.49}$$

By substituting this form into the equation

$$-\square A_\mu = 0,$$ (2.50)

one obtains a so-called dispersion relation for the field:

$$k_\mu k^\mu = k^2 = \mathbf{k}^2 - \omega^2/c^2 = 0.$$ (2.51)

This equation may be thought of as a constraint on the allowed values of k. The total field may be written compactly in the form

$$A_\mu(x) = \int \frac{\mathrm{d}^{n+1}k}{(2\pi)^{n+1}} e^{ik_\mu x^\mu} A_\mu(k)\, \delta(k^2),$$ (2.52)

where $A(k)_\mu$ represents the amplitude of the wave with wavenumber k_i, and the vector index specifies the polarization of the wave modes. From the gauge condition in eqn. (2.23), we have

$$k_\mu A(k)^\mu = 0.$$ (2.53)

The delta-function constraint in eqn. (2.52) ensures that the combination of waves satisfies the dispersion relation in eqn. (2.51). If we use the property of the delta function expressed in Appendix A, eqn. (A.15), then eqn. (2.52) may be written

$$A_\mu(x) = \hat{\epsilon}_\mu \int \frac{\mathrm{d}^{n+1}k}{(2\pi)^{n+1}} e^{i(k_i x^i - \omega t)} A(k) \frac{1}{c\,k_i} \left(\frac{\partial \omega}{\partial k^i} \right)$$
$$\times \left(\delta(k_0 - \sqrt{\mathbf{k}^2}) + \delta(k_0 + \sqrt{\mathbf{k}^2}) \right).$$ (2.54)

The delta functions ensure that the complex exponentials are waves travelling at the so-called phase velocity

$$v^i_{\text{ph}} = \pm \frac{\omega}{k_i}$$ (2.55)

where ω and k_i satisfy the dispersion relation. The amplitude of the wave clearly changes at the rate

$$v^i_{\text{gr}} = \frac{\partial \omega}{\partial k_i},$$ (2.56)

known as the group velocity. By choosing the coefficient $C(k)$ for each frequency and wavelength, the super-position principle may be used to build up any complementary (steady state) solution to the Maxwell field. We shall use this approach for other fields in chapter 5.

2.2 Conservation laws

The simple observation of 'what goes in must come out' applies to many
physical phenomena, including electromagnetism, and forms a predictive framework with rich consequences. Conservation is a physical fact, which must be
reflected in dynamical models. Just as economics uses money as a book-keeping
parameter for transactions, so physics accounts for transactions (interactions)
with energy, charge and a variety of similarly contrived labels which have proven
useful in keeping track of 'stock' in the physical world.

2.2.1 Current conservation

Perhaps the central axiom of electromagnetism is the conservation of total
electric charge. An algebraic formulation of this hypothesis provides an
important relationship, which will be referred to many times. Consider the
electric current I, defined in terms of the rate of flow of charge:

$$I = \int d\sigma \cdot \mathbf{J} = \frac{dQ}{dt}. \tag{2.57}$$

Expressing the charge Q as the integral over the charge density, one has

$$\int \nabla \cdot \mathbf{J} d\sigma = -\partial_t \int \rho_e d\sigma. \tag{2.58}$$

Comparing the integrand on the left and right hand sides gives

$$\frac{\partial \rho_e}{\partial t} + \vec{\nabla} \cdot \mathbf{J} = 0, \tag{2.59}$$

or, in index notation,

$$\partial_i J^i = -\partial_t \rho_e. \tag{2.60}$$

This may now be expressed in 4-vector language (or $(n+1)$-vector language),
and the result is:

$$\partial_\mu J^\mu = 0. \tag{2.61}$$

This result is known as a *continuity condition* or a *conservation law*. All
conservation laws have this essential form, for some $(n+1)$ dimensional current
vector J^μ. The current is then called a conserved current. In electromagnetism
one can verify that the conservation of current is compatible with the equations
of motion very simply in eqn. (2.32) by taking the 4-divergence of the equation:

$$\partial_\mu \partial_\nu F^{\mu\nu} = \mu_0 \, \partial_\mu J^\mu = 0. \tag{2.62}$$

The fact that the left hand size is zero follows directly from the anti-symmetrical
and non-singular nature of $F_{\mu\nu}$.

2.2.2 Poynting's vector and energy conservation

The electromagnetic field satisfies a continuity relation of its own. This relation is arrived at by considering the energy flowing though a unit volume of the field. The quantities defined below will later re-emerge in a more general form as the so-called field theoretical energy–momentum tensor.

The passage of energy through an electromagnetic system may be split up into two contributions. The first is the work done on any electric charges contained in the volume. This may be expressed in terms of the current density and the electric field as follows. The rate at which work is done on a moving charge is given by the force vector dotted with the rate of change of the displacement (i.e. the velocity), $\mathbf{F} \cdot \mathbf{v}$. The force, in turn, is given by the charge multiplied by the electric field strength $q\mathbf{E}$, which we may write in terms of the charge density ρ_e inside a spatial volume $d\sigma$ as $\rho_e \mathbf{E} d\sigma$. The rate at which work is done on charges may now be expressed in terms of an external source or current, by identifying the external current to be the density of charge which is flowing out of some volume of space with a velocity $\mathbf{v}$

$$\mathbf{J}_{\text{ext}} = \rho_e \mathbf{v}. \tag{2.63}$$

We have

$$\text{Rate of work} = \mathbf{E} \cdot \mathbf{J}_{\text{ext}} d\sigma. \tag{2.64}$$

The second contribution to the energy loss from a unit volume is due to the flux of radiation passing through the surface (S) which bounds the infinitesimal volume (σ). This flux of radiation is presently unknown, so we shall refer to it as $\mathcal{S}$. If we call the total energy density $\mathcal{H}$, then we may write that the loss of energy from an infinitesimal volume is the sum of these two contributions:

$$-\partial_t \int_\sigma \mathcal{H} \, d\sigma = \int_S \mathcal{S} \cdot d\mathbf{S} + \int_\sigma \mathbf{E} \cdot \mathbf{J}_{\text{ext}} d\sigma. \tag{2.65}$$

In 1884, Poynting identified $\mathcal{H}$ and $\mathbf{S}$ using Maxwell's equations. We shall now do the same. The aim is to eliminate the current $\mathbf{J}_{\text{ext}}$ from this relation in order to express $\mathcal{H}$ and $\mathbf{S}$ in terms of $\mathbf{E}$ and $\mathbf{B}$ only. We begin by using the fourth Maxwell equation (2.1d) to replace $\mathbf{J}_{\text{ext}}$ in eqn. (2.65):

$$\mathbf{E} \cdot \mathbf{J}_{\text{ext}} = \frac{\mathbf{E} \cdot (\nabla \times \mathbf{B})}{\mu_0} - \epsilon_0 \, \mathbf{E} \cdot \partial_t \mathbf{E}. \tag{2.66}$$

Using the vector identity in Appendix A, eqn. (A.71), we may write

$$\mathbf{E} \cdot (\nabla \times \mathbf{B}) = \nabla \cdot (\mathbf{B} \times \mathbf{E}) + \mathbf{B} \cdot (\nabla \times \mathbf{E}). \tag{2.67}$$

The second Maxwell eqn. (2.1b) may now be used to replace $\nabla \times \mathbf{E}$, giving

$$\mathbf{E} \cdot (\nabla \times \mathbf{B}) = \nabla \cdot (\mathbf{B} \times \mathbf{E}) - \mathbf{B}\partial_t \mathbf{B}. \tag{2.68}$$

Finally, noting that

$$\frac{1}{2}\partial_t(\mathbf{X}\cdot\mathbf{X}) = \mathbf{X}\partial_t\mathbf{X}, \tag{2.69}$$

and using this with $\mathbf{X} = \mathbf{E}$ and $\mathbf{X} = \mathbf{B}$ in eqns. (2.66) and (2.68), we may write:

$$\mathbf{E}\cdot\mathbf{J}_{\text{ext}} = \frac{\nabla\cdot(\mathbf{B}\times\mathbf{E})}{\mu_0} - \frac{1}{2}\partial_t\left(\epsilon_0\mathbf{E}\cdot\mathbf{E} + \frac{1}{\mu_0}\mathbf{B}\cdot\mathbf{B}\right). \tag{2.70}$$

This equation has precisely the form of eqn. (2.65), and the pieces can now be identified:

$$S^0 = \mathcal{H} = \frac{1}{2}\left(\epsilon_0\mathbf{E}\cdot\mathbf{E} + \frac{1}{\mu_0}\mathbf{B}\cdot\mathbf{B}\right)$$

$$\equiv \frac{1}{2}(\mathbf{E}\cdot\mathbf{D} + \mathbf{B}\cdot\mathbf{H}) \tag{2.71}$$

$$S^i = \mathbf{S} = \frac{\mathbf{E}\times\mathbf{B}}{c\mu_0}$$

$$\equiv \frac{\mathbf{E}\times\mathbf{H}}{c}. \tag{2.72}$$

The new fields $\mathbf{D} = \epsilon_0\mathbf{E}$ and $\mu_0\mathbf{H} = \mathbf{B}$ have been defined. The energy density $\mathcal{H}$ is often referred to as a Hamiltonian for the free electromagnetic field, whereas $\mathbf{S}$ is referred to as the Poynting vector.

$$\partial_\mu S^\mu = \left(F_{0\mu}J_{\text{ext}}^\mu\right) \tag{2.73}$$

is the rate at which work is done by an infinitesimal volume of the field. It is clear from the appearance of an explicit zero component in the above that this argument cannot be the whole story. One expects a generally covariant expression. The expression turns out to be

$$\partial_\mu\theta_{\text{Maxwell}}^{\mu\nu} = F^{\mu\nu}J_\mu, \tag{2.74}$$

where $\theta_{\mu\nu}$ is the energy–momentum tensor. Notice how it is a surface integral which tells us about flows in and out of a volume of space. One meets this idea several times, in connection with boundary conditions and continuity.

2.3 Electromagnetism in matter

To describe the effect of matter on the electromagnetic field in a covariant way, one may use either a microscopic picture of the field interacting with matter at the molecular level, or a macroscopic, effective field theory, which hides the details of these interactions by defining equivalent fields.

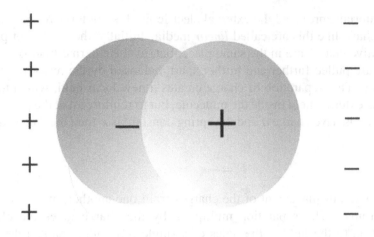

Fig. 2.1.   Matter is not electrically neutral at the microscopic level.

2.3.1 Dielectrics

One tends to think of ordinary matter as being electrically neutral, but of course it is composed of atoms and molecules, which have a centre of positive charge and a centre of negative charge – and these two centres do not necessarily lie at the same place. The more symmetrical a molecule is, the more neutral it is: for instance, the noble gases have highly symmetrical electron orbits and thus have almost no polarizability on average; the water molecule, on the other hand, has an asymmetry which allows a trickle of water to be attracted by a charged comb.

When an electric field is switched on in the vicinity of a dielectric material, the centres of positive and negative charge in each molecule are forced apart slightly (see figure 2.1) in a substance-dependent way. We say that such a molecule has a certain polarizability.

For classical external fields, atoms and molecules behave like dipoles, i.e. there is a clear separation of the charge into two parts: a positive pole and a negative pole. But we would be doing a disservice to the radiation field (not to mention the quantum theory) if we did not recall that the field has a wave nature and a characteristic wavelength. Molecules behave like dipoles if the wavelength of the external field is large compared to the size of the molecule – since then there is a clear direction to the field at every point inside the molecule's charge cloud. If, on the other hand, the wavelength of the field is of the order of the size of the molecule or less, then the field can reverse direction inside the molecule itself. The charge then gets re-distributed into a more complex pattern, and so-called quadrapole moments and perhaps higher 'pole moments' must be taken into account. In this text, we shall only consider the dipole approximation.

The simplest way to model the polarization of an atom or molecule is to view it as opposite charges coupled together by a spring. This model is adequate for many materials, provided the external electric field is not too strong. Materials which behave like this are called *linear* media. Initially, the centres of positive and negative charge are in the same place, but, as the external field is switched on, they are pulled further and further apart, balanced by the restoring force of the spring. This separation of charge creates a new local field, which tends to cancel the external field inside the molecule, but to reinforce it at the poles. If the charge clouds have charge q and the spring constant is κ then the force equation is simply

$$F = -\kappa s = Eq, \tag{2.75}$$

where s is the displacement of the charges from one another, in the rest frame of the atoms. The separation multiplied by the charge gives the effective contribution to the field at the poles of a single molecule, denoted the *dipole moment* $\mathbf{d}$:

$$\mathbf{d} = |\mathbf{s}|q = \frac{q^2}{\kappa}\mathbf{E}. \tag{2.76}$$

The quantity q^2/κ is denoted by α and is called the *polarizability*; it denotes the effective strength of the resistance to polarization. The *polarization* field is

$$\mathbf{P} = \rho_N \mathbf{d} = \rho_N \alpha \mathbf{E} \tag{2.77}$$

where ρ_N is the total number of molecules per unit volume. It is proportional to the field of particles displacements $s^i(x)$ and it hides some non-covariant assumptions (see the next section). Normally speaking, one writes $q = -e$, where $-e$ is the charge on the electron. Then,

$$\alpha_{\text{static}} = \frac{-q^2}{\kappa}. \tag{2.78}$$

If one considers time-varying fields, or specifically waves of the form

$$\mathbf{E} = \mathbf{E}_0 e^{i(\mathbf{k}\cdot\mathbf{x}-\omega t)}, \tag{2.79}$$

it is found that, for a single *optically active* electron (i.e. one in an orbit which can be affected by an external field), the equation of motion is now that of a damped harmonic oscillator:

$$m(\omega_0 + i\gamma\omega - \omega^2)\mathbf{s} = -e\mathbf{E}_0, \tag{2.80}$$

where $\omega_0^2 = \kappa/m$ and γ is a damping term. Using this equation to replace for $\mathbf{s}$ in eqn. (2.76), we get

$$\alpha(\omega) = \frac{q^2/m}{(\omega_0^2 + i\gamma\omega - \omega^2)}. \tag{2.81}$$

Thus the polarizability is a frequency-dependent quantity. This explains why a prism can split a mixture of white light into its component frequencies. A further definition is of interest, namely the *electric susceptibility* $\chi_e = N\alpha(\omega)/\epsilon_0$. For ρ_N particles per unit volume, this is often expressed in terms of the plasma frequency $\omega_p^2 = Ne^2/m$. Thus,

$$\mathbf{P} = \epsilon_0 \chi_e \mathbf{E}. \tag{2.82}$$

This is closely related to the change in the refractive index, $n^2 = 1 + \chi_e$, of a material due to polarization, when $\mu_r = 1$ (which it nearly always is). In real space, we note from eqn. (2.80) that the polarization satisfies a differential equation

$$(\partial_t^2 - \gamma \partial_t + \omega_0^2)\mathbf{P} = \frac{q^2}{m}\rho_N \mathbf{E} \tag{2.83}$$

and thus the real space susceptibility can be thought of as a Green function for the differential operator and $\mathbf{E}$ may be thought of as a source.

$$\mathbf{P}(t) = \epsilon_0 \int dt\, \chi(t - t')\mathbf{E}. \tag{2.84}$$

$\chi(t - t')$ is taken to satisfy retarded boundary conditions, which, in turn, implies that its real and imaginary parts in terms of ω are related. The relationship is referred to as a *Kramers–Kronig relation*, and is simply a consequence of the fact that a retarded Green function is *real*.

2.3.2 Covariance and relative motion: the Doppler effect

The frequency-dependent expressions above are true only in the rest frame of the atoms. The results are not covariant with respect to moving frames of reference. When one studies solid state systems, such as glasses and crystals, these expressions are quite adequate, because the system has a naturally preferred rest frame and the atoms in the material do not move relative to one another, on average. However, in gaseous matter or plasmas, this is not the case: the thermal motion of atoms relative to one another can be important, because of the Doppler effect. This fact can be utilized to good effect; for example, in laser cooling the motion of atoms relative to a laser radiation field can be used to bring all of the atoms into a common rest frame by the use of a resonant, frequency-dependent interaction. A Galilean-covariant expression can be written by treating the field as one co-moving mass, or as a linear super-position of co-moving masses. With only one component co-moving, the transformation of the position of an atom in the displacement field can be written

$$\mathbf{x}(t) \rightarrow \mathbf{x} + \mathbf{v}t, \tag{2.85}$$

where **v** is the velocity of the motion relative to some reference frame (usually the container of the gaseous matter, or the laboratory frame). This means that the invariant form $(kx - \omega t)$ is transformed into

$$\mathbf{k} \cdot \mathbf{x} - \omega t \rightarrow \mathbf{k} \cdot (\mathbf{x} + \mathbf{v}t) - \omega t = \mathbf{k} \cdot \mathbf{x} - \omega_\beta t, \qquad (2.86)$$

where

$$\omega_\beta = \omega(1 - \hat{\mathbf{k}} \cdot \beta) = \omega(1 - \hat{\mathbf{k}} \cdot \mathbf{v}/c). \qquad (2.87)$$

Thus, the expressions above can be used, on replacing ω with a sum over all ω_β, and integrated over all the values of the velocity vector β^i of which the field is composed. The polarizability takes the form

$$\alpha(\omega) = \frac{q^2/m}{(\omega_0^2 + i\gamma\omega - \omega_\beta^2)}. \qquad (2.88)$$

where

$$\omega_\beta = (1 - \hat{k}^i \beta_i)\omega. \qquad (2.89)$$

2.3.3 Refractive index in matter

It appears that the introduction of a medium destroys the spacetime covariance of the equations of motion. In fact this is false. What is interesting is that covariance can be restored by re-defining the $(n + 1)$ dimensional vectors so as to replace the speed of light in a vacuum with the *effective* speed of light in a medium. The speed of light in a dielectric medium is

$$v = \frac{c}{n} \qquad (2.90)$$

where $n = \epsilon_r \mu_r > 1$ is the refractive index of the medium.

Before showing that covariance can be restored, one may consider the equation of motion for waves in a dielectric medium from two perspectives. The purpose is to relate the multifarious fields to the refractive index itself. It is also to demonstrate that the polarization terms can themselves be identified as currents which are set in motion by the electric field. In other words, we will show the equivalence of (i) $\mathbf{P} \neq 0$, but $J_\mu = 0$, and (ii) $\mathbf{P} = 0$ with J_μ given by the current resulting from charges on springs! Taking

$$J^0 = c\rho_e J^i = -\rho_N ec\frac{ds^i}{dt}, \qquad (2.91)$$

the current is seen to be a result of the net charge density set in motion by the field. This particular derivation applies only in $3 + 1$ dimensions.

To obtain the wave equation we begin with the Bianchi identity

$$\epsilon_{ijk}\partial_j E_k + \partial_t B_i = 0, \tag{2.92}$$

and then operate from the left with $\epsilon_{ilm}\partial_l$. Using the identity (A.38) (see Appendix A) for the product of two anti-symmetric tensors, we obtain

$$\left[\nabla^2 E_i - \partial_i (\partial^j E_j)\right] + \epsilon_{ilm}\partial_l\partial_t B_i = 0. \tag{2.93}$$

Taking ∂_t of the fourth Maxwell equation, one obtains

$$\frac{1}{\mu_0\mu_r}\epsilon_{ijk}\partial_j\partial_t B_k = \partial_t J_i + \epsilon_0\epsilon_r\frac{\partial^2 E_i}{\partial t^2}. \tag{2.94}$$

These two equations can be used to eliminate B_i, giving an equation purely in terms of the electric field. Choosing the charge distribution to be isotropic (uniform in space), we have $\partial_i\rho_e = 0$, and thus

$$\left[\nabla^2 - \frac{n^2}{c^2}\frac{\partial^2}{\partial t^2}\right] E_i = \mu_0\mu_r\partial_t J_i. \tag{2.95}$$

In this last step, we used the definition of the refractive index in terms of ϵ_r:

$$n^2 = \epsilon_r\mu_r = (1 + \chi_e)\mu_r. \tag{2.96}$$

This result is already suggestive of the fact that Maxwell's equations in a medium can be written in terms of an effective speed of light.

We may now consider the two cases: (i) $\mathbf{P} \neq 0$, but $J_\mu = 0$,

$$\left[\nabla^2 - \frac{n^2}{c^2}\frac{\partial^2}{\partial t^2}\right] E_i = 0; \tag{2.97}$$

and (ii) $\mathbf{P} = 0$ $(n = 1)$, $J_\mu \neq 0$.

$$\left[\nabla^2 - \frac{1}{c^2}\frac{\partial^2}{\partial t^2}\right] E_i = \mu_0\mu_r\frac{-\rho_N e^2\omega^2/m \cdot E_i}{(\omega_0^2 + i\gamma\omega - \omega^2)}. \tag{2.98}$$

The differential operators on the left hand side can be replaced by $\mathbf{k}^2$ and ω^2, by using the wave solution (2.79) for the electric field to give a 'dispersion relation' for the field. This gives:

$$\begin{aligned}
\frac{\mathbf{k}^2}{\omega^2} &= \frac{1}{c^2}\left(1 + \frac{\mu_r}{\epsilon_0}\frac{\rho_N e^2\omega^2/m}{(\omega_0^2 + i\gamma\omega - \omega^2)}\right) \\
&= \frac{n^2}{c^2}.
\end{aligned} \tag{2.99}$$

So, from this slightly cumbersome expression for the motion of charges, one derives the microscopic form of the refractive index. In fact, comparing eqns. (2.99) and (2.98), one sees that

$$n^2 = 1 + \frac{\rho_N \alpha(\omega) \mu_r}{\epsilon_0}. \tag{2.100}$$

Since μ_r is very nearly unity in all materials that waves penetrate, it is common to ignore this and write

$$n^2 \sim 1 + \chi_e. \tag{2.101}$$

The refractive index is a vector in general, since a material could have a different index of refraction in different directions. Such materials are said to be anisotropic. One now has both microscopic and macroscopic descriptions for the interaction of radiation with matter, and it is therefore possible to pick and choose how one wishes to represent this physical system. The advantage of the microscopic formulation is that it can easily be replaced by a quantum theory at a later stage. The advantage of the macroscopic field description is that it is clear why the form of Maxwell's equations is unaltered by the specific details of the microscopic interactions.

2.4 Aharonov–Bohm effect

The physical significance of the vector potential A_μ (as opposed to the field $F_{\mu\nu}$) was moot prior to the arrival of quantum mechanics. For many, the vector potential was merely an artifice, useful in the computation of certain boundary value problems. The formulation of quantum mechanics as a local field theory established the vector potential as the fundamental *local* field, and the subsequent attention to gauge symmetry fuelled pivotal developments in the world of particle physics. Today, it is understood that there is no fundamental difference between treating the basic electromagnetic interaction as a rank 2 anti-symmetric tensor $F_{\mu\nu}$, or as a vector with the additional requirement of gauge invariance. They are equivalent representations of the problem. In practice, however, the vector potential is the easier field to work with, since it couples locally. The price one pays lies in ensuring that gauge invariance is maintained (see chapter 9).

The view of the vector potential as a mathematical construct was shaken by the discovery of the Aharonov–Bohm effect. This was demonstrated is a classic experiment of electron interference through a double slit, in which electrons are made to pass through an area of space in which $A_\mu \neq 0$ but where $F_{\mu\nu} = 0$. The fact that a change in the electron interference pattern was produced by this configuration was seen as direct evidence for the physical reality of A_μ. Let us examine this phenomenon.

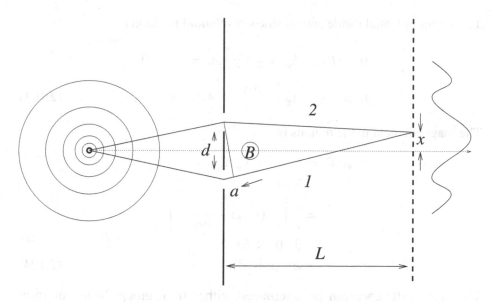

Fig. 2.2.   The Aharonov–Bohm experiment.

The physical layout of the double-slit experiment is shown in figure 2.2. An electron source fires electrons at the slits, these pass through the slits and interfere in the usual way, forming an interference pattern on the screen at the end of their path. In order to observe the Aharonov–Bohm effect, one places a solenoid on the far side of the slits, whose magnetic field is constrained within a cylinder of radius R. The vector potential arising from the solenoid geometry is not confined to the inside of the solenoid however. It also extends outside of the solenoid, but in such a way as to produce no magnetic field.

What is remarkable is that, when the solenoid is switched on, the interference pattern is shifted by an amount x. This indicates that a phase shift $\Delta\theta$ is introduced between the radiation from the two slits, and is caused by the presence of the solenoid. If the distance L is much greater than x and a then we have

$$a \sim \frac{x}{L}\, d$$

$$\Delta\theta = 2\pi \left(\frac{L_1 - L_2}{\lambda} \right) = \frac{2\pi a}{\lambda}$$

$$x = \left(\frac{L\lambda}{2\pi d} \right) \Delta\theta. \tag{2.102}$$

The phase difference can be accounted for by the gauge transformation of the electron field by the vector potential. Although the absolute value of the vector potential is not gauge-invariant, the potential difference between the paths is.

The vector potential inside and outside the solenoid position is

$$(r < R): \quad A_\phi = \frac{1}{2}Br, \quad A_r = A_z = 0$$

$$(r > R): \quad A_\phi = \frac{BR^2}{2r}, \quad A_r = A_z = 0. \tag{2.103}$$

The magnetic field in the regions is

$$\begin{aligned} B_z &= \nabla_r A_\phi - \nabla_\phi A_r \\ &= \nabla_r A_\phi \\ &= \frac{1}{r}\left[\frac{\partial}{\partial_r}(r A_\phi) - \frac{\partial}{\partial\phi}A_r\right] \\ &= 0 \quad (r < R) \\ &= B \quad (r > R). \end{aligned} \tag{2.104}$$

The phase difference can be determined, either from group theory, or from quantum mechanics to be

$$\exp(i\theta) = \exp\left(i\frac{e}{\hbar}\int_P A^i dx_i\right), \tag{2.105}$$

where '*P*' indicates the integral along a given path. Around the closed loop from one slit to the screen and back to the other slit, the phase difference is (using Stokes' theorem)

$$\begin{aligned} \Delta\theta &= \theta_1 - \theta_2 \\ &\sim \frac{e}{\hbar}\oint A_\phi \, dr \\ &= \frac{e}{\hbar}\int (\vec{\nabla} \times \mathbf{B}) \cdot d\mathbf{S} \\ &= \frac{e}{\hbar}\int \mathbf{B} \cdot d\mathbf{S}. \end{aligned} \tag{2.106}$$

The phase shift therefore results from the paths having to travel around the solenoid, i.e. in a loop where magnetic flux passes through a small part of the centre. Note, however, that the flux does not pass through the path of the electrons, only the vector potential is non-zero for the straight-line paths.

There are two ways of expressing this: (i) electrons must be affected by the vector potential, since the field is zero for any classical path from the slits to the screen; or (ii) electrons are stranger than we think: they seem to be affected by a region of space which is classically inaccessible to them. The viewpoints are really equivalent, since the vector potential is simply an analytic extension of the field strength, but the result is no less amazing. It implies a non-locality in the action of the magnetic field: action at a distance, and not only at a distance, but from within a container. If one chooses to believe in the vector potential as a fundamental field, the behaviour seems less objectionable: the interaction is then local. There is no action at a distance, and what happens inside the solenoid is of less interest.

Whether one chooses to view this as evidence for the physical reality of the vector potential or of the strangeness of quantum mechanics is a matter of viewpoint. Indeed, the reality of any field is only substantiated by the measurable effect it has on experiments. However, there are deeper reasons for choosing the interpretation based on the reality of the vector potential, which have to do with locality and topology, so at the very least this gives us a new respect for the utility of the vector potential. In view of the utility of A_μ and its direct appearance in dynamical calculations, it seems reasonable to accept it as the fundamental field in any problem which is simplified by that assumption.

3

Field parameters

The parameters which measure change in dynamical systems have a unique importance: they describe both the layout and the development of a system. Space (position) and time are the most familiar parameters, but there are other possibilities, such as Fourier modes.

In the previous chapter, it was seen how the unification of spatial and temporal parameters, in electromagnetism, led to a tidier and deeper form of the Maxwell equations. It also made the equations easier to transform into other relativistic frames. In the covariant approach to physics one is concerned with what does and does not change, when shifting from one perspective to another, i.e. with the properties of a system which are dependent and independent of the circumstances of observation. In a continuous, holonomic system, this is summarized by two independent concepts: parameter spaces and coordinates.

- **Parameter space** (manifold). This represents the stage for physical reality. A parameter space has coordinate-independent properties such as topology and curvature.

- **Coordinates**. These are arbitrary labels used to mark out a reference scheme, or measurement scheme, in parameter space. There is no unique way to map out a parameter space, e.g. Cartesian or polar coordinates. If there is a special symmetry, calculations are often made easier by choosing coordinates which match this symmetry.

Coordinates are labels which mark a scale on a parameter space. They measure a distance in a particular direction from an arbitrary origin. Clearly, there is nothing fundamental about coordinates: by changing the arbitrary origin, or orientation of measurement, all coordinate labels are changed, but the underlying reality is still the same. They may be based on flat Cartesian (x, y, z) or polar (r, θ, ϕ) conventions; they can be marked on flat sheets or curved shells.

Underneath the details of an arbitrary system of measurement is a physical system which owes nothing to those details.

The invariant properties or *symmetries* of parameter spaces have many implicit consequences for physical systems; not all are immediately intuitive. For this reason, it is useful to study these invariant properties in depth, to see how they dictate the possibilities of behaviour (see chapter 9). For now it is sufficient to define a notation for coordinates on the most important parameter spaces.

This chapter summarizes the formulation of $(n + 1)$ dimensional vectors in Minkowski spacetime and in its complementary space of wavevectors k, usually called *momentum space* or *reciprocal lattice space*.

3.1 Choice of parametrization

The dynamical variables, in field theory, are the fields themselves. They are functions of the parameters which map out the background space or spacetime; e.g.

$$\psi(t), \quad \phi(t, \mathbf{x}), \quad \chi(t, r, \theta, \phi). \tag{3.1}$$

Field variables are normally written as functions of spacetime positions, but other decompositions of the field are also useful. Another ubiquitous choice is to use a complementary set of variables based upon a decomposition of the field into a set of basis functions, a so-called *spectral decomposition*. Given a complete set of functions $\psi_i(x)$, one can always write an arbitrary field as a linear super-position:

$$\phi(x) = \sum_i c_i \psi_i(x). \tag{3.2}$$

Since the functions are fixed and known, a knowledge of the coefficients c_i in this decomposition is equivalent to a knowledge of $\phi(x)$, i.e. as a function of x. However, the function may also be written in a different parametrization:

$$\phi(c_1, c_2, c_3 \ldots). \tag{3.3}$$

This is a shorthand for the decomposition above, just as $\phi(x)$ is a shorthand for a polynomial or series in x. Usually, an infinite number of such coefficients is needed to prescribe a complete decomposition of the field, as, for instance, in the Fourier expansion of a function, described below.

Spacetime is an obvious parameter space for a field theory since it comprises the world around us and it includes laboratories where experiments take place, but other basis functions sometimes reveal simpler descriptions. One important example is the complementary Fourier transform of spacetime. The Fourier transform is important in situations where one suspects a translationally invariant, homogeneous system. The Fourier transform of a function of x is defined to be a new function of the wavenumber k (and the inverse transform) by the relations:

$$f(x) = \int \frac{\mathrm{d}k}{2\pi}\, \mathrm{e}^{\mathrm{i}kx}\, f(k)$$

$$f(k) = \int \mathrm{d}x\, \mathrm{e}^{\mathrm{i}kx}\, f(x). \tag{3.4}$$

k is a continuous label on a continuous set of functions $\exp(\mathrm{i}kx)$, not a discrete set of c_i, for integer i. In solid state physics, the space parametrized by k is called the reciprocal lattice space. Fourier transform variables are useful for many purposes, such as revealing hidden periodicities in a function, since the expansion is based on periodic functions. The Fourier transform is also a useful calculational aid.

Spacetime (configuration space) and the Fourier transform are two complementary ways of describing the basic evolution of most systems. These two viewpoints have advantages and disadvantages. For example, imagine a two-state system whose behaviour in time can be drawn as a square wave. To represent a square wave in Fourier space, one requires either an infinite number of Fourier waves of different frequencies, or merely two positions over time. In that case, it would be cumbersome to use a Fourier representation of the time evolution.

3.2 Configuration space

The four-dimensional vectors used to re-write electromagnetism are easily generalized to $(n+1)$ spacetime dimensions, for any positive n. They place time and space on an *almost* equal footing. In spite of the notational convenience of unified spacetime, some caution is required in interpreting the step. Time is *not* the same as space: formally, it distinguishes itself by a sign in the metric tensor; physically, it plays a special role in determining the dynamics of a system.

3.2.1 Flat and curved space

Physical systems in constrained geometries, such as on curved surfaces, or within containers, are best described using curvilinear coordinates. Experimental apparatus is often spherical or toroidal; shapes with a simple symmetry are commonly used when generating electromagnetic fields; rectangular fields with sharp corners are less common, since these require much higher energy to sustain.

Studies of what happens within the volumes of containers, and what happens on their surface boundaries, are important in many situations [121]. When generalizing, to study systems in $(n + 1)$ dimensions, the idea of surfaces and volumes also has to be generalized. The distinction becomes mainly one of convenience: $(n + 1)$ dimensional curved surfaces are curved spacetimes. The fact that they enclose a volume or partition a space which is $(n + 2)$ dimensional is not always germane to the discussion at hand. This is particularly true in cosmology.

It is important to distinguish between curvilinear coordinates in flat space and coordinate labellings of curved space. An example of the former is the use of polar (r, θ) coordinates to map out a plane. The plane is flat, but the coordinates span the space in a set of curved rings. An example of the latter is (θ, ϕ) coordinates (at fixed r), mapping out the surface of a sphere. Over very short distances, (θ, ϕ) can be likened to a tiny planar patch with Cartesian coordinates (x, y).

Einstein's contribution to the theory of gravity was to show that the laws of gravitation could be considered as an intrinsic curvature of a $(3+1)$ dimensional spacetime. Einstein used the idea of covariance to argue that one could view gravity in one of two equivalent ways: as forced motion in a flat spacetime, or as free-fall in a curved spacetime. Using coordinates and metric tensors, gravitation could itself be described as a field theory, in which the field $g_{\mu\nu}(x)$ was the shape of spacetime itself.

Gravitational effects may be built into a covariant formalism to ensure that every expression is general enough to be cast into an arbitrary scheme of coordinates. If one allows for general coordinates (i.e. general covariance), one does not assume that all coordinates are orthogonal Cartesian systems, and gravity and curvature are not excluded from the discussion.

Spacetime curvature will not be treated in detail here, since this topic is widely discussed in books on relativity. However, we take the issue of curvature 'under advisement' and construct a formalism for dealing with arbitrary coordinates, assured that the results will transform correctly even in a curved environment.

3.2.2 Vector equations

Vector methods express spatial relationships, which remain true regardless of the system of coordinates used to write them down. They thus play a central role in covariant formulation. For example, the simple vector equation

$$\mathbf{A} \cdot \mathbf{B} = 0 \tag{3.5}$$

expresses the fact that two vectors $\mathbf{A}$ and $\mathbf{B}$ are orthogonal. It says nothing about the orientation of the vectors relative to a coordinate system, nor their position relative to an origin; rather, it expresses a relationship of more intrinsic value between the vectors: their relative orientation. Vector equations and covariance are natural partners.

Vector equations are form-invariant under changes of coordinates, but the details of their components do change. For instance, in the above equation, if one fixes a coordinate system, then the components of the two vectors take on definite values. If one then rotates or translates the coordinates, the values of the components change, but the equation itself remains true.

3.2.3 Coordinate bases

A coordinate basis is a set of $(n + 1)$ linearly independent reference vectors $\mathbf{e}_\mu$, used to provide a concise description of any vector within a vector space. They are 'standard arrows'; without them, every direction would need to have a different name.[1]

In index notation, the components of a vector $\mathbf{a}$ are written, relative to a basis or set of axes $\mathbf{e}_i$, as $\{a^i\}$, i.e.

$$\mathbf{a} = \sum_\mu a^\mu \, \mathbf{e}_\mu \equiv a^\mu \, \mathbf{e}_\mu. \tag{3.6}$$

Note that, as usual, there is an implied summation convention over repeated indices throughout this book. The subscript μ runs over the number of dimensions of the space.

Linearity is a central concept in vector descriptions. One does not require what happens within the space to be linear, but the basis vectors must be locally linear in order for the vector description to be single-valued. Consider, then, the set of all linear scalar functions of vectors. Linearity implies that a linear combination of arguments leads to a linear combination of the functions:

$$\omega(c^\mu \mathbf{e}_\mu) = c^\mu \omega(\mathbf{e}_\mu). \tag{3.7}$$

Also, the linear combination of different functions results in new linear functions:

$$\omega'(\mathbf{v}) = \sum_\mu c_\mu \omega^\mu(\mathbf{v}). \tag{3.8}$$

The space of these functions is therefore also a vector space V^*, called the dual space. It has the same dimension as the vector space (also called the tangent space). The duality refers to the fact that one may consider the 1-forms to be linear functions of the basis vectors, or vice versa, i.e.

$$\omega(\mathbf{v}) = \mathbf{v}(\omega). \tag{3.9}$$

---

[1] In terms of information theory, the vector basis provides a systematic $(n+1)$-tuple of numbers, which in turn provides an optimally compressed coding of directional information in the vector space. Without such a system, we would be stuck with names like north, south, east, west, north-north-west, north-north-north-west etc. for each new direction.

Vector components v^i are written

$$\mathbf{v} = v^\mu \mathbf{e}_\mu, \tag{3.10}$$

and dual vector (1-form) components are written

$$\mathbf{v} = v_\mu \omega^\mu. \tag{3.11}$$

The scalar product is

$$
\begin{aligned}
\mathbf{v} \cdot \mathbf{v} = \mathbf{v}^* \mathbf{v} &= (v_\mu \omega^\mu)(v^\nu \mathbf{e}_\nu) \\
&= v_\mu v^\nu \, (\omega^\mu \mathbf{e}_\nu) \\
&= v_\mu v^\nu \, \delta^\nu_\mu \\
&= v_\mu v^\mu,
\end{aligned} \tag{3.12}
$$

where

$$(\omega^\mu \mathbf{e}_\nu) = \delta^\mu_\nu. \tag{3.13}$$

The metric tensor $g_{\mu\nu}$ maps between these equivalent descriptions:

$$
\begin{aligned}
v_\mu &= g_{\mu\nu} v^\nu \\
v^\mu &= g^{\mu\nu} v_\nu,
\end{aligned} \tag{3.14}
$$

and

$$\mathbf{e}_\mu \cdot \mathbf{e}_\nu = g_{\mu\nu} \tag{3.15a}$$
$$\omega^\mu \cdot \omega^\nu = g^{\mu\nu}. \tag{3.15b}$$

When acting on scalar functions, the basis vectors $\mathbf{e}_\mu \to \partial_\mu$ are tangential to the vector space; the 1-forms $\omega^\mu \to \mathrm{d}x^\mu$ lie along it.

In general, under an infinitesimal shift of the coordinate basis by an amount $\mathrm{d}x^\mu$, the basis changes by an amount

$$\mathrm{d}\mathbf{e}_\mu = \Gamma_{\mu\nu}{}^\lambda \, \mathbf{e}_\lambda \, \mathrm{d}x^\nu. \tag{3.16}$$

The symbol $\Gamma_{\mu\nu}{}^\lambda$ is called the affine connection, or Christoffel symbol. From this, one determines that

$$\partial_\nu \mathbf{e}_\mu = \Gamma_{\mu\nu}{}^\lambda \, \mathbf{e}_\lambda, \tag{3.17}$$

and by differentiating eqn. (3.13), one finds that

$$\partial_\nu \omega^\lambda = -\Gamma_{\nu\mu}{}^\lambda \, \omega_\mu. \tag{3.18}$$

The connection can be expressed in terms of the metric, by differentiating eqn. (3.15a):

$$
\begin{aligned}
\partial_\lambda g_{\mu\nu} &= \partial_\lambda \mathbf{e}_\mu \cdot \mathbf{e}_\nu + \mathbf{e}_\mu \cdot \partial_\lambda \mathbf{e}_\nu \\
&= \Gamma_{\mu\lambda}{}^\lambda g_{\rho\nu} + g_{\rho\mu} \Gamma_{\nu\lambda}{}^\lambda.
\end{aligned} \tag{3.19}
$$

By permuting indices in this equation, one may show that

$$\Gamma_{\lambda\mu}{}^{\sigma} = \frac{1}{2}g^{\nu\sigma}\left\{\partial_\lambda g_{\mu\nu} + \partial_\mu g_{\lambda\nu} - \partial_\nu g_{\mu\lambda}\right\}. \tag{3.20}$$

The connection is thus related to cases where the metric tensor is not constant. This occurs in various contexts, such when using curvilinear coordinates, and when fields undergo conformal transformations, such as in the case of gauge transformations.

3.2.4 Example: Euclidean space

In n-dimensional Euclidean space, the spatial indices i of a vector's components run from 1 to n except where otherwise stated. The length of a vector interval **ds** is an invariant quantity, which is defined by the inner product. This may be written

$$\mathbf{ds} \cdot \mathbf{ds} = \mathrm{d}x^2 + \mathrm{d}y^2 + \mathrm{d}z^2 \tag{3.21}$$

in a Cartesian basis. In the index notation (for $n = 3$) this may be written,

$$\mathbf{ds} \cdot \mathbf{ds} = \mathrm{d}x^i \mathrm{d}x_i. \tag{3.22}$$

Repeated indices are summed over, unless otherwise stated. We distinguish, in general, between vector components with raised indices (called *contravariant* components) and those with lower indices (called, confusingly, *covariant* components,[2] and 'normal' components, which we shall almost never use. In a Cartesian basis $(x, y, z \ldots)$ there is no difference between these components. In other coordinate systems, such as polar coordinates however, they are different.

Results which are independent of coordinate basis always involve a sum over one raised index and one lower index. The length of the vector interval above is an example. We can convert an up index into a down index using a matrix (actually a tensor) called the metric tensor g_{ij},

$$a_i = g_{ij}a^j. \tag{3.23}$$

The inverse of the metric g_{ij} is written g^{ij} (with indices raised), and it serves to convert a lower index into an upper one:

$$a^i = g^{ij}a_j. \tag{3.24}$$

The metric and its inverse satisfy the relation,

$$g_{ij}g^{jk} = g_i{}^k = \delta_i{}^k. \tag{3.25}$$

---

[2] There is no connection between this designation and the usual meaning of covariant.

In Cartesian components, the components of the metric are trivial. It is simply the identity matrix, or Kronecker delta:

$$(\text{Cartesian}) : \quad g_{ij} = g^{ij} = \delta_{ij}. \tag{3.26}$$

To illustrate the difference between covariant, contravariant and normal components, consider two-dimensional polar coordinates as an example. The vector interval, or line element, is now written

$$\mathbf{ds} \cdot \mathbf{ds} = dr^2 + r^2 d\theta^2. \tag{3.27}$$

The normal components of the vector **ds** have the dimensions of length in this case, and are written

$$(dr, r d\theta). \tag{3.28}$$

The contravariant components are simply the coordinate intervals,

$$ds^i = (dr, d\theta), \tag{3.29}$$

and the covariant components are

$$ds_i = (dr, r^2 d\theta). \tag{3.30}$$

The metric tensor is then defined by

$$g_{ij} = \begin{pmatrix} 1 & 0 \\ 0 & r^2 \end{pmatrix}, \tag{3.31}$$

and the inverse tensor is simply

$$g^{ij} = \begin{pmatrix} 1 & 0 \\ 0 & r^{-2} \end{pmatrix}. \tag{3.32}$$

The covariant and contravariant components are used almost exclusively in the theory of special relativity.

Having introduced the metric tensor, we may define the scalar product of any two vectors **a** and **b** by

$$\mathbf{a} \cdot \mathbf{b} = a^i b_i = a^i g_{ij} b^j. \tag{3.33}$$

The definition of the vector product and the curl are special to three space dimensions. We define the completely anti-symmetric tensor in three dimensions by

$$\epsilon^{ijk} = \begin{cases} +1 & ijk = 123 \text{ and even permutations} \\ -1 & ijk = 321 \text{ and other odd permutations} \\ 0 & \text{otherwise.} \end{cases} \tag{3.34}$$

This is also referred to as the three-dimensional Levi-Cevita tensor in some texts. Since its value depends on permutations of 123, and its indices run only over these values, it can only be used to generate products in three dimensions. There are generalizations of this quantity for other numbers of dimensions, but the generalizations must always have the same number of indices as spatial dimensions, thus this object is unique in three dimensions. More properties of anti-symmetric tensors are described below.

In terms of this tensor, we may write the ith covariant component of the three-dimensional vector cross-product as

$$(\mathbf{b} \times \mathbf{c})_i = \epsilon_{ijk} b^j c^k. \tag{3.35}$$

Contracting with a scalar product gives the volume of a parallelepiped spanned by vectors $\mathbf{a}$, $\mathbf{b}$ and $\mathbf{c}$,

$$\mathbf{a} \cdot (\mathbf{b} \times \mathbf{c}) = \epsilon_{ijk} a^i b^j c^k, \tag{3.36}$$

which is basis-independent.

3.2.5 Example: Minkowski spacetime

The generalization of Euclidean space to relativistically motivated spacetime is called Minkowski spacetime. Close to the speed of light, the lengths of n-dimensional spatial vectors are not invariant under boosts (changes of speed), due to the Lorentz length contraction. From classical electromagnetism, one finds that the speed of light in a vacuum must be constant for all observers:

$$c^2 = \frac{1}{\epsilon_0 \mu_0}, \tag{3.37}$$

and one deduces from this that a new quantity is invariant; we refer to this as the invariant line element

$$ds^2 = -c^2 \, dt^2 + dx^2 + dy^2 + dz^2 = -c^2 \, d\tau^2, \tag{3.38}$$

where $d\tau$ is referred to as the proper time. By comparing the middle and rightmost terms in this equation, it may be seen that the proper time is the time coordinate in the rest frame of a system, since there is no change in the position variables. The negative sign singles out the time contribution as special. The nomenclature 'timelike separation' is used for intervals in which $ds^2 < 0$, 'spacelike separation' is used for $ds^2 > 0$, and 'null' is used for $ds^2 = 0$.

In terms of $(n + 1)$ dimensional vectors, one writes:

$$ds^2 = dx^\mu dx_\mu = dx^\mu g_{\mu\nu} dx^\nu \tag{3.39}$$

where $\mu, \nu = 0, 1, 2, \ldots, n$ In a Cartesian basis, the contravariant and covariant components of the spacetime interval are defined, respectively, by

$$dx^\mu = (ct, x, y, z, \ldots)$$
$$dx_\mu = (-ct, x, y, z, \ldots), \tag{3.40}$$

and the metric tensor in this Cartesian basis, or locally inertial frame (LIF), is the constant tensor

$$\eta_{\mu\nu} \equiv g_{\mu\nu}\big|_{\text{LIF}} = \begin{pmatrix} -1 & 0 & 0 \cdots & 0 \\ 0 & 1 & 0 \cdots & 0 \\ 0 & 0 & 1 \cdots & 0 \\ \vdots & \vdots & \vdots & \vdots \\ 0 & 0 & 0 \cdots & 1 \end{pmatrix}. \tag{3.41}$$

This is a special case of a metric in a general frame $g_{\mu\nu}$.

This placement of signs in the metric is arbitrary, and two other conventions are found in the literature: the opposite sign for the metric, with corresponding movement of the minus sign from the time to the space parts in the covariant and contravariant components; and a Euclidean formulation, in which the metric is entirely positive (positive definite), and the time components of the components are symmetrically *ict*. This last form, called a Euclidean formulation (or Riemannian in curved spacetime), has several uses, and thus we adopt conventions in this text in which it is trivial to convert to the Euclidean form and back.

Contravariant vectors describe regular parametrizations of the coordinates. In order to define a frame-invariant derivative, we need to define partial derivatives by the requirement that the partial derivative of x^1 with respect to x_1 be unity:

$$\frac{\partial}{\partial x^1} x^1 = \partial_1 x^1 = 1. \tag{3.42}$$

Notice that 'dividing by' an upper index makes it into an object with an effectively lower index. More generally, we require:

$$\frac{\partial}{\partial x^\mu} x^\nu = \partial_\mu x^\nu = \delta_\mu{}^\nu. \tag{3.43}$$

From this, one sees that the Cartesian components of the derivative must be

$$\partial_\mu = \left(\frac{1}{c}\partial_t, \partial_x, \partial_y, \partial_z, \ldots\right)$$
$$\partial^\mu = \left(-\frac{1}{c}\partial_t, \partial_x, \partial_y, \partial_z, \ldots\right). \tag{3.44}$$

Velocity is a relative concept, by definition. It is intimately associated with a choice of Lorentz frame. The relative velocity is defined as the time derivative of the position

$$\beta^\mu = \frac{1}{c}\frac{dx^\mu}{dt} = \frac{dx^\mu}{dx^0}. \tag{3.45}$$

Unfortunately, because both x^μ and t are frame-dependent, this quantity does not transform like a vector. To obtain a vector, we choose to look at

$$U^\mu = \frac{1}{c}\frac{x^\mu}{d\tau}. \tag{3.46}$$

The components of the relative velocity are as follows:

$$\beta^\mu = (\beta^0, \beta^i) = (1, v^i/c). \tag{3.47}$$

The relationship to the velocity vector is given by

$$U^\mu = \gamma c \beta^\mu. \tag{3.48}$$

Hence,

$$U^\mu U_\mu = -c^2. \tag{3.49}$$

3.3 Momentum space and waves

The reciprocal wavevector space of k_μ plays a complementary role to that of spacetime. It measures changes in waves when one is not interested in spacetime locations. Pure harmonic (sinusoidal) waves are spread over an infinite distance. They have no beginning or end, only a definite wavelength.

In the quantum theory, energy and momentum are determined by the operators

$$E \to i\hbar \partial_t, \quad p_i \to -i\hbar \partial_i, \tag{3.50}$$

which have pure values when acting on plane wave states

$$\psi \sim \exp i(k_i x^i - \omega t). \tag{3.51}$$

In $(n + 1)$ dimensional notation, the wavevector becomes:

$$k_\mu = \left(-\frac{\omega}{c}, k_i\right), \tag{3.52}$$

so that plane waves take the simple form

$$\psi \sim \exp(ik_\mu x^\mu). \tag{3.53}$$

The energy and momentum are therefore given by the time and space eigenvalues of the operator

$$p_\mu = -i\hbar \partial_\mu, \tag{3.54}$$

respectively, as they act upon a plane wave. This leads to the definition of an $(n + 1)$ dimensional energy–momentum vector,

$$p_\mu = \hbar k_\mu = \left(-\frac{E}{c}, p_i \right). \tag{3.55}$$

The identification $p_\mu = \hbar k_\mu$ is the de Broglie relation for matter waves. This is one of the most central and important relations in the definition of the quantum theory of matter.

In discussing wavelike excitations, it is useful to resolve the components of vectors along the direction of motion of the wave (longitudinal) and perpendicular (transverse) to the direction of motion. A longitudinal vector is one proportional to a vector in the direction of motion of a wave k^μ. A transverse vector is orthogonal to this vector. The longitudinal and transverse components of a vector are defined by

$$V_{\mathrm{L}}^\mu \equiv \frac{k^\mu k_\nu}{k^2} V^\nu$$
$$V_{\mathrm{T}}^\mu \equiv \left(g_{\mu\nu} - \frac{k^\mu k_\nu}{k^2} \right) V^\nu. \tag{3.56}$$

It is straightforward to verify that the two projection operators

$$P_{\mathrm{L}}{}^\mu{}_\nu = \frac{k^\mu k_\nu}{k^2}$$
$$P_{\mathrm{T}}{}^\mu{}_\nu = \left(g_{\mu\nu} - \frac{k^\mu k_\nu}{k^2} \right) \tag{3.57}$$

are orthogonal to one another:

$$(P_{\mathrm{L}})^\mu{}_\nu (P_{\mathrm{T}})^\nu{}_\lambda = 0. \tag{3.58}$$

3.4 Tensor transformations

Vector equations remain true in general coordinate frames because the components of a vector transform according to specific rules under a coordinate transformation U:

$$\mathbf{v}' = U \mathbf{v}, \tag{3.59}$$

or

$$v'^i = U^i{}_j \, v^j, \qquad (3.60)$$

where the components of the matrix U are fixed by the requirement that the equations remain true in general coordinate systems. This is a valuable property, and we should be interested in generalizations of this idea which might be useful in physics.

Tensors are objects with any number of indices, which have the same basic transformation properties as vectors. The number of indices on a tensor is its rank. Each free index in a tensor equation requires a transformation matrix under changes of coordinates; free indices represent the components in a specific coordinate basis, and each summed index is invariant since scalar products are independent of basis.

Under a change of coordinates, $x \rightarrow x'$, a scalar (rank 0-tensor) transforms simply as

$$\phi(x) \rightarrow \phi(x'). \qquad (3.61)$$

For a vector (rank 1-tensor), such a simple rule does make sense. If one rotates a coordinate system, for instance, then all the components of a vector must change, since it points in a new direction with respect to the coordinate axes. Thus, a vector's components must transform separately, but as linear combinations of the old components. The rule for a vector with raised index is:

$$V^\mu(x') = \frac{\partial x'^\mu}{\partial x^\nu} \, V^\nu(x) = (\partial_\nu x'^\mu) \, V^\nu(x). \qquad (3.62)$$

For a vector with lowered index, it is the converse:

$$V_\mu(x') = \frac{\partial x^\nu}{\partial x'^\mu} \, V_\nu(x) = (\partial'_\mu x^\nu) \, V_\nu(x). \qquad (3.63)$$

Here we have used two notations for the derivatives: the longhand notation first for clarity and the shorthand form which is more compact and is used throughout this book.

The metric tensor is a tensor of rank 2. Using the property of the metric in raising and lowering indices, one can also deduce its transformation rule under the change of coordinates from x to x'. Starting with

$$V^\mu(x') = g^{\mu\nu}(x')V_\nu(x'), \qquad (3.64)$$

and expressing it in the x coordinate system, using the transformation above, one obtains:

$$(\partial_\nu x'^\mu)V^\nu(x) = g^{\mu\sigma}(x')(\partial'_\sigma x^\rho)V_\rho(x). \qquad (3.65)$$

However, it is also known that, in the unprimed coordinates,

$$V^\nu(x) = g^{\nu\sigma}(x)V_\sigma(x). \tag{3.66}$$

Comparing eqns. (3.65) and (3.66), it is possible to deduce the transformation rule for the inverse metric $g^{\mu\nu}$. To do this, one rearranges eqn. (3.65) by multiplying by $(\partial'_\mu x^\tau)$ and using the chain-rule:

$$(\partial_\nu x'^\mu)(\partial'_\mu x^\tau) = \delta_\nu^\tau. \tag{3.67}$$

Being careful to re-label duplicate indices, this gives

$$\delta\nu\,\tau\,V^\nu(x) = g^{\mu\sigma}(x')(\partial'_\mu x^\tau)(\partial'_\sigma x^\rho)\,V_\rho(x), \tag{3.68}$$

which is

$$V^\tau(x) = g^{\mu\rho}(x')(\partial'_\mu x^\tau)(\partial'_\rho x^\sigma)V_\sigma(x). \tag{3.69}$$

Comparing this with eqn. (3.66), one finds that

$$g^{\rho\mu}(x')(\partial'_\mu x^\tau)(\partial'_\rho x^\sigma) = g^{\tau\sigma}(x), \tag{3.70}$$

or, equivalently, after re-labelling and re-arranging once more,

$$g^{\mu\nu}(x') = (\partial_\rho x'^\mu)(\partial_\sigma x'^\nu)g^{\rho\sigma}(x). \tag{3.71}$$

One sees that this follows the same pattern as the vector transformation with raised indices. The difference is that there is now a partial derivative matrix $(\partial_\sigma x'^\nu)$ for each index. In fact, this is a general feature of tensors. Each raised index transforms with a factor like $(\partial_\sigma x'^\nu)$ and each lowered index transforms with a factor like $\partial'_\mu x^\nu$. For instance,

$$T^{\mu\nu}_{\ \ \rho\sigma}(x') = (\partial_\alpha x'^\mu)(\partial_\beta x'^\nu)(\partial'_\rho x^\gamma)(\partial'_\sigma x^\delta)T^{\alpha\beta}_{\ \ \gamma\delta}. \tag{3.72}$$

3.5 Properties

The following properties of tensors are instructive and useful.

(1) Any matrix T may be written as a sum of a symmetric part $\overline{T}_{ij} = \frac{1}{2}(T_{ij} + T_{ji})$ and an anti-symmetric part $\tilde{T}_{ij} = \frac{1}{2}(T_{ij} - T_{ji})$. Thus one may write any 2×2 matrix in the form

$$T = \begin{pmatrix} \overline{T}_{11} & \overline{T}_{12} + \tilde{T}_{12} \\ \overline{T}_{12} - \tilde{T}_{12} & \overline{T}_{22} \end{pmatrix} \tag{3.73}$$

(2) It may be shown that the trace of the product of a symmetric matrix with an anti-symmetric matrix is zero, i.e. $\overline{S}^{ij}\tilde{T}_{ij} = 0$.

(3) By considering similarity transformations of the form $T \to \Lambda^{-1} T \Lambda$, one may show that the trace of any matrix is an invariant, equal to the sum of its eigenvalues.

(4) By definition, a rank 2-tensor T transforms by the following matrix multiplication rule:

$$T \to \Lambda^{\mathrm{T}} T \Lambda, \tag{3.74}$$

for some transformation matrix Λ. Consider a general 2×2 tensor

$$T = \begin{pmatrix} \frac{1}{2} t + \Delta T_{11} & \overline{T}_{12} + \tilde{T}_{12} \\ \overline{T}_{12} + \tilde{T}_{12} & \frac{1}{2} t + \Delta T_{22} \end{pmatrix},$$

where t is the trace $t = (\overline{T}_{11} + \overline{T}_{22})$, and consider the effect of the following matrices on T:

$$\Lambda_0 = \begin{pmatrix} a & 0 \\ 0 & d \end{pmatrix}$$

$$\Lambda_1 = \begin{pmatrix} 0 & i \\ -i & 0 \end{pmatrix}$$

$$\Lambda_2 = \begin{pmatrix} 0 & 1 \\ 1 & 0 \end{pmatrix}$$

$$\Lambda_3 = \frac{1}{\sqrt{2}} \begin{pmatrix} 1 & 1 \\ 1 & -1 \end{pmatrix}. \tag{3.75}$$

For each of these matrices, compute:

(a) $\Lambda^{\mathrm{T}} \Lambda$,

(b) $\Lambda^{\mathrm{T}} T \Lambda$.

It may be shown that, used as a transformation on T:

(a) the anti-symmetric matrix Λ_1 leaves anti-symmetric terms invariant and preserves the trace of T;

(b) the off-diagonal symmetric matrix Λ_2 leaves the off-diagonal symmetric terms invariant and preserves the trace of T;

(c) the symmetrical, traceless matrix Λ_3, preserves only the trace of T.

It may thus be concluded that a tensor T in n dimensions has three separately invariant parts and may be written in the form

$$T_{ij} = \frac{1}{n} T^k_k \delta_{ij} + \overline{T}_{ij} + \left(\tilde{T}_{ij} - \frac{1}{n} T^k_k \delta_{ij} \right). \tag{3.76}$$

3.6 Euclidean and Riemannian spacetime

Minkowski spacetime has an indefinite metric tensor signature. In Euclidean and Riemannian spacetime, the metric signature is definite (usually positive definite). General curved spaces with a definite signature are referred to as Riemannian manifolds. Multiplying the time components of vectors and tensors by the square-root of minus one (i) allows one to pass from Minkowski spacetime to Euclidean spacetime and back again. This procedure is known as *Wick rotation* and is encountered in several contexts in quantum theory. For instance, it serves a regulatory role: integrals involving the Lorentzian form $(k^2 + m^2)^{-1}$ are conveniently evaluated in Euclidean space, where $k^2 + m^2$ has no zeros. Also, there is a convenient relationship between equilibrium thermodynamics and quantum field theory in Euclidean space.

We shall use subscripts and superscripts 'E' to indicate quantities in Euclidean space; 'M' denotes Minkowski space, for this section only. Note that the transformation affects only the time or zeroth components of tensors; the space parts are unchanged.

The transformation required to convert from Minkowski spacetime (with its indefinite metric) to Euclidean spacetime (with its definite metric) is motivated by the appearance of plane waves in the Fourier decomposition of field variables. Integrals over plane waves of the form $\exp\ \mathrm{i}(\mathbf{k} \cdot \mathbf{x} - \omega t)$ have no definite convergence properties, since the complex exponential simply oscillates for all values of $\mathbf{k}$ and ω. However, if one adds a small imaginary part to time $t \to t - \mathrm{i}\tau$, then we turn the oscillatory behaviour into exponential decay:

$$e^{\mathrm{i}(\mathbf{k}\cdot\mathbf{x}-\omega t)} \to e^{\mathrm{i}(\mathbf{k}\cdot\mathbf{x}-\omega t)}e^{-\omega\tau}. \tag{3.77}$$

The requirement of decay rather than growth chooses the sign for the Wick rotation. An equivalent motivation is to examine the Lorentzian form:

$$\frac{1}{k^2 + m^2} = \frac{1}{-k_0^2 + \mathbf{k}^2 + m^2} = \frac{1}{(-k_0 + \sqrt{\mathbf{k}^2 + m^2})(k_0 + \sqrt{\mathbf{k}^2 + m^2})}. \tag{3.78}$$

This is singular and has poles on the real k_0 axis at $k_0 = \pm\sqrt{\mathbf{k}^2 + m^2}$. This makes the integral of k_0 non-analytical, and a prescription must be specified for integrating around the poles. The problem can be resolved by adding a small (infinitesimal) imaginary part to the momenta:

$$\frac{1}{k^2 + m^2 - \mathrm{i}\epsilon} = \frac{1}{(-k_0 - \mathrm{i}\epsilon + \sqrt{\mathbf{k}^2 + m^2})(k_0 - \mathrm{i}\epsilon + \sqrt{\mathbf{k}^2 + m^2})}. \tag{3.79}$$

This effectively shifts the poles from the real axis to above the axis for negative k_0 and below the axis for positive k_0. Since it is possible to rotate the contour

90 degrees onto the imaginary axis without having to pass through any poles, by defining (see section 6.1.1)

$$k_0^E = ik_0, \tag{3.80}$$

this once again chooses the sign of the rotation. The contour is rotated clockwise by 90 degrees, the integrand is positive definite and no poles are encountered in an integral over κ_0:

$$\frac{1}{-k_0^2 + \mathbf{k}^2 + m^2 - i\epsilon} \rightarrow \frac{1}{k_{0E}^2 + \mathbf{k}^2 + m^2}. \tag{3.81}$$

All variables in a field theory must be rotated consistently:

$$x_E^0 = -ix^0 \tag{3.82}$$
$$x_0^E = ix_0 \tag{3.83}$$
$$k_0^E = ik_0 = -i\omega/c. \tag{3.84}$$

The inner product

$$k_\mu x^\mu = \mathbf{k} \cdot \mathbf{x} + k_0 x^0 \rightarrow \mathbf{k} \cdot \mathbf{x} + \kappa_0 x^0 \tag{3.85}$$

is consistent with

$$\partial_0 x^0 = \partial_0^E x_E^0 = 1 \tag{3.86}$$

where

$$\partial_0^E = i\partial_0, \tag{3.87}$$

since $\partial_0^E \rightarrow i\kappa_0$. Since the Wick transformation affects derivatives and vectors, it also affects Maxwell's equations. From

$$\partial^\nu F_{\mu\nu} = \mu_0 J_\mu, \tag{3.88}$$

we deduce that

$$J_0^E = iJ_0 \tag{3.89}$$
$$A_0^E = iA_0, \tag{3.90}$$

which are necessary in view of the homogeneous form of the field strength:

$$-iF_{0i}^E = \partial_0 A_i - \partial_i A_0 = F_{0i}. \tag{3.91}$$

Notice that, in $(3 + 1)$ dimensions, this means that

$$\frac{1}{2} F^{\mu\nu} F_{\mu\nu} = \left(\mathbf{B}^2 - \frac{\mathbf{E}^2}{c^2}\right) = \left(\mathbf{B}^2 + \frac{\mathbf{E}_E^2}{c^2}\right). \tag{3.92}$$

Notice how the Euclideanized Lagrangian takes on the appearance of a Hamiltonian. This result is the key to relating Wick-rotated field theory to thermodynamical partition functions. It works because the quantum phase factor $\exp(iS_M/\hbar)$ looks like the partition function, or statistical weight factor $\exp(-\beta H_M)$ when Wick-rotated:

$$S_E = -iS_M, \tag{3.93}$$

since the volume measure $dV_x^E = -idV_x$. The superficial form of the Lagrangian density is unchanged in theories with only quadratic derivatives provided everything is written in terms of summed indices, but internally all of the time-summed terms have changed sign. Thus, one has that

$$\exp\left(i\frac{S_M}{\hbar}\right) = \exp\left(-\frac{S_E}{\hbar}\right) \sim \exp\left(-\frac{1}{\hbar}\int dV_E\, \mathcal{H}_M\right). \tag{3.94}$$

A Euclideanized invariant becomes something which looks like a Minkowski space non-invariant. The invariant F^2, which is used to deduce the dynamics of electromagnetism, transformed into Euclidean space, resembles a non-invariant of Minkowski space called the *Hamiltonian*, or total energy function (see eqn. (2.70)). This has physical as well as practical implications for field theories at finite temperature. If one takes the Euclidean time to be an integral from zero to $\hbar\beta$ and take $H = \int d\sigma\mathcal{H}$,

$$\exp\left(i\frac{S_M}{\hbar}\right) = \exp\left(-\frac{1}{\beta}H_M\right), \tag{3.95}$$

then a Euclidean field theory phase factor resembles a Minkowski space, finite-temperature Boltzmann factor. This is discussed further in chapter 6.

In a Cartesian basis, one has

$$g_{\mu\nu} \rightarrow g_{\mu\nu}^E = \delta_{\mu\nu}. \tag{3.96}$$

4

The action principle

The variational principle is central to covariant field theory. It displays symmetries, field equations and continuity conditions on an equal footing. It can be used as the starting point for every field theoretical analysis. In older books, the method is referred to as Hamilton's principle. In field theory it is referred to more colloquially as the *action principle*. Put plainly, it is a method of *generating functionals*; it compresses all of the kinematics and dynamics of a physical theory into a single integral expression S called the action.

The advantage of the action principle is that it guarantees a well formulated dynamical problem, assuming only the existence of a set of parameters on which the dynamical variables depends. Any theory formulated as, and derived from an action principle, automatically leads to a complete dynamical system of equations with dynamical variables which play the roles of positions and momenta, by analogy with Newtonian mechanics. To formulate a new model in physics, all one does is formulate invariant physical properties in the form of an action, and the principle elucidates the resulting kinematical and dynamical structure in detail.

4.1 The action in Newtonian particle mechanics

Consider a system consisting of a particle with position $q(t)$ and momentum $p(t)$. The kinetic energy of the particle is

$$T = \frac{1}{2}m\dot{q}^2, \tag{4.1}$$

and the potential energy is simply denoted $V(q)$. The 'dot' over the q denotes the time derivative, or

$$\dot{q} = \frac{\mathrm{d}q}{\mathrm{d}t}. \tag{4.2}$$

Classical mechanics holds that the equation of motion for a classical particle is Newton's law:

$$F = m\ddot{q} = -\frac{dV}{dq}, \tag{4.3}$$

but it is interesting to be able to derive this equation from a general principle. If many equations of motion could be derived from a common principle, it would represent a significant compression of information in physics. This is accomplished by introducing a generating function L called the Lagrangian. For a conservative system, the Lagrangian is defined by

$$L = T - V, \tag{4.4}$$

which, in this case, becomes

$$L = \frac{1}{2}m\dot{q}^2 - V(q). \tag{4.5}$$

This form, kinetic energy minus potential energy, is a coincidence. It does not apply to all Lagrangians. In relativistic theories, for instance, it is not even clear what one should refer to as the kinetic and potential energies. The Lagrangian is a generating function; it has no unique physical interpretation.

The Lagrangian is formally a function of q and $\dot{q}$. The general rule for obtaining the equations of motion is the well known Euler–Lagrange equations. They are

$$\frac{\partial L}{\partial q} - \frac{d}{dt}\left(\frac{\partial L}{\partial \dot{q}}\right) = 0. \tag{4.6}$$

If the physical system is changed, one only has to change the Lagrangian: the general rule will remain true. Evaluating, in this case,

$$\frac{\partial L}{\partial q} = -\frac{dV}{dq}$$
$$\frac{\partial L}{\partial \dot{q}} = m\dot{q}, \tag{4.7}$$

one obtains the field equations (4.3), as promised.

Is this approach better than a method in which one simply writes down the field equations? Rather than changing the field equations for each case, one instead changes the Lagrangian. Moreover, eqn. (4.6) was pulled out of a hat, so really there are two unknowns now instead of one! To see why this approach has more to offer, we introduce the *action*.

4.1.1 Variational principle

The fact that one can derive known equations of motion from an arbitrary formula involving a constructed function L is not at all surprising – there are hundreds of possibilities; indeed, the motivation for such an arbitrary procedure is not clear. The fact that one can obtain them from a function involving only the potential and kinetic energies of the system, for any conservative system, is interesting. What is remarkable is the fact that one can derive the Euler–Lagrange equations (i.e. the equations of motion), together with many other important physical properties for any system, from one simple principle: the action principle.

Consider the action S from the Lagrangian by

$$S_{12} = \int_{t_1}^{t_2} L(q, \dot{q}) \mathrm{d}t. \tag{4.8}$$

The action has (naturally) dimensions of *action* or 'energy $\times$ time', and is thought of as being a property of the path $q(t)$ of our particle between the fixed points $q(t_1)$ and $q(t_2)$. The action has no physical significance in itself. Its significance lies instead in the fact that it is a *generating functional* for the dynamical properties of a physical system.

When formulating physics using the action, it is not necessary to consider the fact that q and $\dot{q}$ are independent variables: that is taken care of automatically. In fact, the beauty of the action principle is that all of the useful information about a physical system falls out of the action principle more or less automatically.

To extract information from S, one varies it with respect to its dynamical variables, i.e. one examines how the integral changes when the key variables in the problem are changed. The details one can change are t_1 and t_2, the end-points of integration, and $q(t)$, the path or world-line of the particle between those two points (see figure 4.1). Note however that $Q(t)$ is the path the particle would take from A to B, and that is not arbitrary: it is determined by, or determines, physical law, depending on one's view. So, in order to make the variational principle a useful device, we have to be able to select the correct path by some simple criterion.

Remarkably, the criterion is the same in every case: one chooses the path which minimizes (or more correctly: makes stationary) the action; i.e. we look for paths $q(t)$ satisfying

$$\frac{\delta S}{\delta q(t)} = 0. \tag{4.9}$$

These are the *stable* or *stationary* solutions to the variational problem. This tells us that most physical laws can be thought of as regions of stability in a space of all solutions. The action behaves like a potential, or stability measure, in this space.

It is an attractive human idea (Occam's razor) that physical systems do the 'least action' possible; however, eqn. (4.9) is clearly no ordinary differentiation. First of all, S is a scalar number – it is integrated over a dummy variable t, so t is certainly not a variable on which S depends. To distinguish this from ordinary differentiation of a function with respect to a variable, it is referred to as *functional differentiation* because it is differentiation with respect to a function.

The functional variation of S with respect to $q(t)$ is defined by

$$\delta S = S[q + \delta q] - S[q], \tag{4.10}$$

where $\delta q(t)$ is an infinitesimal change in the form of the function q at time t. Specifically, for the single-particle example,

$$\delta S = \int dt \left\{ \frac{1}{2} m(\dot{q} + \delta \dot{q})^2 - V(q + \delta q) \right\} - \int dt \left\{ \frac{1}{2} m\dot{q}^2 - V(q) \right\}. \tag{4.11}$$

Now, since δq is infinitesimal, we keep only the first-order contributions, so on expanding the potential to first order as a Taylor series about $q(t)$,

$$V(q + \delta q) = V(q) + \frac{dV}{dq} \delta q + \cdots, \tag{4.12}$$

one obtains the first-order variation of S,

$$\delta S = \int dt \left\{ m\dot{q}(\partial_t \delta q) - \frac{dV}{dq} \delta q \right\}. \tag{4.13}$$

A 'dot' has been exchanged for an explicit time derivative to emphasize the time derivative of δq. Looking at this expression, one notices that, if the time derivative did not act on δq, we would be able to take out an overall factor of δq, and we would be almost ready to move δq to the left hand side to make something like a derivative. Since we are now operating under the integral sign, it is possible to integrate by parts, using the property:

$$\int dt\, A(\partial_t B) = \left[AB \right]_{t_1}^{t_2} - \int dt\, (\partial_t A) B, \tag{4.14}$$

so that the time derivative can be removed from δq, giving:

$$\delta S = \int dt \left\{ -m\ddot{q}(t) - \frac{dV}{dq(t)} \right\} \delta q(t) + \left[m\dot{q} \cdot \delta q(t) \right]_{t_1}^{t_2}. \tag{4.15}$$

The stationary action criterion tells us that $\delta S = 0$. Assuming that $q(t)$ is not always zero, one obtains a restriction on the allowed values of $q(t)$. This result must now be interpreted.

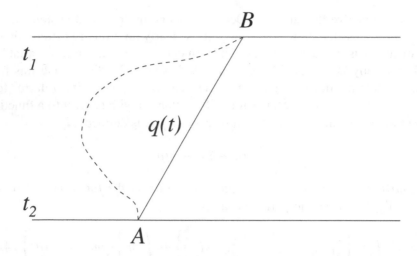

Fig. 4.1. The variational formula selects the path from *A* to *B* with a *stationary* value of the action. Stationary or *minimum* means that the solution is stable on the surface of all field solutions. Unless one adds additional perturbations in the action, it will describe the 'steady state' behaviour of the system.

4.1.2 δS: equation of motion

The first thing to notice about eqn. (4.15) is that it is composed of two logically separate parts. The first term is an integral over all times which interpolate between t_1 and t_2, and the second is a term which lives only at the end-points. Now, suppose we ask the question: what path $q(t)$ is picked out by the action principle, if we consider all the possible variations of paths $q(t) + \delta q(t)$, given that the two end-points are always fixed, i.e. $\delta q(t_1) = 0$ and $\delta q(t_2) = 0$?

The requirement of fixed end-points now makes the second term in eqn. (4.15) vanish, so that $\delta S = 0$ implies that the contents of the remaining curly braces must vanish. This gives precisely the equation of motion

$$m\ddot{q} = -\frac{dV}{dq}. \qquad (4.16)$$

The action principle delivers the required formula as promised. This arises from an equation of *constraint* on the path $q(t)$ – a constraint which forces the path to take a value satisfying the equation of motion. This notion of constraint recurs later, in more advanced uses of the action principle.

4.1.3 The Euler–Lagrange equations

The Euler–Lagrange equations of motion are trivially derived from the action principle for an arbitrary Lagrangian which is a function of q and $\dot{q}$. The action one requires is simply

$$S = \int dt\, L(q(t), \dot{q}(t)), \tag{4.17}$$

and its variation can be written, using the functional chain-rule,

$$\delta S = \int dt \left\{ \frac{\delta L}{\delta q} \delta q + \frac{\delta L}{\delta(\partial_t q)} \delta(\partial_t q) \right\} = 0. \tag{4.18}$$

The variation of the path commutes with the time derivative (trivially), since

$$\delta(\partial_t q) = \partial_t q(\tau + \delta\tau) - \partial_t q(\tau) = \partial_t(\delta q). \tag{4.19}$$

Thus, one may re-write eqn. (4.18) as

$$\delta S = \int dt \left\{ \frac{\delta L}{\delta q} \delta q + \frac{\delta L}{\delta(\partial_t q)} \partial_t(\delta q) \right\} = 0. \tag{4.20}$$

Integrating the second term by parts, one obtains

$$\delta S = \int dt \left\{ \frac{\delta L}{\delta q} \delta q - \partial_t \left(\frac{\delta L}{\delta(\partial_t q)} \right) (\delta q) \right\} + \int d\sigma \left[\frac{\delta L}{\delta(\partial_t q)} \delta q \right] = 0. \tag{4.21}$$

The second term vanishes independently (since its variation is zero at the fixed end-points), and thus one obtains the Euler–Lagrange equations (4.6).

4.1.4 δS: continuity

Before leaving this simple world of classical particles, there is one more thing to remark about eqn. (4.21). Consider the second term; when one asks the question: what is the condition on $q(t)$ for the classical trajectories with stationary action and fixed end-points? – this term drops out. It vanishes by assumption. It contains useful information however. If we consider the example of a single particle, the surface term has the form

$$m\dot{q} \cdot \delta q = p\delta q. \tag{4.22}$$

This term represents the momentum of the particle. For a general Lagrangian, one can use this fact to define a 'generalized momentum'. From eqn. (4.21)

$$p = \frac{\delta L}{\delta(\partial_t q)} \equiv \Pi. \tag{4.23}$$

Traditionally, this quantity is called the canonical momentum, or conjugate momentum, and is denoted generically as Π.

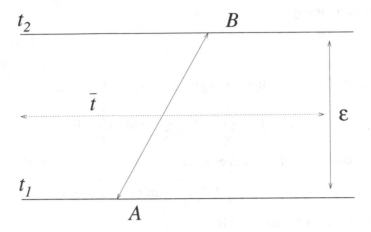

Fig. 4.2. The continuity of paths obeying the equations of motion, over an infinitesimal interval is assured by the null variation of the action over that interval.

Suppose one asks a different question of the variation. Consider only an infinitesimal time period $t_2 - t_1 = \epsilon$, where $\epsilon \to 0$. What happens between the two limits of integration in eqn. (4.21) is now less important. In fact, it becomes decreasingly important as $\epsilon \to 0$, since

$$\delta S_{12} = [\, p\delta q\,]_{t_1}^{t_2} + \mathrm{O}(\epsilon). \tag{4.24}$$

What infinitesimal property of the action ensures that $\delta S = 0$ for all intermediate points between the limits t_1 and t_2? To find out, we relax the condition that the end-points of variation should vanish. Then, over any infinitesimal interval ϵ, the change in $\delta q(t)$ can itself only be infinitesimal, unless $q(t)$ is singular, but it need not vanish. However, as $\epsilon \to 0$, the change in this quantity must also vanish as long as $q(t)$ is a smooth field, so one must take $\Delta(\delta q) = 0$.[1] This means that

$$\Delta p \equiv p(t_2) - p(t_1) = 0; \tag{4.25}$$

i.e. the change in momentum across any infinitesimal surface is zero, or momentum is conserved at any point. This is a continuity condition on $q(t)$. To see this, ask what would happen if the potential $V(q)$ contained a singular term at the surface:

$$V(q, t) = \delta(t - \bar{t})\Delta V + V(q), \tag{4.26}$$

---

[1] Note that we are assuming that the field is a continuous function, but the momentum need not be strictly continuous if there are impulsive forces (influences) on the field. This is fully consistent with our new philosophy of treating the 'field' q as a fundamental variable, and p as a derived quantity.

where $\frac{1}{2}(t_1 + t_2)$ is the mid-point of the infinitesimal interval. Here, the delta function integrates out immediately, leaving an explicit surface contribution from the potential, in addition to the term from the integration by parts:

$$\delta S_{12} = \frac{d\Delta V}{dq}\delta q + [\,p\delta q\,]_{t_1}^{t_2} + O(\epsilon) = 0, \tag{4.27}$$

Provided ΔV is finite, using the same argument as before, one obtains,

$$\Delta p = -\frac{d\Delta V}{dq}, \tag{4.28}$$

i.e. the change in momentum across any surface is a direct consequence of the impulsive force $d\Delta V/dq$ at that surface.

We thus have another facet of the action: it evaluates relationships between dynamical variables which satisfy the constraints of stable behaviour. This property of the action is very useful: it generates standard continuity and boundary conditions in field theory, and is the backbone of the canonical formulation of both classical and quantum mechanics. For instance, in the case of the electromagnetic field, we can generate all of the 'electromagnetic boundary conditions' at interfaces using this technique (see section 21.2.2). This issue occurs more generally in connection with the energy–momentum tensor, in chapter 11, where we shall re-visit and formalize this argument.

4.1.5 Relativistic point particles

The relativistically invariant form of the action for a single point particle is

$$S = \int dt\sqrt{-g_{00}}\left\{-\frac{1}{2}m\frac{dx^i(t)}{dt}g_{ij}\frac{dx^j(t)}{dt} + V\right\}. \tag{4.29}$$

The particle positions trace out world-lines $q(\tau) = \mathbf{x}(\tau)$. If we re-express this in terms of the proper time τ of the particle, where

$$\tau = t\gamma^{-1}$$
$$\gamma = 1/\sqrt{(1 - \beta^2)}$$
$$\beta^2 = \frac{\mathbf{v}^2}{c^2} = \frac{1}{c^2}\left(\frac{d\mathbf{x}}{dt}\right)^2, \tag{4.30}$$

then the action may now be written in the frame of the particle,

$$dt \to \gamma d\tau$$
$$\sqrt{g} \to \gamma\sqrt{g}, \tag{4.31}$$

giving

$$S = \int d\tau \sqrt{g_{00}} \left\{ -\frac{1}{2}m \left(\frac{d\mathbf{x}(\tau)}{d\tau} \right)^2 + V\gamma^{-2} \right\}. \tag{4.32}$$

The field equations are therefore

$$\frac{\delta S}{\delta \mathbf{x}} = m\frac{d^2\mathbf{x}}{d\tau^2} + \frac{\partial V'}{\partial \mathbf{x}} = 0, \tag{4.33}$$

i.e.

$$\mathbf{F} = m\mathbf{a}, \tag{4.34}$$

where

$$\mathbf{F} = -\nabla V'$$
$$\mathbf{a} = \frac{d^2\mathbf{x}}{d\tau^2}. \tag{4.35}$$

The conjugate momentum from the continuity condition is

$$\mathbf{p} = m\frac{d\mathbf{x}}{d\tau}, \tag{4.36}$$

which is simply the relativistic momentum vector $\mathbf{p}$. See section 11.3.1 for the energy of the classical particle system.

In the above derivation, we have treated the metric tensor as a constant, but in curved spacetime $g_{\mu\nu}$ depends on the coordinates. In that case, the variation of the action leads to the field equation

$$\frac{d}{d\tau}\left(g_{\mu\nu}\frac{dx^\nu}{d\tau} \right) - \frac{1}{2}(\partial_\mu g_{\rho\nu})\frac{dx^\nu}{d\tau}\frac{dx^\rho}{d\tau} = 0. \tag{4.37}$$

The equation of a free particle on a curved spacetime is called the geodesic equation. After some manipulation, it may be written

$$\frac{d^2x^\mu}{d\tau^2} + \Gamma^\mu{}_{\nu\rho}\frac{dx^\nu}{d\tau}\frac{dx^\rho}{d\tau} = 0. \tag{4.38}$$

Interestingly, this equation can be obtained from the absurdly simple variational principle:

$$\delta \int ds = 0, \tag{4.39}$$

where ds is the line element, described in section 3.2.5. See also section 25.4.

4.2 Frictional forces and dissipation

In many branches of physics, phenomenological equations are used for the dissipation of energy. Friction and ohmic resistance are two common examples. Empirical frictional forces cannot be represented by a microscopic action principle, since they arise physically only through time-dependent boundary conditions on the system. No fundamental dynamical system is dissipative at the microscopic level; however, fluctuations in dynamical variables, averaged over time, can lead to a re-distribution of energy within a system, and this is what leads to dissipation of energy from one part of a system to another. More advanced statistical notions are required to discuss dissipation fully, but a few simple observations can be made at the level of the action.

Consider the example of the frictional force represented by Langevin's equation:

$$m \frac{d^2 x}{dt} + \alpha \dot{x} = F(t).$$ (4.40)

Initially it appears as though one could write the action in the following way:

$$S = \int dt \left\{ \frac{1}{2} m \left(\frac{dx}{dt} \right)^2 + \frac{1}{2} \alpha x \frac{dx}{dt} \right\}.$$ (4.41)

However, if one varies this action with respect to x, the term proportional to α gives

$$\int dt \, \alpha \left(\delta x \frac{d}{dt} x + x \frac{d}{dt} \delta x \right).$$ (4.42)

But this term is a total derivative. Integrating by parts yields

$$\int dt_a^b \frac{d}{dt} (x^2) = x^2 \Big|_a^b = 0,$$ (4.43)

which may be ignored, since it exists only on the boundary. Because of the reversibility of the action principle, one cannot introduce terms which pick out a special direction in time. The only place where such terms can appear is through boundary conditions. For the same reason, it is impossible to represent Ohm's law

$$J^i = \sigma E^i$$ (4.44)

in an action principle. An ohmic resistor has to dissipate heat as current passes through it.

In some cases, the action principle can tricked into giving a non-zero contribution from velocity-dependent terms by multiplying the whole Lagrangian with an 'integrating factor' $\exp(\gamma(t))$, but the resulting field equations require $\gamma(t)$ to make the whole action decay exponentially, and often the results are ambiguous and not physically motivated.

We shall return to the issue of dissipation in detail in chapter 6 and show the beginnings of how physical boundary conditions and statistical averages can be incorporated into the action principle, in a consistent manner, employing the principle of causality. It is instructive to show that it is not possible to write down a gauge-invariant action for the equation

$$J^i = \sigma E^i. \tag{4.45}$$

i.e. Ohm's law, in terms of the vector potential A_μ. The equation is only an effective representation of an averaged statistical effect, because it does provide a reversible description of the underlying physics.

(1) By varying with respect to A_μ, one may show that the action

$$S = \int (\mathrm{d}x) \left\{ J^i A_i - \sigma_{ij} A^i E^j \right\} \tag{4.46}$$

with $E_i = -\partial_t A_i - \partial_i A_0$, does not give eqn. (4.45). If one postulates that E^i and J^i may be replaced by their steady state (time-independent) averages $\langle E^i \rangle$ and $\langle J^i \rangle$, then we can show that this does give the correct equation. This is an indication that some averaging procedure might be the key to representing dissipative properties of bulk matter.

(2) Consider the action

$$S = \int (\mathrm{d}x) \left\{ J^\mu A_\mu - \sigma_{ij} A^i E^j e^{-\gamma^\mu x_\mu} \right\}. \tag{4.47}$$

This may be varied with respect to A_0 and A_i to find the equations of motion; gauge invariance requires the equations to be independent of the vector potential A_μ. On taking $\sigma_{ij} = \sigma \delta_{ij}$, one can show that gauge invariance requires that the vector potential decay exponentially. Readers are encouraged to check whether the resulting equations of motion are a satisfactory representation of Ohm's law.

4.3 Functional differentiation

It is useful to define the concept of functional differentiation, which is to ordinary differentiation what $\delta q(t)$ is to $\mathrm{d}q$. Functional differentiation differs from normal differentiation in some important ways.

The ordinary derivative of a function with respect to its control variable is defined by

$$\frac{\mathrm{d}f(t)}{\mathrm{d}t} = \lim_{\delta t \to 0} \frac{f(t + \delta t) - f(t)}{\delta t}. \tag{4.48}$$

It tells us about how a function changes with respect to the value of its control variable at a given point. Functional differentiation, on the other hand, is something one does to an integral expression; it is performed with respect to a function of some variable of integration. The 'point of differentiation' is now a function $f(t)$ evaluated at a special value of its control variable t'. It takes some value from within the limits of the integral. So, whereas we start with a quantity which is not a function of t or t', the result of the functional derivation is a function which is evaluated at the point of differentiation. Consider, as an example, the arbitrary functional

$$F[f] = \int dt \sum_n a_n (f(t))^n. \tag{4.49}$$

This is clearly not a function of t due to the integral. The variation of such a functional $F[f]$ is given by

$$\delta F[f] = F[f(t) + \delta f(t)] - F[f(t)]. \tag{4.50}$$

We define the functional derivative by

$$\frac{\delta F}{\delta f(t')} = \lim_{\epsilon \to 0} \frac{F[f(t) + \epsilon \delta(t - t')] - F[f(t)]}{\epsilon}. \tag{4.51}$$

This is a function, because an extra variable t' has been introduced. You can check that this has the unusual side effect that

$$\frac{\delta q(t)}{\delta q(t')} = \delta(t - t'), \tag{4.52}$$

which is logical (since we expect the derivative to differ from zero only if the function is evaluated at the same point), but unusual, since the right hand side is not dimensionless – in spite of the fact that the left hand side seems to be. On the other hand, if we define a functional

$$Q = \int dt\, q(t) \tag{4.53}$$

then we have

$$\frac{\delta Q}{\delta q(t')} = \int dt \frac{\delta q(t)}{\delta q(t')} = \int \delta(t - t') = 1. \tag{4.54}$$

Thus, the integral plays a key part in the definition of differentiation for functionals.

4.4 The action in covariant field theory

The action principle can be extended to generally covariant field theories. This generalization is trivial in practice. An important difference is that field theories are defined in terms of variables which depend not only on time but also on space; $\phi(\mathbf{x}, t) = \phi(x)$. This means that the action, which must be a scalar, without functional dependence, must also be integrated over space in addition to time. Since the final action should have the dimensions of energy $\times$ time, this means that the Lagrangian is to be replaced by a Lagrangian *density* $\mathcal{L}$

$$S = \int_\sigma^{\sigma'} (\mathrm{d}x)\mathcal{L}(\phi(\mathbf{x}, t), \partial_\mu\phi(\mathbf{x}, t), x). \tag{4.55}$$

The integral measure is $(\mathrm{d}x) = \mathrm{d}V_x/c$, where $\mathrm{d}V_x = c\mathrm{d}t\mathrm{d}^n\mathbf{x}\sqrt{g} = \mathrm{d}x^0\mathrm{d}^n\mathbf{x}\sqrt{g}$. Although it would be nice to use $\mathrm{d}V_x$ here (since this is the Minkowski space volume element), this is not possible if $\mathcal{L}$ is an energy density and S is to have the dimensions of action.[2] The non-relativistic action principle has already chosen this convention for us. The special role played by time forces is also manifest in that the volume is taken between an earlier time t and a later time t' – or, more correctly, from one spacelike hyper-surface, σ, to another, σ'.

The classical interpretation of the action as the integral over $T - V$, the kinetic energy minus the potential energy, does not apply in the general case. The Lagrangian density has no direct physical interpretation, it is merely an artefact which gives the correct equations of motion. What is important, however, is how one defines a Hamiltonian, or energy functional, from the action. The Hamiltonian is related to measurable quantities, namely the total energy of the system at a given time, and it is responsible for the time development of the system. One must be careful to use consistent definitions, e.g. by sticking to the notation and conventions used in this book.

Another important difference between field theory and particle mechanics is the role of position. Particle mechanics describes the trajectories of particles, $q(t)$, as a function of time. The position was a function with time as a parameter. In field theory, however, space and time are independent parameters, on a par with one another, and the ambient field is a function which depends on both of them. In particle mechanics, the action principle determines the equation for a constrained path $q(t)$; the field theoretical action principle determines an equation for a field which simultaneously exists at all spacetime points, i.e. it does not single out any trajectory in spacetime, but rather a set of allowed solutions for an omnipresent field space. In spite of this difference, the formal properties of the action principle are identical, but for an extra integration:

---

[2] One could absorb a factor of c into the definition of the field $\phi(x)$, since its dimensions are not defined, but this would then mean that the Lagrangian and Hamiltonian would not have the dimensions of energy. This blemish on the otherwise beautiful notation is eliminated when one chooses natural units in which $c = 1$.

4.4.1 Field equations and continuity

For illustrative purposes, consider the following action:

$$S = \int (\mathrm{d}x) \left\{ \frac{1}{2} (\partial^\mu \phi)(\partial_\mu \phi) + \frac{1}{2} m^2 \phi^2 - J\phi \right\}, \tag{4.56}$$

where $\mathrm{d}V_x = c\mathrm{d}t \, \mathrm{d}\mathbf{x}$. Assuming that the variables $\phi(x)$ commute with one another, the variation of this action is given by

$$\delta S = \int (\mathrm{d}x) \left\{ (\partial^\mu \delta \phi)(\partial_\mu \phi) + m^2 \phi \delta \phi - J \delta \phi \right\}. \tag{4.57}$$

Integrating this by parts and using the commutativity of the field, one has

$$\delta S = \int (\mathrm{d}x) \left\{ -\Box \phi + m^2 \phi - J \right\} + \int \mathrm{d}\sigma^\mu \, \delta \phi (\partial_\mu \phi). \tag{4.58}$$

From the general arguments given earlier, one recognizes a piece which is purely a surface integral and a piece which applies the field in a general volume of spacetime. These terms vanish separately. This immediately results in the field equations of the system,

$$(-\Box + m^2)\phi(x) = J(x), \tag{4.59}$$

and a continuity condition which we shall return to presently.

The procedure can be reproduced for a general Lagrangian density $\mathcal{L}$ and gives the Euler–Lagrange equations for a field. Taking the general form of the action in eqn. (4.55), one may write the first variation

$$\delta S = \int (\mathrm{d}x) \left\{ \frac{\partial \mathcal{L}}{\partial \phi} \delta \phi + \frac{\partial \mathcal{L}}{\partial (\partial^\mu \phi)} \delta (\partial_\mu \phi) \right\}. \tag{4.60}$$

Now, the variation symbol and the derivative commute with one another since they are defined in the same way:

$$\partial_\mu \delta \phi = \partial_\mu \phi (x + \Delta x) - \partial_\mu \phi(x)$$
$$= \delta(\partial_\mu \phi); \tag{4.61}$$

thus, one may integrate by parts to obtain

$$\delta S = \int (\mathrm{d}x) \left\{ \frac{\partial \mathcal{L}}{\partial \phi} - \partial_\mu \left(\frac{\partial \mathcal{L}}{\partial (\partial^\mu \phi)} \right) \right\} + \frac{1}{c} \int \mathrm{d}\sigma^\mu \, \delta \phi \left(\frac{\partial \mathcal{L}}{\partial (\partial^\mu \phi)} \right) \tag{4.62}$$

The first of these terms exists for every spacetime point in the volume of integration, whereas the second is restricted only to the bounding hyper-surfaces σ and σ'. These two terms must therefore vanish independently in general.

The vanishing integrand of the first term gives the Euler–Lagrange equations of motion for the field

$$\frac{\partial \mathcal{L}}{\partial \phi} - \partial_\mu \left(\frac{\partial \mathcal{L}}{\partial (\partial^\mu \phi)} \right) = 0, \tag{4.63}$$

and the vanishing of the second term leads to the boundary continuity condition,

$$\Delta \left(\delta\phi \frac{\partial \mathcal{L}}{\partial (\partial^\mu \phi)} \right) = 0. \tag{4.64}$$

If this result is compared with eqns. (4.22) and (4.23), an analogous 'momentum', or conjugate variable to the field $\phi(x)$, can be defined. This conjugate variable is unusually denoted $\Pi(x)$:

$$\Pi(x) = \frac{\delta L}{\partial (\partial^0 \phi)}, \tag{4.65}$$

and is derived by taking the canonical spacelike hyper-surface with $\sigma = 0$. Note the position of indices such that the variable transforms like a covariant vector $p = \partial_0 q$. The covariant generalization of this is

$$\Pi_\sigma(x) = \frac{\delta L}{\partial (\partial^\sigma \phi)}. \tag{4.66}$$

4.4.2 Uniqueness of the action

In deriving everything from the action principle, one could gain the impression that there is a unique prescription at work. This is not the case. The definition of the action itself is not unique. There is always an infinity of actions which generates the correct equations of motion. This infinity is obtained by multiplying the action by an arbitrary complex number. In addition to this trivial change, there may be several actions which give equivalent results depending on (i) what we take the object of variation to be, and (ii) what we wish to deduce from the action principle. For example, we might choose to re-parametrize the action using new variables. The object of variation and its conjugate are then re-defined.

It is clear from eqn. (4.21) that the field equations and boundary conditions would be the same if one were to re-define the Lagrangian by multiplying by a general complex number:

$$S \to (a + ib)S. \tag{4.67}$$

The complex factor would simply cancel out of the field equations and boundary conditions. Moreover, the Lagrangian itself has no physical meaning, so there is no physical impediment to such a re-definition. In spite of this, it is normal to choose the action to be *real*. The main reason for this is that this choice makes for a clean relationship between the Lagrangian and a new object, the Hamiltonian, which is related to the energy of the system and is therefore, by assumption, a real quantity.

Except in the case of the gravitational field, one is also free to add a term on to the action which is independent of the field variables, since this is always zero with respect to variations in the fields:

$$S \rightarrow S + \int (\mathrm{d}x)\, \Lambda. \qquad (4.68)$$

Such a term is often called a cosmological constant, because it was introduced by Einstein into the theory of relativity in order to create a static (non-expansive) cosmology. Variations of the action with respect to the metric are not invariant under the addition of this term, so the energy–momentum tensor in chapter 11 is not invariant under this change, in general. Since the Lagrangian density is an energy density (up to a factor of c), the addition of this arbitrary term in a flat (gravitation-free) spacetime simply reflects the freedom one has in choosing an origin for the scale of energy density for the field.[3]

Another way in which the action can be re-defined is by the addition of a total derivative,

$$S \rightarrow S + \int (\mathrm{d}x)\partial^{\mu} F_{\mu}[\phi]$$

$$= S + \int \mathrm{d}\sigma^{\mu} F_{\mu}[\phi]. \qquad (4.69)$$

The additional term exists only on the boundaries σ of the volume integral. By assumption, the surface term vanishes independently of the rest, thus, since the field equations are defined entirely from the non-surface contributions, they will never be affected by the addition of such a total derivative. However, the boundary conditions or continuity will depend on this addition. This has a physical interpretation: if the boundary of a physical system involves a discontinuous change, it implies the action of an external agent at the boundary. Such a jump is called a *contact potential*. It might signify the connection of a system to an external potential source (a battery attached by leads, for instance). The connection of a battery to a physical system clearly does not change the laws of physics (equations of motion) in the system, but it does change the boundary conditions.

In light of this observation, we must be cautious to write down a 'neutral', or unbiased action for free systems. This places a requirement on the action,

---

[3] Indeed, the action principle $\delta S = 0$ can be interpreted as saying that only potential differences are physical. The action potential itself has no unique physical interpretation.

namely that the action must be *Hermitian*, time-reversal-invariant, or symmetrical with respect to the placement of derivatives, so that, if we let $t \to -t$, then nothing is changed. For instance, one writes

$$(\partial^\mu \phi)(\partial_\mu \phi) \quad \text{instead of} \quad \phi(-\Box\, \phi), \tag{4.70}$$

for quadratic derivatives, and

$$\frac{1}{2}(\phi^* \stackrel{\leftrightarrow}{\partial_t} \phi) = \frac{1}{2}(\phi^*(\partial_t \phi) - (\partial_t \phi^*)\phi) \quad \text{instead of} \quad \phi^* \partial_t \phi, \tag{4.71}$$

in the case of linear derivatives. These alternatives differ only by an integration by parts, but the symmetry is essential for the correct interpretation of the action principle as presented. This point recurs in more detail in section 10.3.1.

4.4.3 Limitations of the action principle

In 1887, Helmholtz showed that an equation of motion can only be derived from Lagrange's equations of motion (4.6) if the generalized force can be written

$$F_i = -\partial_i V + \frac{\mathrm{d}}{\mathrm{d}t} \frac{\partial V}{\partial \dot{q}_i}, \tag{4.72}$$

where $V = V(q, \dot{q}, t)$ is the potential $L = T - V$, and the following identities are satisfied:

$$\frac{\partial F_i}{\partial \ddot{q}_j} = \frac{\partial F_j}{\partial \ddot{q}_i}$$

$$\frac{\partial F_i}{\partial \dot{q}_j} + \frac{\partial F_j}{\partial \dot{q}_i} = \frac{\mathrm{d}}{\mathrm{d}t}\left(\frac{\partial F_i}{\partial \ddot{q}_j} + \frac{\partial F_j}{\partial \ddot{q}_i}\right)$$

$$\partial_j F_i - \partial_i F_j = \frac{\mathrm{d}}{\mathrm{d}t}\left(\frac{\partial F_i}{\partial \dot{q}_j} - \frac{\partial F_j}{\partial \dot{q}_i}\right) \tag{4.73}$$

For a review and discussion of these conditions, see ref. [67]. These relations lie at the core of Feynman's 'proof' of Maxwell's equations [42, 74]. Although they are couched in a form which derives from the historical approach of varying the action with respect to the coordinate q_i and its associated velocity, $\dot{q}_i$, separately, their covariant generalization effectively summarizes the limits of generalized force which can be derived from a local action principle, even using the approach taken here. Is this a significant limitation of the action principle?

Ohm's law is an example where a Lagrangian formulation does not work convincingly. What characterizes Ohm's law is that it is a substantive relationship between large-scale averages, derived from a deeper theory, whose actual dynamics are hidden and approximated at several levels. The relation summarizes a coarse average result of limited validity. Ohm's law cannot be derived from symmetry principles, only from a theory with complex hidden variables. The deeper theory from which it derives (classical electrodynamics and linear response theory) does have an action principle formulation however.

Ohm's law is an example of how *irreversibility* enters into physics. The equations of fundamental physics are reversible because they deal only with infinitesimal changes. An infinitesimal interval, by assumption, explores so little of its surrounding phase space that changes are trivially reversed. This is the main reason why a generating functional (action) formulation is so successful at generating equations of motion: it is simply a mechanism for exploring the differential structure of the action potential-surface in a local region; the action is a definition of a conservation book-keeping parameter (essentially energy), parametrized in terms of field variables. The reversible, differential structure ensures conservation and generates all of the familiar quantities such as momentum. Irreversibility arises only when infinitesimal changes are compounded into significant changes; i.e. when one is able to explore the larger part of the phase space and take account of long-term history of a system. The methods of statistical field theory (closed time path [116] and density matrices [49]) may be used to study long-term change, based on sums of differential changes. Only in this way can one relate differential law to macroscopic change.

Another way of expressing the above is that the action principle provides a concise formulation of Markov processes, or processes whose behaviour now is independent of what happened in their past. Non-Markov processes, or processes whose behaviour now depends on what happened to them earlier, require additional long-term information, which can only be described by the combination of many infinitesimal changes.

Clearly, it is possible to write down equations which cannot be easily derived from an action principle. The question is whether such equations are of interest to physics. Some of them are (such as Ohm's law), but these only fail because, employing an action principle formulation of a high-level emergent phenomenon ignores the actual energy accounting taking place in the system. If one jumps in at the level of an effective field theory, one is not guaranteed an effective energy parameter which obeys the reversible accounting rules of the action principle. If an action principle formulation fails to make sense, it is possible to go to a deeper, more microscopic theory and re-gain an action formulation, thereby gaining a more fundamental (though perhaps more involved) understanding of the problem.

So are there any fundamental, elementary processes which cannot be derived from an action principle? The answer is probably not. Indeed, today all formulations of elementary physics assume an action principle formulation at the outset. What one can say in general is that any theory derived from an action principle, based on local fields, will lead to a well defined problem, within a natural, covariant formulation. This does not guarantee any prescription understanding physical phenomena, but it does faithfully generate differential formulations which satisfy the symmetry principle.

4.4.4 *Higher derivatives*

Another possibility which is not considered in this book is that of higher derivative terms. The actions used here are at most quadratic in the derivatives. Particularly in speculative gravitational field theories, higher derivative terms do occur in the literature (often through terms quadratic in the curvature, such as Gauss–Bonnet terms or Weyl couplings); these are motivated by geometrical or topological considerations, and are therefore 'natural' to consider. Postulating higher order derivative terms is usually not useful in other contexts.

Higher derivative terms are often problematic, for several reasons. The main reason is that they lead to acausal solutions and 'ghost' excitations, or to field modes which appear to be solutions, but which actually do not correspond to physical propagations. In the quantum field theory, they are non-renormalizable. Although none of these problems is itself sufficient to disregard higher derivatives entirely, it limits their physical significance and usefulness. Some higher derivative theories can be factorized and expressed as coupled local fields with no more than quadratic derivatives; thus, a difficult action may be re-written as a simpler action, in a different formulation. This occurs, for instance, if the theories arise from non-local self-energy terms.

4.5 Dynamical and non-dynamical variations

It is convenient to distinguish between two kinds of variations of tensor quantities. These occur in the derivation of field equations and symmetry generators, such as energy and momentum, from the action.

4.5.1 *Scalar fields*

The first kind of variation is a *dynamical* variation; it has been used implicitly up to now. A dynamical variation of an object q is defined by

$$\delta q = q'(x) - q(x). \qquad (4.74)$$

This represents a change in the *function $q(x)$* at constant position x. It is like the 'rubber-banding' of a function into a new function: a parabola into a cubic curve, and so on.

The other kind of variation is a coordinate variation, or *kinematical* variation, which we denote

$$\delta_x q(x) = q(x') - q(x). \qquad (4.75)$$

This is the apparent change in the height of the function when making a shift in the coordinates x, or perhaps some other parameter which appears either explicitly or implicitly in the action. More generally, the special symbol δ_ξ is used for a variation with respect to the parameter ξ. By changing the coordinates in successive variations, δ_x, one could explore the entire function $q(x)$ at different points. This variation is clearly related to the partial (directional) derivative of q. For instance, under a shift

$$x^\mu \rightarrow x^\mu + \epsilon^\mu, \tag{4.76}$$

i.e. $\delta x^\mu = \epsilon^\mu$, we have

$$\delta_x q(x) = (\partial_\mu q)\epsilon^\mu. \tag{4.77}$$

One writes the total variation in the field q as

$$\delta_T \equiv \delta + \sum_i \delta_{\xi^i}. \tag{4.78}$$

4.5.2 Gauge and vector fields

The coordinate variation of a vector field is simply

$$\begin{aligned} \delta_x V_\mu &= V_\mu(x') - V_\mu(x) \\ &= (\partial_\lambda V_\mu)\epsilon^\lambda. \end{aligned} \tag{4.79}$$

For a gauge field, the variation is more subtle. The field at position x' need only be related to the Taylor expansion of the field at x up to a gauge transformation, so

$$\begin{aligned} \delta_x A_\mu &= A_\mu(x') - A_\mu(x) \\ &= (\partial_\lambda A_\mu)\epsilon^\lambda + \partial_\lambda(\partial_\mu s)\epsilon^\lambda. \end{aligned} \tag{4.80}$$

The gauge transformation s is important because $\delta_x A_\mu(x)$ is a *potential difference*, and we know that potential differences are observable as the electric and magnetic fields, so this variation should be gauge-invariant. To make this so, one identifies the arbitrary gauge function s by $\partial_\lambda s = -A_\lambda$, which is equally arbitrary, owing to the gauge symmetry. Then one has

$$\begin{aligned} \delta_x A_\mu &= (\partial_\lambda A_\mu - \partial_\mu A_\lambda)\epsilon^\lambda \\ &= F_{\lambda\mu}\epsilon^\lambda. \end{aligned} \tag{4.81}$$

Neglect of the gauge freedom has led to confusion over the definition of the energy–momentum tensor for gauge fields; see section 11.5.

The dynamical variation of a vector field follows from the general tensor transformation

$$V'(x') = \frac{\partial x^\rho}{\partial x'^\mu} V_\rho(x). \tag{4.82}$$

From this we have

$$
\begin{aligned}
\delta V_\mu(x) &= V'_\mu(x) - V_\mu(x) \\
&= V'_\mu(x') - (\partial_\lambda V_\mu)\epsilon^\lambda - V_\mu(x) \\
&= \frac{\partial x^\rho}{\partial x'^\mu} V_\rho(x) - (\partial_\lambda V_\mu)\epsilon^\lambda - V_\mu(x) \\
&= -(\partial_\nu \epsilon_\mu)V^\nu - (\partial_\lambda V_\mu)\epsilon^\lambda.
\end{aligned} \tag{4.83}
$$

For the gauge field, one should again be wary about the implicit coordinate variation. The analogous derivation gives

$$
\begin{aligned}
\delta A_\mu(x) &= A'_\mu(x) - A_\mu(x) \\
&= A'_\mu(x') - F_{\lambda\mu}\epsilon^\lambda - A_\mu(x) \\
&= \frac{\partial x^\rho}{\partial x'^\mu} A_\rho(x) - F_{\lambda\mu}\epsilon^\lambda - A_\mu(x) \\
&= -(\partial_\nu \epsilon_\mu)A^\nu - F_{\lambda\mu}\epsilon^\lambda.
\end{aligned} \tag{4.84}
$$

4.5.3 The metric and second-rank tensors

The coordinate variation of the metric is obtained by Taylor-expanding the metric about a point x,

$$
\begin{aligned}
\delta_x g_{\mu\nu} &= g_{\mu\nu}(x') - g_{\mu\nu}(x) \\
&= (\partial_\lambda g_{\mu\nu}(x))\epsilon^\lambda.
\end{aligned} \tag{4.85}
$$

To obtain the dynamical variation, we must use the tensor transformation rule

$$g'_{\mu\nu}(x') = \frac{\partial x^\rho}{\partial x'^\mu} \frac{\partial x^\sigma}{\partial x'^\nu} g_{\rho\sigma}(x), \tag{4.86}$$

where

$$\frac{\partial x^\rho}{\partial x'^\mu} = \delta^\rho{}_\mu - (\partial_\mu \epsilon^\rho) + \cdots + O(\epsilon^2). \tag{4.87}$$

Thus,

$$
\begin{aligned}
\delta g_{\mu\nu} &= g'_{\mu\nu}(x) - g_{\mu\nu}(x) \\
&= \frac{\partial x^\rho}{\partial x'^\mu} \frac{\partial x^\sigma}{\partial x'^\nu} g_{\rho\sigma}(x) - (\partial_\rho g'_{\mu\nu})\epsilon^\rho - g_{\mu\nu}(x)
\end{aligned}
$$

$$= -(\partial_\lambda g_{\mu\nu})\epsilon^\lambda - (\partial_\mu\epsilon^\lambda)g_{\lambda\nu} - (\partial_\nu\epsilon^\lambda)g_{\lambda\mu}$$
$$= -(\partial_\lambda g_{\mu\nu})\epsilon^\lambda - \{\partial_\mu\epsilon_\nu + \partial_\nu\epsilon_\mu\}, \tag{4.88}$$

where one only keeps terms to first order in ϵ^μ.

4.6 The value of the action

There is a frequent temptation to assign a physical meaning to the action, beyond its significance as a generating functional. The differential structure of the action, and the variational principle, give rise to canonical systems obeying conservation laws. This is the limit of the action's physical significance. The impulse to deify the action should be stifled.

Some field theorists have been known to use the value of the action as an argument for the triviality of a theory. For example, if the action has value zero, when evaluated on the constraint shell of the system, one might imagine that this is problematic. In fact, it is not. It is not the numerical value of the action but its differential structure which is relevant.

The vanishing of an action on the constraint shell is a trivial property of any theory which is linear in the derivatives. For instance, the Dirac action and the Chern–Simons [12] action have this property. For example:

$$S = \int (\mathrm{d}x)\overline{\psi}(\mathrm{i}\gamma^\mu\partial_\mu + m)\psi$$

$$\frac{\delta S}{\delta\overline{\psi}} = (\mathrm{i}\gamma^\mu\partial_\mu + m)\psi = 0$$

$$S\Big|_\psi = 0. \tag{4.89}$$

The scalar value of the action is irrelevant, even when evaluated on some specified constraint surface. Whether it is zero, or non-zero, it has no meaning. The only exception to this is in the Wick-rotated theory, where a serendipitous link to finite temperature physics relates the Wick-rotated action to the Hamiltonian or energy operator of the non-rotated theory.

5

Classical field dynamics

A field is a dynamically changing potential $V(\mathbf{x}, t)$, which evolves in time according to an equation of motion. The equation of motion is a constraint on the allowed behaviour of the field. It expresses the dynamical content of the theory. The solution of that constraint, called the physical field, is the pivotal variable from which we glean all of the physical properties of the system. In addition to dynamical equations, a field theory has a conceptual basis composed of physical assumptions, interpretations and boundary conditions.

The familiar equations of motion, in classical field dynamics, include the Schrödinger equation, Maxwell's equations, Dirac's relativistic equation and several others. In the context of field theory, we call such equations *classical* as long as we are not doing *quantum field theory* (see chapter 15), since the method of solution is directly analogous to that of classical electrodynamics. In spite of this designation, we know that the solutions of Schrödinger's field equation are wavefunctions, i.e. the stuff of quantum mechanics. Whole books have been written about these solutions and their interpretation, but they are not called field theory; they use a different name.

Field theory embraces both quantum mechanics and classical electrodynamics, and goes on to describe the most fundamental picture of matter and energy known to physics. Our aim here is to seek a unified level of description for matter and radiation, by focusing on a field theoretical formulation. This approach allows a uniquely valuable perspective, which forms the basis for the full quantum theory. The equations presented 'classically' in this book have many features in common, although they arise from very different historical threads, but – as we shall see in this chapter – the completeness of the field theoretical description of matter and radiation can only be appreciated by introducing further physical assumptions brought forcefully to bear by Einsteinian relativity. This is discussed in chapter 15.

5.1 Solving the field equations

A solution is a mathematical expression of the balance between the *freedom* expressed by the variables of a theory and the constraints which are implicitly imposed upon them by symmetries and equations of motion.

Each physical model has a limited validity, and each has a context into which one builds its interpretation. Some solutions must be disregarded on the basis of these physical assumptions. Sometimes, additional constraints, such as boundary conditions, are desirable to make contact with the real world. The basic vocabulary of solutions involves some common themes.

5.1.1 Free fields

Free particles or fields do not interact. They experience no disturbances and continue in a fixed state of motion for ever. Free particles are generally described by plane wave fields or simple combinations of plane waves, which may be written as a Fourier transform,

$$\Phi(x) = \int \frac{d^{n+1}k}{(2\pi)^{n+1}} e^{ikx} \Phi(k), \tag{5.1}$$

or, using Schwinger's compact notation for the integration measure, as

$$\Phi(x) = \int (dk) \, e^{ikx} \Phi(k). \tag{5.2}$$

For this combination to satisfy the field equations, we must add a condition $\chi(k) = 0$, which picks out a hyper-surface (a sub-set) of all of the k_μ which actually satisfy the equations of motion:

$$\Phi(x) = \int (dk) e^{ikx} \Phi_\chi(k) \delta(\chi), \tag{5.3}$$

where $\chi = 0$ is the constraint imposed by the equations of motion on k. Without such a condition, the Fourier transform can represent an arbitrary function. Notice that $\Phi(k)$ and $\Phi_\chi(k)$ have different dimensions by a factor of k due to the delta function. This condition χ is sometimes called the mass shell in particle physics. Elsewhere it is called a dispersion relation. Fields which satisfy this condition (i.e. the equations of motion) are said to be *on shell*, and values of k which do not satisfy this condition are *off shell*. For free fields we have

$$\chi_R = \hbar^2(-\omega^2 + \mathbf{k}^2 c^2) + m^2 c^4 = 0$$

$$\chi_{NR} = \frac{\hbar^2 \mathbf{k}^2}{2m} - \omega = 0, \tag{5.4}$$

for the relativistic and non-relativistic scalar fields, respectively. The delta-function constraint ensures that the combinations of plane waves obey the field equations. It has the additional side effect that one component of the wavenumber k_μ is not independent and can be eliminated. It is normal to integrate over the zeroth (energy) component to eliminate the delta function. From Appendix A, eqn. (A.15), we have

$$\Phi(x) = \int (dk) \left| \frac{\partial \chi}{\partial k_0} \right|^{-1} e^{i(\mathbf{k}\cdot\mathbf{x} - \omega(\mathbf{k})t)} \Phi(\mathbf{k}, \omega(\mathbf{k})). \tag{5.5}$$

Travelling waves carry momentum $k_i > 0$ or $k_i < 0$, while stationary waves carry no momentum, or rather both k_i and $-k_i$ in equal and opposite amounts.

5.1.2 Boundary conditions and causality I

A common strategy for simplifying the analysis of physical systems is to assume that they are infinitely large, or that they are uniform in space and/or time, or that they have been running uniformly in a steady state for ever. Assumptions like this allow one to do away with the complicated behaviour which is associated with the starting up or shutting down of a dynamical process. It also allows one to consider bulk behaviour without dealing with more difficult effects in the vicinity of the edges of a system. Some of the effects of finite size and starting up/shutting down can be dealt with by imposing *boundary conditions* on the behaviour of a system. The term *boundary conditions* is used with a variety of meanings.

- Boundary conditions can be a specification of the absolute value of the field at some specific spacetime points, e.g.

$$\phi(x)\Big|_{x=x_0} = 0. \tag{5.6}$$

  This indicates a constraint associated with some inhomogeneity in spacetime.

- A corollary to the above is the specification of the value of the field on the walls of a container in a finite system.

- At junctions or interfaces, one is interested in continuity conditions, like those derived in section 4.1.4 and generalizations thereof. Here, one matches the value of the field, perhaps up to a symmetry transformation, across the junction, e.g.

$$\Delta\phi(x_0) = 0, \tag{5.7}$$

  meaning that the field does not change discontinuously across a junction. Conditions of this type are sometimes applied to fields, but usually it is more correct to apply them to conserved quantities such as invariant products of fields, probabilities

$$\Delta \left(\psi^\dagger \psi \right) = 0, \tag{5.8}$$

etc. since fields can undergo discontinuous phase changes at boundaries when the topology of spacetime allows or demands it.

- Related to the last case is the issue of spatial topology. Some boundary conditions tell us about the connectivity of a system. For example, a field in a periodic lattice or circle of length L could satisfy

$$\phi(x + L) = U(L)\phi(x). \tag{5.9}$$

In other words, the value of the field is identical, up to a possible phase or symmetry factor $U(L)$, on translating a distance L.

- Another kind of condition which one can impose on a reversible physical system is a direction for causal development. The keywords here are *advanced*, *retarded* and *Feynman boundary conditions* or fluctuations. They have to do with a freedom to change perspective between cause and effect in time-reversible systems. Is the source switched on/off before or after a change in the field? In other words, does the source cause the effect or does it absorb and dampen the effect? This is a matter of viewpoint in reversible systems. The boundary conditions known as Feynman boundary conditions mix these two causal perspectives and provide a physical model for fluctuations of the field or 'virtual particles': a short-lived effect which is caused and then absorbed shortly afterwards.

5.1.3 Positive and negative energy solutions

The study of fields in relativistic systems leads to solutions which can be interpreted as having both positive and negative energy. Free relativistic field equations are all transcriptions of the energy relation

$$E = \pm\sqrt{p^2c^2 + m^2c^4}, \tag{5.10}$$

with the operator replacement $p_\mu = -i\hbar\partial_\mu$ and a field on which the operators act. This is most apparent in the case of the Klein–Gordon equation,

$$(-\hbar^2c^2\Box + m^2c^4)\phi(x) = 0. \tag{5.11}$$

Clearly, both signs for the energy are possible from the square-root in eqn. (5.10). The non-relativistic theory does not suffer from the same problem, since the Schrödinger equation is linear in the energy and the sign is defined to be positive:

$$\frac{\mathbf{p}^2}{2m} = E. \tag{5.12}$$

The field $\phi(x)$ can be expanded as a linear combination of a complete set of plane wavefunctions satisfying the equation of motion. The field can therefore be written

$$\phi(x) = \int (dk)\phi(k)e^{ikx}\delta\left(\hbar^2 c^2 k^2 + m^2 c^4\right), \tag{5.13}$$

where $\phi(k)$ are arbitrary coefficients, independent of x. The integral ranges over all energies, but one can separate the positive and negative energy solutions by writing

$$\phi(x) = \phi^{(+)}(x) + \phi^{(-)}(x), \tag{5.14}$$

where

$$\phi^{(+)}(x) = \int (dk)\phi(k)e^{ikx}\theta(k_0)\delta\left(\hbar^2 c^2 k^2 + m^2 c^4\right)$$

$$\phi^{(-)}(x) = \int (dk)\phi(k)e^{ikx}\theta(-k_0)\delta\left(\hbar^2 c^2 k^2 + m^2 c^4\right). \tag{5.15}$$

The symmetry of the energy relation then implies that

$$\phi^{(+)}(x) = \left(\phi^{(-)}(x)\right)^*. \tag{5.16}$$

The physical interpretation of negative energy solutions is an important issue, not because negative energy is necessarily unphysical (energy is just a label which embraces a variety of conventions), but rather because there are solutions with arbitrarily large negative energy. A transition from any state to a state with energy $E = -\infty$ would produce an infinite amount of real energy for free. This is contrary to observations and is, presumably, nonsense.

The positive and negative energy solutions to the free relativistic field equations form independently complete sets, with respect to the scalar product,

$$(\phi^{(+)}(x), \phi^{(+)}(x)) = \text{const.}$$
$$(\phi^{(-)}(x), \phi^{(-)}(x)) = \text{const.}$$
$$(\phi^{(+)}(x), \phi^{(-)}(x)) = 0. \tag{5.17}$$

In the search for physically meaningful solutions to the free relativistic equations, it might therefore be acceptable to ignore the negative energy solutions on the basis that they are just the mirror image of the positive energy solutions, describing the same physics with a different sign.

This is the case for plane waves, or any solutions which are translationally invariant in time. Such a wave has a time dependence of the form,

$$\phi(t) \sim \exp\left(-i\frac{E}{\hbar}(t - t_0)\right), \tag{5.18}$$

where t_0 is an arbitrary origin for time. If $E < 0$, one can simply recover a positive energy description by moving the origin for time t_0 into the far future, $t_0 \to \infty$, which essentially switches $t \to -t$. Since a free particle cannot change its energy by interaction, it will always have a definite energy, either positive or negative. It cannot therefore extract energy from the field by making a transition.

The real problem with negative energies arises in interacting theories. It is not clear how to interpret these solutions from the viewpoint of classical field theory. An extra assumption is needed. This assumption is more clearly justified in the quantum theory of fields (see chapter 15), but is equally valid in the classical theory. The assumption is that there exists a physical state of lowest energy (called the vacuum state) and that states below this energy are interpreted as anti-matter states.

It is sometimes stated that relativistic quantum mechanics (prior to second quantization) is sick, and that quantum field theory is required to make sense of this problem. This is not correct, and would certainly contradict modern thinking about effective field theories.[1] All that is required is a prescription for interpreting the negative energies. The assumptions of quantum field theory, although less well justified, are equally effective and no more arbitrary here. In fact, they are essential since the classical field theory is a well defined limit to the fully quantized field theory.

5.1.4 Sources

The terms *source* and *current* are often used interchangeably in field theory, but they refer to logically distinct entities. Sources (sometimes referred to emphatically as *external sources*) are infinitesimal perturbations to a physical system; currents represent a transfer between one part of a system and another. In an isolated (closed) system, matter and energy can flow from one place to another, and such currents are conserved. There is a close formal similarity between sources and currents, which is no accident. Sources – and their opposites: sinks – can be thought of as infinitesimal currents which are not conserved. They represent the flow of something into or out of a physical system, and thus a perturbation to it. Sources are also the generators of infinitesimal field changes, called virtual processes or fluctuations.

---

[1] Certain specific Lagrangians lead to unphysical theories, but this is only a reason to reject certain models, not the quantum theory itself.

In mathematics, any quantity on the 'right hand side' of a field equation is called a source, 'forcing term' or 'driving term'. A source perturbs or drives the field linearly. For example, consider the Klein–Gordon equation

$$\left(-\Box + \frac{m^2 c^2}{\hbar^2}\right)\phi(x) = J. \tag{5.19}$$

One says that $J(x)$ is a source for the field $\phi(x)$. J is sometimes also referred to as a *generalized force*. Sources are included in the action in the form

$$S \rightarrow S + \int (dx) J\phi(x). \tag{5.20}$$

For example, the Klein–Gordon action with a source term becomes

$$S = \int (dx) \left\{ \frac{1}{2}\hbar^2 c^2 (\partial^\mu \phi)(\partial_\mu \phi) + \frac{1}{2}m^2 c^4 \phi^2 - J\phi \right\}. \tag{5.21}$$

When this action is varied, one obtains

$$\frac{\delta S}{\delta \phi} = \left(-\hbar^2 c^2 \Box + m^2 c^4\right)\phi - J = 0, \tag{5.22}$$

which leads directly to eqn. (5.19). Other source terms include

$$S_{\text{Maxwell}} \rightarrow S_{\text{Maxwell}} + \int (dx) J^\mu A_\mu \tag{5.23}$$

for the electromagnetic field, and

$$S_{\text{complex}} \rightarrow S_{\text{complex}} + \int (dx) \left\{ J\phi^* + J^*\phi \right\} \tag{5.24}$$

for a complex scalar field. Most interactions with the field do not have the form of an infinitesimal perturbation. For instance, the interaction with a Schrödinger field, in quantum mechanics, has the form $\psi^* V \psi$, making $J = V\psi$, which is not infinitesimal. However, if one assumes that V is small, or infinitesimal, then this may be expanded around the field ψ for a free theory in such a way that it appears to be a series of infinitesimal impulsive sources; see section 17.5. In this way, the source is the basic model for causal change in the field.

Another definition of the source is by functional differentiation:

$$\frac{\delta S}{\delta \phi_A} = J_A, \tag{5.25}$$

where ϕ is a generic field. This is a generic definition and it follows directly from eqn. (5.20), where one does not treat the source term as part of the action S.

A current represents a flow or transport. To define current, one looks to the only example of current known prior to field theory, namely the electric current. Recall Maxwell's equation

$$\partial_\mu F^{\nu\mu} = \mu_0 J^\nu. \tag{5.26}$$

The quantity J_μ is the $(n + 1)$ dimensional current vector. It is known, from the microscopics of electromagnetism, that this is the electric current: electric currents and electric charges are responsible for the electromagnetic field. However, one may also say that J_μ is a source for the electromagnetic field, because it prevents the left hand side of this equation from being equal to zero. It perturbs the equation of motion. In electromagnetism the current is a source for the field $F_{\mu\nu}$ or A_μ, so it is common to treat source and current as being the same thing. This tendency spills over for other fields too, and one often defines a generic current by eqn. (5.25). Of course, normally one imagines a current as being a vector, whereas the quantity in eqn. (5.25) is a scalar, but this may be used as a definition of 'current'. The notion of conserved currents and their relation to symmetries recurs in chapter 9.

5.1.5 Interactions and measurements

Fields undergo interactions with other fields, and perhaps with themselves (self-interaction). When fields interact with other fields or potentials (either static or dynamical), the state of the field is modified. Classically, the field responds deterministically according to a well defined differential equation (the equation of motion), and interactions apply new constraints. One way to understand weakly interacting systems is to imagine them to be assemblies of weakly-coupled oscillators. In special circumstances, it is possible to construct models with interactions which can be solved exactly. Often, however, approximate methods are required to unravel the behaviour of interacting fields.

In quantum mechanics the act of measurement itself is a kind of temporary interaction, which can lead to a discontinuous change of state. It is not fundamentally different from switching on a potential in field theory. The 'collapse of the wavefunction' thus occurs as a transition resulting from an interaction with a measurement apparatus. This collapse has no detailed description in the theory.

5.2 Green functions and linear response

5.2.1 The inverse problem

Consider an equation of the form

$$\mathcal{D} \, y(t) = f(t), \tag{5.27}$$

where $\mathcal{D}$ is a differential operator, $y(t)$ is a variable we seek to determine, and $f(t)$ is some forcing term, or 'source'. We meet this kind of equation repeatedly in field theory, and $\mathcal{D}$ is often an operator of the form $\mathcal{D} = -\Box + m^2$.

Normally, one would attempt to solve a differential equation either by integrating it directly, or by 'substituting in' a trial solution and looking for consistency. An alternative method is the method of Green functions. The idea can be approached in a number of ways. Let us first take a naive approach.

If $\mathcal{D}$ is an operator, then, if a unique solution to the above equation exists, it must have an inverse. We can therefore write the solution to this equation formally (because the following step has no meaning until we have defined the inverse) by

$$y(t) = (\mathcal{D})^{-1} f(t) = \frac{f(x)}{\mathcal{D}}. \tag{5.28}$$

This is much like the approach used to solve matrix equations in linear algebra. Both the notations in the equation above are to be found in the literature. If the inverse exists, then it must be defined by a relation of the form

$$\frac{\mathcal{D}}{\mathcal{D}} = \mathcal{D}\mathcal{D}^{-1} = I, \tag{5.29}$$

where I is the identity operator.[2] We do not yet know what these quantities are, but if an inverse exists, then it must be defined in this way. An obvious thing to notice is that our eqn. (5.27) is a differential equation, so the solution involves some kind of integration of the right hand side. Let us now postpone the remainder of this train of thought for a few lines and consider another approach.

The second way in which we can approach this problem is to think of eqn. (5.27) as a 'linear response' equation. This means that we think of the right hand side as being a forcing term which perturbs the solution $y(t)$ by kicking it over time into a particular shape. We can decompose the force $f(t)$ into a set of delta-function impulse forces over time,

$$f(t) = \int dt' \delta(t, t') f(t'). \tag{5.30}$$

This equation, although apparently trivial (since it defines the delta function), tells us that we can think of the function $f(t)$ as being a sum of delta functions at different times, weighted by the values of $f(t')$. We can always build up a function by summing up delta functions at different times. In most physical problems we expect the value of $y(t)$ to depend on the past history of all the kicks it has received from the forcing function $f(t)$. This gives us a clue as to how we can define an inverse for the differential operator $\mathcal{D}$.

---

[2] Note that the ordering of the operator and inverse is an issue for differential operators. We require a 'right-inverse', but there may be no left inverse satisfying $\mathcal{D}^{-1}\mathcal{D} = I$.

Suppose we introduce a bi-local function $G(t, t')$, such that

$$y(t) = \int dt' \, G(t, t') f(t'); \qquad (5.31)$$

i.e. when we sum up the contributions to the force over time with this weight, it gives us not the force itself at a later time, but the solution. This, in fact, is the way we define the inverse $\mathcal{D}^{-1}$. It has to be a bi-local function, as we shall see below, and it involves an integration, in spite of the purely formal notation in eqn. (5.29).

Substituting this trial solution into the equation of motion, we have

$$\mathcal{D} \int dt' \, G(t, t') f(t') = f(t), \qquad (5.32)$$

where the operator $\mathcal{D}$ acts on the variable t only, since the dummy variable t' is integrated out from minus to plus infinity. Thus, we may write,

$$\int dt' \, \overset{t}{\mathcal{D}} \, G(t, t') f(t') = f(t). \qquad (5.33)$$

This equation becomes the defining equation for the delta function (5.30) if and only if

$$\overset{t}{\mathcal{D}} \, G(t, t') = \delta(t, t'), \qquad (5.34)$$

and this equation is precisely of the form of an inverse relation, where the delta function is the identity operator. We have therefore obtained a consistent set of relations which allow us to write a formal solution $y(t)$ in terms of an inverse for the operator $G(t, t')$; we also have an equation which this inverse must satisfy, so the problem has been changed from one of finding the solution $y(t)$ to one of calculating the inverse function. It turns out that this is often an easier problem than trying to integrate eqn. (5.27) directly.

The function $G(t, t')$ goes by several names. It is usually referred to as the *Green('s) function* for the operator $\mathcal{D}$, but it is also called the *kernel* for $\mathcal{D}$ and, in quantum field theory, the *propagator*.

We can, of course, generalize this function for differential operators which act in an $(n + 1)$ dimensional spacetime. The only difference is that we replace t, t' by x, x' in the above discussion:

$$\mathcal{D} \, y(x) = f(x)$$
$$\mathcal{D} G(x, x') = c\delta(x, x')$$
$$y(x) = \int (dx') G(x, x') f(x'). \qquad (5.35)$$

Or, equivalently,

$$\mathcal{D}G(x, x') = \delta(\mathbf{x}, \mathbf{x}')\delta(t, t')$$
$$y(x) = \int (\mathrm{d}x') G(x, x') f(x'). \tag{5.36}$$

We are not quite finished with Green functions yet, however: we have skirted around an important issue above, which is described in the next section.

5.2.2 Boundary conditions and causality II

The discussion above is not quite complete: we have written down a function which relates the solution at x to a forcing term at x' via a bi-local function $G(x, x')$. The inverse relation involves an integral over all intermediate times and positions x', but over what values does this integral run? And over what values of x' was the force defined? Was it switched on suddenly at some time in the past (giving an integral from a fixed time in the past to the present), or has it always existed (giving an integral from minus infinity)? Moreover, why should x' be in the past? We know that physics is usually time-reversible, so why could we not run time backwards and relate a solution in the past to a value of the force in the future, or perhaps a combination of the past and future?

All of these things are possible using different Green functions. We therefore see that the inverse is not unique, and it is not unique because the definition of the inverse involves an integration, and integrals have limits. Physically we are talking about the need to specify initial or boundary conditions on our physical system.

The commonly used Green functions are as follows.

- **Retarded** Green function $G_r(x, x')$. This relates a solution at the present to forces strictly in the past. It is the basis of linear response theory. Due to its origins in electromagnetism, it is often referred to as the *susceptibility* $\chi(x, x') \equiv \chi' + i\chi''$ in other books, with real and imaginary parts as denoted.

- **Advanced** Green function $G_a(x, x')$. This relates a solution at the present to forces strictly in the future.

- **Feynman** Green function $G_F(x, x')$. This relates a solution at the present to forces disposed equally in the past and the future. Its interpretation is rather subtle, since it turns real fields into complex fields as they propagate. The Feynman Green function is a correlation function, and a model for fluctuations in a system. It is sometimes denoted $\Delta(x, x')$, $C(x, x')$ or $S(x, x')$ in other books.

- **Wightman functions**. The positive and negative frequency Wightman functions $G^{(\pm)}(x, x')$ may be thought of as building blocks out of which all the other Green functions may be constructed.

5.2.3 Green functions in Fourier momentum space[3]

A useful way of calculating quantities is to use an integral transformation, usually the Fourier transformation on the Green functions. The purpose of this step is to turn an operator equation into an ordinary algebraic equation, plus a single integral. This is often referred to as transforming into 'momentum space', since the choice of units makes the Fourier transform variables equivalent to momenta.

We shall focus largely on the Green functions for the scalar field, since most of the Green functions for other fields can be obtained from this by differentiation. We are looking to solve an equation of the form

$$(-\Box + M^2)G(x, x') = \delta(x, x'), \tag{5.37}$$

where M^2 is some real mass term. We define the Fourier transforms of the Green function by the mutually inverse relations,

$$G(r) = \int (\mathrm{d}k)e^{ikr} G(k) \tag{5.38a}$$

$$G(k) = \int (\mathrm{d}r)e^{-ikr} G(x, x'), \tag{5.38b}$$

where we have assumed that $G(r) = G(x, x')$ is a translationally invariant function of the coordinates (a function only of the difference $x - x'$), which is reasonable since M^2 is constant with respect to x. We shall also have use for the Fourier representation of the delta function, defined in Appendix A, eqn. (A.10). Notice how the Fourier integral is a general linear combination of plane waves $\exp(ik(x - x'))$, with coefficients $G(k)$. Using this as a solution is just like substituting complex exponentials into differential equations. Substituting these transformed quantities into eqn. (5.37), and comparing the integrands on the left and right hand sides, we obtain

$$(k^2 + M^2)G(k) = 1. \tag{5.39}$$

This is now an algebraic relation which may be immediately inverted and substituted back into eqn. (5.38b) to give

$$G(x, x') = \int (\mathrm{d}k)\frac{e^{ik(x-x')}}{k^2 + M^2}. \tag{5.40}$$

---

[3] In this section we set $\hbar = c = 1$ for convenience.

In addition to this 'particular integral', one may add to this any linear combination of plane waves which satisfies the mass shell constraint $k^2 + M^2 = 0$. Thus the general solution to the Green function is

$$G_X(x, x') = \int (\mathrm{d}k) e^{ik(x-x')} \left[\frac{1}{k^2 + M^2} + X(k, \overline{x}) \, \delta(k^2 + M^2) \right], \quad (5.41)$$

where $X(k, \overline{x})$ is an arbitrary function of k, and in the unusual case of inhomogeneous systems it can also depend on the average position $\overline{x} = \frac{1}{2}(x + x')$. This arbitrariness in the complementary function is related to the issue of boundary conditions in the previous section and the subsequent discussion in the remainder of this chapter, including the choice of integration path for the Green function. In most cases studied here, $X(k, \overline{x}) = 0$, and we choose a special solution (retarded, advanced, etc.) for the Green function. This term becomes important in satisfying special boundary conditions, and occurs most notably in statistical 'many-particle' systems, which vary slowly with $\overline{t}$ away from equilibrium.

We are therefore left with an integral which looks calculable, and this is correct. However, its value is ambiguous for the reason mentioned above: we have not specified any boundary conditions. The ambiguity in boundary conditions takes on the form of a division by zero in the integrand, since

$$k^2 + M^2 = -k_0^2 + \mathbf{k}^2 + M^2 = (\omega_k - k_0)(\omega_k + k_0), \quad (5.42)$$

where $\omega_k = \sqrt{\mathbf{k}^2 + M^2}$. This $G(k)$ has simple poles at

$$k_0 = \pm \omega_k. \quad (5.43)$$

In order to perform the integral, we need to define it unambiguously in the complex plane, by choosing a prescription for going around the poles. It turns out that this procedure, described in many texts, is equivalent to choosing boundary conditions on the Green function.

5.2.4 Limitations of the Green function method

The Green function method nearly always works well in field theory, but it is not without its limitations. The limitations have to do with the order of the differential operator, $\mathcal{D}$, the number of spacetime dimensions and whether or not the operator contains a mass term. For a massive operator

$$(-\Box + M^2)\phi(x) = J(x), \quad (5.44)$$

the general solution is given by

$$\phi(x) = \int (\mathrm{d}x) \, G(x, x') J(x'). \quad (5.45)$$

For a massless field, it is clear that one can always add to this a polynomial of order lower than the order of the differential operator. In the example above, setting $M = 0$ allows us to add

$$\phi(x) = \int (\mathrm{d}x)\, G(x, x') J(x') + \alpha(x - x') + \beta. \qquad (5.46)$$

A more serious limitation of the Green function method arises when the order of the differential operator exceeds the number of spacetime dimensions involved in the operator. This leads to non-simple poles in the Green function, which presents problems for the evaluation of the Green function. For example, a second-order operator in one dimension

$$\partial_t^2 G(t, t') = \delta(t, t'). \qquad (5.47)$$

If we try to solve this using the Fourier method, we end up with an integral of the form

$$G(t, t') = \int \frac{\mathrm{d}\omega}{2\pi} \frac{e^{-i\omega(t-t')}}{-(\omega \pm i\epsilon)^2}. \qquad (5.48)$$

This integral has a second-order pole and cannot be used to solve an equation involving ∂_t^2. For example, the equation for the position of a Newtonian body

$$\partial_t^2 x(t) = F/m, \qquad (5.49)$$

cannot be solved in this way since it is not homogeneous in the source F/m. The solution is easily obtained by integration

$$x(t) = \frac{1}{2} \frac{F}{m} t^2 + vt + x_0. \qquad (5.50)$$

Since there are terms in this solution which are not proportional to F/m, it is clear that the Green function method cannot provide this full answer. However, the equation can still be solved by the Green function method in two stages.

5.2.5 Green functions and eigenfunction methods

In introductory quantum mechanics texts, the usual approach to solving the system is based on the use of the eigenfunctions of a Hamiltonian operator. This is equivalent to the use of Green functions. The Fourier space expressions given thus far assume that an appropriate expansion can be made in terms of plane wave eigenfunctions:

$$u_k(x) = e^{ikx}. \qquad (5.51)$$

Written in this notation, the Green functions have the form

$$G(x, x') = \sum_n G_n \, u_n(x) u_n^*(x') , \qquad (5.52)$$

where the u_n are a complete set of eigenfunctions, or solutions of the field equations, and the G_n are a set of constants in this new expansion. The labels n are sometimes discrete (as in bound state problems) and sometimes continuous, as in the case $n = k$, $G(k)$ and so on. In addition to the above expansion, the question of boundary conditions must be addressed. This can be accomplished by multiplying the coefficients by step functions:

$$G_n(x, x') \propto \left(\alpha_n \, \theta(t - t') + \beta_n \, \theta(t' - t) \right). \qquad (5.53)$$

This is true in many situations, at least when the system concerned is translationally invariant. However, in bound state problems and situations of special symmetry, this expansion leads to an inefficient and sometimes pathological approach.

Consider the relativistic scalar field as an example. The complex scalar field satisfies the equation

$$\left(-\Box + m^2 + V \right) \phi(x) = J(x). \qquad (5.54)$$

Now let φ_n be a complete set of eigenfunctions of the operator in this equation, such that a general wavefunction $\phi(x)$ may be expanded in terms of a complete set of these with coefficients c_n,

$$\phi(x) = \sum_n c_n \varphi_n(x), \qquad (5.55)$$

such that

$$\int d\sigma_x (\varphi_n, \varphi_m) \Bigg|_{t=t'} = \delta_{nm}. \qquad (5.56)$$

The wavefunction $\phi(x)$ and the eigenfunctions $\varphi_n(x)$ are assumed to be one-particle wavefunctions. The discrete indices n, m denote any bound state quantum numbers which the wavefunction might have. The eigenfunctions satisfy

$$\left(-\Box + m^2 + V \right) \varphi_n(x) = 0. \qquad (5.57)$$

The eigenfunctions can also be expressed in terms of their positive and negative frequency parts,

$$\varphi_n(x) = \varphi_n^{(+)}(x) + \varphi_n^{(-)}(x), \qquad (5.58)$$

where $\varphi_n^{(+)}(x) = (\varphi_n^{(-)}(x))^*$,

$$\phi_n^{(+)}(x) = \int (dk) e^{ikx} \theta(-k_0) \delta(k^2 + m^2 + V) a_n(k), \qquad (5.59)$$

and $a_n(k)$ is a c-number. The Green function for the field (wavefunction) $\phi(x)$ is the inverse of the operator in eqn. (5.54), satisfying,

$$\left(-\Box + m^2 + V\right) G_{nm}(x, x') = \delta_{nm} \delta(x, x'). \qquad (5.60)$$

Using eqn. (5.57) and eqn. (A.21) from Appendix A, we can solve this equation with an object of the form

$$G_{nm} = \left(\alpha\, \theta(t - t') + \beta\, \theta(t' - t)\right) \sum_{n,m} \varphi_n(x) \varphi_m^*(x'), \qquad (5.61)$$

where α and β are to be fixed by the choice of boundary conditions on the Green function.

5.3 Scalar field Green function

The Green function for the scalar field is defined by the relation

$$(-\hbar^2 c^2 \Box + m^2 c^4) G(x, x') = \delta(\mathbf{x}, \mathbf{x}') \delta(t, t'). \qquad (5.62)$$

It is often convenient to express this in terms of the $(n + 1)$ dimensional delta function

$$\delta(\mathbf{x}, \mathbf{x}') \delta(t, t') = c \delta(\mathbf{x}, \mathbf{x}') \delta(x^0, x^{0'}) = c \delta(x, x'). \qquad (5.63)$$

The right hand side of eqn. (5.62) differs from an $(n + 1)$ dimensional delta function by a factor of c because the action is defined as an integral over $dV_t = (dx)$ rather than dV_x. This convention is chosen because it simplifies the coupling between matter and radiation, and because it makes the Lagrangian density have the dimensions of an energy density. In natural units, $\hbar = c = 1$, this distinction does not arise. The formal expression for the scalar Green function on solving this equation is

$$G(x, x') = c \int (dk) \frac{e^{ik(x-x')}}{p^2 c^2 + m^2 c^4}, \qquad (5.64)$$

where $p_\mu = \hbar k_\mu$. Thus, $G(x, x')$ has the dimensions of $\phi^2(x)$. This Green function can be understood in a number of ways. For the remainder of this section, we shall explore its structure in terms of the free-field solutions and the momentum-space constraint surface $p^2 c^2 + m^2 c^4 = 0$, which is referred to in the literature as the 'mass shell'.

5.3.1 The Wightman functions

It is useful to define two quantities, known in quantum field theory as the positive and negative frequency Wightman functions, since all the Green functions can be expressed in terms of these. The Wightman functions are the solutions to the free differential equation,[4]

$$(-\hbar^2 c^2 \Box + m^2 c^4) G^{(\pm)}(x, x') = 0. \tag{5.65}$$

For convenience, it is useful to separate the solutions of this equation into those which have positive frequency, $k_0 = |\omega_k|$, and those which have negative frequency, $k_0 = -|\omega_k|$. They may be written by inspection as a general linear combination of plane waves, using a step function, $\theta(\pm k_0)$, to restrict the sign of the frequency, and a delta function to ensure that the integral over all k is restricted only to those values which satisfy the equations of motion,

$$G^{(+)}(x, x') = -2\pi i\, c \int (dk) e^{ik(x-x')} \theta(-k_0) \delta(p^2 c^2 + m^2 c^4)$$

$$G^{(-)}(x, x') = 2\pi i\, c \int (dk) e^{ik(x-x')} \theta(k_0) \delta(p^2 c^2 + m^2 c^4). \tag{5.66}$$

Because of unitarity,[5] these two functions are mutually conjugate (adjoint) in the relativistic theory.

$$G^{(+)}(x, x') = \left[G^{(-)}(x, x') \right]^* = -G^{(-)}(x', x). \tag{5.67}$$

In the non-relativistic limit, field theory splits into a separate theory for particles (which have positive energy) and for anti-particles (which have negative energy). Although this relation continues to be true, when comparing the particle theory with the anti-particle theory, it is not true for straightforward Schrödinger theory where the negative frequency Wightman function is zero at zero temperature.

The delta function in the integrands implies that one of the components of the momentum is related to all the others,[6] thus we may integrate over one of them, k_0, in order to eliminate this and express it in terms of the others. The equations of motion tell us that $ck_0 = \pm\omega_k$, where

$$\hbar\omega_k = \sqrt{\hbar^2 \mathbf{k}^2 c^2 + m^2 c^4}, \tag{5.68}$$

i.e. there are two solutions, so we may use the identity proven in eqn. (A.15) to write

$$\delta(p^2 c^2 + m^2 c^4) = \frac{1}{2\hbar^2 c^2 |\omega_k|} \left\{ \delta\left(-k_0 + \frac{|\omega_k|}{c}\right) + \delta\left(k_0 + \frac{|\omega_k|}{c}\right) \right\} \tag{5.69}$$

---

[4] They are analogous to the complementary function in the theory of linear partial differential equations.

[5] Unitarity is the property of field theories which implies conservation of energy and probabilities.

[6] The momentum is said to be 'on shell' since the equation, $k^2 + m^2 = 0$, resembles the equation of a spherical shell in momentum space with radius im.

This relation is valid under the integral sign for k_0. Noting that the step functions, $\theta(\pm k_0)$, pick out only one or the other delta function on the right hand side, we have

$$G^{(+)}(x, x') = -2\pi i \, (\hbar^2 c)^{-1} \int \frac{(dk)}{2\pi} \frac{1}{2\omega_k} e^{i(k \cdot (x-x') - |\omega_k|(t-t'))}$$

$$G^{(-)}(x, x') = 2\pi i \, (\hbar^2 c)^{-1} \int \frac{(dk)}{2\pi} \frac{1}{2\omega_k} e^{i(k \cdot (x-x') + |\omega_k|(t-t'))}$$

$$= 2\pi i \, (\hbar^2 c)^{-1} \int \frac{(dk)}{2\pi} \frac{1}{2\omega_k} e^{-i(k \cdot (x-x') - |\omega_k|(t-t'))}.$$

(5.70)

Before leaving this section, we define two further symbols which appear in field theory,

$$\tilde{G}(x, x') = G^{(+)}(x, x') + G^{(-)}(x, x')$$
$$\overline{G}(x, x') = G^{(+)}(x, x') - G^{(-)}(x, x').$$

(5.71)

$\overline{G}(x, x')$ is the sum of all solutions to the free-field equations and, in quantum field theory, becomes the so-called *anti-commutator* function.[7] Note that this quantity is explicitly the sum of $G^{(+)}(x, x')$ and its complex conjugate $G^{(-)}(x, x')$ and is therefore real in the relativistic theory.[8]

The symmetric and anti-symmetric combinations satisfy the identities

$$\overset{x'}{\partial_t} \overline{G}(x, x')\bigg|_{t=t'} = 0$$

(5.72)

and

$$\overset{x'}{\partial_t} \tilde{G}(x, x')\bigg|_{t=t'} = \delta(\mathbf{x}, \mathbf{x}').$$

(5.73)

The latter turns out to be equivalent to the fundamental commutation relations in the quantum theory of fields. $\tilde{G}(x, x')$ becomes the *commutator* function in the quantum theory of fields.

---

[7] This looks wrong from the definitions in terms of Green functions, but recall the signs in the definitions of the Green functions. The tilde denotes the fact that it is a commutator of the quantum fields in the quantum theory.

[8] This symmetry is broken by the non-relativistic theory as $G^{(-)}(x, x')$ vanishes at the one-particle level.

Finally, we may note that ω_k is always positive, since it is the square-root of a positive, real quantity, so we may drop the modulus signs in future and take this as given.

5.3.2 Boundary conditions and poles in the k_0 plane

When solving differential equations in physics, the choice of boundary conditions normally determines the appropriate mixture of particular integral and complementary functions. The same is true for the Green function approach, but here the familiar procedure is occluded by the formalism of the Green function.

The Wightman functions are the general solutions of the free-field equations: they are the complementary functions, which one may always add to any particular integral. There are two ways to add them to a special solution. One is to use the term X in eqn. (5.41); the other is to deform the complex contour around the poles. This deformation accomplishes precisely the same result as the addition of complementary solutions with complex coefficients. Let us now consider how the deformation of the complex contour leads to the choice of boundary conditions for the field.

The retarded, advanced and Feynman Green functions solve the equations of motion in the presence of a source, with specific boundary conditions as mentioned in section 5.2.2. In this section, we shall impose those boundary conditions and show how this leads to an automatic prescription for dealing with the complex poles in the integrand of eqn. (5.40). The most intuitive way of imposing the boundary conditions is to write the Green functions in terms of the step function:

$$G_r(x, x') = -\theta(\sigma, \sigma')\tilde{G}(x, x') \tag{5.74a}$$

$$G_a(x, x') = \theta(\sigma', \sigma)\tilde{G}(x, x') \tag{5.74b}$$

$$G_F(x, x') = -\theta(\sigma, \sigma')G^{(+)}(x, x') + \theta(\sigma', \sigma)G^{(-)}(x, x'). \tag{5.74c}$$

Note that, since the retarded and advanced Green functions derive from $\tilde{G}(x, x')$, they are real in x, x' space (though this does not mean that their Fourier transforms are real in k space), except in the non-relativistic theory. When we write $\theta(\sigma, \sigma')$ in this way, the σ's usually refer to two time coordinates $\theta(t, t')$, but in general we may be measuring the development of a system with respect to more general spacelike hyper-surfaces, unconnected with the Cartesian coordinate t or x^0. For simplicity, we shall refer to t and t' in future. The physical meaning of these functions is as advertised: the retarded function propagates all data from earlier times to later times, the advanced function propagates all data from future times to past times, and the Feynman function takes positive frequency data and propagates them forwards in time, while propagating negative frequency data backwards in time.

To convert these expressions into momentum-space integrals, we make use of the integral representations of the step function,

$$\theta(t - t') = i \lim_{\epsilon \to 0} \int_\infty^\infty \frac{d\alpha}{2\pi} \frac{e^{-i\alpha(t-t')}}{\alpha + i\epsilon}$$

$$\theta(t' - t) = -i \lim_{\epsilon \to 0} \int_\infty^\infty \frac{d\alpha}{2\pi} \frac{e^{-i\alpha(t-t')}}{\alpha - i\epsilon}. \tag{5.75}$$

Writing $\Delta x \equiv x - x'$ for brevity, we can now evaluate these expressions using the momentum-space forms for the Wightman functions in eqn. (5.70).

To evaluate the Green functions in momentum-space, it is useful to employ Cauchy's residue theorem, which states that the integral around a closed (anti-clockwise) circuit of a function equals 2π times the sum of the residues of the function. Suppose the function $\phi(z)$ has simple poles in the complex plane at z_i, then, assuming that the closed contour is in the anti-clockwise (positive) sense, we have

$$\oint_C \phi(z)dz = 2\pi i \sum_i (z - z_i)\phi(z)\bigg|_{z=z_i}. \tag{5.76}$$

If the contour C is in the clockwise sense, the sign is reversed.

The complex contour method for evaluating integrals is a useful tool for dealing with Green functions, but one should not confuse the contours with the Green functions themselves. The Green functions we seek are only defined on the real axis, but Cauchy's formula only works for a closed contour with generally complex pieces. We can evaluate integrals over any contour, in order to use Cauchy's formula, provided we can extract the value purely along the real axis at the end. The general strategy is to choose a contour so that the contributions along uninteresting parts of the curve are zero.

5.3.3 Retarded Green function

Let us begin with the retarded (causal) Green function, sometimes called the susceptibility χ, and write it as an integral expression in k space. We substitute the integral expressions in eqn. (5.75) into eqn. (5.70) and eqn. (5.74a), giving

$$G_r(x, x') = -\frac{2\pi}{\hbar^2 c} \int \frac{d\alpha}{2\pi} \frac{e^{-i\alpha\Delta t}}{\alpha + i\epsilon} \int \frac{(dk)}{2\pi} \left[\frac{e^{i(k\Delta x - \omega_k \Delta t)}}{2\omega_k} - \frac{e^{i(k\Delta x + \omega_k \Delta t)}}{2\omega_k} \right]$$

$$= -\frac{1}{\hbar^2 c} \int \frac{(dk)d\alpha}{(2\pi)} \left[\frac{e^{i(k\Delta x - (\omega_k + \alpha)\Delta t)}}{2\omega_k(\alpha + i\epsilon)} - \frac{e^{i(k\Delta x - (\alpha - \omega_k)\Delta t)}}{2\omega_k(\alpha + i\epsilon)} \right]. \tag{5.77}$$

We now shift $\alpha \to \alpha - \omega_k$ in the first term and $\alpha \to \alpha + \omega_k$ in the second term. This gives

$$G_{\rm r}(x, x') = - (\hbar^2 c)^{-1} \int \frac{{\rm d}^n {\bf k}\, {\rm d}\alpha}{(2\pi)^{n+1}} \frac{{\rm e}^{{\rm i}({\bf k}\Delta{\bf x} - \alpha \Delta t)}}{2\omega_k}$$

$$\times \left[\frac{1}{(\alpha - \omega_k + {\rm i}\epsilon)} - \frac{1}{(\alpha + \omega_k + {\rm i}\epsilon)} \right]. \qquad (5.78)$$

Re-labelling $\alpha \to k_0$ and combining the partial fractions on the right hand side, we are left with,

$$G_{\rm r}(x, x') = (\hbar^2 c)^{-1} \int ({\rm d}k)\, {\rm e}^{{\rm i}k\Delta x} \frac{1}{-(k_0 + {\rm i}\epsilon)^2 + \omega_k^2}, \qquad (5.79)$$

or to first order, re-defining $\epsilon \to \epsilon/2$,

$$G_{\rm r}(x, x') = c \int ({\rm d}k)\, {\rm e}^{{\rm i}k\Delta x} \frac{1}{p^2 c^2 + m^2 c^4 - {\rm i}p_0\epsilon}. \qquad (5.80)$$

This is the significant form we have been looking for. It may be compared with the expression in eqn. (5.40), and we notice that it reduces to eqn. (5.40) in the limit $\epsilon \to 0$. What is important is that we now have an unambiguous prescription for dealing with the poles: they no longer lie in the real k_0 axis. If we examine the poles of the integrand in eqn. (5.79) we see that they have been shifted below the axis to

$$ck_0 = \pm\omega_k - {\rm i}\epsilon; \qquad (5.81)$$

see figure 5.1. An alternative and completely equivalent contour is shown in figure 5.2. In this approach, we bend the contour rather than shift the poles; the end result is identical.

This $i\epsilon$ prescription tells us how to avoid the poles on the real axis, but it does not tell us how to complete the complex contour. Although the result we are looking for is equal to the value of the integral along the real axis only, Cauchy's theorem only gives us a prescription for calculating an integral around a closed contour, so we must complete the contour by joining the end of the real axis at $+\infty$ and $-\infty$ with a loop. After that, we extract the value of the portion which lies along the real axis.

The simplest way to evaluate the contribution to such a loop is to make it a semi-circle either in the upper half-plane or in the lower half-plane (see figure 5.2). But which do we choose? In fact, the choice is unimportant as long as we can extract the part of integral along the real axis.

Evaluation around two closed loops We begin by writing the integrals piece-wise around the loop in the complex k_0 plane. It is convenient to use $\omega = k_0 c$ as

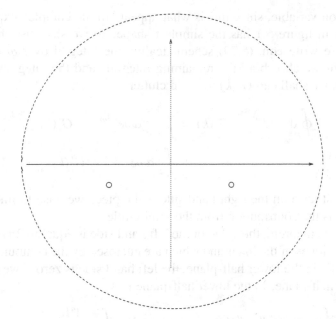

Fig. 5.1. Contour in the complex plane for the retarded Green function with poles shifted using the iϵ prescription.

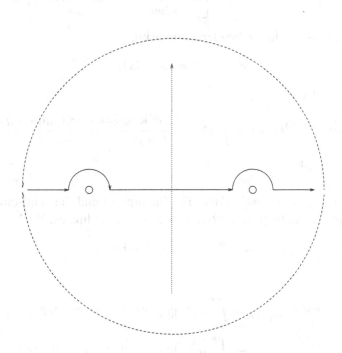

Fig. 5.2. Contour in the complex plane for the retarded Green function.

the integration variable, since this is what appears in the complex exponential. The contour in figure 5.1 has the simplest shape, so we shall use this as our template. We write eqn. (5.79) schematically: the integral over ω is written explicitly, but we absorb all the remaining integrals and the integrand into an object which we shall call $G'_r(k)$ to avoid clutter;

$$
\oint d\omega e^{-i\omega(t-t')} G'_r(k) = \int_{-\infty}^{+\infty} d\omega e^{-i\omega(t-t')} G'_r(k)
$$
$$
+ \int_{SC} d\omega e^{-i\omega(t-t')} G'_r(k), \qquad (5.82)
$$

where the first term on the right hand side is the piece we wish to find and the second term is the contribution from the semi-circle.

By Cauchy's theorem, the value of the left hand side is equal to $2\pi i$ times the sum of the residues of the integrand which are enclosed by the contour. Since all of the poles lie in the lower half-plane, the left hand side is zero if we complete in the upper half-plane. In the lower half-plane it is

$$
\oint d\omega e^{-i\omega(t-t')} G'_r(k) = -2\pi i \, (\hbar^2 c)^{-1} \int \frac{d^n k}{(2\pi)^{n+1}} \times
$$
$$
\left[\frac{e^{i(\mathbf{k}\cdot\Delta\mathbf{x}+\omega_k\Delta t)}}{-2\omega_k} + \frac{e^{i(\mathbf{k}\cdot\Delta\mathbf{x}-\omega_k\Delta t)}}{2\omega_k}. \right] \qquad (5.83)
$$

Re-labelling $k \to -k$ in the first term and using

$$
e^{ix} - e^{-ix} = 2i \sin(x), \qquad (5.84)
$$

we have ($\Delta t > 0$)

$$
\oint d\omega e^{-i\omega(t-t')} G'_r(k) = \int (\hbar^2 c)^{-1} \frac{d^n k}{(2\pi)^n} \frac{\cos(\mathbf{k}\cdot\Delta\mathbf{x}) \sin(\omega_k \Delta t)}{\omega_k}. \qquad (5.85)
$$

This is clearly real.

Semi-circle in the upper half-plane The integral around the semi-circle in the upper half-plane can be parametrized using polar coordinates. We let

$$
\omega = r e^{i\theta} = r(\cos\theta + i \sin\theta), \qquad (5.86)
$$

so that,

$$
\int_{SC} d\omega e^{-i\omega(t-t')} G'_r(k) = \int_0^{\pi} i r e^{i\theta} d\theta \, e^{-ir(\cos\theta+i\sin\theta)(t-t')} G'_r(r e^{i\theta})
$$
$$
= \int_0^{\pi} i r e^{i\theta} d\theta \, e^{-ir\cos\theta(t-t')} e^{r\sin\theta(t-t')} G'_r(r e^{i\theta}).
$$
$$
\qquad (5.87)
$$

Note what has happened here. The imaginary component from the semi-circle (the contribution involving $\sin \theta (t - t')$) has created a real exponential. This real exponential causes the integrand to either blow up or decay to zero at $r = \infty$, depending on the sign of the $\sin \theta (t - t')$ term. So we have two cases:

$$\int_{SC} d\omega e^{-i\omega(t-t')} G'_r(k) = 0 \qquad (t - t' < 0)$$

$$= ? \qquad (t - t' > 0). \qquad (5.88)$$

In the first case, in which we do not expect the retarded function to be defined, the integral over the semi-circle vanishes. Since the complete integral around the loop also vanishes here, the real axis contribution that we are looking for (looking at eqn. (5.82)), must also be zero. In the second case, the contribution from the loop is difficult to determine, so the contribution to the real axis part, from eqn. (5.82) is also difficult to determine. In fact, we cannot derive any useful information from this, so for $t - t' > 0$, we cannot determine the value of the integral. In order to evaluate the integral for $t - t' > 0$ we close the contour in the lower half-plane where the semi-circle contribution is again well behaved.

Semi-circle in the lower half-plane The integral around the semi-circle in the lower half-plane can also be parametrized using polar coordinates,

$$\int_{SC} d\omega e^{-i\omega(t-t')} G'_r(k) = - \int_0^{-\pi} i r e^{i\theta} d\theta \, e^{-ir(\cos\theta + i\sin\theta)(t-t')} G'_r(re^{i\theta})$$

$$= - \int_0^{-\pi} i r e^{i\theta} d\theta \, e^{-ir\cos\theta(t-t')} e^{-r|\sin\theta|(t-t')} G'_r(re^{i\theta}). \qquad (5.89)$$

Now the opposite happens:

$$\int_{SC} d\omega e^{-i\omega(t-t')} G'_r(k) = ? \qquad (t - t' < 0)$$

$$= 0 \qquad (t - t' > 0). \qquad (5.90)$$

This time the situation is reversed. The value of the integral tells us nothing for $t - t' < 0$. In the second case, however, the contribution to the loop goes to zero, making the integral along the real axis equal to the loop integral result in eqn. (5.85).

Piece-wise definition Because of the infinite pieces, we must close the contour for the retarded Green function separately for $t - t' > 0$ (lower half-plane, non-zero result) and $t - t' < 0$ (upper half-plane, zero result). This is not a serious problem for evaluating single Green functions, but the correct choice of contour becomes more subtle when calculating products of Green functions using the momentum-space forms. We have nonetheless established that these momentum-space prescriptions lead to a Green function which propagates from the past into the future:

$$G_{\rm r}(x, x') = (\hbar^2 c)^{-1} \int \frac{{\rm d}^n {\bf k}}{(2\pi)^n} \frac{\cos({\bf k} \cdot \Delta {\bf x}) \sin(\omega_k \Delta t)}{\omega_k} \quad (t - t' > 0)$$
$$= 0 \qquad (t - t' < 0). \tag{5.91}$$

5.3.4 Advanced Green function

The treatment of this function is identical in structure to that for the retarded propagator. The only difference is that the poles lie in the opposite half-plane, and thus the results are reversed:

$$G_{\rm a}(x, x') = - (\hbar^2 c)^{-1} \int ({\rm d}k) \, {\rm e}^{ik\Delta x} \frac{1}{-(k_0 - i\epsilon)^2 + \omega_k^2}. \tag{5.92}$$

We see that the poles are shifted above the axis and that the complex contour may now be completed in the opposite manner to the retarded Green function. The result is

$$G_{\rm a}(x, x') = - (\hbar^2 c)^{-1} \int \frac{{\rm d}^n {\bf k}}{(2\pi)^n} \frac{\sin({\bf k} \cdot \Delta {\bf x} - \omega_k \Delta t)}{\omega_k} \quad (t - t' < 0)$$
$$= 0 \qquad (t - t' > 0). \tag{5.93}$$

5.3.5 Feynman Green function

$$G_{\rm F}(x, x') = - \frac{2\pi}{\hbar^2 c} \int \frac{{\rm d}\alpha \, ({\rm d}k)}{2\pi \, (2\pi)} \left[\frac{{\rm e}^{i({\bf k}\Delta{\bf x} - (\omega_k + \alpha)\Delta t)}}{(\alpha + i\epsilon)2\omega_k} - \frac{{\rm e}^{i({\bf k}\Delta{\bf x} - (\alpha - \omega_k)\Delta t)}}{(\alpha - i\epsilon)2\omega_k} \right]. \tag{5.94}$$

Shifting $\alpha \to \alpha - \omega_k$ in the first fraction and $\alpha \to \alpha + \omega_k$ in the second fraction, and re-labelling $\alpha \to k_0$ we obtain,

$$G_{\rm F}(x, x') = (\hbar^2 c)^{-1} \int ({\rm d}k) \frac{{\rm e}^{ik\Delta x}}{2\omega_k} \left[\frac{1}{(k_0 + \omega_k - i\epsilon)} - \frac{1}{(k_0 - \omega_k + i\epsilon)} \right]. \tag{5.95}$$

It is normal to re-write this in the following way. Remember that we are interested in the limit $\epsilon \to 0$. Combining the partial fractions above, we get

$$G_{\rm F}(x, x') = (\hbar^2 c)^{-1} \int ({\rm d}k) \, {\rm e}^{ik\Delta x} \left[\frac{-1}{(k_0 + \omega - i\epsilon)(k_0 - \omega + i\epsilon)} + {\rm O}(\epsilon) \right]. \tag{5.96}$$

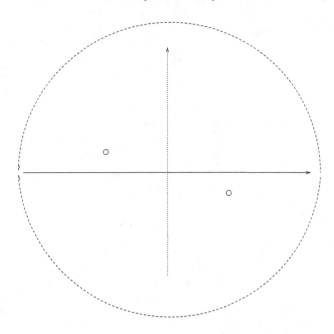

Fig. 5.3. Contour in the complex plane for the Feynman Green function. This shows how the $i\epsilon$ prescription moves the poles effectively from the real axis.

From this expression, we see that the poles have been shifted from the real axis to

$$ck_0 = \omega_k - i\epsilon$$
$$ck_0 = -\omega_k + i\epsilon, \tag{5.97}$$

i.e. the negative root is shifted above the axis and the positive root below the axis in the k_0 plane (see figure 5.4). An equivalent contour is shown in figure 5.3. Although it does not improve one's understanding in any way, it is normal in the literature to write the Feynman Green function in the following way. Re-writing the denominator, we have

$$(ck_0 + \omega - i\epsilon)(ck_0 - \omega + i\epsilon) = c^2 k_0^2 - \omega_k^2 + 2i\epsilon\omega_k + \epsilon^2. \tag{5.98}$$

Now, since ϵ is infinitesimal and $\omega_k > 0$, we may drop ϵ^2, and write $2i\epsilon\omega_k = i\epsilon'$. This allows us to write

$$G_F(x, x') = c \int (dk) \frac{e^{ik\Delta x}}{p^2 c^2 + m^2 c^4 - i\epsilon'}. \tag{5.99}$$

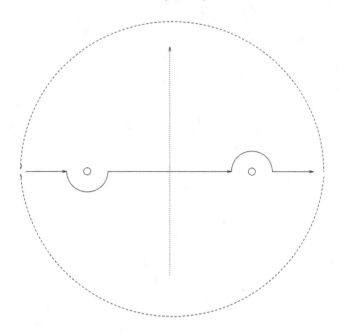

Fig. 5.4. Contour in the complex plane for the Feynman Green function. Here we bend the contour rather than moving the poles. The result is identical.

5.3.6 Comment on complex contours

The procedure described by Green functions is a formalism for extracting solutions to the inverse-operator problem. It has a direct analogy in the theory of matrices or linear algebra. There the issue concerns the invertibility of matrices and the determinant of the matrix operator. Suppose we have a matrix equation

$$M \cdot \mathbf{x} = \mathbf{J}, \tag{5.100}$$

with a matrix M given by

$$M = \begin{pmatrix} a & b \\ c & d \end{pmatrix}. \tag{5.101}$$

If this matrix has an inverse, which is true if the determinant $ad - bc$ does not vanish,

$$M^{-1} = \frac{1}{ad - bc} \begin{pmatrix} d & -b \\ -c & a \end{pmatrix}, \tag{5.102}$$

then eqn. (5.100) has a unique solution. We would not expect this case to correspond to the solution of a differential equation such as the one we are considering, since we know that the general solution to second-order differential equations usually involves a linear super-position of many solutions.

If the determinant of M does vanish, then it means that there is an infinite number of solutions, which corresponds to a sub-space of x (a hyper-surface which is determined by a constraint linking the coordinates). In this case, the inverse defined above in eqn. (5.102) has a pole. For example, suppose we take M to be the matrix

$$M = \begin{pmatrix} 1 & 2 & 1 \\ 1 & 1 & 0 \\ 4 & 8 & 4 \end{pmatrix}, \tag{5.103}$$

and

$$J = \begin{pmatrix} 4 \\ 2 \\ 16 \end{pmatrix}. \tag{5.104}$$

This matrix clearly has no inverse, since the third row is a multiple of the first. The determinant vanishes, but in this trivial case we can solve the equations directly. Since there are only two independent equations and three unknowns, it is not possible to find a unique solution. Instead, we eliminate all but one of the variables, leaving

$$x_2 + x_3 = 2. \tag{5.105}$$

This is the equation of a straight line, or a sub-space of the full three-dimensional solution space. We regard this as an incomplete constraint on the solution space rather than a complete solution.

This is analogous to the situation we have with the Green functions. The poles indicate that the solution to the differential equation which we are trying to solve is not unique. In fact, there is an infinite number of plane wave solutions which lie on the hyper-surface $k^2 + m^2 = 0$, called the mass shell.

5.4 Scalar Green functions in real space

Although the momentum-space representations of the Green functions are useful for calculations, we are usually interested in their forms in real space. For general fields with a mass, these can be quite complicated, but in the massless limit the momentum-space integrals can be straightforwardly evaluated.

Again, since the other relativistic Green functions can be expressed in terms of that for the scalar field, we shall focus mainly on this simple case.

5.4.1 The retarded Green function for $n = 3$ as $m \rightarrow 0$

From Cauchy's residue theorem in eqn. (5.76), we have

$$
G_{\mathrm{r}}(x, x') = -2\pi \mathrm{i} \, (\hbar^2 c)^{-1} \int \frac{\mathrm{d}^3 k}{(2\pi)^4} \left[\frac{\mathrm{e}^{\mathrm{i}(\mathbf{k}\cdot\Delta\mathbf{x}-\omega_k\Delta t)}}{2\omega_k} - \frac{\mathrm{e}^{\mathrm{i}(\mathbf{k}\cdot\Delta\mathbf{x}+\omega_k\Delta t)}}{2\omega_k} \right].
$$

$$(5.106)$$

For general $m \neq 0$, this integral defines Bessel functions. For $m = 0$, however, the integral is straightforward and can be evaluated by going to three-dimensional polar coordinates in momentum space:

$$
\omega_k = |r|c
$$

$$
\int \mathrm{d}^3 k = \int_0^\infty r^2 \, \mathrm{d}r \int_0^\pi \sin\theta \mathrm{d}\theta \int_0^{2\pi} \mathrm{d}\phi
$$

$$
\mathbf{k} \cdot \mathbf{x} = |r|\Delta X \cos\theta,
$$

$$(5.107)$$

where $\Delta X = |\Delta\mathbf{x}|$, so that

$$
G_{\mathrm{r}}(x, x') = \frac{-\mathrm{i}}{16\pi^3} \, (\hbar^2 c)^{-1} \int_0^\infty 2\pi r^2 \, \mathrm{d}r
$$

$$
\times \int_0^\pi \sin\theta \, \mathrm{d}\theta \, \frac{\mathrm{e}^{\mathrm{i}r\Delta x \cos\theta}}{r} \left[\mathrm{e}^{-\mathrm{i}rc\Delta t} - \mathrm{e}^{\mathrm{i}rc\Delta t} \right].
$$

$$(5.108)$$

The integral over $\mathrm{d}\theta$ may now be performed, giving

$$
G_{\mathrm{r}}(x, x') = \frac{-1}{8\pi^2 \Delta X} \, (\hbar^2 c)^{-1} \int_0^\infty \mathrm{d}r \left\{ \mathrm{e}^{-\mathrm{i}r(\Delta X+c\Delta t)} \right.
$$

$$
\left. - \mathrm{e}^{\mathrm{i}r(\Delta t-c\Delta X)} - \mathrm{e}^{\mathrm{i}r(\Delta X-c\Delta t)} + \mathrm{e}^{\mathrm{i}r(\Delta X+c\Delta t)} \right\}.
$$

$$(5.109)$$

Note that both Δt and Δx are positive by assumption. From the definition of the delta function, we have

$$
2\pi \delta(x) = \int_{-\infty}^{+\infty} \mathrm{d}k \, \mathrm{e}^{\mathrm{i}kx}
$$

$$
= \int_0^\infty \left[\mathrm{e}^{\mathrm{i}kx} + \mathrm{e}^{-\mathrm{i}kx} \right].
$$

$$(5.110)$$

Using this result, we see that the first and last terms in eqn. (5.109) vanish, since Δx can never be equal to $-\Delta t$ as both Δx and Δt are positive. This leaves us with

$$
G_{\mathrm{r}}(x, x') = \frac{1}{4\hbar^2 c\pi \Delta X} \delta(ct - \Delta X)
$$

$$
= \frac{1}{4\pi\hbar^2 c|\mathbf{x} - \mathbf{x}'|} \, \delta\left(c(t - t') - |\mathbf{x} - \mathbf{x}'|\right).
$$

$$(5.111)$$

5.4.2 The $G^{(\pm)}$ and G_F for $n = 3$ as $m \to 0$

From eqn. (5.74c) we can see that the Feynman propagator is manifestly equal to $-G^{(+)}$ for $t > t'$ and is equal to $G^{(-)}$ for $t' < t$. The calculation of all three quantities can therefore be taken together. We could, in fact, write this down from the definitions, but it is useful to use the residue theorem on eqn. (5.95) to show the consistency of the procedure with the definitions we have already given. In fact, we shall see that the Wightman functions *are* just the residues, up to a sign which depends on the orientation of the closed contour.

For $t - t' > 0$, we complete the contour in the lower half-plane, creating an anti-clockwise contour. The residue theorem then tells us that

$$\oint dk_0 G'_F(k_0) = (\hbar^2 c)^{-1} \int \frac{d(\mathbf{k})}{(2\pi)} \frac{1}{2\omega_k} e^{ik\Delta x} \times -2\pi i \{-1\} \Big|_{k_0 = \omega_k}. \quad (5.112)$$

Comparing this equation with eqns. (5.66), we see that this is precisely equal to $-G^{(+)}(x, x')$. In the massless limit with $n = 3$, we may therefore write

$$G^{(+)} = -i (\hbar^2 c)^{-1} \frac{d^3 \mathbf{k}}{(2\pi)^4} \frac{e^{ik\Delta x}}{2|\mathbf{k}|c^2}$$

$$= \frac{-1}{8\pi^2 \hbar^2 c |\mathbf{x}' - \mathbf{x}|} \int_0^\infty dr \left\{ e^{-ir(\Delta X + c\Delta t)} - e^{ir(\Delta X - c\Delta t)} \right\}. \quad (5.113)$$

Similarly, for $t - t' > 0$, we complete the contour in the upper half-plane, creating a clockwise contour. This gives

$$\oint dk_0 G'_F(k_0) = (\hbar^2 c)^{-1} \int \frac{d(\mathbf{k})}{(2\pi)} \frac{1}{2\omega_k} e^{i(\mathbf{k}\cdot\Delta\mathbf{x} + \omega_k \Delta t)} \times 2\pi i \{+1\} \Big|_{k_0 = \omega_k} \quad (5.114)$$

Comparing this equation with eqn. (5.66), we see that this is precisely equal to $G^{(-)}(x, x')$, and

$$G^{(-)} = \frac{1}{8\pi^2 \hbar^2 c |\mathbf{x}' - \mathbf{x}|} \int_0^\infty dr \left\{ e^{-ir(\Delta X - c\Delta t)} - e^{ir(\Delta X + c\Delta t)} \right\}. \quad (5.115)$$

It may be checked that these expressions satisfy eqn. (5.67). Finally, we may piece together the Feynman Green function from $G^{(\pm)}$. Given that the Δt are assumed positive, we have

$$G_F(x, x') = \frac{1}{8\pi^2 \hbar^2 c |\mathbf{x} - \mathbf{x}'|} \int_0^\infty dr \left\{ e^{-ir(\Delta X)} - e^{ir(\Delta X)} e^{-irc\Delta t} \right\}$$

$$= \frac{-i}{4\pi^2 \hbar^2 c |\mathbf{x} - \mathbf{x}'|} \int_0^\infty dr \, \sin(r|\mathbf{x}' - \mathbf{x}|) e^{-irc|t-t'|}. \quad (5.116)$$

We may note that the difference between the retarded and Feynman Green functions is

$$G_F(x, x') - G_r(x, x') = \lim_{\alpha \to 0} 2 \int_0^\infty dr \; e^{ir(c\Delta t - \Delta X + i\alpha)} - e^{ir(c\Delta t + \Delta X + i\alpha)}$$

$$= \frac{i}{|\mathbf{x} - \mathbf{x}'| - c|t - t'|} - \frac{i}{|\mathbf{x} + \mathbf{x}'| - c|t - t'|},$$
(5.117)

where α is introduced to define the infinite limit of the complex exponential. This difference is a purely imaginary number, which diverges on the light cone.

5.4.3 Frequency-dependent form of G_F and G_r in $n = 3$

In atomic physics and optics, one usually deals in implicitly translational invariant systems, in the rest frames of an atom, where the frequency ω and time are the only variables entering physical models. To use standard field theoretical methods in these cases, it is useful to have the Green functions in such a form, by integrating over spatial wavenumbers leaving only the Fourier transform over time. These are obtained trivially by re-writing the non-zero contributions to eqns. (5.109) and (5.116) with $r \to \omega/c$:

$$G_F(x, x') = \frac{-i}{4\pi^2 \hbar^2 c^2 |\mathbf{x} - \mathbf{x}'|} \int_0^\infty d\omega \; \sin\left(\frac{\omega}{c}|\mathbf{x} - \mathbf{x}'|\right) e^{-i\omega|t - t'|}$$

$$G_r(x, x') = \frac{1}{4\pi^2 \hbar^2 c^2 |\mathbf{x} - \mathbf{x}'|} \int_0^\infty d\omega \; \cos\left(\frac{\omega}{c}|\mathbf{x} - \mathbf{x}'| - \omega|t - t'|\right).$$
(5.118)

5.4.4 Euclidean Green function in $2 + 0$ dimensions

In the special case of a space-only Green function (the inverse of the Laplacian operator), there is no ambiguity in the boundary conditions, since the Green function is time-independent and there are no poles in the integrand. Let us define the inverse Laplacian by

$$(-\nabla^2 + m^2)g(x, x') = \delta(x, x').$$
(5.119)

To evaluate this function, we work in Fourier space and write

$$g(x, x') = \int \frac{d^2k}{(2\pi)^2} \frac{e^{ik(x-x')}}{k^2 + m^2},$$
(5.120)

where $k^2 = k_1^2 + k_2^2$. Expressing this in polar coordinates, we have

$$g(x, x') = \int_0^{2\pi} \int_0^\infty \frac{r\,dr\,d\theta}{(2\pi)^2} \frac{e^{ir|x-x'|\cos\theta}}{r^2 + m^2}.$$
(5.121)

Massless case In the massless limit, this integral can be evaluated straightforwardly using a trick which is frequently useful in the evaluation of Fourier integrals. The trick is only valid strictly when $x \neq x'$, but we shall leave it as an exercise to show what happens in that case. The integral is then evaluated by setting m to zero in eqn. (5.121) and cancelling a factor of r from the integration measure. To evaluate the expression, we differentiate under the integral sign with respect to the quantity $|x - x'|$:

$$\frac{d}{d|x - x'|} g(x - x') = \int_0^{2\pi} \int_0^\infty \frac{i \cos \theta}{(2\pi)^2} e^{ir|x-x'| \cos \theta} dr \, d\theta. \quad (5.122)$$

Notice that this step cancels a factor of r in the denominator, which means that the integral over r is now much simpler. Formally, we have

$$\frac{d}{d|x - x'|} g(x - x') = \int_0^{2\pi} \frac{d\theta}{(2\pi)^2} e^{ir|x-x'| \cos \theta} \Big|_0^\infty. \quad (5.123)$$

There is still a subtlety remaining, however: since we are integrating a complex, multi-valued function, the limit at infinity has an ambiguous limit. The limit can be defined uniquely (analytically continued) by adding an infinitesimal positive imaginary part to r, so that $r \to r(i + \epsilon)$ and letting $\epsilon \to 0$ afterwards. This makes the infinite limit converge to zero, leaving only a contribution from the lower limit:

$$\frac{d}{d|x - x'|} g(x - x') = \lim_{\epsilon \to 0} \int_0^{2\pi} \frac{d\theta}{(2\pi)^2} \frac{1}{1 - i\epsilon} e^{(ir - \epsilon r)|x-x'| \cos \theta} \Big|_0^\infty$$

$$= -\int_0^{2\pi} \frac{d\theta}{(2\pi)^2} \frac{1}{|x - x'|}. \quad (5.124)$$

To complete the evaluation, we evaluate the two remaining integrals trivially, first the anti-derivative with respect to $|x - x'|$, which gives rise to a logarithm, and finally the integral over θ, giving:

$$g(x, x') = -\frac{1}{2\pi} \ln |x - x'|, \quad (5.125)$$

where it is understood that $x \neq x'$.

5.4.5 Massive case

In the massive case, we can write down the result in terms of Bessel functions J_ν, K_ν, by noting the following integral identities [63]:

$$J_\nu(z) = \frac{(z/2)^\nu}{\Gamma(\nu + \frac{1}{2})\Gamma(\frac{1}{2})} \int_0^\pi e^{\pm iz \cos \theta} \sin^{2\nu} \theta \, d\theta \quad (5.126)$$

$$K_{\nu-\mu}(ab) = \frac{2^\mu \Gamma(\mu + 1)}{a^{\nu-\mu} b^\mu} \int_0^\infty \frac{J_\nu(bx) \, x^{\nu+1}}{(x^2 + a^2)} dx. \quad (5.127)$$

From the first of these, we can choose $\nu = 0$ and use the symmetry of the cosine function to write

$$J_0(z) = \frac{1}{2\pi} \int_0^{2\pi} e^{iz\cos\theta} d\theta. \tag{5.128}$$

Eqn. (5.121) may now be expressed in the form

$$g(x, x') = \int_0^{\infty} \frac{r dr}{2\pi} \frac{J_0(r|x - x'|)}{r^2 + m^2}, \tag{5.129}$$

and hence

$$g(x, x') = \frac{1}{2\pi} K_0(m|x - x'|). \tag{5.130}$$

The massless limit is singular, but with care can be inferred from the small argument expansion

$$K_0(m(x - x')) = \lim_{m \to 0} -\ln\left(\frac{m(x - x')}{2}\right) \sum_{k=0}^{\infty} \frac{\left(\frac{m(x-x')}{2}\right)^{2k}}{(k!)^2}. \tag{5.131}$$

5.5 Schrödinger Green function

Being linear in the time derivative, the solutions of the Schrödinger equation have positive definite energy. The Fourier transform may therefore be written as,

$$\psi(x) = \int_0^{\infty} \frac{d\tilde{\omega}}{2\pi} \int_{-\infty}^{+\infty} (dk) \, e^{i(\mathbf{k}\cdot\Delta\mathbf{x} - \tilde{\omega}\Delta t)} \psi(\mathbf{k}, \tilde{\omega}) \theta(\tilde{\omega}) \delta\left(\frac{\hbar^2 \mathbf{k}^2}{2m} - \hbar\tilde{\omega}\right). \tag{5.132}$$

This singles out the Schrödinger field amongst the other relativistic fields which have solutions of both signs. Correspondingly, the Schrödinger field has only a positive energy Wightman function, the negative energy function vanishes from the particle theory.[9] The positive frequency Wightman function is

$$G_{NR}^{(+)}(x, x') = -2\pi i \int_0^{\infty} \frac{d\tilde{\omega}}{2\pi} \int_{-\infty}^{+\infty} (dk) e^{i(\mathbf{k}\cdot\Delta\mathbf{x} - \tilde{\omega}\Delta t)} \theta(\tilde{\omega}) \delta\left(\frac{\hbar^2 \mathbf{k}^2}{2m} - \hbar\tilde{\omega}\right). \tag{5.133}$$

The negative frequency Wightman function vanishes now,

$$G_{NR}^{(-)}(x, x') = 0, \tag{5.134}$$

---

[9] This does not remain true at finite temperature or in interacting field theory, but there remains a fundamental asymmetry between positive and negative energy Green functions in the non-relativistic theory.

since there is no pole in the negative $\tilde{\omega}$ plane to enclose. Moreover, this means that there is no Feynman Green function in the non-relativistic theory, only a retarded one. In the non-relativistic limit, both the Feynman Green function and the retarded Green function for relativistic particles reduce to the same result, which has poles only in the lower half complex $\tilde{\omega}$ plane. This non-relativistic Green function satisfies the equation

$$\left(-\frac{\hbar^2\nabla^2}{2m} - i\hbar\partial_t\right) G_{\mathrm{NR}}(x, x') = \delta(\mathbf{x}, \mathbf{x}')\delta(t, t').\tag{5.135}$$

This Green function can be evaluated from the expression corresponding to those in eqns. (5.74):

$$G_{\mathrm{NR}}(x, x') = -\theta(t - t')G_{\mathrm{NR}}^{(+)}(x, x').\tag{5.136}$$

Using eqn. (5.75) in eqn. (5.133), we have

$$G_{\mathrm{NR}}(x, x') = -\int_{-\infty}^{+\infty} d\alpha \int_0^\infty \frac{d\tilde{\omega}}{2\pi} \int_{-\infty}^{+\infty} (dk)$$
$$\times \frac{e^{i(\mathbf{k}\cdot\Delta\mathbf{x}-(\tilde{\omega}+\alpha)\Delta t)}}{(\alpha + i\epsilon)}\delta\left(\frac{\hbar^2\mathbf{k}^2}{2m} - \hbar\tilde{\omega}\right).\tag{5.137}$$

The integral over α can be shifted, $\alpha \to \alpha - \tilde{\omega}$, without consequences for the limits or the measure, giving

$$G_{\mathrm{NR}}(x, x') = -\int_{-\infty}^{+\infty} d\alpha \int_0^\infty \frac{d\tilde{\omega}}{2\pi} \int_{-\infty}^{+\infty} (dk)$$
$$\times \frac{e^{i(\mathbf{k}\cdot\Delta\mathbf{x}-\alpha\Delta t)}}{(\alpha - \tilde{\omega}) + i\epsilon}\delta\left(\frac{\hbar^2\mathbf{k}^2}{2m} - \hbar\tilde{\omega}\right).\tag{5.138}$$

We may now integrate over $\tilde{\omega}$ to invoke the delta function. Noting that the argument of the delta function is defined only for positive $\tilde{\omega}$, and that the integral is also over this range, we have simply

$$G_{\mathrm{NR}}(x, x') = -\int_{-\infty}^{+\infty} d\alpha \int_{-\infty}^{+\infty} (dk) \frac{e^{i(\mathbf{k}\cdot\Delta\mathbf{x}-\alpha\Delta t)}}{\left(\hbar\alpha - \frac{\hbar^2\mathbf{k}^2}{2m}\right) + i\epsilon},\tag{5.139}$$

or, re-labelling $\alpha \to \tilde{\omega}$,

$$G_{\mathrm{NR}}(x, x') = \int_{-\infty}^{+\infty} d\tilde{\omega} \int_{-\infty}^{+\infty} (dk) \frac{e^{i(\mathbf{k}\cdot\Delta\mathbf{x}-\tilde{\omega}\Delta t)}}{\left(\frac{\hbar^2\mathbf{k}^2}{2m} - \hbar\tilde{\omega}\right) - i\epsilon}.\tag{5.140}$$

In spite of appearances, the parameter $\tilde{\omega}$ is not really the energy of the system, since it runs from minus infinity to plus infinity. It should properly be regarded only as a variable of integration. It is clear from this expression that the Schrödinger field has a single pole in the lower half complex plane. It therefore satisfies purely retarded boundary conditions. We shall see in section 13.2.2 how the relativistic Feynman Green function reduces to a purely retarded one in the non-relativistic limit.

5.6 Dirac Green functions

The Dirac Green function satisfies an equation which is first order in the derivatives, but which is matrix-valued. The equation of motion for the Dirac field,

$$(-i\gamma^\mu \partial_\mu + m)\psi = J, \tag{5.141}$$

tells us that a formal solution may be written as

$$\psi = \int dV_{x'} S(x, x') J(x'), \tag{5.142}$$

where the spinor Green function is defined by

$$(-i\hbar c\gamma^\mu \partial_\mu + mc^2) S(x, x') = \delta(x, x'). \tag{5.143}$$

Although this looks rather different to the scalar field case, $S(x, x')$ can be obtained from the expression for the scalar propagator by noting that

$$(-i\hbar c\gamma^\mu \partial_\mu + mc^2)(i\hbar c\gamma^\mu \partial_\mu + mc^2)$$
$$= -\hbar^2 c^2 \Box + m^2 c^4 + \frac{1}{2}[\gamma^\mu, \gamma^\nu]\partial_\mu \partial_\nu, \tag{5.144}$$

and the latter term vanishes when operating on non-singular objects. It follows for the free field that

$$(i\hbar c\gamma^\mu \partial_\mu + mc^2)G^{(\pm)}(x, x') = S^{(\pm)}(x, x') \tag{5.145}$$
$$(i\hbar c\gamma^\mu \partial_\mu + mc^2)G_F(x, x') = S_F(x, x') \tag{5.146}$$
$$(-i\hbar c\gamma^\mu \partial_\mu + mc^2)S^{(\pm)}(x, x') = 0 \tag{5.147}$$
$$(-i\hbar c\gamma^\mu \partial_\mu + mc^2)S_F(x, x') = \delta(x, x'). \tag{5.148}$$

5.7 Photon Green functions

The Green function for the Maxwell field satisfies the $(n+1)$ dimensional vector equation

$$\left[-\Box \, \delta_\mu^\nu + \partial_\mu \partial^\nu \right] A^\mu(x) = \mu_0 J^\nu. \tag{5.149}$$

As usual, we look for the inverse of the operator,[10] which satisfies

$$\left[-\Box \, \delta_\mu^{\,\nu} + \partial_\mu \partial^\nu\right] D_\nu^\rho(x, x') = \mu_0 c \delta_\mu^{\,\rho} \delta(x, x'). \tag{5.150}$$

Formally, it can be written as a Fourier transform:

$$D_{\mu\nu}(x, x') = \mu_0 c \int (dk) e^{ik(x-x')} \left[\frac{g_{\mu\nu}}{k^2} - \frac{k^\mu k^\nu}{k^4}\right]. \tag{5.151}$$

In this case, however, there is a problem. In inverting the operator, we are looking for a constraint which imposes the equations of motion. For scalar particles, this is done by going to momentum space and constructing the Green function, which embodies the equations of motion in the dispersion relation $k^2 + m^2 = 0$ (see eqn. (5.40)). In this case, that approach fails.

The difficulty here is the gauge symmetry. Suppose we consider the determinant of the operator in eqn. (5.149). A straightforward computation shows that this determinant vanishes:

$$\begin{vmatrix} -\Box + \partial_0 \partial^0 & \partial_0 \partial^i \\ \partial_i \partial^0 & -\Box + \partial_i \partial^i \end{vmatrix} = 0. \tag{5.152}$$

In linear algebra, this would be a signal that the matrix was not invertible, the matrix equivalent of dividing by zero. It also presents a problem here. The problem is not that the operator is not invertible (none of the Green function equations are invertible when the constraints they impose are fulfilled, since they correspond precisely to a division by zero), but rather that it implies no constraint at all. In the case of a scalar field, we have the operator constraint, or its momentum-space form:

$$-\hbar^2 c^2 \Box + m^2 c^4 = 0$$
$$p^2 c^2 + m^2 c^4 = 0. \tag{5.153}$$

In the vector case, one has

$$\det\left[-\Box \, \delta_\mu^{\,\nu} + \partial_\mu \partial^\nu\right] = 0, \tag{5.154}$$

but this is an identity which is solved for every value of the momentum. Thus, the Green function in eqn. (5.151) supplies an infinite number of solutions for A_μ for every J, one for each unrestricted value of k, which makes eqn. (5.151) singular.

The problem can be traced to the gauge symmetry of the field $A_\mu(x)$. Under a gauge transformation, $A_\mu \to A_\mu + \partial_\mu s$, but

$$\left[-\Box \, \delta_\mu^{\,\nu} + \partial_\mu \partial^\nu\right] (\partial_\nu s) = 0 \tag{5.155}$$

---

[10] Note that the operator has one index up and one index down, thereby mapping contravariant eigenvectors to contravariant eigenvectors

for any function $s(x)$. It can be circumvented by breaking the gauge symmetry in such a way that the integral over k in eqn. (5.151) is restricted. A convenient choice is the so-called Lorentz gauge condition

$$\partial_\mu A^\mu = 0. \tag{5.156}$$

This can be enforced by adding a Lagrange multiplier to the Maxwell action,

$$S \to \int (\mathrm{d}x) \left\{ \frac{1}{4\mu_0} F^{\mu\nu} F_{\mu\nu} - J^\mu A_\mu + \frac{1}{2\alpha} \mu_0^{-1} (\partial^\mu A_\mu)^2 \right\}, \tag{5.157}$$

so that eqn. (5.149) is modified to

$$\left[-\Box \, \delta_\mu^{\ \nu} + \left(1 - \frac{1}{\alpha} \right) \partial_\mu \partial^\nu \right] A^\mu(x) = J^\nu. \tag{5.158}$$

It may now be verified that the determinant of the operator no longer vanishes for all α; thus, a formal constraint is implied over the k_μ, and the Green function may be written

$$D_{\mu\nu}(x, x') = c\mu_0 \int (\mathrm{d}k) e^{ik(x-x')} \left[\frac{g_{\mu\nu}}{k^2} + (\alpha - 1) \frac{k_\mu k_\nu}{k^4} \right]. \tag{5.159}$$

This constraint is not a complete breakage of the gauge symmetry, since one may gauge transform eqn. (5.156) and show that

$$\partial_\mu A^\mu \to \partial_\mu A^\mu + \Box \, s(x) = 0. \tag{5.160}$$

Thus, the gauge condition still admits restricted gauge transformations such that

$$\Box \, s(x) = 0. \tag{5.161}$$

However, this modification is sufficient to obtain a formal Green function, and so the additional gauge multi-valuedness is often not addressed.

5.8 Principal values and Kramers–Kronig relations

Green functions which satisfy retarded (or advanced) boundary conditions satisfy a special pair of Fourier frequency-space relations, called the Kramers–Kronig relations (these are also referred to as Bode's law in circuit theory), by virtue of the fact that all of their poles lie in one half-plane (see figure 5.5). These relations are an indication of purely causal or purely acausal behaviour. In particular, physical response functions satisfy such relations, including the refractive index (or susceptibility, in non-magnetic materials) and the conductivity.

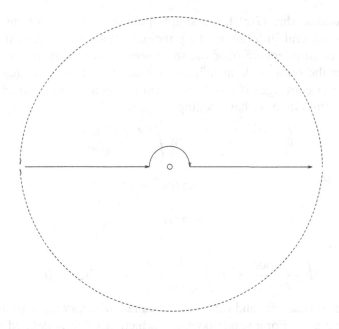

Fig. 5.5.   Contour in the complex plane for the Kramers–Kronig relations.

Cauchy's integral formula states that the value of a function $G(\omega)$, which is analytic at every point within and on a closed curve C, and is evaluated at a point $\omega = z$, is given by the integral around the closed loop C of

$$\oint_C \frac{G(\omega)}{\omega - z} = 2\pi i G(z). \qquad (5.162)$$

If a point P lies outside the closed loop, the value of the integral at that point is zero. Consider then a field $G(t-t')$ which satisfies retarded boundary conditions

$$G(t - t') = \int \frac{d\omega}{2\pi} e^{-i\omega(t-t')} G(\omega). \qquad (5.163)$$

The Fourier transform $G(\omega)$, where

$$G(\omega) = \int d(t - t') e^{i\omega(t-t')} G(t - t') \qquad (5.164)$$

is analytic in the upper half-plane, as in figure 5.5, but has a pole on the real axis. In the analytic upper region, the integral around a closed curve is zero, by Cauchy's theorem:

$$\oint_C \frac{G(\omega)d\omega}{\omega - z} = 0, \qquad (5.165)$$

where we assume that $G(z)$ has a simple pole at $\omega = z$. We can write the parts of this integral in terms of the principal value of the integral along the real axis, plus the integral around the small semi-circle enclosing the pole. The integral over the semi-circle at infinity vanishes over the causal region, since $\exp(i\omega(t - t'))$ converges if $t - t' > 0$ and ω has a positive imaginary part. Around the semi-circle we have, letting $\omega - z = \epsilon\, e^{i\theta}$,

$$\oint_{SC} \frac{G(\omega)d\omega}{\omega - z} = -\lim_{\epsilon \to 0} \int_0^\pi \frac{G(\epsilon e^{i\theta})i\epsilon e^{i\theta}d\theta}{\epsilon e^{i\theta}}$$

$$= -i\pi(\epsilon e^{i\theta} + z)\bigg|_{\epsilon \to 0}$$

$$= -i\pi G(z). \tag{5.166}$$

Then we have

$$\oint_C \frac{G(\omega)d\omega}{\omega - z} = P\int_{-\infty}^\infty \frac{G(\omega)d\omega}{\omega - z} - i\pi G(z) = 0. \tag{5.167}$$

The first term on the left hand side is the so-called *principal value* of the integral along the real axis. For a single pole, the principal value is defined strictly by the limit

$$P\int_{-\infty}^{+\infty} \equiv \lim_{\epsilon \to 0} \left\{ \int_{-\infty}^{p_i - \epsilon} + \int_{p_i + \epsilon}^\infty \right\}, \tag{5.168}$$

which approaches the singularity from equal distances on both sides. The expression may be generalized to two or more poles by arranging the limits of the integral to approach all poles symmetrically. Thus, if we now write the real and imaginary parts of $G(\omega)$ explicitly as

$$G(z) \equiv G_R(z) + iG_I(z), \tag{5.169}$$

and substitute this into eqn. (5.167), then, comparing real and imaginary parts we have:

$$P\int_{-\infty}^\infty \frac{G_R(\omega)d\omega}{\omega - z} = -\pi G_I(z)$$

$$P\int_{-\infty}^\infty \frac{G_I(\omega)d\omega}{\omega - z} = \pi G_R(z). \tag{5.170}$$

These are the so-called Kramers–Kronig relations. They indicate that the analyticity of $G(t - t')$ implies a relationship between the real and imaginary parts of $G(t - t')$.

The generalization of these expressions to several poles along the real axis may be written

$$P\int_{-\infty}^\infty \frac{G_{I/R}(\omega)d\omega}{\omega - z} = \sum_{\text{poles}} \pm\pi G_{R/I}(z). \tag{5.171}$$

The integral along the real axis piece of the contour may be used to derive an expression for the principal value of $1/\omega$. From eqn. (5.167), we may write

$$\frac{1}{\omega - z} = P\frac{1}{\omega - z} - i\pi\delta(\omega - z).$$ (5.172)

This relation assumes that we have integrated along the real axis in a positive direction, avoiding a single pole on the real axis by passing *above* it (or, equivalently, by pushing the pole into the lower half-plane by an infinitesimal amount $i\epsilon$). Apart from these assumptions, it is quite general. It does not make any other assumptions about the nature of $G(\omega)$, nor does it depend on the presence of any other poles which do not lie on the real axis. It is a property of the special contour segment which passes around one pole. Had the contour passed under the pole instead of over it, the sign of the second term would have been changed. These results can be summarized and generalized to several poles on the real axis, by writing

$$\frac{1}{\omega - z \pm i\epsilon_j} = P\frac{1}{\omega - z} \mp \sum_j i\pi\delta(\omega - z_j),$$ (5.173)

where z is a general point in the complex plane, z_i are the poles on the real axis and $\epsilon \to 0$ is assumed. The upper sign is that for passing over the poles, while the lower sign is for passing under.

5.9 Representation of bound states in field theory

Bound states are states in which 'particles' are completely confined by a potential $V(x)$. Confinement is a simple interaction between two different fields: a dynamical field $\psi(x)$ and a static confining field $V(x)$. The way in which one represents bound states in field theory depends on which properties are germane to the description of the physical system. There are two possibilities.

The first alternative is the approach traditionally used in quantum mechanics. Here one considers the potential $V(x)$ to be a fixed potential, which breaks translational symmetry, e.g.

$$\left(-\frac{\hbar^2}{2m}\nabla^2 + V(\mathbf{x})\right)\psi(x) = i\partial_t\psi(x).$$ (5.174)

One then considers the equation of motion of $\psi(x)$ in the rest frame of this potential and solves it using whatever methods are available. A Green function formulation of this problem leads to the Lippman–Schwinger equation for example (see section 17.5). In this case, the dynamical variable is the field, which moves in an external potential and is confined by it, e.g. electrons moving in the spherical hydrogen atom potential.

A second possibility is to consider bound states as multi-level, internal properties of the dynamical variables in question. For instance, instead of formulating the motion of electrons in a hydrogen atom, one formulates the motion of hydrogen atoms with internal electron levels which can be excited. To do this, one introduces multiplet states (an index A on the field and on the constant potential), e.g.

$$\left(-\frac{\hbar^2}{2m} \nabla^2 + V_A \right) \psi_A(x) = i\partial_t \psi_A(x). \tag{5.175}$$

This is an effective theory in which one takes the average value of the potential V_A at N different levels, where $A = 1, \ldots, N$. The values of V_A signify the energy differences between levels in the atom. The field ψ_A now represents the whole atom, not the electron within in. Clearly, all the components of ψ_A move together, according to the same equation of motion. The internal indices have the character of a broken internal 'symmetry'. This approach allows one to study the dynamics and kinematics of hydrogen atoms in motion (rather than the behaviour of electrons in the rest frame of the atom). Such a study is of interest when considering how transitions are affected by sources outside the atom. An example of this is provided by the classic interaction between two levels of a neutral atom and an external radiation field (see section 10.6.3). This approach is applicable to laser cooling, for instance, where radiation momentum has a breaking effect on the kinetic activity of the atoms.

6

Statistical interpretation of the field

6.1 Fluctuations and virtual processes

Although it arises naturally in quantum field theory from unitarity, the Feynman Green function does not arise naturally in classical field theory. It contains explicitly acausal terms which defy our experience of mechanics. It has special symmetry properties: it depends only on $|x - x'|$, and thus distinguishes no special direction in space or time. It seems to characterize the uniformity of spacetime, or of a physical system in an unperturbed state.

The significance of the Feynman Green function lies in the *effective* understanding of complex systems, where Brownian fluctuations in bulk have the macroscopic effect of mixing or stirring. In field theory, its use as an intuitive model for fluctuations allows the analysis of population distributions and the simulation of field decay, by spreading an energy source evenly about the possible modes of the system.

6.1.1 Fluctuation generators: $G_F(x, x')$ and $G_E(x, x')$

The Feynman Green function is related to the Green function for Euclidean space. Beginning with the expression in eqn. (5.99), one performs an anti-clockwise rotation of the integration contour (see figure 5.3):

$$k_0^E = ik_0. \tag{6.1}$$

There are no obstacles (poles) which prevent this rotation, so the two expressions are completely equivalent. With this contour definition the integrand is positive definite and no poles are encountered in an integral over $\hat{k}_{0E}$:

$$\frac{1}{-k_0^2 + \mathbf{k}^2 + m^2 - i\epsilon} \rightarrow \frac{1}{k_{0E}^2 + \mathbf{k}^2 + m^2}. \tag{6.2}$$

There are several implications to this equivalence between the Feynman Green function and the Euclidean Green function. The first is that Wick rotation to

Euclidean space is a useful technique for evaluating Green function integrals, without the interference of poles and singularities. Another is that the Euclidean propagator implies the same special causal relationship between the source and the field as does the Feynman Green function. In quantum field theory, one would say that these Green functions formed time-ordered products.

In the classical theory, the important point is the spacetime symmetry of the Green functions. Owing to the quadratic nature of the integral above, it is clear that both the Feynman and Euclidean Green functions depend only on the absolute value of $|x - x'|$. They single out no special direction in time. Physically they represent processes which do not develop in time, or whose average effect over an infinitesimal time interval is zero.

These Green functions are a differential representation of a cycle of emission and absorption (see below). They enable one to represent *fluctuations* or *virtual processes* in the field which do not change the overall state. These are processes in which an excitation is emitted from a source and is absorbed by a sink over a measurable interval of time.[1] This is a doorway to the study of statistical equilibria.

Statistical (many-particle) effects are usually considered the domain of quantum field theory. their full description, particularly away from equilibrium, certainly requires the theory of interacting fields, but the essence of statistical mechanics is contained within classical concepts of ensembles. The fact that a differential formulation is possible through the Green function has profound consequences for field theory. Fluctuations are introduced implicitly through the boundary conditions on the Green functions. The quantum theory creates a more elaborate framework to justify this choice of boundary conditions, and takes it further. However, when it comes down to it, the idea of random fluctuations in physical systems is postulated from experience. It does not follow from any deeper physical principle, nor can it be derived. Its relationship to Fock space methods of counting states is fascinating though. This differential formulation of statistical processes is explored in this chapter.[2]

6.1.2 Correlation functions and generating functionals

The Feynman (time-ordered) Green function may be obtained from a generating functional W which involves the action. From this generating functional it is possible to see that a time-translation-invariant field theory, expressed in terms of the Feynman Green function, is analytically related to a statistically

---

[1] Actually, almost all processes can be studied in this way by assuming that the field tends to a constant value (usually zero) at infinity.

[2] In his work on source theory, Schwinger [118, 119] constructs quantum transitions and statistical expectation values from the Feynman Green function Δ_+, using the principle of spacetime uniformity (the Euclidean hypothesis). The classical discussion here is essentially equivalent to his treatment of weak sources.

weighted ensemble of static systems. The action $S[\phi(x)]$ is already a generating functional for the mechanics of the system, as noted in chapter 4. The additional generating functional $W[J]$ may be introduced in order to study the statistical correlations in the field. This is a new concept, and it requires a new generating functional, the effective action. The effective action plays a central role in quantum field theory, where the extension to interacting fields makes internal dynamics, and thence the statistical interpretation, even more pressing.

We begin by defining averages and correlated products of the fields. This is the route to a statistical interpretation. Consider a field theory with fields ϕ^A, $\phi^{\dagger B}$ and action $S^{(2)}$. The superscript here denotes the fact that the action is one for free fields and is therefore of purely quadratic order in the fields. In the following sections, we use the complex field $\phi(x)$ to represent an arbitrary field. The same argument applies, with only irrelevant modifications, for general fields. We may write

$$S^{(2)} = \int (\mathrm{d}x)\, \phi^{\dagger A} \hat{\mathcal{O}}_{AB} \phi^B, \tag{6.3}$$

where the Gaussian weighted average, for statistical weight $\rho = \exp(\mathrm{i}S/s)$ is then defined by

$$\langle F[\phi] \rangle = \frac{\mathrm{Tr}(\rho F)}{\mathrm{Tr}\,\rho}$$
$$= \frac{\int \mathrm{d}\mu[\phi] F[\phi] \mathrm{e}^{\frac{\mathrm{i}}{s} S^{(2)}}}{\int \mathrm{d}\mu[\phi] \mathrm{e}^{\frac{\mathrm{i}}{s} S^{(2)}}}. \tag{6.4}$$

where s is an arbitrary scale with the dimensions of action. In quantum field theory, it is normal to use $s = \hbar$, but here we keep it general to emphasize that the value of this constant cancels out of relevant formulae at this classical level. Do not be tempted to think that we are now dealing with quantum field theory, simply because this is a language which grew up around the second quantization. The language is only a convenient mathematical construction, which is not tied to a physical model. In this section, we shall show that the Gaussian average over pairs of fields results in the classical Feynman Green function. Consider the generating functional

$$Z\left[J, J^{\dagger}\right] = \int \mathrm{d}\mu\left[\phi, \phi^{\dagger}\right] \mathrm{e}^{\frac{\mathrm{i}}{s} \int (\mathrm{d}x)\left[\phi^{\dagger A} \hat{\mathcal{O}}_{AB} \phi^B - \phi^{\dagger A} J_A - J_B^{\dagger} \phi^B\right]}, \tag{6.5}$$

which bears notable similarities to the classical thermodynamical partition function. From the definitions above, we may write

$$\frac{Z\left[J, J^{\dagger}\right]}{Z\left[0, 0\right]} = \left\langle \exp\left(-\frac{\mathrm{i}}{s} \int (\mathrm{d}x) \phi^{\dagger A} J_A - \frac{\mathrm{i}}{s} \int (\mathrm{d}x) J_A^{\dagger} \phi^A\right) \right\rangle, \tag{6.6}$$

where the currents J^A and $J^{\dagger B}$ are of the same type as ϕ^A and $\phi^{\dagger B}$, respectively. The effective action, as a function of the sources $W[J, J^{\dagger}]$, is defined by

$$\exp\left(\frac{i}{s}W[J, J^{\dagger}]\right) = Z[J, J^{\dagger}], \qquad (6.7)$$

thus $W[J, J^{\dagger}]$ is like the average value of the action, where the average is defined by the Gaussian integral. Now consider a shift of the fields in the action, which diagonalizes the exponent in eqn. (6.6):

$$(\phi^{\dagger A} + K^A)\hat{O}_{AB}(\phi^B + L^B) - K^A\hat{O}_{AB}L^B$$
$$= \phi^{\dagger A}\hat{O}_{AB}\phi^B + \phi^{\dagger A}\hat{O}_{AB}L^B + K^A\hat{O}_{AB}\phi^B. \qquad (6.8)$$

The right hand side of this expression is the original exponent in eqn. (6.5), provided we identify

$$\hat{O}_{AB}L^B(x) = J^A(x) \qquad (6.9)$$

$$\Rightarrow\; L^A(x) = \int (dx')(\hat{O}^{-1})^{AB}(x, x')J_B(x') \qquad (6.10)$$

and

$$K^A(x)\hat{O}_{AB} = J^{\dagger}_B(x) \qquad (6.11)$$

$$\Rightarrow\; K^A(x) = \int (dx')J^{\dagger}_B(x')(\hat{O}^{-1})^{AB}(x, x'), \qquad (6.12)$$

where $\int (dx')\hat{O}^{-1\,AB}\hat{O}_{BC} = \delta^A_C$. With these definitions, it follows that

$$K^A\hat{O}_{AB}L^B = \int (dx)(dx')\, J^{\dagger}_A(\hat{O}^{-1})^{AB}J_B \qquad (6.13)$$

and so

$$Z[J, J^{\dagger}] = \int d\mu[\phi, \phi^{\dagger}]\, e^{\frac{i}{s}\int (dx)\left[(\phi^{\dagger A}+K^A)\hat{O}_{AB}(\phi^B+L^B)-J^{\dagger}_A(\hat{O}^{-1})^{AB}J_B\right]}. \qquad (6.14)$$

We may now translate away L^A and K^A, assuming that the functional measure is invariant. This leaves

$$Z[J, J^{\dagger}] = \exp\left(-\frac{i}{s}\int (dx)(dx')\, J^{\dagger}_A(\hat{O}^{-1})^{AB}J_B\right) Z[0, 0] \qquad (6.15)$$

or

$$W[J, J^{\dagger}] = -\int (dx)(dx')\, J^{\dagger}_A(\hat{O}^{-1})^{AB}J_B + \text{const.} \qquad (6.16)$$

By differentiating $W[J, J^\dagger]$ with respect to the source, we obtain

$$\langle \phi^{\dagger A} \rangle = \frac{\delta W}{\delta J_A^\dagger(x)} = is \int (\mathrm{d}x') J_A^\dagger (\hat{O}^{-1})^{AB} \tag{6.17}$$

$$\langle \phi^B \rangle = \frac{\delta W}{\delta J_B(x)} = is \int (\mathrm{d}x')(\hat{O}^{-1})^{AB} J_B \tag{6.18}$$

$$\langle \phi^{\dagger A} \phi^{\dagger B} \rangle = is \frac{\delta^2 W}{\delta J^\dagger \delta J_B^\dagger} = 0 \tag{6.19}$$

$$\langle \phi^A \phi^B \rangle = is \frac{\delta^2 W}{\delta J_A \delta J_B} = 0 \tag{6.20}$$

$$\langle \phi^A \phi^{\dagger B} \rangle = is \frac{\delta^2 W}{\delta J_A^\dagger \delta J_B} = is(\hat{O}^{-1})^{AB} \tag{6.21}$$

One may now identify $(\hat{O}^{-1})^{AB}$ as the inverse of the operator in the quadratic part of the action, which is clearly a Green function, i.e.

$$\langle \phi^A \phi^{\dagger B} \rangle = is G^{AB}(x, x'). \tag{6.22}$$

Moreover, we have evaluated the generator for correlations in the field $W[J]$. Returning to real scalar fields, we have

$$W[J] = -\frac{1}{2} \int (\mathrm{d}x)(\mathrm{d}x') \, J_A(x) G^{AB}(x, x') J_B(x'). \tag{6.23}$$

We shall use this below to elucidate the significance of the Green function for the fluctuations postulated in the system. Notice that although the generator, in this classical case, is independent of the scale s, the definition of the correlation function in eqn. (6.21) does depend on this scale. This tells us simply the magnitude of the fluctuations compared with the scale of $W[J]$ (the typical energy scale or rate of work in the system over a time interval). If one takes $s = \hbar$, we place the fluctuations at the quantum level. If we take $s \sim \beta^{-1}$, we place fluctuations at the scale of thermal activity kT.[3] Quantum fluctuations become unimportant in the classical limit $\hbar \to 0$; thermal fluctuations become unimportant in the low-temperature limit $\beta \to \infty$. At the level of the present discussion, the results we can derive from the correlators are independent of this scale, so a macroscopic perturbation would be indistinguishable from a microscopic perturbation. It would be a mistake to assume that this scale were unimportant however. Changes in this scaling factor can lead to changes in the correlation lengths of a system and phase transitions. This, however, is the domain of an interacting (quantum) theory.

---

[3] These remarks reach forward to quantum field theories; they cannot be understood from the simple mechanical considerations of the classical field. However they do appeal to one's intuition and make the statistical postulate more plausible.

We have related the generating functional $W[J]$ to weighted-average products over the fields. These have an automatic symmetry in their spacetime arguments, so it is clear that the object $G^{AB}(x, x')$ plays the role of a correlation function for the field. The symmetry of the generating functional alone implies that $(\hat{O}^{-1})^{ij}$ must be the Feynman Green function. We shall nevertheless examine this point more closely below.

A note of caution to end this section: the spacetime symmetry of the Green function follows from the fact that the integrand in

$$G(x, x') = \int (\mathrm{d}k) \frac{e^{ik(x-x')}}{k^2 + m^2} \tag{6.24}$$

is a purely quadratic quantity. A correlator must depend only on the signless difference between spacetime events $|x - x'|$, if it is to satisfy the relations in the remainder of this section on dissipation and transport. If the spectrum of excitations were to pick up, say, an absorbative term, which singled out a special direction in time, this symmetry property would be spoiled, and, after an infinitesimal time duration, the Green functions would give the wrong answer for the correlation functions. In that case, it would be necessary to analyse the system more carefully using methods of non-equilibrium field theory. In practice, the simple formulae given in the rest of this chapter can only be applied to derive instantaneous tendencies of the field, never prolonged instabilities.

6.1.3 Symmetry and causal boundary conditions

There are two Green functions which we might have used in eqn. (6.21) as the inverse of the Maxwell operator; the retarded Green function and the Feynman Green function. Both satisfy eqn. (5.62). The symmetry of the expression

$$W = -\frac{1}{2} \int (\mathrm{d}x)(\mathrm{d}x') J(x) G(x, x') J(x') \tag{6.25}$$

precludes the retarded function however. The integral is spacetime-symmetrical, thus, only the symmetrical part of the Green function contributes to the integral. This immediately excludes the retarded Green function, since

$$
\begin{aligned}
W_{\mathrm{r}} &= -\frac{1}{2} \int (\mathrm{d}x)(\mathrm{d}x') J(x) G_{\mathrm{r}}(x, x') J(x') \\
&= -\frac{1}{2} \int (\mathrm{d}x)(\mathrm{d}x') J(x) [G^{(+)}(x, x') + G^{(-)}(x, x')] J(x') \\
&= -\frac{1}{2} \int (\mathrm{d}x)(\mathrm{d}x') J(x) [G^{(+)}(x, x') - G^{(+)}(x', x)] J(x') \\
&= 0,
\end{aligned}
\tag{6.26}
$$

where the last line follows on re-labelling x, x' in the second term. This relation tells us that there is no dissipation in one-particle quantum theory. As we shall see, however, it does not preclude dissipation by re-distribution of energy in 'many-particle' or statistical systems coupled to sources. See an example of this in section 7.4.1. Again, the link to statistical systems is the Feynman Green function or correlation function. The Feynman Green function is symmetrical in its spacetime arguments. It is straightforward to show that

$$W = -\frac{1}{2} \int (\mathrm{d}x)(\mathrm{d}x') J(x) G_{\mathrm{F}}(x, x') J(x')$$

$$= -\frac{1}{2} \int (\mathrm{d}x)(\mathrm{d}x') J(x) \overline{G}(x, x') J(x'). \qquad (6.27)$$

The imaginary part of $\overline{G}(x, x')$ is

$$\mathrm{Im}\, \overline{G}(x, x') = 2\, \mathrm{Im}\, G^{(+)}(x, x'). \qquad (6.28)$$

6.1.4 Work and dissipation at steady state

Related to the idea of transport is the idea of energy dissipation. In the presence of a source J, the field can decay due to work done on the source. Of course, energy is conserved within the field, but the presence of fluctuations (briefly active sources) allows energy to be transferred from one part of the field to another; i.e. it allows energy to be mixed randomly into the system in a form which cannot be used to do further work. This is an increase in entropy.

The instantaneous rate at which the field decays is proportional to the imaginary part of the Feynman Green function. In order to appreciate why, we require a knowledge of the energy–momentum tensor and Lorentz transformations, so we must return to this in section 11.8.2. Nonetheless, it is possible to gain a partial understanding of the problem by examining the Green functions, their symmetry properties and the roles they play in generating solutions to the field equations. This is an important issue, which is reminiscent of the classical theory of hydrodynamics [53].

The power dissipated by the field is the rate at which field does work on external sources,[4]

$$P = \frac{\mathrm{d}w}{\mathrm{d}t}. \qquad (6.29)$$

Although we cannot justify this until chapter 11, let us claim that the energy of the system is determined by

$$\text{energy} = -\left.\frac{\delta S[\phi]}{\delta t}\right|_{\phi=\int GJ}. \qquad (6.30)$$

---

[4] Note that we use a small w for work since the symbol W is generally reserved to mean the value of the action, evaluated at the equations of motion.

So, the rate of change of energy in the system is equal to minus the rate at which work is done by the system:

$$\frac{d}{dt} \text{ energy} = -\frac{dw}{dt}. \tag{6.31}$$

Let us define the action functional W by

$$\frac{\delta W}{\delta J} = \frac{\delta S[\phi, J]}{\delta J}\bigg|_{\phi = \int GJ}, \tag{6.32}$$

where the minus sign is introduced so that this represents the work done by the system rather than the energy it possesses. The object W clearly has the dimensions of action, but we shall use it to identify the rate at which work is done. Eqn. (6.32) is most easily understood with the help of an example. The action for a scalar field is

$$\delta_J S = -\int (dx) \, \delta J \phi(x), \tag{6.33}$$

so, evaluating this at

$$\phi(x) = \int (dx') \, G(x, x') J(x'), \tag{6.34}$$

one may write, up to source-independent terms,

$$W[J] = -\frac{1}{2} \int (dx)(dx') \, J(x) G_F(x, x') J(x'). \tag{6.35}$$

This bi-linear form recurs repeatedly in field theory. Schwinger's source theory view of quantum fields is based on this construction, for its spacetime symmetry properties. Notice that it is based on the Feynman Green function. Eqn. (6.34) could have been solved by either the Feynman Green function or the retarded Green function. The explanation follows shortly. The work done over an infinitesimal time interval is given by

$$\Delta w = \text{Im} \frac{dW}{dt}. \tag{6.36}$$

Expressed in more useful terms, the instantaneous decay rate of the field is

$$\int dt \, \gamma(t) = -\frac{2}{\chi_h} \text{Im} W. \tag{6.37}$$

The sign, again, indicates the difference between work done and energy lost. The factor of χ_h is included because we need a scale which relates energy and time (frequency). In quantum mechanics, the appropriate scale is $\chi_h = \hbar$. In fact, any constant with the dimensions of action will do here. There is nothing specifically quantum mechanical about this relation. The power is proportional to the rate of work done. The more useful quantity, the power spectrum $P(\omega)$ or power at frequency ω, is

$$\int d\omega \frac{P(\omega, t)}{\hbar \omega} = -\gamma(t), \qquad (6.38)$$

giving the total power

$$P = \int d\omega P(\omega). \qquad (6.39)$$

We speak of the instantaneous decay rate because, in a real analysis of dissipation, the act of work being done acts back on all time varying quantities. Taking the imaginary part of W to be the decay rate for the field assumes that the system changes only adiabatically, as we shall see below.

6.1.5 Fluctuations

The interpretation of the field as a statistical phenomenon is made plausible by considering the effect of infinitesimal perturbations to the field. This may be approached in two equivalent ways: (i) through the introduction of linear perturbations to the action, or sources

$$S \to S - \int (dx) \, J\phi, \qquad (6.40)$$

where J is assumed to be weak, or (ii) by writing the field in terms of a fluctuating 'average' part $\langle\phi\rangle$ and a remainder part φ,

$$\phi(x) = \langle\phi(x)\rangle + \varphi(x). \qquad (6.41)$$

These two constructions are equivalent for all dynamical calculations. This can be confirmed by the use of the above generating functionals, and a change of variable.

It is worth spending a moment to consider the meaning of the function $W[J]$. Although originally introduced as part of the apparatus of quantum field theory [113], we find it here completely divorced from such origins, with no trace of quantum field theoretical states or operators (see chapter 15). The structure of this relation is a direct representation of our model of fluctuations or virtual processes. $W[J]$ is the generator of fluctuations in the field. The Feynman Green function, in eqn. (6.25), is sandwiched between two sources symmetrically. The Green function itself is symmetrical: for retarded times, it propagates a field radiating from a past source to the present, and for advanced times it propagates the field from the present to a future source, where it is absorbed.

The symmetry of advanced and retarded boundary conditions makes $W[J]$ an explicit representation of a virtual process, at the purely classical level.[5]

The first derivative of the effective action with respect to the source is

$$\frac{\delta W}{\delta J(x)} = \langle \phi(x) \rangle, \qquad (6.42)$$

which implies that, for the duration of an infinitesimal fluctuation $J \neq 0$, the field has an average value. If it has an average value, then it also deviates from this value, thus we may write

$$\phi(x) = \frac{\delta W}{\delta J(x)} + \varphi(x), \qquad (6.43)$$

where $\varphi(x)$ is the remainder of the field due to J. The average value vanishes once the source is switched off, meaning that the fluctuation is the momentary appearance of a non-zero average in the local field. This is a smearing, stirring or mixing of the field by the infinitesimal generalized force J. The rate of change of this average is

$$(i\hbar)\frac{\delta^2 W[J]}{\delta^2 J} = \langle \phi(x)\phi(x') \rangle - \langle \phi(x) \rangle \langle \phi(x') \rangle. \qquad (6.44)$$

This is the correlation function $C_{AB}(x, x')$, which becomes the Feynman Green function as $J \to 0$. It signifies the response of the field to its own fluctuations nearby, i.e. the extent to which the field has become mixed. The correlation functions become large as the field becomes extremely uniform. This is called (off-diagonal[6]) long-range order.

The correlation function interpretation is almost trivial at the classical (free-field) level, but becomes enormously important in the interacting quantum theory.

Instantaneous thermal fluctuations Fluctuations have basically the same form regardless of their origin. If we treat all thermal fluctuations as instantaneous, then we may account for them by a Euclidean Green function; the fluctuations of the zero-temperature field are generated by the Feynman Green function. In an approximately free theory, these two are the same thing. Consider a thermal Boltzmann distribution

$$\mathrm{Tr}(\rho(x, x')\phi(x)\phi(x')) = \mathrm{Tr}(\mathrm{e}^{-\beta\hbar\omega}\phi(\omega)\phi(-\omega)). \qquad (6.45)$$

---

[5] For detailed discussions of these points in the framework of quantum field theory, see the original papers of Feynman [46, 47, 48] and Dyson [41]. The generator $W[J]$ was introduced by Schwinger in ref. [113].

[6] 'Off-diagonal' refers to $x \neq x'$.

Since the average weight is $e^{iS/\hbar}$, and the Green function in momentum space involves a factor $\exp(-i\omega(t - t'))$, one can form a representation of the Boltzmann exponential factor $\exp(-\beta E)$ by analytically continuing

$$t \to t - i\hbar\beta \qquad (6.46)$$

or

$$t' \to t + i\hbar\beta. \qquad (6.47)$$

This introduces an imaginary time element such as that obtained by Wick rotating to Euclidean space. It also turns the complex exponential into a real, decaying exponential. If the real part of the time variable plays no significant role in the dynamics (a static system), then it can be disregarded altogether. That is why Euclidean spacetime is essentially equivalent to equilibrium thermodynamics. However, from the spacetime symmetry of the correlation functions, we should have the same result if we re-label t and t' so

$$G(t - t' + i\hbar\beta) = G(t' - t + i\hbar\beta) \qquad (6.48)$$

or, in the Wick-rotated theory,

$$G(t_{\mathrm{E}} - t'_{\mathrm{E}} + \hbar\beta) = G(t'_{\mathrm{E}} - t_{\mathrm{E}} + \hbar\beta). \qquad (6.49)$$

This is only possible if

$$e^{i\omega_{\mathrm{E}}(t_{\mathrm{E}} - \hbar\beta - t'_{\mathrm{E}})} = e^{i\omega_{\mathrm{E}}(t'_{\mathrm{E}} - \hbar\beta - t_{\mathrm{E}})} \qquad (6.50)$$

or

$$e^{i\hbar\beta\omega_{\mathrm{E}}} = 1. \qquad (6.51)$$

From this; we deduce that the Euclidean Green function must be periodic in imaginary time and that the Euclidean frequency

$$\omega_{\mathrm{E}}(n) = \frac{2n\pi}{\beta}, \qquad n = 0, \pm 1, \pm 2, \ldots, \qquad (6.52)$$

where $\omega_{\mathrm{E}}(n)$ are called the Matsubara frequencies.

Thermal fluctuations in time Using the fluctuation model, we may represent a system in thermal equilibrium by the same idealization as that used in thermodynamics. We may think of the source and sink for thermal fluctuations as being a large reservoir of heat, so large that its temperature remains constant at $T = 1/k\beta$, even when we couple it to our system. The coupling to the heat bath is by sources. Consider the fluctuation model as depicted in figure 6.1.

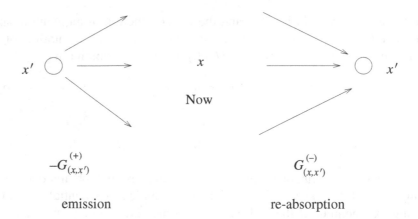

Fig. 6.1.   Thermal fluctuations occur when the source is a heat reservoir at constant temperature.

Since the fluctuation generator is $W[J]$, which involves

$$W[J] = -\frac{1}{2} \int (\mathrm{d}x)(\mathrm{d}x') \; J(x) G_\mathrm{F}(x, x') J(x')$$

$$\sim J(x) \left[-G^{(+)}(\omega)\theta(\text{past}) + G^{(-)}(\omega)\theta(\text{future}) \right] J(x'), \quad (6.53)$$

then, during a fluctuation, the act of emission from the source is represented by $-G^{(+)}(\omega)$ and the act of re-absorption is represented by $G^{(-)}(\omega)$. In other words, these are the susceptibilities for thermal emission and absorption. In an isolated system in thermal equilibrium, we expect the number of fluctuations excited from the heat bath to be distributed according to a Boltzmann probability factor [107]:

$$\frac{\text{emission}}{\text{absorption}} = \frac{-G^{(+)}(\omega)}{G^{(-)}(\omega)} = e^{\hbar\beta|\omega|}. \quad (6.54)$$

We use $\hbar\omega$ for the energy of the mode with frequency ω by tradition, though $\hbar$ could be replaced by any more appropriate scale with the dimensions of action. This is a classical understanding of the well known Kubo–Martin–Schwinger relation [82, 93] from quantum field theory. In the usual derivation, one makes use of the quantum mechanical time-evolution operator and the cyclic property of the trace in eqn. (6.45) to derive this relation for a thermal equilibrium. What makes these two derivations equivalent is the principle of spacetime uniformity of fluctuations. The argument given here is identical to Einstein's argument for stimulated and spontaneous emission in a statistical two-state system, and the derivation of the well known A and B coefficients. It can be interpreted as the relative occupation numbers of particles with energy $\hbar\omega$. Here, the two states

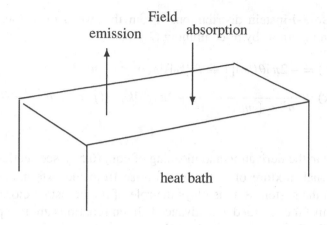

Fig. 6.2. Contact with a thermodynamic heat bath. Fluctuations represent emission and absorption from a large thermal field.

are the heat bath and the system (see figure 6.2). We can use eqn. (6.54) to find the thermal forms for the Wightman functions (and hence all the others). To do so we shall need the extra terms $X(k)$ mentioned in eqn. (5.41). Generalizing eqns. (5.66), we write,

$$G^{(+)}(k) = -2\pi i[\theta(k_0) + X]\delta(p^2c^2 + m^2c^2)$$
$$G^{(-)}(k) = 2\pi i[\theta(-k_0) + Y]\delta(p^2c^2 + m^2c^4) \qquad (6.55)$$

with X and Y to be determined. The commutator function $\tilde{G}(x, x')$ represents the difference between the emission and absorption processes, which cannot depend on the average state of the field since it represents the completeness of the dynamical system (see section 14.1.8 and eqn. (5.73)). It follows that $X = Y$. Then, using eqn. (6.54), we have

$$\theta(\omega) + X = e^{\hbar\omega\beta}[\theta(-\omega) + X] \qquad (6.56)$$

and hence

$$X(e^{\beta\hbar|\omega|} - 1) = \theta(\omega), \qquad (6.57)$$

since $\theta(-\omega)e^{\beta\omega} = 0$. Thus, we have

$$X = \theta(\omega)f(\omega), \qquad (6.58)$$

where

$$f(\omega) = \frac{1}{e^{\beta\hbar\omega} - 1}. \qquad (6.59)$$

This is the Bose–Einstein distribution. From this we deduce the following thermal Green functions by re-combining $G^{(\pm)}(k)$:

$$G^{(+)}(k) = -2\pi i\theta(k_0)[1 + f(|k_0|)]\delta(p^2c^2 + m^2c^4) \tag{6.60}$$

$$G_F(k) = \frac{1}{p^2c^2 + m^2c^4 - i\epsilon} + 2\pi i f(|k_0|)\delta(p^2c^2 + m^2c^4)\theta(k_0). \tag{6.61}$$

For a subtlety in the derivation and meaning of eqn. (6.59), see section 13.4.

The additional mixture of states which arise from the external (boundary) conditions on the system X thus plays the role of a macrostate close to steady state. Notice that the retarded and advanced Green functions are independent of X. This must be the case for unitarity to be preserved.

6.1.6 Divergent fluctuations: transport

The fluctuations model introduced above can be used to define instantaneous transport coefficients in a statistical system. Long-term, time-dependent expressions for these coefficients cannot be obtained because of the limiting assumptions of the fluctuation method. However, such a non-equilibrium situation could be described using the methods of non-equilibrium field theory.

Transport implies the propagation or flow of a physical property from one place to another over time. Examples include

- thermal conduction,

- electrical conduction (current),

- density conduction (diffusion).

The conduction of a property of the field from one place to another can only be accomplished by dynamical changes in the field. We can think of conduction as a persistent fluctuation, or a fluctuation with very long wavelength, which never dies. All forms of conduction are essentially equivalent to a diffusion process and can be analysed hydrodynamically, treating the field as though it were a fluid.

Suppose we wish to consider the transport of a quantity X: we are therefore interested in fluctuations in this quantity. In order to generate such fluctuations, we need a source term in the action which is generically conjugate to the fluctuation (see section 14.5). We add this as follows:

$$S \to S - \int (\mathrm{d}x)X \cdot F, \tag{6.62}$$

and consider the generating functional of fluctuations $W[F]$ as a function of the infinitesimal source $F(x)$; Taylor-expanding, one obtains

$$\delta W[F] = W[0] + \int (\mathrm{d}x) \frac{\delta W[0]}{\delta F(x)} F(x)$$
$$+ \int (\mathrm{d}x)(\mathrm{d}x') \frac{\delta^2 W[0]}{\delta F(x) \delta F(x')} F(x) \delta F(x') + \cdots . \quad (6.63)$$

Now, since

$$W[F] = \int (\mathrm{d}x)(\mathrm{d}x') \, F(x) \langle X(x) X(x') \rangle F(x'), \quad (6.64)$$

we have the first few terms of the expansion

$$W[0] = 0$$
$$\frac{\delta W[0]}{\delta F(x)} = \langle X(x) \rangle$$
$$\frac{\delta^2 W[0]}{\delta F(x) \delta F(x')} = -\frac{\mathrm{i}}{\hbar} \langle X(x) X(x') \rangle \quad (6.65)$$

Thus, linear response theory gives us, generally,

$$\langle X(x) \rangle = \frac{\mathrm{i}}{\hbar} \int (\mathrm{d}x') \langle X(x) X(x') \rangle F(x'), \quad (6.66)$$

or

$$\frac{\delta \langle X(x) \rangle}{\delta F(x')} = \frac{\mathrm{i}}{\hbar} \langle X(x) X(x') \rangle. \quad (6.67)$$

Since the correlation functions have been generated by the fluctuation generator W, they satisfy Feynman boundary conditions; however, in eqn. (6.81) we shall derive a relation which may be used to relate this to the linear response of the field with retarded boundary conditions. It remains, of course, to express the correlation functions of the sources in terms of known quantities. Nevertheless, the above expression may be used to determine the transport coefficients for a number of physical properties. As an example, consider the electrical conductivity, as defined by Ohm's law,

$$J_i = \sigma_{ij} E_i. \quad (6.68)$$

If we write $E_i \sim \partial_t A_i$ in a suitable gauge, then we have

$$J_i(\omega) = \sigma_{ij} \omega A_j(\omega), \quad (6.69)$$

or

$$\frac{\delta J_i}{\delta A_j} = \sigma_{ij} \omega. \quad (6.70)$$

From eqn. (6.67), we may therefore identify the transport coefficient as the limit in which the microscopic fluctuations' wavelength tends to infinity and leads to a lasting effect across the whole system,

$$\sigma_{ij}(\omega) = \lim_{\mathbf{k} \to 0} \frac{i}{\hbar\omega} \langle J_i(\omega) J_j(-\omega) \rangle. \tag{6.71}$$

The evaluation of the right hand side still needs to be performed. To do this, we need to know the dynamics of the sources J_i and the precise meaning of the averaging process, signified by $\langle \ldots \rangle$. Given this, the source method provides us with a recipe for calculating transport coefficients.

In most cases, one is interested in calculating the average transport coefficients in a finite temperature ensemble. Thermal fluctuations may be accounted for simply by noting the relationship between the Feynman boundary conditions used in the generating functional above and the retarded boundary conditions, which are easily computable from the mechanical response. We make use of eqn. (6.54) to write

$$\begin{aligned} G_r(t, t') &= -\theta(t - t') \left[G^{(+)} + G^{(-)} \right] \\ &= -\theta(t - t') G^{(+)} \left[1 - e^{-\hbar\beta\omega} \right]. \end{aligned} \tag{6.72}$$

The retarded part of the Feynman Green function is

$$G_F = -\theta(t - t') G^{(+)} \theta(t - t'), \tag{6.73}$$

so, over the retarded region,

$$G_r(x, x') = (1 - e^{-\hbar\beta\omega}) G_F(x, x'), \tag{6.74}$$

giving

$$\sigma_{ij}(\omega) = \lim_{\mathbf{k} \to 0} \frac{(1 - e^{-\hbar\beta\omega})}{\hbar\omega} \langle J_i(\omega) J_j(-\omega) \rangle, \tag{6.75}$$

for the conductivity tensor, assuming a causal response between source and field. The formula in eqn. (6.75) is one of a set of formulae which relate the fluctuations in the field to transport coefficients. The strategy for finding such relations is to identify the source which generates fluctuations in a particular quantity. We shall return to this problem in general in section 11.8.5.

6.1.7 Fluctuation dissipation theorem

In a quasi-static system, the time-averaged field may be defined by

$$\langle \phi \rangle = \frac{1}{T} \int_{\bar{t}-T/2}^{\bar{t}+T/2} \phi(x) dt. \tag{6.76}$$

From the generating functional in eqn. (6.5), we also have

$$\langle \phi(x) \rangle = i\hbar \frac{\delta W[J]}{\delta J(x)}, \tag{6.77}$$

and further

$$\frac{\delta \langle \phi \rangle}{\delta J}\bigg|_{J=0} = -\frac{i}{\hbar} \langle \phi(x)\phi(x') \rangle = \mathrm{Im}\; G_{\mathrm{F}}(x, x'). \tag{6.78}$$

The field may always be causally expressed in terms of the source, using the retarded Green function in eqn. (6.76), provided the source is weak so that higher terms in the generating functional can be neglected; thus

$$\langle \phi \rangle = \frac{1}{T} \int_{\bar{t}-T/2}^{\bar{t}+T/2} \int (\mathrm{d}x') G_{\mathrm{r}}(x, x') J(x')\mathrm{d}t. \tag{6.79}$$

Now, using eqns. (6.78) and (6.79), we find that

$$\frac{\delta}{\delta J(x')} \frac{\delta}{\delta t} \phi(x) = -\mathrm{Im}\; \partial_t G_{\mathrm{F}}(x, x') = \frac{1}{T} G_{\mathrm{r}}(x, x'). \tag{6.80}$$

Thus, on analytically continuing to Euclidean space,

$$G_{\mathrm{r}}(\omega) = -\hbar\beta\omega\, G_{\mathrm{E}}(\omega). \tag{6.81}$$

This is the celebrated fluctuation dissipation theorem. It is as trivial as it is profound. It is clearly based on assumptions about the average behaviour of a statistical system over macroscopic times T, but also refers to the effects of microscopic fluctuations over times contained in $x - x'$. It relates the Feynman Green function to the retarded Green function for a time-averaged field; i.e. it relates the correlation function, which measures the spontaneous fluctuations

$$\varphi = \phi - \langle \phi \rangle \tag{6.82}$$

in the field, to the retarded Green function, which measures the purely mechanical response to an external source. The fluctuations might be either thermal or quantum in origin, it makes no difference. their existence is implicitly postulated through the use of the correlation function. Thus, the truth or falsity of this expression lies in the very assumption that microscopic fluctuations are present, even though the external source $J \to 0$ on average (over macroscopic time T). This requires further elaboration.

In deriving the above relation, we have introduced sources and then taken the limit in which they tend to zero. This implies that the result is only true for an infinitesimal but non-zero source J. The source appears and disappears, so that it is zero on average, but it is present long enough to change the distribution of modes in the system, little by little. An observer who could resolve only macroscopic behaviour would therefore be surprised to see the system changing, apparently without cause. This theorem is thus about the mixing of scales.

The fluctuation dissipation theorem tells us that an infinitesimal perturbation to the field, $J \to 0$, will lead to microscopic fluctuations, which can decay by mechanical response (mixing or diffusion). The decay rate may be related to the imaginary part of the correlation function, but this gives only an instantaneous rate of decay since the assumptions we use to derive the expression are valid only for the brief instant of the fluctuation.[7]

The Feynman Green function seems to have no place in a one-particle mechanical description, and yet here it is, at the classical level. But we have simply introduced it *ad hoc*, and the consequences are profound: we have introduced fluctuations into the system. This emphasizes the importance of boundary conditions and the generally complex nature of the field.

6.2 Spontaneous symmetry breaking

Another aspect of fluctuating statistical theories, which arises in connection with symmetry, is the extent to which the average state of the field, $\langle \phi \rangle$, displays the full symmetry afforded it by the action. In interacting theories, collective effects can lead to an average ordering of the field, known as long-range order. The classic example of this is the ferromagnetic state in which spin domains line up in an ordered fashion, even though the action allows them to point in any direction, and indeed the fluctuations in the system occur completely at random. However, it is energetically favourable for fluctuations to do this close to an average state in which all the spins are aligned, provided the fluctuations are small. Maximum stability is then achieved by an ordered state. As fluctuations grow, perhaps by increasing temperature, the stability is lost and a phase transition can occur. This problem is discussed in section 10.7, after the chapters on symmetry.

---

[7] The meaning of this 'theorem' for Schwinger's source theory viewpoint is now clear [119]. Spacetime uniformity in the quantum transformation function tells us that the Green function we should consider is the Feynman Green function. The symmetry of the arguments tells us that this is a correlation function and it generates fluctuations in the field. The infinitesimal source is a model for these fluctuations. Processes referred to as the decay of the vacuum in quantum field theory are therefore understood in a purely classical framework, by understanding the meaning of the Feynman boundary conditions.

7
Examples and applications

To expose the value of the method in the foregoing chapters, it is instructive to apply it to a number of important and well known physical problems. Through these examples we shall see how a unified methodology makes the solution of a great many disparate systems essentially routine. The uniform approach does not necessarily convey with it any automatic physical understanding, but then no approach does. What we learn from this section is how many problems can be reduced to the basics of 'cause followed by effect', or, here, 'source followed by field'.

7.1 Free particles

Solving Newton's law $F = ma$ using a Green function approach is hardly to be recommended for any practical purpose; in fact, it is a very inefficient way of solving the problem. However, it is useful to demonstrate how the Green function method can be used to generate the solution to this problem. This simple test of the theory helps to familiarize us with the working of the method in practice. The action for a one-dimensional-particle system is

$$S = \int dt \ \left\{ -\frac{1}{2}m\dot{x}^2 - Fx \right\}. \tag{7.1}$$

The variation of the action leads to

$$\delta S = \int dt \ \{m\ddot{x} - F\} \delta x + \Delta(m\dot{x})\delta x = 0, \tag{7.2}$$

which gives us the equation of motion

$$F = m\ddot{x} \tag{7.3}$$

and the continuity condition

$$\Delta(m\dot{x}) = 0, \tag{7.4}$$

131

which is the conservation of momentum. The equation of motion can be written in the form of 'operator acting on field equals source',

$$\mathcal{D}x = J, \tag{7.5}$$

by rearranging

$$\partial_t^2 \, x(t) = F/m. \tag{7.6}$$

Clearly, we can integrate this equation directly with a proper initial condition $x(t_0) = x_0$, $\dot{x}(t_0) = v$, to give

$$x(t) - x_0 = \frac{F}{2m}(t - t_0) + v(t - t_0). \tag{7.7}$$

But let us instead try to use the Green function method to solve the problem. There are two ways to do this: the first is quite pointless and indicates a limitation of the Green function approach, mentioned in section 5.2.4. The second approach demonstrates a way around the limitation and allows us to see the causality principle at work.

Method 1 The operator on the left hand side of eqn. (7.6) is ∂_t^2, so we define a Green function

$$\partial_t^2 \, G(t, t') = \delta(t, t'). \tag{7.8}$$

As usual, we expect to find an integral expression by Fourier transforming the above equation:

$$G(t - t') = \int \frac{d\omega}{2\pi} \frac{e^{-i\omega(t-t')}}{-\omega^2}. \tag{7.9}$$

This expression presents us with a problem, however: it has a non-simple pole, which must be eliminated somehow. One thing we can do is to re-write the integral as follows:

$$G(t - t') = \int d\tilde{t} \int d\tilde{t} \int \frac{d\omega}{2\pi} \, e^{-i\omega\tilde{t}},$$

$$= \int d\tilde{t} \int d\tilde{t} \, \delta(\tilde{t}), \tag{7.10}$$

where $\tilde{t} = t - t'$. It should be immediately clear that this is just telling us to replace the Green function with a double integration (which is how one would normally solve the equation). We obtain two extra, unspecified integrals:

$$x(t) = \int dt'\, G(t, t') F/m$$

$$= \int d\tilde{t} d\tilde{t} dt'\, \delta(t - t') F/m$$

$$= \int d\tilde{t} d\tilde{t}\, F/m$$

$$= \int d\tilde{t}\, [F/m(t - t') + v]$$

$$= \frac{F}{2m}(t - t_0)^2 + v(t - t_0) + x_0. \tag{7.11}$$

So, the result is the same as that obtained by direct integration and for the same reason: the Green function method merely adds one extra (unnecessary) integration and re-directs us to integrate the equation directly. The problem here was that the denominator contained a non-simple pole. We can get around this difficulty by integrating it in two steps.

Method 2 Suppose we define a Green function for the linear differential operator

$$\partial_t\, g(t, t') = \delta(t, t'). \tag{7.12}$$

From section A.2, in Appendix A, we immediately recognize this function as the Heaviside step function. (We could take the Fourier transform, but this would only lead to an integral representation of the step function.) The solution has advanced and retarded forms

$$g_r(t, t') = \theta(t - t')$$
$$g_a(t, t') = -\theta(t' - t). \tag{7.13}$$

Now we have an integrable function, which allows us to solve the equation in two steps:

$$\partial_t\, x(t) = \int dt'\, g_r(t, t')\, F/m$$

$$= \frac{F}{m}(t - t') + \partial_t\, x(t') \qquad (t > t'). \tag{7.14}$$

Then, applying the Green function again,

$$x(t) = \int dt'\, g_r(t - t')\left[\frac{F}{m}(t - t') + \partial_t\, x(t')\right]$$

$$= \frac{F}{2m}(t - t_0)^2 + v(t - t_0) + x_0. \tag{7.15}$$

Again we obtain the usual solution, but this time we see explicitly the causality inferred by a linear derivative. The step function tells us that the solution only exists for a causal relationship between force F and response $x(t)$.

7.1.1 Velocity distributions

In a field of many particles, there is usually a distribution of velocities or momenta within the field. In a particle field this refers to the momenta p_i of individual localizable particles. In other kinds of field there is a corresponding distribution of wavenumbers k_i of the wave modes which make up the field. The action describes the dynamics of a generic particle, but it does not capture the macroscopic state of the field. The macrostate is usually described in terms of the numbers of components (particles or modes) with a given momentum or energy (the vector nature of momentum is not important in an isotropic plasma).

The distribution function f is defined so that its integral with respect to the distribution parameter gives the number density or particles per unit volume. We use a subscript to denote the control variable:

$$N = \int d^n \mathbf{k} \; f_k(\mathbf{k})$$

$$= \int d^n \mathbf{p} \; f_p(\mathbf{p})$$

$$= \int d^n \mathbf{v} \; f_v(\mathbf{v}). \tag{7.16}$$

This distribution expresses averages of the field. For example, the average energy is the weighted average of the energies of the different momenta:

$$\langle E \rangle = \frac{1}{N} \int d^n \mathbf{k} \; f_k(\mathbf{k}) E(k). \tag{7.17}$$

7.2 Fields of bound particles

A field of particles, sometimes called a plasma when charged, is an *effective* field, formed from the continuum approximation of discrete particles. Its purpose is to capture some of the bulk dynamics of material systems; it should not be confused with the deeper description of the atoms and their sub-atomic components in terms of fundamental fields which might focus on quite different properties, not relevant for the atoms in a bulk context. The starting point for classical analyses of atomic systems coupled to an electromagnetic field is the idea that matter consists of billiard-ball atoms with some number density ρ_N, and that the wavelength of radiation is long enough to be insensitive to the particle nature of the atoms. The only important fact is that there are many particles whose combined effect in space is to act like a smooth field. When perturbed by radiation (which we shall represent as an abstract source J_i) the particles are displaced by a spatial vector s^i where $i = 1, 2, \ldots, n$. The action for this system may be written

$$S_{\text{eff}} = \frac{1}{\sigma_x} \int (\mathrm{d}x) \left\{ -\frac{1}{2} m \dot{s}^2 + \frac{1}{2} \kappa s^2 - m \gamma s \dot{s} - J^i s_i \right\}. \qquad (7.18)$$

This requires some explanation. The factor of the spatial volume of the total system, σ_x, reflects the fact that this is an effective average formulation. Dividing by a total scale always indicates an averaging procedure. As an alternative to using this explicit value, we could use the average density, $\rho = m/\sigma_x$, and other parameter densities to express the action in appropriate dimensions. The first term is a kinetic energy term, which will describe the acceleration of particles in response to the forcing term J^i. The second term is a harmonic oscillator term, which assumes that the particles are bound to a fixed position $s_i = 0$, just as electrons are bound to atoms or ions are bound in a lattice. The effective spring constant of the harmonic interaction is κ. Because $s^i(x)$ represents the displacement of the particles from their equilibrium position, we use the symbol s^i rather than x^i, since it is not the position which is important, but the deviation from equilibrium position. The dimensions of $s^i(x)$ are position divided by the square-root of the density because of the volume integral in the action, and $s^i(x)$ is a function of x^μ because the displacement could vary from place to place and from time to time in the system. The final term in eqn. (7.18) is a term which will provide a phenomenological damping term for oscillations, as though the system were leaky or had friction. As we have already discussed in section 4.2, this kind of term is not well posed unless there is some kind of boundary in the system which can leak energy. The term is actually a total derivative. Nevertheless, since this is not a microscopic fundamental theory, it is possible to make sense of this as an effective theory by 'fiddling' with the action. This actually forces us to confront the reason why such terms cannot exist in fundamental theories, and is justifiable so long as we are clear about the meaning of the procedure.

The variation of the action is given, after partial integration, by

$$\delta S = \frac{1}{\sigma_x} \int (\mathrm{d}x) \left\{ m \ddot{s}_i + \kappa s_i - m \gamma \dot{s}_i + m \gamma \dot{s}_i - J_i \right\} \delta s^i$$

$$+ \frac{1}{\sigma_x} \int \mathrm{d}\sigma \, [m \dot{s}_i + m \gamma s_i] \, \delta s^i. \qquad (7.19)$$

The terms containing γ clearly cancel, leaving only a surface term. But suppose we divide the source into two parts:

$$J^i = J^i_\gamma + J^i_s, \qquad (7.20)$$

where J^i_γ is postulated to satisfy the equation

$$-m \gamma \dot{s}^i = J^i_\gamma. \qquad (7.21)$$

This then has the effect of preventing the frictional terms from completely disappearing. Clearly this is a fiddle, since we could have simply introduced a source in the first place, with a velocity-dependent nature. However, this is precisely the point. If we introduce a source or sink for the energy of the system, then it is possible to violate the conservational properties of the action by claiming some behaviour for J^i which is not actually determined by the action principle. The lesson is this: *if we specify the behaviour of a field rather than deriving it from the action principle, we break the closure of the system and conservation laws.* What this tells us is that dissipation in a system has to come from an external agent; it does not arise from a closed mechanical theory, and hence this description of dissipation is purely phenomenological. Taking eqn. (7.21) as given, we have the equation of motion for the particles

$$m\ddot{s}^i - m\gamma\dot{s}^i + \kappa s = J_s^i, \tag{7.22}$$

with continuity condition

$$\Delta\,(m\dot{s} + m\gamma s) = 0. \tag{7.23}$$

It is usual to define the natural frequency $\omega_0^2 = \kappa/m$ and write

$$(\partial_t^2 - \gamma\partial_t + \omega_0^2)s^i(x) = \frac{J_s^i}{m}. \tag{7.24}$$

If we consider a plane wave solution of the form

$$s(x) = \int (\mathrm{d}k)\,\mathrm{e}^{\mathrm{i}(k_i x^i - \omega t)} s(k), \tag{7.25}$$

then we may write

$$(-\omega^2 + \mathrm{i}\gamma\omega + \omega_0^2)s^i(k) = \frac{J_s^i(k)}{m}. \tag{7.26}$$

From this we see that the Green function $G_{ij}(x, x')$ for $s^i(x)$ is

$$G_{ij}(x, x') = \delta_{ij}\int (\mathrm{d}k)\,\frac{\mathrm{e}^{\mathrm{i}(k_i x^i - \omega t)}}{(-\omega^2 + \mathrm{i}\gamma\omega + \omega_0^2)}. \tag{7.27}$$

As long as the integral contains both positive and negative frequencies, this function is real and satisfies retarded boundary conditions. It is often referred to as the susceptibility, χ_{ij}. In a quantum mechanical treatment, $\hbar\omega_0 = E_2 - E_1$ is the difference between two energy levels.

Notice that the energy density

$$P^i E_i = \int E_i(x)G_{ij}(x, x')E_j(x')\,(\mathrm{d}x') \tag{7.28}$$

cannot be expressed in terms of a retarded Green function, since the above expression requires a spacetime symmetrical Green function. The Feynman Green function is then required. This indicates that the energy of the field is associated with a statistical balance of virtual processes of emission and absorption, rather than simply being a process of emission. In general, the interaction with matter introduces an imaginary part into the expression for the energy, since the Green function idealizes the statistical processes by treating them as steady state, with no back-reaction. It thus implicitly assumes the existence of an external source whose behaviour is unaffected by the response of our system. The energy density reduces to E^2 in the absence of material interactions and the result is then purely real.

7.3 Interaction between matter and radiation

Classical field theory is normally only good enough to describe non-interacting field theories. A complete description of interactions requires the quantum theory. The exception to this rule is the case of an external source. In electromagnetism we are fortunate in having a system in which the coupling between matter and radiation takes on the form of a linear external source J^μ, so there are many systems which behave in an essentially classical manner.

7.3.1 Maxwell's equations

The interaction between matter and radiation begins with the relativistically invariant Maxwell action

$$S = \int (\mathrm{d}x) \left\{ \frac{1}{4\mu_0} F^{\mu\nu} F_{\mu\nu} - J^\mu A_\mu \right\}. \tag{7.29}$$

The variation of the action,

$$
\begin{aligned}
\delta S &= \int (\mathrm{d}x) \left\{ (\partial^\mu \delta A^\nu) F_{\mu\nu} - J^\mu \delta A_\mu \right\} \\
&= \int (\mathrm{d}x) \left\{ \delta A^\nu (-\partial^\mu F_{\mu\nu}) - J^\mu \delta A_\mu \right\} + \int \mathrm{d}\sigma^\mu \left\{ \delta A^\nu F_{\mu\nu} \right\} \\
&= 0,
\end{aligned}
\tag{7.30}
$$

leads immediately to the field equations for the electromagnetic field interacting with charges in an ambient vacuum:

$$\partial_\mu F^{\mu\nu} = -\mu_0 J^\nu. \tag{7.31}$$

The spatial continuity conditions are

$$\Delta F_{i\mu} = 0, \tag{7.32}$$

or

$$\Delta E_i = 0$$
$$\Delta B_i = 0. \tag{7.33}$$

7.3.2 Electromagnetic waves

In the Lorentz gauge, $\partial^\mu A_\mu = 0$, Maxwell's equations (7.31) reduce to

$$-\Box\, A_\mu = J_\mu. \tag{7.34}$$

The solution to this equation is a linear combination of a particular integral with non-zero J_μ and a complementary function with $J_\mu = 0$. The free-field equation,

$$-\Box\, A_\mu = 0, \tag{7.35}$$

is solved straightforwardly by taking the Fourier transform:

$$A_\mu(x) = \int \frac{\mathrm{d}^{n+1}k}{(2\pi)^{n+1}}\, e^{ik_\mu x^\mu} A_\mu(k). \tag{7.36}$$

Substituting into the field equation, we obtain the constraint

$$\chi(k) = k^2 = k^\mu k_\mu = \left(-\frac{\omega^2}{c^2} + k^i k_i\right) = 0. \tag{7.37}$$

This is the result we found in eqn. (2.52), obtained only slightly differently. The retarded and Feynman Green functions for the field clearly satisfy

$$-\Box\, D_{\mu\nu}(x, x') = g_{\mu\nu}\, c\delta(x, x'). \tag{7.38}$$

Thus, the solution to the field in the presence of the source is, by analogy with eqn. (5.41),

$$A_\mu(x) = \int (\mathrm{d}x')\, D_{\mu\nu}(x, x') J^\nu(x')$$
$$= \int (\mathrm{d}k)\, e^{ik_\mu(x-x')^\mu} \left[\frac{1}{k^2} + X(k)\delta(k^2)\right] J(x'), \tag{7.39}$$

where $X(k)$ is an arbitrary and undetermined function. In order to determine this function, we need to make some additional assumptions and impose some additional constraints on the system.

7.3.3 Dispersion and the Faraday effect

When linearly polarized electromagnetic waves propagate through a magnetized medium, in the direction of the applied magnetization, the plane of polarization becomes rotated in an anti-clockwise sense about the axis of propagation by an amount proportional to z, where z is the distance travelled through the medium. The angle of rotation

$$\psi = VBz, \tag{7.40}$$

where B is the magnetic field and V is Verdet's constant, the value of which depends upon the dielectric properties of the material. This phenomenon is important in astronomy in connection with the polarization of light from distant stars. It is also related to optical activity and the Zeeman effect.

Classical descriptions of this effect usually involve a decomposition of the electric field vector into two contra-rotating vectors which are then shown to rotate with different angular velocities. The sum of these two vectors represents the rotation of the polarization plane. An alternative description can be formulated in complex coordinates to produce the same result more quickly and without prior assumptions about the system.

Let us now combine some of the above themes in order to use the action method to solve the Faraday system. Suppose we have a particle field, $s^i(x)$, of atoms with number density ρ_N, which measures the displacement of optically active electrons $-e$ from their equilibrium positions, and a magnetic field $B = B_3$, which points along the direction of motion for the radiation. In the simplest approximation, we can represent the electrons as being charges on springs with spring constant κ. As they move, they generate an electric current density

$$J_i = -e\rho_N \dot{s}_i. \tag{7.41}$$

Since the Faraday effect is about the rotation of radiation's polarization vector (which is always perpendicular to the direction of motion x_3), we need only s^i for $i = 1, 2$. The action then can be written

$$S = \frac{1}{\sigma_x} \int (\mathrm{d}x) \left\{ -\frac{1}{2}m(\partial_t s)(\partial_t s) + \frac{1}{2}eB\epsilon_{ij}s^i(\partial_t s) + ks^i s_i - J^i s_i \right\}. \tag{7.42}$$

Here, J^i is an external source which we identify with the radiation field

$$J^i(x) = -eE^i(x) = -\frac{e}{c}F^{0i}(x). \tag{7.43}$$

As is often the case with matter–radiation interactions, the relativistically invariant electromagnetic field is split into E^i, B^i by the non-relativistically invariant matter field s^i. The field equations are now obtained by varying the action with respect to δs^i:

$$\delta S = \int (dx) \left\{ m\ddot{s}_i + eB\epsilon_{ij}\dot{s}^j + \kappa s_i - J_i \right\} \delta s^i$$
$$+ \int d\sigma \left[m\dot{s}_i + eB\epsilon_{ij}s^j \right] \delta s^i. \tag{7.44}$$

Thus, the field equations are

$$m\ddot{s}_i + eB\epsilon_{ij}\dot{s}^j + \kappa s_i = J_i = -eE_i, \tag{7.45}$$

and continuity of the field requires

$$\Delta(m\dot{s}_i) = 0$$
$$\Delta(eB\epsilon_{ij}s^j) = 0. \tag{7.46}$$

The first of these is simply the conservation of momentum for the electrons. The latter tells us that any sudden jumps in the magnitude of magnetic field must be compensated for by a sudden jump in the amplitude of the transverse displacement.

If we compare the action and the field equations with the example in section 7.2, it appears as though the magnetic field has the form of a dissipative term. In fact this is not the case. Magnetic fields do no work on particles. The crucial point is the presence of the anti-symmetric matrix ϵ_{ij} which makes the term well defined and non-zero.

Dividing eqn. (7.45) through by the mass, we can defined the Green function for the $s^i(x)$ field:

$$\left[\left(\frac{d^2}{dt^2} + \omega_0^2 \right) \delta_{ij} + \frac{eB}{m}\epsilon_{ij} \right] G_{jk}(x, x') = \delta_{ik}(x, x'), \tag{7.47}$$

where $\omega_0^2 = \kappa/m$, so that the formal solution for the field is

$$s_i(x) = \int (dx') G_{ij}(x, x') J^j(x'). \tag{7.48}$$

Since we are interested in coupling this equation to an equation for the radiation field J_i, we can go no further. Instead, we turn to the equation of motion (7.34) for the radiation. Because of the gauge freedom, we may use a gauge in which $A_0 = 0$, this simplifies the equation to

$$-\Box A_i = \mu_0 J_i^e$$
$$E_i = -\partial_t A_i. \tag{7.49}$$

Thus, using the Green function $D_{ij}(x, x')$,

$$-\Box D_{ij}(x, x') = \delta_{ij} c\delta(x, x'), \tag{7.50}$$

for $A_i(x)$, we may write the solution for the electric field formally as

$$E_i(x) = -\mu_0 \partial_t \int (\mathrm{d}x') D_{ij}(x, x')(-e\rho_N \dot{s}_j(x')) = -J_i/e. \qquad (7.51)$$

This result can now be used in eqn. (7.45), giving

$$\left[\left(\frac{\mathrm{d}^2}{\mathrm{d}t^2} + \omega_0^2 \right) \delta_{ij} + \frac{eB\omega}{m} \epsilon_{ij} \right] s_j(x) =$$
$$-\frac{e^2}{m} \rho_N \mu_0 \, \partial_t \int (\mathrm{d}x') D_{jk}(x, x') \dot{s}^k. \qquad (7.52)$$

Operating from the left with $-\overset{x}{\Box}$, we have

$$(-\Box) \left[\left(\frac{\mathrm{d}^2}{\mathrm{d}t^2} + \omega_0^2 \right) \delta_{ij} + \frac{eB\omega}{m} \epsilon_{ij} \right] s_j(x) = -\frac{e^2}{m} \rho_N \mu_0 \ddot{s}_i. \qquad (7.53)$$

This is a matrix equation, with a symmetric part proportional to δ_{ij} and an anti-symmetric part proportional to ϵ_{ij}. If we take plane wave solutions moving along the $x_3 = z$ axis,

$$s^i(x) = \int \frac{\mathrm{d}^{n+1}k}{(2\pi)^{n+1}} \, \mathrm{e}^{\mathrm{i}(k_z z - \omega t)} s^i(k) \delta(\chi)$$

$$E^i(x) = \int \frac{\mathrm{d}^{n+1}k}{(2\pi)^{n+1}} \, \mathrm{e}^{\mathrm{i}(k_z z - \omega t)} E^i(k) \delta(\chi), \qquad (7.54)$$

for the dispersion relation χ implied by eqn. (7.53), eqn. (7.53) implies that the wavenumber k_z must be a matrix in order to find a solution. This is what will lead to the rotation of the polarization plane for E_i. Substituting the above form for $s^i(x)$ into eqn. (7.53) leads to the replacements $\partial_z \rightarrow \mathrm{i}k_z$ and $\partial_t \rightarrow -\mathrm{i}\omega$. Thus the dispersion relation is

$$\chi = \left(k_z^2 - \frac{\omega^2}{c^2} \right) \left[(-\omega^2 + \omega_0^2)\delta_{ij} + \frac{eB\omega}{m} \epsilon_{ij} \right] - \frac{e^2}{m} \rho_N \mu_0 \omega^2 \, \delta_{ij} = 0,$$
$$\qquad (7.55)$$

or re-arranging,

$$k_{z\,ij}^2 = \frac{\omega^2}{c^2} \left[\delta_{ij} + \frac{\frac{e^2}{m\epsilon_0} \rho_N \left[(\omega^2 + \omega_0^2)\delta_{ij} + \frac{eB\omega}{m} \epsilon_{ij} \right]}{(-\omega^2 + \omega_0^2)^2 - (\frac{eB\omega}{m})^2} \right]. \qquad (7.56)$$

This only makes sense if the wavenumber k_z is itself a matrix with a symmetric and anti-symmetric part:

$$k_{z\,ij} = \bar{k} \delta_{ij} + \tilde{k} \epsilon_{ij}. \qquad (7.57)$$

It is the anti-symmetric part which leads to a rotation of the plane of polarization. In fact, k_z has split into a generator of linear translation $\bar{k}$ plus a generator or rotations $\tilde{k}$ about the z axis:

$$k_z = \bar{k} \times \begin{pmatrix} 1 & 0 \\ 0 & 1 \end{pmatrix} + \tilde{k} \times \begin{pmatrix} 0 & 1 \\ -1 & 0 \end{pmatrix}. \tag{7.58}$$

The exponential of the second term is

$$\begin{pmatrix} \cos(\tilde{k}z) & \sin(\tilde{k}z) \\ -\sin(\tilde{k}z) & \cos(\tilde{k}z) \end{pmatrix}, \tag{7.59}$$

so $\tilde{k}$ is the rate of rotation. Using a binomial approximation for small B, we can write simply

$$\tilde{k}_{z\,ij} = \frac{\frac{e^3 B}{2m\epsilon_0}\rho_N}{(-\omega^2 + \omega_0^2)^2 - (\frac{eB\omega}{m})^2}. \tag{7.60}$$

Verdet's constant is defined by the phenomenological relation,

$$\tilde{k}z = VBz, \tag{7.61}$$

so we have

$$V = \frac{Ne^3\omega^2}{2m^2c\epsilon_0|(\omega_o^2 - \omega^2)^2 - (eB\omega/m)^2|}. \tag{7.62}$$

7.3.4 Radiation from moving charges in $n = 3$: retardation

The derivation of the electromagnetic field emanating from a charged particle in motion is one of the classic topics of electrodynamics. It is an important demonstration of the Green function method for two reasons. First of all, the method of Green functions leads quickly to the answer using only straightforward algebraic steps. Prior to the Green function method, geometrical analyses were carried out with great difficulty. The second reason for looking at this example here is that it brings to bear the causal or retarded nature of the physical field, i.e. the property that the field can only be generated by charge disturbances in the past. This retardation property quickly leads to algebraic pitfalls, since the dynamical variables become defined recursively in term of their own motion in a strange-loop. Unravelling these loops demonstrates important lessons.

We begin by choosing the Lorentz gauge for the photon propagator with $\alpha = 1$. This choice will give the result for the vector potential in a form which is most commonly stated in the literature. Our aim is to compute the vector potential $A_\mu(x)$, and thence the field strength $F_{\mu\nu}$, for a particle at position $\mathbf{x}_p(t)$ which is in motion with speed $\mathbf{v} = \partial_t \mathbf{x}_p(t)$. The current distribution for a point particle is singular, and may be written

$$J^\mu = qc\beta^\mu \, \delta^n(\mathbf{x} - \mathbf{x}_p(t)). \qquad (7.63)$$

The vector potential is therefore, in terms of the retarded propagator,

$$A_\mu(x) = \mu_0 \int (dx') \, G_r(x, x') J_\mu(x')$$

$$= \frac{q}{4\pi\epsilon_0 c} \int (dx') \, \beta^\mu(t') \frac{\delta\left(c(t' - t_{\text{ret}})\right)}{|\mathbf{x} - \mathbf{x}'|} \delta\left(\mathbf{x}' - \mathbf{x}_p(t)\right), \qquad (7.64)$$

where the retarded time is defined by $t_{\text{ret}} = t - |\mathbf{x} - \mathbf{x}'|/c$. Performing the integral over $x^{0'}$ in the presence of the delta function sets $t' \to t_{\text{ret}}$:

$$A_\mu(x) = \frac{q}{4\pi\epsilon_0 c} \int d\sigma_{x'} \frac{\beta_\mu(t_{\text{ret}})\delta\left(\mathbf{x}' - \mathbf{x}_p(t_{\text{ret}})\right)}{|\mathbf{x} - \mathbf{x}'|}. \qquad (7.65)$$

Here $\mathbf{x}$ is a free continuous coordinate parameter, which varies over all space around the charge, and $\mathbf{x}_p(t_{\text{ret}})$ is the retarded trajectory of the charge q. We may now perform the remaining integral. Here it is convenient to change variables. Let

$$\int d\sigma_{x'} \delta(\mathbf{x}' - \mathbf{x}_p(t_{\text{ret}})) = \int d\sigma_r \delta(\mathbf{r})|J|, \qquad (7.66)$$

where $J = \det J_{ij}$ and

$$J_{ij}^{-1} = \partial_i' r_j = \partial_i'(x' - x_p(t_{\text{ret}}))_j$$

$$= g_{ij} - \frac{\partial x_p^i}{\partial t_{\text{ret}}} \frac{\partial t_{\text{ret}}}{\partial x^{i'}}, \qquad (7.67)$$

is the Jacobian of the transformation. At this stage, t_{ret} is given by $t_{\text{ret}} = t - |\mathbf{x} - \mathbf{x}'|/c$, i.e. it does not depend implicitly on itself. After the integration we are about to perform, it will. We complete the integration by evaluating the Jacobian:

$$\partial_i' t_{\text{ret}} = \frac{\hat{r}_i}{c}$$

$$J_{ij}^{-1} = g_{ij} - \frac{v_j}{c}\hat{r}_i$$

$$\det J_{ij}^{-1} = (1 - \beta^i \hat{r}_i)\Big|_{t_{\text{ret}}}. \qquad (7.68)$$

The last line uses the fact that r_i depends only on x_i', not on x_j for $i \neq j$. In this instance, the determinant becomes $1 + \mathrm{Tr}(J_{ij}^{-1})$, giving

$$A_\mu(x) = \frac{q}{4\pi\epsilon_0 c} \int d\sigma_\mathrm{r} \frac{\beta^\mu(t_\mathrm{ret})\delta(\mathbf{r})}{|\mathbf{x} - \mathbf{x}_p(t_\mathrm{ret}) - \mathbf{r}|}$$

$$= \frac{q\beta_\mu(t_\mathrm{ret})}{4\pi\epsilon_0 c\kappa|\mathbf{x} - \mathbf{x}_p|}, \tag{7.69}$$

where $\kappa \equiv (1 - \beta \cdot \hat{\mathbf{r}})$, and all quantities (including κ itself) are evaluated at t_ret. If we define the light ray r^μ as the vector from $\mathbf{x}_p$ to $\mathbf{x}$, then $r^\mu = (r, \mathbf{r})$ and $r = |\mathbf{r}|$, since, for a ray of light, $r = c\Delta t = c \times r/c$. Finally, noting that $r\kappa = -r^\mu \beta_\mu$, we have the Liénard–Wiechert potential in the Lorentz gauge,

$$A_\mu(x) = \frac{-q}{4\pi\epsilon_0 c} \left(\frac{\beta_\mu}{r^\mu \beta_\mu}\right)_{t_\mathrm{ret}}. \tag{7.70}$$

To proceed with the evaluation of the field strength $F_{\mu\nu}$, or equivalently the electric and magnetic fields, it is useful to derive a number of relations which conceal subtleties associated with the fact that the retarded time now depends on the position evaluated at the retarded time. In other words, the retarded time t_ret satisfies an implicit equation

$$t_\mathrm{ret} = t - \frac{|\mathbf{x} - \mathbf{x}_p(t_\mathrm{ret})|}{c} = t - \frac{r}{c}. \tag{7.71}$$

The derivation of these relations is the only complication to this otherwise purely algebraic procedure. Differentiating eqn. (7.71) with respect to t_ret, we obtain

$$1 = \frac{\partial t}{\partial t_\mathrm{ret}} + \hat{r}^i \beta_i(t_\mathrm{ret})$$

$$\partial_t(t_\mathrm{ret}) = \kappa^{-1}\Big|_{t_\mathrm{ret}}. \tag{7.72}$$

Moreover,

$$(\partial_i t_\mathrm{ret}) = -\frac{1}{c}(\partial_i r), \tag{7.73}$$

$$(\partial_i r) = \partial_i \sqrt{r^j r_j}$$

$$= \hat{r}^j(\partial_i r_j), \tag{7.74}$$

$$(\partial_i r_j) = g_{ij} - \frac{\partial x_{pj}}{\partial t_\mathrm{ret}}(\partial_i t_\mathrm{ret})$$

$$= g_{ij} + \beta_j(\partial_i r) \tag{7.75}$$

$$(\partial_i r_j) = g_{ij} + \beta_j \hat{r}^k(\partial_i r_k). \tag{7.76}$$

The last line cannot be simplified further; however, on substituting eqn. (7.76) into eqn. (7.74), it is straightforward to show that

$$(\partial_i r)(1 - (\hat{\mathbf{r}} \cdot \beta)^2) = \hat{r}_i(1 + \hat{\mathbf{r}} \cdot \beta), \tag{7.77}$$

and thus

$$(\partial_i r) = \frac{\hat{r}_i}{\kappa}. \tag{7.78}$$

This may now be substituted back into eqn. (7.75) to give

$$(\partial_i r_j) = g_{ij} + \frac{\beta_j \hat{r}_i}{\kappa}. \tag{7.79}$$

Continuing in the same fashion, one derives the following relations:

$$(\partial_0 r) = -\frac{\hat{r}^i \beta_i}{\kappa}$$

$$(\partial_0 r^i) = -\frac{\beta^i}{\kappa}$$

$$(\partial_i r) = \frac{\hat{r}_i}{\kappa}$$

$$(\partial_i r_j) = g_{ij} + \frac{\beta_j \hat{r}_i}{\kappa}.$$

$$(\partial_0 \beta_i) = \frac{\alpha_i}{\kappa}$$

$$(\partial_i \beta_j) = -\frac{\hat{r}_i \alpha_j}{\kappa}$$

$$\partial_0 (r\kappa) = \frac{1}{\kappa} \left(\beta^2 - \hat{\mathbf{r}} \cdot \beta - \alpha \cdot \mathbf{r} \right)$$

$$\partial_i (r\kappa) = \frac{\hat{r}_i}{\kappa} \left(1 - \beta^2 + \alpha \cdot \mathbf{r} \right) - \beta_i, \tag{7.80}$$

where we have defined $\alpha_\mu = \partial_0 \beta_\mu = (0, \dot{\mathbf{v}}/c^2)$. The field strength tensor may now be evaluated. From eqn. (7.70) one has

$$F_{\mu\nu} = \partial_\mu A_\nu - \partial_\nu A_\mu$$
$$= \frac{q}{4\pi \epsilon_0 c} \left[\frac{\partial_\mu \beta_\nu - \partial_\nu \beta_\mu}{r\kappa} - \frac{(\beta_\nu \partial_\mu - \beta_\mu \partial_\nu)(r\kappa)}{r^2 \kappa^2} \right]. \tag{7.81}$$

And, noting that $\beta_0 = -1$ is a constant, we identify the three-dimensional electric and magnetic field vectors:

$$E_i = cF_{i0}$$
$$= \frac{-q}{4\pi \epsilon_0} \left[\frac{\partial_0 \beta_i}{r\kappa} - \frac{(\beta_i \partial_0 - \beta_0 \partial_i)(r\kappa)}{r^2 \kappa^2} \right]$$
$$= \frac{-q}{4\pi \epsilon_0} \left[\frac{\alpha_i}{\kappa^2 r} + \frac{(\beta_i - \hat{r}_i)}{r^2 \kappa^3} \left(\alpha \cdot \mathbf{r} + (1 - \beta^2) \right) \right]. \tag{7.82}$$

$$B_i = \frac{1}{2}\epsilon_{ijk}F_{jk}$$

$$= \frac{q}{4\pi\epsilon_0 c}\epsilon_{ijk}\left[\frac{\partial_j \beta_k}{r\kappa} - \frac{(\beta_k \partial_j)(r\kappa)}{r^2\kappa^2}\right]$$

$$= \frac{-q}{4\pi\epsilon_0 c}\epsilon_{ijk}\hat{r}_j\left[\frac{\alpha_k}{\kappa} + \frac{\beta_k}{r^2\kappa^3}\left(\alpha\cdot\mathbf{r} + (1-\beta^2)\right)\right] \tag{7.83}$$

$$= \frac{1}{c}\epsilon_{ijk}\hat{r}_j E_k$$

$$= \frac{1}{c}(\hat{\mathbf{r}}\times\mathbf{E})_i. \tag{7.84}$$

From these relations, it is clear that the magnetic field is perpendicular to both the light ray $\mathbf{r}$ and the electric field. The electric field can be written as a sum of two parts, usually called the radiation field and the near field:

$$E_{i\,\text{rad}} = \frac{q}{4\pi\epsilon_0 c}\left[\frac{\alpha_i}{\kappa^2 r} + \frac{(\beta_i - \hat{r}_i)(\alpha\cdot\hat{\mathbf{r}})}{\kappa^3 r}\right] \tag{7.85}$$

$$E_{i\,\text{near}} = \frac{q}{4\pi\epsilon_0 c}\left[\frac{(\beta_i - \hat{r}_i)(1-\beta^2)}{r^2\kappa^3}\right]. \tag{7.86}$$

The near field falls off more quickly than the long-range radiation field. The radiation field is also perpendicular to the light ray $\hat{\mathbf{r}}$. Thus, the far-field electric and magnetic vectors are completely transverse to the direction of propagation, but the near-field electric components are not completely transverse except at very high velocities $\beta \sim 1$. Note that all of the vectors in the above expressions are assumed to be evaluated at the retarded time.

Owing to their special relationship, the magnitude of the magnetic and electric components are equal up to dimensional factors:

$$|\mathbf{E}|^2 = c^2|\mathbf{B}|^2. \tag{7.87}$$

Finally, the rate of work or power expended by the field is given by Poynting's vector,

$$S_i = \epsilon_{ijk}E_j H_k$$

$$= (\mu_0 c)^{-1}\epsilon_{ijk}E_j(\hat{\mathbf{r}}\times\mathbf{E})_k$$

$$= \epsilon_0 c\,\epsilon_{ijk}E_j(\epsilon_{klm}\hat{r}_l E_m)$$

$$\mathbf{S} = -\epsilon_0 c(\mathbf{E}\cdot\mathbf{E})\hat{\mathbf{r}}. \tag{7.88}$$

7.4 Resonance phenomena and dampening fields

In the interaction between matter and radiation, bound state transitions lead to resonances, or phenomena in which the strength of the response to a radiation field is amplified for certain frequencies. Classically these special frequencies are the normal modes of vibration for spring-like systems with natural frequency ω_0; quantum mechanically they are transitions between bound state energy levels with a definite energy spacing $\omega_0 = (E_2 - E - 1)/\hbar$. The examples which follow are all cases of one mathematical phenomenon which manifests itself in several different physical scenarios. We see how the unified approach reveals these similarities.

7.4.1 Cherenkov radiation

The radiation emitted by charged particles which move in a medium where the speed of light is less than the speed of the particles themselves is called Cherenkov radiation. The effect was observed by Cherenkov [25] and given a theoretical explanation by Tamm and Frank [127] within the framework of classical electrodynamics. The power spectrum of the radiation may be calculated with extraordinary simplicity using covariant field theory [122].

Using the covariant formulation in a material medium from section 21.2.4 and adapting the expression in eqn. (5.118) for the Maxwell field, we have the Feynman Green function in the Lorentz–Feynman $\alpha = 1$ gauge, given by

$$
D_F(x, x') = \frac{-i}{4\pi^2 c^2 |\mathbf{x} - \mathbf{x}'|} \int_0^\infty d\omega \, \sin\left(\frac{n\omega}{c}|\mathbf{x} - \mathbf{x}'|\right) e^{-i\omega|t - t'|},
\tag{7.89}
$$

where n is the refractive index of the medium. Note that this index is assumed to be constant here, which is not the case in media of interest. One should really consider $n = n(\omega)$. However, the expressions generated by this form will always be correct in ω space for each value of ω, since the standard textbook assumption is to ignore transient behaviour (t-dependence) of the medium. We may therefore write the dissipation term as

$$
W = \mu_0 \mu_r \int (dx)(dx') \hat{J}^\mu(x) \hat{D}_{F\mu\nu}(x, x') \hat{J}^\nu(x'),
\tag{7.90}
$$

and we are interested in the power spectrum which is defined by

$$
\int d\omega \frac{P(\omega)}{\omega} = \frac{2}{\hbar} \mathrm{Im} \frac{dW}{dt}.
\tag{7.91}
$$

Substituting in expressions for $\hat{J}_\mu$, we obtain

$$
\mathrm{Im} W = -\frac{1}{8\pi^2} \int d\omega (dx)(dx') \frac{\mu_0 \mu_r \sin(\frac{n\omega}{c}|\mathbf{x} - \mathbf{x}'|)}{c^2 |\mathbf{x} - \mathbf{x}'|}
$$
$$
\times \cos(\omega|t - t'|) \hat{J}^\mu \hat{J}_\mu,
\tag{7.92}
$$

from which we obtain

$$P(\omega) = -\frac{\omega\mu_0\mu_r}{4\pi^2 n^2} \int d\sigma_x (dx') \frac{\sin(\frac{n\omega}{c}|\mathbf{x} - \mathbf{x}'|)}{n^2|\mathbf{x} - \mathbf{x}'|} \cos(\omega|t - t'|)$$
$$\times \left[\rho(x)\rho(x') - \frac{n^2}{c^2} J^i(x) J_i(x') \right]. \quad (7.93)$$

The current distribution for charged particles moving at constant velocity is

$$\rho = q\delta(\mathbf{x} - \mathbf{v}t)$$
$$J^i = qv^i \delta(\mathbf{x} - \mathbf{v}t); \quad (7.94)$$

thus we have

$$P(\omega, t) = \frac{q^2}{4\pi^2} \frac{\mu_0\mu_r\omega\beta}{c} \left(1 - \frac{1}{n(\omega)^2\beta^2}\right) \int_{-\infty}^{\infty} \sin(n\beta\omega\tau)\cos(\omega\tau)d\tau$$
$$= \begin{cases} 0 & n\beta < 1 \\ \frac{q^2}{4\pi} \frac{\mu_0\mu_r\omega\beta}{c} \left(1 - \frac{1}{n^2\beta^2}\right) & n\beta > 1. \end{cases} \quad (7.95)$$

This is the power spectrum for Cherenkov radiation, showing the threshold behaviour at $n\beta = 1$. We have derived the Cherenkov resonance condition for charges interacting with electromagnetic radiation. The Cherenkov effect is more general than this, however. It applies to any interaction in which particles interact with waves, either transverse or longitudinal.

7.4.2 Cyclotron radiation

Cyclotron, or synchrotron, radiation is emitted by particles accelerated by a homogeneous magnetic field. Its analysis proceeds in the same manner as that for Cherenkov radiation, but with a particle distribution executing circular rather than linear motion. For the current, one requires

$$\rho = q\delta(\mathbf{x} - \mathbf{x}_0)$$
$$J^i = qv^i \delta(\mathbf{x} - \mathbf{x}_0), \quad (7.96)$$

where $\mathbf{x}_0$ is the position of the charged particle. Since the electromagnetic field is not self-interacting, the Green function for the radiation field is not affected by the electromagnetic field in the absence of a material medium. In the presence of a polarizable medium, there is an effect, but it is small. (See the discussion of Faraday rotation.)

The force on charges is

$$F_i = \frac{dp_i}{dt} = q(\mathbf{v} \times \mathbf{B})_i$$
$$= qF_{ij}v^j, \quad (7.97)$$

and, since this is always perpendicular to the motion, no work is done; thus the energy is a constant of the motion:

$$\frac{dE}{dt} = 0. \tag{7.98}$$

The generic equation of circular motion is

$$\frac{dv_i}{dt} = (\omega \times \mathbf{v})_i, \tag{7.99}$$

which, in this case, may be written as

$$\frac{dv_i}{dt} = \frac{q}{m}\sqrt{1 - \beta^2}(\mathbf{v} \times \mathbf{B})_i, \tag{7.100}$$

where $p_i = mv_i/\sqrt{1 - \beta^2}$ and $\beta_i = v_i/c$. Thus, the angular frequency of orbit is the Larmor frequency,

$$\omega_i = -\frac{qB_i}{m}\sqrt{1 - \beta^2} = -\frac{qB_i c^2}{E}, \tag{7.101}$$

which reduces to the cyclotron frequency, $\omega_c \simeq eB/m$, in the non-relativistic limit $\beta_i \to 0$. The radius of revolution is correspondingly

$$R = \frac{|v|}{\omega} = \frac{mc\beta}{|q|B\sqrt{1 - \beta^2}}. \tag{7.102}$$

The primary difficulty in analysing this problem is a technical one associated with the circular functions. Taking boundary conditions such that the particle position is given by

$$x_1(t) = R\cos(\omega t)$$
$$x_2(t) = R\sin(\omega t)$$
$$x_3(t) = 0, \tag{7.103}$$

one finds the velocity

$$v_1(t) = -R\omega\sin(\omega t)$$
$$v_2(t) = R\omega\cos(\omega t)$$
$$v_3(t) = 0. \tag{7.104}$$

This may be substituted into the current in order to evaluate the power spectrum. This is now more difficult: one can use an integral representation of the delta function, such as the Fourier transform; this inevitably leads to exponentials of sines and cosines, or Bessel functions. We shall not pursue these details of evaluation here. See ref. [121] for further study of this topic.

7.4.3 Landau damping

Landau damping is the name given to the dissipative mixing of momenta in any particle field or plasma which interacts with a wave. The phenomenon of Landau damping is quite general and crops up in many guises, but it is normally referred to in the context of the interaction of a plasma with electromagnetic waves. In a collisionless plasma (no scattering by self-interaction), there is still scattering by the interaction of plasma with the ambient electromagnetic field, similar to the phenomenon of stimulated absorption/emission. However, any linear perturbation or source can cause the energy in one plasma mode to be re-channelled into other modes, thus mixing the plasma and leading to dissipation. All one needs is a linear interaction between the waves and the plasma field, and a resonant amplifier, which tends to exaggerate a specific frequency.

In simple terms, a wave acts like a sinusoidal potential which scatters and drags the particle field. If the phase of the field is such that it strikes the upward slope of the wave, it is damped or reflected, losing energy. If the phase is such that the field 'rolls down' the downward slope of the wave, it is enhanced and gains energy. In a random system, the average effect is to dissipate or to dampen the field so that all particles or field modes tend to become uniform. In short, Landau damping is the re-organization of energy with the modes of a field due to scattering off wavelets of another field.

Let us consider an unbound particle displacement field with action

$$S = \frac{1}{\sigma_x} \int (\mathrm{d}x) \left\{ -\frac{1}{2} m \dot{s}^2 - J^i s_i \right\}, \tag{7.105}$$

coupled through the current J_i to the electromagnetic field. The position of a particle is

$$x^i = \overline{x}^i + \delta x^i = \overline{x}^i + s^i, \tag{7.106}$$

and its velocity is

$$\dot{x}^i = \overline{v}^i + \delta v. \tag{7.107}$$

The velocity of a free particle is constant until the instant of its infinitesimal perturbation by a wave, so we write

$$\overline{x}^i = \overline{v}t, \tag{7.108}$$

so that

$$k_\mu x^\mu = k_i x^i - \omega t = k_i s^i + (k_i \overline{v}^i - \omega)t. \tag{7.109}$$

The perturbation is found from the solution to the equation of motion:

$$s^i = \int (\mathrm{d}x) G^{ij}(x, x') E_j(x'), \tag{7.110}$$

or

$$\delta v^i = \frac{q}{m} \, \mathrm{Re} \, \frac{E_0^i \, \exp \mathrm{i} \left(k_i s^i + (k_i \bar{v}^i - \omega)t \pm \gamma t\right)}{\mathrm{i}(k_i \bar{v}^i - \omega) \pm \gamma}$$

$$s^i = \frac{q}{m} \, \mathrm{Re} \, \frac{E_0^i \, \exp \mathrm{i} \left(k_i s^i + (k_i \bar{v}^i - \omega)t \pm \gamma t\right)}{\left[\mathrm{i}(k_i \bar{v}^i - \omega) \pm \gamma\right]^2}. \tag{7.111}$$

An infinitesimal regulating parameter, γ, is introduced here in order to define a limit in what follows. This has causal implications for the system. It means that the field either grows from nothing in the infinite past or dissipates to nothing in the infinite future. This is reflected by the fact that its sign determines the sign of the work done. Eventually, we shall set γ to zero. The work done by this interaction between the charged particle q and the electric field E^i is $q E_i x^i$. The rate of work is

$$q \frac{\mathrm{d}}{\mathrm{d}t} \left[E_i x^i\right] = q \partial_t E_i \, x^i + E_i v^i. \tag{7.112}$$

The two terms signify action and reaction, so that the total rate of work is zero, expressed by the total derivative. The second term is the rate of work done by the charge on the field. It is this which is non-zero and which leads to the dampening effect and apparent dissipation. Following Lifshitz and Pitaevskii [90], we calculate the rate of work per particle as follows,

$$\begin{aligned} \frac{\mathrm{d}w}{\mathrm{d}t} &= \mathrm{Re} \, q v_i E^i \\ &= \mathrm{Re} \, q(\bar{v}^i + \delta v^i) E(t, \mathbf{x} + \mathbf{s}) \\ &= \mathrm{Re} \, q(\bar{v}^i + \delta v^i)(E_i(t, \mathbf{x}) + \partial_j E_i(t, \mathbf{x}) s^j + \cdots). \end{aligned} \tag{7.113}$$

To first order, the average rate of work is thus

$$\begin{aligned} \left\langle \frac{\mathrm{d}w}{\mathrm{d}t} \right\rangle &= \mathrm{Re} \, q \bar{v}_i \langle (\partial_j E_i) s^i \rangle + \mathrm{Re} \, q \langle \delta v^i E_i \rangle \\ &= \frac{1}{2} q \bar{v}_i (\partial_j E_i^*) s^i + \frac{1}{2} q \delta v^i E_i^*. \end{aligned} \tag{7.114}$$

Here we have used the fact that

$$\mathrm{Re} \, A = \frac{1}{2}(A + A*) \tag{7.115}$$

and

$$\langle \mathrm{Re} \, A \cdot \mathrm{Re} \, B \rangle = \frac{1}{4}(AB^* + A^*B) = \frac{1}{2} \mathrm{Re} \, (AB^*), \tag{7.116}$$

since terms involving A^2 and B^2 contain $e^{2i\omega t}$ average to zero over time (after setting $\gamma \to 0$). Substituting for s^i and δv^i, we obtain

$$\left\langle \frac{dw}{dt} \right\rangle = \frac{q^2}{2m} E^i E^j \left[\frac{i\bar{v}_i k_j}{[i(k_i v^i - \omega) \pm \gamma]^2} + \frac{\delta_{ij}}{[i(k_i v^i - \omega) \pm \gamma]} \right].$$

(7.117)

The tensor structure makes it clear that the alignment, k_i, and polarization, E_0^i, of the electric wave to the direction of particle motion, v_i, is important in deciding the value of this expression. Physically, one imagines a wave (but not a simple transverse wave) moving in direction k_i and particles surfing over the wave in a direction given by v_i. The extent to which the wave offers them resistance, or powers them along, decides what work is done on them. For transverse wave components, $k^i E_i = 0$, the first term vanishes. From the form of eqn. (7.117) we observe that it is possible to write

$$\left\langle \frac{dw}{dt} \right\rangle = \frac{q^2}{2m} E_i E_j \frac{d}{d(k_i v_j)} \left[\frac{\pm \gamma (k_m v^m)}{[i(k_i v^i - \omega) \pm \gamma]} \right],$$

(7.118)

and, using

$$\lim_{\gamma \to 0} \frac{\gamma}{z^2 + \gamma^2} = \pi \, \delta(z)$$

(7.119)

we have

$$\left\langle \frac{dw}{dt} \right\rangle = \pm \frac{q^2 \pi}{m} E_i E_j \frac{d}{d(k_i v_j)} (k_i v^i) \delta(k_j v^j - \omega).$$

(7.120)

To avoid unnecessary complication, let us consider the contribution to this which is most important in the dampening process, namely a one-dimensional alignment of k_i and v_i:

$$\left\langle \frac{dw}{dt} \right\rangle = \pm \frac{q^2 \pi}{2m} |E_{\parallel}|^2 \frac{d}{d(kv)} (kv) \, \delta(kv - \omega).$$

(7.121)

This expression is for one particle. For the whole particle field we must perform the weighted sum over the whole distribution, $f(\omega)$, giving the total rate of work:

$$\left\langle \frac{dW}{dt} \right\rangle = \pm \frac{q^2 \pi}{2m} |E_{\parallel}|^2 \int d\omega \, f(\omega) \frac{d}{d(kv)} (kv) \delta(kv - \omega)$$

$$= \mp \frac{q^2 \pi}{m} |E_{\parallel}|^2 \int d\omega \frac{d f(\omega)}{d\omega} (kv) \delta(kv - \omega)$$

$$= \mp \frac{q^2 \mathbf{p} i \omega}{2m} |E_{\parallel}|^2 \frac{d f(\omega)}{d\omega} \bigg|_{v=\omega/k}.$$

(7.122)

The integral over the delta function picks out contributions when the velocity of particles, v_i, matches the phase velocity of the electromagnetic wave, ω/k_i. This result can now be understood either in real space from eqn. (7.114) or in momentum space from eqn. (7.122). The appearance of the gradient of the electric field in eqn. (7.114) makes it clear that the dissipation is caused as a result of motion in the potential of the electric field. Eqn. (7.122) contains $df/d\omega$, for frequencies where the phase velocity is in resonance with the velocity of the particles within the particle field; this tells us that particles with $v < \omega/k$ gain energy from the wave, whereas $v > \omega/k$ lose energy to it ($\gamma > 0$). The electric field will be dampened if the shape of the distribution is such that there are more particles with $v < \omega/k$ than with $v > \omega/k$. This is typical for long-tailed distributions like thermal distributions.

This can be compared with the discussion in section 6.1.4.

7.4.4 Laser cooling

Another example of resonant scattering with many experimental applications is the phenomenon of laser cooling. This can be thought of as Landau damping for neutral atoms, using the dipole force as the breaking agent. We shall consider only how the phenomenon comes about in terms of classical fields, and sketch the differences in the quantum mechanical formulation. By now, this connection should be fairly familiar. The shift in energy of an electromagnetic field by virtue of its interaction with a field of dipoles moving at fractional speed β^i is the work done in the rest frame of the atom,

$$\Delta W = -\frac{1}{2} \int d\sigma_x \, \mathbf{P}(x) \cdot \mathbf{E}(x)$$

$$= \frac{q^2}{2m} \int (dx') d\sigma_x \, E^i(x) G_{ij}^\beta(x, x') E^j(x'), \qquad (7.123)$$

where

$$((1 - \beta^i)^2 \partial_t^2 - \gamma \partial_t + \kappa) G_{ij}^\beta(x, x') = \delta_{ij} \, c\delta(x, x') \qquad (7.124)$$

(see eqn. (2.88)), and therefore the dipole force $\mathbf{F}$ on each atom may be deduced from $dW = \mathbf{F} \cdot d\mathbf{r}$. The imaginary part of the energy is the power exchanged by the electromagnetic field, which is related to the damping rate or here the cooling rate of the atoms. The force on an atom is the gradient of the real part of the work:

$$F_i^\beta = -\frac{q^2}{2m} \int d\sigma_x \, \overset{x}{\partial_i} \left[E^j(x) \int (dx') \, G_{jk}^\beta(x, x') E^k(x') \right]. \qquad (7.125)$$

If we consider a source of monochromatic radiation interacting with the particle field (refractive index n^i),

$$E^i(x) = E_0^i \, e^{ikx} = E_0^i \, e^{i(\mathbf{k} \cdot \mathbf{x} - \omega t)}, \qquad (7.126)$$

Table 7.1. Doppler effect on momentum.

Resonance-enhanced	Parallel	Anti-parallel
(diagonal)	$\hat{k}^i \beta_i > 0$	$\hat{k}^i \beta_i < 0$
$\omega_\beta > \omega_0$	$F_i \hat{\beta}^i < 0$	$F_i \hat{\beta}^i > 0$
$\omega_\beta < \omega_0$	$F_i \hat{\beta}^i > 0$	$F_i \hat{\beta}^i < 0$

where the frequency ω is unspecified but satisfies $\mathbf{k}^2 c^2 = \mathbf{n} \omega^2$, then we have

$$
\begin{aligned}
F_i^\beta &= -\frac{q^2}{2m} E_0^2 \rho_N \int d\sigma_x \, \partial_i \frac{e^{ix(k+k')+ix'(k'-k)}}{-\omega_\beta^2 + \omega_0^2 + i\gamma\omega} \\
&= -\frac{q^2}{2m} E_0^2 \rho_N \int d\sigma_x \, \partial_i \frac{e^{i2kx}}{-\omega_\beta^2 + \omega_0^2 + i\gamma\omega}.
\end{aligned}
\tag{7.127}
$$

This expression contains forward and backward moving photons of fixed frequency, ω, and wavenumber, k_i. The sign of the force acting on the atoms depends on the frequency relative to the resonant frequency, ω_0, and we are specifically interested in whether the force acts to accelerate the atoms or decelerate them relative to their initial velocity. The fact that atoms in the particle field move in all directions on average means that some will experience Doppler blue-shifted radiation frequencies and others will experience red-shifted frequencies, relative to the direction of photon wavevector, k^i. In effect, the Doppler effect shifts the resonant peak above and below its stationary value making two resonant 'side bands'. These side bands can lead to energy absorption. This is best summarized in a table (see table 7.1).

As the velocity component, $v^i = \beta^i c$, of a particle field increases, the value of $1 - \beta^i \hat{k}_i$ either increases (when $\hat{k}$ and β^i point in opposing directions) or decreases (when $\hat{k}$ and β^i point in the same direction). The component of velocity in the direction of the photons, $\mathbf{E}^i$, is given by $\hat{k}^i \beta_i$, and its sign has two effects. It can bring ω_β closer to or further from the resonant frequency, ω_0, thus amplifying or attenuating the force on the particles. The force is greater for those values which are closest to resonance. It also decides whether the sign of the force is such that it tends to increase the magnitude of β^i or decrease the magnitude of β^i. It may be seen from table 7.1 that the force is always such as to make the velocity tend to a value which makes $\omega_\beta = \omega_0$. Thus by sweeping the value of ω from a value just above resonance to resonance, it should be possible to achieve $\beta^i \to 0$. The lowest attainable temperature according to this simple model is limited by the value of ω_0.

In order to reduce all of the components of the velocity to minimal values, it is desirable to bathe a system in crossed laser beams in three orthogonal directions.

Such laser beams are called *optical molasses*, a kind of quagmire for resonant particle fields. Clearly, systems with a low-frequency resonance are desirable in order to push the magnitude of β^i down to a minimum. The rate of energy loss is simply the damping constant, γ.

7.5 Hydrodynamics

The study of the way in which bulk matter fields spread through a system is called hydrodynamics. Because it deals with bulk matter, hydrodynamics is a macroscopic, statistical discussion. It involves such ideas as flow and diffusion, and is described by a number of essentially classical phenomenological equations.

7.5.1 Navier–Stokes equations

The Navier–Stokes equations are the central equations of fluid dynamics. They are an interesting example of a vector field theory because they can be derived from an action principle in two different ways. Fluid dynamics describes a stationary system with a fluid flowing through it. The velocity is a function of position and time, since the flow might be irregular; moreover, because the fluid flows relative to a fixed pipe or container, the action is not invariant under boosts.

Formulation as a particle field Using a 'microscopic' formulation, we can treat a fluid as a particle displacement field without a restoring force (spring tension zero). We begin by considering such a field at rest:

$$S = \int (\mathrm{d}x) \left\{ \frac{1}{2}\rho \dot{s}^2 - \frac{1}{2}\eta(\partial^i s^j) \overset{\leftrightarrow}{\partial_t} (\partial_i s_{,}) + s^i(F_i - \partial_i P) \right\}. \quad (7.128)$$

Notice the term linear in the derivative which is dissipative and represents the effect of a viscous frictional force (see section 7.2). η is the coefficient of viscosity. In this form, the equations have made an assumption which relates bulk and shear viscosity, leaving only a single effective viscosity. This is the form often used experimentally. Varying the action with respect to s^i leads to the field equation

$$-\rho \ddot{s}_i + \eta \nabla^2 \dot{s}_i + F_i - \partial_i P = 0. \quad (7.129)$$

Or, setting $v_i \equiv \dot{s}_i$,

$$\rho \dot{v}_i - \eta \nabla^2 v_i + \partial_i P = F_i. \quad (7.130)$$

This is the equation of a velocity field at rest. In order to boost it into a moving frame, we could re-define positions by $x_i \rightarrow x_i - v_i t$, but it is more convenient to re-define the time coordinate to so-called retarded time (see section 9.5.2). With this transformation, we simple replace the time derivative for v^i by

$$\frac{d}{dt_{ret}} = \partial_t + v^i \, \partial_j. \tag{7.131}$$

This gives

$$\rho \frac{d}{dt_{ret}} v_i - \eta \, \nabla^2 \, v_i + \partial_i P = F_i. \tag{7.132}$$

In fluid dynamics, this derivative is sometimes called the *substantive derivative*; it is just the total derivative relative to a moving frame. This transformation of perspective introduces a non-linearity into the equation which was not originally present. It arises physically from a non-locality in the system; i.e. the fact that the velocity-dependent forces at a remote point lead to a delayed effect on the velocity at local point. Put another way, the velocity at one point interacts with the velocity at another point because of the flow, just as in a particle scattering problem. In particle theory parlance, we say that the velocity field scatters off itself, or is self-interacting. It would have been incorrect to apply this transformation to the action before variation since the action is a scalar and was not invariant under this transformation, thus it would amount to a change of the physics. Since the action is a generator of constraints, it would have additional consequences for the system, as we shall see below.

Formulation as an effective velocity field The description above is based upon a microscopic picture of a fluid as a collection of particles. We need not think like this, however. If we had never built a large enough microscope to be able to see atoms, then we might still believe that a fluid were a continuous substance. Let us then formulate the problem directly in terms of a velocity field. We may write the action

$$S = \tau \int (dx) \left\{ -\frac{1}{2} \rho \, v^i \stackrel{\leftrightarrow}{\partial_t} v_i + \frac{1}{2} \eta (\partial^i v^j)(\partial_i v_j) - v^i (F_i - \partial_i P) \right\}. \tag{7.133}$$

The constant scale τ has the dimensions of time and is necessary on purely dimensional grounds. The fact that we need such an arbitrary scale is an indication that this is just an average, smeared out field theory rather than a microscopic description. It has no physical effect on the equations of motion unless we later attempt to couple this action to another system where the same scale is absent or different. Such is the nature of dimensional analysis. The linear derivatives in the action are symmetrized for the reasons discussed in section 4.4.2. Varying this action with respect to the velocity v^i, and treating ρ as a constant for the time being, leads to

$$\rho \, \partial_t v_i - \eta \, \nabla^2 \, v_i + \partial_i P = F_i. \tag{7.134}$$

Changing to retarded time, as before, we have the Navier–Stokes equation,

$$\rho\,\partial_t v_i + \rho v^j (\partial_j v_i) - \eta\,\nabla^2 v_i + \partial_i P = F_i. \tag{7.135}$$

Again, it would be incorrect to transform the action before deriving the field equations, since the action is a scalar and it is not invariant under this transformation.

Consider what would have happened if we had tried to account for the retardation terms in the action from the beginning. Consider the action

$$S = \tau \int (dx) \left\{ -\frac{1}{2}\rho\, v^i \stackrel{\leftrightarrow}{\partial_t}\, v_i + \frac{1}{2}\rho\, v^i\,(\partial_i v_j) v^j - \frac{1}{2}\rho\,\partial_i(v^i v^j) v_j \right.$$
$$\left. + \frac{1}{2}\eta(\partial^i v^j)(\partial_i v_j) - v^i(F_i - \partial_i P) \right\}. \tag{7.136}$$

The action is now non-linear from the beginning since it contains the same retardation information as the transformed eqn. (7.132). The derivatives are symmetrized also in spatial directions. The variation of the action is also more complicated. We shall now let ρ depend on x. After some calculation, variation with respect to v^i leads to an equation which can be separated into parts:

$$(\partial_t \rho) v_i + \rho v_i (\partial_j v^j) + \frac{1}{2}(\partial_i \rho) v^2 = 0$$
$$\rho\,(\partial_t v_i) + \rho v^j \partial_i v_j - \eta\,\nabla^2 v_i + \partial_i P = F_i. \tag{7.137}$$

The first of these occurs because the density is no longer constant; it is tantalizingly close to the conservation equation for current

$$-\partial_t \rho = \partial_i(\rho v^i), \tag{7.138}$$

but alas is not quite correct. The equations of motion (7.137) are almost the same as before, but now the derivative terms are not quite correct. Instead of

$$v^j \partial_j v_i \tag{7.139}$$

we have the symmetrical

$$v^j \partial_i v_j. \tag{7.140}$$

This result is significant. The terms are not unrelated. In fact, since we can always add and subtract a term, it is possible to relate them by

$$v^j \partial_j v_i = v^j (\partial_i v_j) + v^j (\partial_j v_i - \partial_i v_j). \tag{7.141}$$

The latter term is the curl of the velocity. What this means is that the two terms are equivalent provided that the curl of the velocity vanishes. It vanishes in the absence of eddies or other phenomena which select a preferred direction in space or time. This is indicative of the symmetry of the action. Since the action was invariant under space and time reversal, it can only lead to equations of motion with the same properties. Physically, this restriction corresponds to purely *irrotational flow*. Notice how the symmetry which is implicit in the action leads directly to a symmetry in the field equations. The situation was different in our first formulation, where we chose to transform the action to retarded time (an intrinsically asymmetrical operation).

The problem of an x-dependent density ρ is not resolvable here. The fundamental problem is that the flow equation is not reversible, whereas the action would like to be. If we omit the non-linear terms, the problem of finding an action which places no restriction on ρ is straightforward, though not particularly well motivated. We shall not pursue this here. The lesson to be learned from this exercise is that, because the action is a scalar, the action principle will always tend to generate field equations consistent with the symmetries of the fields it is constructed from. Here we have tried to generate a term $v^j \partial_j v_i$ from an action principle, but the infinitesimal variation of this term led to new constraints since action is spacetime-reflection-invariant. The problem of accommodating an x-dependent density is confounded by these other problems. In short, non-covariant analyses do not lend themselves to a covariant formulation, but should be obtained as a special case of a more well defined problem as in the first method.

7.5.2 Diffusion

Let us consider the rate at which conserved matter diffuses throughout a system when unencumbered by collisions. Consider a matter current, J_μ, whose average, under the fluctuations of the system, is conserved:

$$\partial_\mu \langle J^\mu \rangle = 0. \tag{7.142}$$

We need not specify the nature of the averaging procedure, nor the origin of the fluctuations here. Phenomenologically one has a so-called constitutive relation [53], which expresses a phenomenological rate of flow in terms of local density gradients:

$$\langle J_i \rangle = -D \partial_i \langle \rho \rangle. \tag{7.143}$$

Substituting this into the conservation equation gives

$$(\partial_t - D\nabla^2)\langle \rho \rangle = 0. \tag{7.144}$$

This is a diffusion equation, with diffusion coefficient D. If we multiply this equation by the positions squared, x^2, and integrate over the entire system,

$$\int d\sigma \, x^2 (\partial_t - D\nabla^2)\langle \rho \rangle = 0, \tag{7.145}$$

we can interpret the diffusion constant in terms of the mean square displacement of the field. Integrating by parts, and assuming that there is no diffusion at the limits of the system, one obtains

$$\partial_t \langle x^2 \rangle - 2D \sim 0, \tag{7.146}$$

or

$$\langle x^2 \rangle \sim 2Dt, \tag{7.147}$$

which indicates that particles diffuse at a rate of $\sqrt{2D}$ metres per unit time. Notice that, since D characterizes the diffusion of averaged quantities, it need not be a constant. We shall think of it as a slowly varying function of space and time. The variation, however, must be so slow that it is effectively constant over the dominant scales of the system. We shall derive a Kubo-type relation for this quantity [53].

From eqn. (7.144), we may solve

$$\langle \rho \rangle (x) = \int \frac{d^n k d\omega}{(2\pi)^{n+1}} e^{i(\mathbf{k} \cdot \mathbf{x} - \omega t)} \rho(k) \, \delta(-i\omega - D\mathbf{k}^2), \tag{7.148}$$

or

$$G^{(\pm)}(k) = \frac{1}{\mp i\omega - D\mathbf{k}^2}. \tag{7.149}$$

Thus

$$\langle \rho \rangle (x) = \int \frac{d^n k d\omega}{(2\pi)^{n+1}} e^{i\mathbf{k} \cdot \mathbf{x} - D\mathbf{k}^2 t} \rho(k). \tag{7.150}$$

To determine the effect of fluctuations in this system, consider adding an infinitesimal source,

$$(\partial_t - D\nabla^2)\langle \rho \rangle = F. \tag{7.151}$$

The purely mechanical retarded response to F gives us the following relation:

$$\langle \rho \rangle (x) = \int (dx') \, G_r(x, x') F(x'), \tag{7.152}$$

where the retarded Green function may be evaluated by analogy with eqn. (5.77)

$$\begin{aligned} G_r(x, x') &= \int \frac{d^n k d\omega}{(2\pi)^{n+1}} e^{i(\mathbf{k} \cdot \mathbf{x} - \omega t)} \left[\frac{1}{\omega + iD\mathbf{k}^2 - i\epsilon} - \frac{1}{\omega - iD\mathbf{k}^2 + i\epsilon} \right] \\ &= \int \frac{d^n k d\omega}{(2\pi)^{n+1}} e^{i(\mathbf{k} \cdot \mathbf{x} - \omega t)} \frac{-2iD\mathbf{k}^2}{(\omega - i\epsilon)^2 + D^2 \mathbf{k}^4}. \end{aligned} \tag{7.153}$$

From eqn. (6.67) we have

$$\langle \rho \rangle = \frac{i}{\hbar} \int (dx')\langle \rho(x)\rho(x')\rangle F(x'), \tag{7.154}$$

where

$$\frac{i}{\hbar}\langle \rho(x)\rho(x')\rangle = \frac{\delta^2 W}{\delta F^2} = -i\mathrm{Im}G_F(x, x'). \tag{7.155}$$

The Feynman Green function may be evaluated using the phase, or weight $\exp(iS/\hbar)$, by analogy with eqn. (5.95):

$$\begin{aligned}
G_F(x, x') &= \int \frac{d^n\mathbf{k}d\omega}{(2\pi)^{n+1}} e^{i(\mathbf{k}\cdot\mathbf{x}-\omega t)} \left[\frac{1}{\omega + iD\mathbf{k}^2 - i\epsilon} - \frac{1}{\omega - iD\mathbf{k}^2 + i\epsilon} \right] \\
&= \int \frac{d^n\mathbf{k}d\omega}{(2\pi)^{n+1}} e^{i(\mathbf{k}\cdot\mathbf{x}-\omega t)} \frac{-2iD\mathbf{k}^2}{\omega^2 + D^2\mathbf{k}^4 - i\epsilon}.
\end{aligned} \tag{7.156}$$

For thermal or other distributions it will be somewhat different. We may now compare this (in momentum space) with the linear response equation:

$$\langle \rho \rangle(k) = \mathrm{Im}G_F(k)F = \frac{2D\mathbf{k}^2}{\omega^2 + D^2\mathbf{k}^4} F. \tag{7.157}$$

Thus, eliminating the source from both sides of this equation, we may define the instantaneous 'D.C.' ($\omega \to 0$) diffusion constant, given by the Kubo-type relation,

$$\langle D(\omega \to 0) \rangle = \lim_{\omega \to 0} \left(\lim_{\mathbf{k}\to 0} \frac{\omega^2}{\mathbf{k}^2} G_F(k) \right). \tag{7.158}$$

If we take G_F from eqn. (7.156), we see the triviality of this relation for purely collisionless quantum fluctuations of the field, $\langle \rho \rangle$. By taking the fluctuation average to be $\exp(iS/\hbar)$, we simply derive a tautology. However, once we switch on thermal fluctuations or quantum interactions (for which we need to know about quantum field theory), the Feynman Green function picks up a temperature dependence and a more complicated analytical structure, and this becomes non-trivial; see eqn. (6.61). Then it becomes possible to express D in terms of independent parameters, rather than as the phenomenological constant in eqn. (7.143).

7.5.3 Forced Brownian motion

A phenomenological description of Brownian motion for particles in a field is given by the Langevin model. Newton's second law for a particle perturbed by random forces may be written in the form

$$m\frac{dv^i}{dt} = F^i - \alpha v^i, \tag{7.159}$$

where v^i is the velocity of a particle in a field and α is a coefficient of friction, by analogy with Stokes' law. This equation clearly expresses only a statistical phenomenology, and it cannot be derived from an action principle, since it contains explicitly velocity-dependent terms, which can only arise from statistical effects in a real dynamical system. The forcing term, F^i, is a random force. By this, we mean that the time average of this force is zero, i.e. it fluctuates in magnitude and direction in such a way that its time average vanishes:

$$\langle F(t) \rangle = \frac{1}{T} \int_{\bar{t}-T/2}^{\bar{t}+T/2} F(t)\,\mathrm{d}t = 0. \tag{7.160}$$

We may solve this equation simply, in the following ways.

Green function approach Consider the general solution of

$$a\frac{\mathrm{d}u}{\mathrm{d}t} + bu = f(t), \tag{7.161}$$

where a and b are positive constants. Using the method of Green functions, we solve this in the usual way. Writing this in operator/source form,

$$\left(a\frac{\mathrm{d}}{\mathrm{d}t} + b\right)u = f(t), \tag{7.162}$$

we have the formal solution in terms of the retarded Green function

$$u(t) = \int \mathrm{d}t'\, G_{\mathrm{r}}(t, t')\, f(t'), \tag{7.163}$$

where

$$\left(a\frac{\mathrm{d}}{\mathrm{d}t} + b\right)G_{\mathrm{r}}(t, t') = \delta(t, t'). \tag{7.164}$$

Taking the Fourier transform, we have

$$G_{\mathrm{r}}(t - t') = \int \frac{\mathrm{d}\omega}{2\pi} \frac{\mathrm{e}^{-\mathrm{i}\omega(t-t')}}{(-\mathrm{i}a\omega + b)}. \tag{7.165}$$

This Green function has a simple pole for $t - t' > 0$ at $\omega = -\mathrm{i}b/a$, and the contour is completed in the lower half-plane for ω, making the semi-circle at infinity vanish. The solution for the field $u(t)$ is thus

$$
\begin{aligned}
u(t) &= \int \mathrm{d}\tau \int \frac{\mathrm{d}\omega}{2\pi} \frac{\mathrm{e}^{-\mathrm{i}\omega(t-\tau)}}{(-\mathrm{i}a\omega + b)} f(\tau) \\
&= \int_{-\infty}^{t} \mathrm{d}\tau \frac{1}{2\pi} - 2\pi\mathrm{i}\left(\frac{1}{-\mathrm{i}a}\mathrm{e}^{\frac{b}{a}(\tau-t)} f(\tau)\right) \\
&= \frac{1}{a}\int_{-\infty}^{t} \mathrm{d}\tau\, f(\tau)\, \mathrm{e}^{\frac{b}{a}(\tau-t)}.
\end{aligned}
\tag{7.166}
$$

The lower limit of the integral is written as minus infinity since we have not specified the time at which the force was switched on, but we could replace this by some finite time in the past by specifying boundary conditions more fully.

Differential equation approach Although the Green function method is straightforward and quite simple, this eqn. (7.161) can also be solved by an alternative method. When $f(t) = 0$ it is solved by separation of variables, giving

$$\frac{du}{dt} = -\frac{b}{a}u$$

$$u(t) = u_0 \, e^{-\frac{b}{a}t}, \tag{7.167}$$

for some constant u_0. This is therefore the complementary function for the differential equation. If the forcing term $f(t)$ is non-zero, this hints that we can make the equation integrable by multiplying through by the integrating factor $\exp(-bt/a)$.

$$\frac{d}{dt}\left(e^{\frac{b}{a}t}u(t)\right) = \frac{1}{a}\left(a\frac{du}{dt} + b\,u(t)\right)e^{\frac{b}{a}t}$$

$$e^{\frac{b}{a}t}u(t) = \frac{1}{a}\int_0^t d\tau \; f(\tau)e^{\frac{b}{a}\tau}$$

$$u(t) = \frac{1}{a}\int_0^t d\tau \; f(\tau)e^{\frac{b}{a}(\tau-t)}. \tag{7.168}$$

This is exactly analogous to making a gauge transformation in electrodynamics. Note that, since the integral limits are from 0 to t, $u(t)$ cannot diverge unless $f(t)$ diverges. The lower limit is by assumption. The general solution to eqn. (7.161) is therefore given by the particular integral in eqn. (7.168) plus an arbitrary constant times the function in eqn. (7.167). The solutions are typically characterized by exponential damping. This reproduces the answer in eqn. (7.166) marginally more quickly than the tried and trusted method of Green functions. This just goes to show that it never does any harm to consider alternative methods, even when in possession of powerful methods of general applicability.

Diffusion and mobility Langevin's equation plays a central role in the kinetic theory of diffusion and conduction. Let $\dot{x}^i = v^i$, then, multiplying through by x, we have

$$mx\frac{d\dot{x}}{dt} = m\left[\frac{d}{dt}(x\dot{x}) - \dot{x}^2\right] = -\alpha x\dot{x} + xF(t). \tag{7.169}$$

Taking the kinetic (ensemble) average of both sides, and recalling that the fluctuating force has zero average, we have that

$$m \left\langle \frac{d}{dt}(x\dot{x}) \right\rangle = m \frac{d}{dt} \langle x\dot{x} \rangle = kT - \alpha \langle x\dot{x} \rangle, \qquad (7.170)$$

where we have used the result from kinetic theory (the equi-partition theorem) that $\frac{1}{2}m\langle \dot{x}^2 \rangle = \frac{1}{2}kT$. We can solve this to give

$$\langle x\dot{x} \rangle = C e^{-\alpha t/m} + \frac{kT}{\alpha}. \qquad (7.171)$$

At large times, the first of these terms decays and the system reaches a steady state. We may integrate this to give

$$\langle x^2 \rangle = \frac{2kT}{\alpha} t. \qquad (7.172)$$

This tells us the mean square position. By comparing this to the diffusion equation in eqn. (7.146) we find the effective diffusion coefficient

$$D = \frac{kT}{\alpha}. \qquad (7.173)$$

A related application is that of electrical conduction. Consider the same diffusion process for charges e in a uniform electric field E. The average of the Langevin equation is now

$$m \frac{d\langle v^i \rangle}{dt} = eE^i - \alpha \langle v^i \rangle, \qquad (7.174)$$

since $\langle F \rangle = 0$. In a steady state, the average acceleration is also zero, even though microscopically there might be collisions which cause fluctuations in the velocity. Thus we have, at steady state,

$$eE^i = \alpha \langle v^i \rangle. \qquad (7.175)$$

We define the *mobility*, μ, of the charges, for an isotropic system, as

$$\mu = \frac{\langle v^i \rangle}{E^i} = \frac{e}{\alpha}. \qquad (7.176)$$

The mobility is related to the diffusion constant by the Einstein relation

$$\frac{\mu}{D} = \frac{e}{kT}. \qquad (7.177)$$

In an anisotropic system, there might be different coefficients for diffusion and mobility in different directions. Then, eqn. (7.176) would become a tensor relation, $\mu_{ij} E v^i / E_j$.

7.6 Vortex fields in 2 + 1 dimensions

Although one generally avoids speaking of particulate matter in field theory, since classically it is used to describe mainly smooth, continuous fields, there are occasions on which the solutions to the equations of motion lead unambiguously to pointlike objects. One such situation is the case of vortices.

Vortices are charged, singular objects which arise in some physical systems such as the non-linear Schrödinger equation. Vortices have the property that they acquire a phase factor, by an Aharonov–Bohm-like effect, when they wind around one another. They can usually be thought of as pointlike objects which are penetrated by an infinitely thin line of magnetic flux. In 2 + 1 dimensions, vortices are also referred to as *anyons*, and have a special relationship with Chern–Simons field theories. It might seem strange that a field variable $\phi(x)$, which covers all of space and time, could be made to represent such singular objects as vortices. As we shall see in the following example, this is made possible precisely by the singular nature of Green functions.

Consider a field, $\phi(x)$, representing pointlike objects in two spatial dimensions with coordinates denoted for simplicity by $r = (x, y)$. We define the winding angle, θ, between any two pointlike objects in the field by

$$\theta(r - r') = \tan^{-1} \frac{\Delta y}{\Delta x} = \tan^{-1} \frac{y - y'}{x - x'}. \tag{7.178}$$

Notice that $\theta(r - r')$ is a function of coordinate differences between pairs of points. We shall, in fact, relate this winding angle to the Green function $g(x, x')$, for the Laplacian in two dimensions, which was calculated in section 5.4.4.

7.6.1 A vortex model

The study of Chern–Simons theories is motivated principally by two observations: namely that important aspects of the quantum Hall effect are described by a Chern–Simons theory, and that a viable theory of high-temperature superconductivity should be characterized by a parity-violating, anti-ferromagnetic state. Symmetry considerations lead to an action which does not possess space-reflection symmetry. The Chern–Simons action fits this prescription. These two physical systems are also believed to be essentially two-dimensional, planar systems.

In its most primitive form, the action for the Chern–Simons model may be written in (2 + 1) dimensional flat spacetime as

$$S = \int dt\, d^2x \left((D^\mu \Phi)^\dagger (D_\mu \Phi) + m^2 \Phi^2 + \frac{\lambda}{6} \Phi^4 + \frac{1}{2} \mu \epsilon^{\mu\nu\lambda} A_\mu \partial_\nu A_\lambda \right). \tag{7.179}$$

The equation of motion is thus

$$\frac{1}{2}\mu\epsilon^{\mu\nu\lambda}F_{\nu\lambda} = J^{\mu}. \tag{7.180}$$

The gauge-invariant current, J^{μ}, is introduced for convenience and represents the interaction with the matter fields arising from the gauge-covariant derivatives in eqn. (7.179). We shall not consider the full dynamics of this theory here; rather, it is interesting to see how the singular vortex phenomenon is reflected in the field variables.

7.6.2 Green functions

The basic Green function we shall use in the description of two-dimensional vortices is the inverse Laplacian which was derived in section 5.4.4, but it is also useful to define and elaborate on some additional symbols which are encountered in the literature. We shall use the symbol r^i as an abbreviation for the coordinate difference $\Delta r^i = \Delta x^i = x^i - x^{i\prime}$, and the symbol Δr for the scalar length of this vector. Some authors define a Green function vector by

$$G^i(r - r') = \epsilon^{ij}\partial_j\, g(r - r')$$

$$= -\frac{1}{2\pi}\epsilon^{ij}\frac{\hat{r}^j}{r - r'}, \tag{7.181}$$

where $\hat{r}$ is a unit vector along $r - r'$. The two-dimensional curl of this function is thus

$$\nabla \times \mathbf{G}(r) = \epsilon^{ij}\partial_i G_j(r - r')$$

$$= \epsilon^{ij}\epsilon_{jk}\partial_i\partial_k g(r - r')$$

$$= -\nabla^2 g(r - r')$$

$$= \delta(r - r'). \tag{7.182}$$

In other words, $G^i(r - r')$ is the inverse of the curl operator.

7.6.3 Relationship between $\theta(r - r')$ and $g(r - r')$

To obtain a relationship between the coordinates and the winding function $\theta(r)$, we note that

$$\partial_i \tan\theta(r - r') = \partial_i\left(\frac{\sin\theta(r - r')}{\cos\theta(r - r')}\right)$$

$$= \partial_i\theta(r - r')\, \sec^2\theta(r - r')$$

$$= \partial_i\theta(r - r')(1 + \tan^2\theta(r - r')). \tag{7.183}$$

From eqn. (7.178), this translates into

$$\partial_i \theta = \frac{\partial_i \left(\frac{\Delta y}{\Delta x} \right)}{1 + \left(\frac{\Delta y}{\Delta x} \right)^2}$$

$$= \frac{\Delta x (\partial_i \Delta y) - \Delta y (\partial_i \Delta x)}{r^2}$$

$$= -\epsilon_{ij} \frac{\hat{r}^j}{r}. \qquad (7.184)$$

This last form is significant since the logarithm has a similar property, namely

$$\epsilon^{ij} \partial_j \ln |r - r'| = \epsilon^{ij} \frac{\hat{r}_j}{r - r'}, \qquad (7.185)$$

and thus we immediately have the relationship:

$$-\frac{1}{2\pi} (\partial_i \theta (r - r')) = G(r) = -\epsilon_{ij} \partial_j g(r - r'). \qquad (7.186)$$

It is understood that partial derivatives acting on $r - r'$ act on the first argument r.

7.6.4 Singular nature of $\theta(r - r')$

The consistency of the above relations supplies us with an unusual, and perhaps somewhat surprising relation, namely

$$\epsilon^{ij} \partial_i \partial_j \theta (r - r') = 2\pi \, \delta(r - r') \qquad (7.187)$$

or

$$[\partial_1, \partial_2] \theta (r - r') = 2\pi \delta(r - r'). \qquad (7.188)$$

This relation tells us that the partial derivatives do not commute when acting on the function $\theta(r)$. This is the manifestation of a logarithmic singularity in the field, or, physically, the non-triviality of the phase accrued by winding vortices around one another. Although the field is formally continuous, it has this non-analytical property at every point.

Using complex coordinates $z = x^1 + ix^2$ and conjugate variables $\bar{z}$, the above discussion leads to the relations in complex form:

$$\partial_z (\bar{z})^{-1} = \partial_z \partial_{\bar{z}} \ln |z|^2$$

$$= \pi \delta(|z|). \qquad (7.189)$$

Part 2
Groups and fields

8

Field transformations

The previous chapters take a pragmatic, almost engineering, approach to the solution of field theories. The recipes of chapter 5 are invaluable in generating solutions to field equations in many systems, but the reason for their effectiveness remains hidden. This chapter embarks upon a train of thought, which lies at the heart of the theory of dynamical systems, which explain the fundamental reasons why field theories look the way they do, how physical quantities are related to the fields in the action, and how one can construct theories which give correct answers regardless of the perspective of the observer. Before addressing these issues directly, it is necessary to understand some core notions about symmetry on a more abstract level.

8.1 Group theory

To pursue a deeper understanding of dynamics, one needs to know the language of transformations: *group theory*. Group theory is about families of transformations with special symmetry. The need to *parametrize* symmetry groups leads to the idea of *algebras*, so it will also be necessary to study these.

Transformations are central to the study of dynamical systems because all changes of variable, coordinates or measuring scales can be thought of as transformations. The way one parametrizes fields and spacetime is a matter of convenience, but one should always be able to transform any results into a new perspective whenever it might be convenient. Even the dynamical development of a system can be thought of as a series of transformations which alter the system's state progressively over time. The purpose of studying groups is to understand the implications posed by constraints on a system: the field equations and any underlying symmetries – but also the rules by which the system unfolds on the background spacetime. In pursuit of this goal, we shall find universal themes which enable us to understand many structures from a few core principles.

169

8.1.1 Definition of a group

A group is a set of objects, usually numbers or matrices, which satisfies the following conditions.

(1) There is a rule of composition for the objects. When two objects in a group are combined using this rule, the resulting object also belongs to the group. Thus, a group is *closed* under the action of the composition rule. If a and b are two matrices, then $a \cdot b \neq b \cdot a$ is not necessarily true. If $a \cdot b = b \cdot a$, the group is said to be Abelian, otherwise it is non-Abelian.

(2) The combination rule is associative, i.e. $(a \cdot b) \cdot c = a \cdot (b \cdot c)$.

(3) The identity element belongs to the set, i.e. an object which satisfies $a \cdot I = a$.

(4) Every element a in the set has a right-inverse a^{-1}, such that $a^{-1} \cdot a = I$.

A group may contain one or more *sub-groups*. These are sub-sets of the whole group which also satisfy all of the group axioms. Sub-groups always overlap with one another because they must all contain the identity element. Every group has two trivial or improper sub-groups, namely the identity element and the whole group itself. The *dimension* of a group d_G is defined to be the number of independent degrees of freedom in the group, or the number of generators required to represent it. This is most easily understood by looking at the examples in the next section. The *order* of a group O_G is the number of distinct elements in the group. In a continuous group the order is always infinite.

If the ordering of elements in the group with respect to the combination rule matters, i.e. the group elements do not commute with one another, the group is said to be non-Abelian. In that case, there always exists an Abelian sub-group which commutes with *every* element of the group, called the *centre*. *Schur's lemma* tells us that any element of a group which commutes with every other must be a multiple of the identity element. The centre of a group is usually a discrete group, Z_N, with a finite number, N, of elements called the *rank* of the group.

8.1.2 Group transformations

In field theory, groups are used to describe the relationships between components in a multi-component field, and also the behaviour of the field under spacetime transformations. One must be careful to distinguish between two vector spaces in the discussions which follow. It is also important to be very clear about what is being transformed in order to avoid confusion over the names.

- *Representation space.* This is the space on which the group transformations act, or the space in which the objects to be transformed live. In field theory, when transformations relate to internal symmetries, the components of field multiplets $(\phi_1, \phi_2, \ldots, \phi_{d_R})$ are the coordinates on representation space. When transformations relate to changes of spacetime frame, then spacetime coordinates are the representation space.

- *Group space.* This is an abstract space of dimension d_G. The dimension of this space is the number of independent transformations which the group is composed of. The coordinates $(\theta_1, \theta_2, \ldots, \theta_{d_G})$ in this space are measured with respect to a set of basis matrices called the *generators* of the group.

Since fields live on spacetime, the full representation space of a field consists of spacetime (μ, ν indices) combined with any hidden degrees of freedom: spin, charge, colour and any other hidden labels or indices (all denoted with indices $A, B, a, b, \alpha, \beta$) which particles might have. In practice, some groups (e.g. the Lorentz group) act only on spacetime, others (e.g. $SU(3)$) act only on hidden indices. In this chapter, we shall consider group theory on a mainly abstract level, so this distinction need not be of concern.

A field, $\phi(x)$, might be a spacetime-scalar (i.e. have no spacetime indices), but also be vector on representation space (have a single group index).

$$\phi(x)_A = \begin{pmatrix} \phi_1(x) \\ \phi_2(x) \\ \vdots \\ \phi_{d_R}(x) \end{pmatrix}. \tag{8.1}$$

The transformation rules for fields with spacetime (coordinate) indices are therefore

$$\phi \rightarrow \phi'$$
$$A_\mu \rightarrow U_\mu{}^\nu A_\nu$$
$$g_{\mu\nu} \rightarrow U_\mu{}^\rho U_\nu{}^\lambda g_{\rho\lambda}, \tag{8.2}$$

and for multiplet transformations they are

$$\phi^A \rightarrow U_{AB} \phi^B$$
$$A_\mu^a \rightarrow U_{ab} A_\mu^b$$
$$g_{\mu\nu}^A \rightarrow U_{AB} g_{\mu\nu}^B. \tag{8.3}$$

All of the above have the generic form of a vector **v** with Euclidean components $v^A = v_A$ transforming by matrix multiplication:

$$\mathbf{v} \rightarrow U\mathbf{v}, \tag{8.4}$$

or

$$v^{A'} = U^A_{\ B}\, v^B. \tag{8.5}$$

The label $A = 1, \ldots, d_R$, where d_R is the dimension of the representation. Thus, the transformation matrix U is a $d_R \times d_R$ matrix and $\mathbf{v}$ is a d_R-component column vector. The group space is Euclidean, so raised and lowered A, B indices are identical here.

Note that multiplet indices (those which do not label spacetime coordinates) for general group representations G_R are labelled with upper case Latin characters $A, B = 1, \ldots, d_R$ throughout this book. Lower case Latin letters $a, b = 1, \ldots, d_G$ are used to distinguish the components of the adjoint representation G_{adj}.

In general, the difference between a representation of a group and the group itself is this: while a group might have certain unique abstract properties which define it, the realization of those properties in terms of numbers, matrices or functions might not be unique, and it is the explicit representation which is important in practical applications. In the case of Lie groups, there is often a variety of possible locally isomorphic groups which satisfy the property (called the Lie *algebra*) that defines the group.

8.1.3 Use of variables which transform like group vectors

The property of transforming a dynamical field by simple matrix multiplication is very desirable in quantum theory where symmetries are involved at all levels. It is a direct representation of the Markov property of physical law. In chapter 14, it becomes clear that invariances are made extremely explicit and are algebraically simplest if transformation laws take the multiplicative form in eqn. (8.5).

An argument against dynamical variables which transform according to group elements is that they cannot be observables, because they are non-unique. Observables can only be described by invariant quantities. A vector is, by definition, not invariant under transformations; however, the scalar product of vectors is invariant.

In classical particle mechanics, the dynamical variables $q(t)$ and $p(t)$ do not transform by simple multiplication of elements of the Galilean symmetry. Instead, there is a set of eqns. (14.34) which describes how the variables change under the influence of group generators. Some would say that such a formulation is most desirable, since the dynamical variables are directly observable, but the price for this is a more complicated set of equations for the symmetries.

As we shall see in chapter 14, the quantum theory is built upon the idea that the dynamical variables should transform like linear combinations of vectors on some group space. Observables are extracted from these vectors with the help of operators, which are designed to pick out actual data as eigenvalues of the operators.

8.2 Cosets and the factor group

8.2.1 Cosets

Most groups can be decomposed into non-overlapping sub-sets called *cosets*. Cosets belong to a given group and one if its sub-groups. Consider then a group G of order O_G, which has a sub-group H of order O_H. A coset is defined by acting with group elements on the elements of the sub-group. In a non-Abelian group one therefore distinguishes between *left* and *right* cosets, depending on whether the group elements pre- or post-multiply the elements of the sub-group. The left coset of a given group element is thus defined by

$$GH \equiv \left\{ GH_1, GH_2, \ldots, GH_{d_H} \right\} \tag{8.6}$$

and the right coset is defined by

$$HG = \left\{ H_1G, H_2G, \ldots, H_{d_H}G \right\}. \tag{8.7}$$

The cosets have order O_H and one may form a coset from every element of G which is not in the sub-group itself (since the coset formed by a member of the coset itself is simply that coset, by virtue of the group axioms). This means that cosets do not overlap.

Since cosets do not overlap, one can deduce that there are $O_G - O_H$ distinct cosets of the sub-group. It is possible to go on forming cosets until all these elements are exhausted. The full group can be written as a sum of a sub-group and all of its cosets.

$$G = H + G_1H + G_2H + \cdots + G_pH, \tag{8.8}$$

where p is some integer. The value of p can be determined by counting the orders of the elements in this equation:

$$O_G = O_H + O_H + O_H + \cdots + O_H = (p+1)O_H. \tag{8.9}$$

Thus,

$$O_G = (p+1)O_H. \tag{8.10}$$

Notice that the number of elements in the sub-group must be a factor of the number of elements in the whole group. This is necessarily true since all cosets are of order O_H.

8.2.2 Conjugacy and invariant sub-groups

If g_1 is an element of a group G, and g_2 is another element, then g_c defined by

$$g_c = g_2 \, g \, g_2^{-1} \tag{8.11}$$

is said to be an element of the group G which is conjugate to g_1. One can form conjugates from every other element in the group. Every element is conjugate to itself since

$$g = I \, g \, I^{-1}. \tag{8.12}$$

Similarly, all elements in an Abelian group are conjugate only to themselves. Conjugacy is a mutual relationship. If g_1 is conjugate to g_2, then g_2 is conjugate to g_1, since

$$\begin{aligned} g_1 &= g \, g_2 \, g^{-1} \\ g_2 &= g^{-1} \, g_1 \, (g^{-1})^{-1}. \end{aligned} \tag{8.13}$$

If g_1 is conjugate to g_2 and g_2 is conjugate to g_3, then g_1 and g_3 are also conjugate. This implies that conjugacy is an equivalence relation.

Conjugate elements of a group are similar in the sense of similarity transformations, e.g. matrices which differ only by a change of basis:

$$A' = \Lambda \, M \, \Lambda^{-1}. \tag{8.14}$$

The *conjugacy class* of a group element g is the set of all elements conjugate to g:

$$\left\{ I \, g \, I^{-1}, \; g_1 \, g \, g_1^{-1}, \; g_2 \, g \, g_2^{-1}, \dots \right\}. \tag{8.15}$$

A sub-group H of G is said to be an *invariant sub-group* if every element of the sub-group is conjugate to another element in the sub-group:

$$H_c = G \, H \, G^{-1} = H. \tag{8.16}$$

This means that the sub-group is invariant with respect to the action of the group, or that the only action of the group is to permute elements of the sub-group. It follows trivially from eqn. (8.16) that

$$GH = HG, \tag{8.17}$$

thus the left and right cosets of an invariant sub-group are identical. This means that all of the elements within H commute with G. H is said to belong to the *centre* of the group.

8.2.3 Schur's lemma and the centre of a group

Schur's lemma states that any group element which commutes with every other element of the group must be a multiple of the identity element. This result proves to be important in several contexts in group theory.

8.2.4 The factor group G/H

The *factor group*, also called the group of cosets is formed from an invariant sub-group H of a group G. Since each coset formed from H is distinct, one can show that the set of cosets of H with G forms a group which is denoted G/H. This follows from the Abelian property of invariant sub-groups. If we combine cosets by the group rule, then

$$H g_1 \cdot H g_2 = H H g_1 g_2 = H(g_1 \cdot g_2,) \tag{8.18}$$

since $H \cdot H = H$. The group axioms are satisfied.

(1) The combination rule is the usual combination rule for the group.

(2) The associative law is valid for coset combination:

$$(H g_1 \cdot H g_2) \cdot H g_3 = H(g_1 \cdot g_2) \cdot H g_3 = H((g_1 \cdot g_2) \cdot g_3). \tag{8.19}$$

(3) The identity of G/H is $H \cdot I$.

(4) The inverse of $H g$ is $H g^{-1}$.

The number of independent elements in this group (the order of the group) is, from eqn. (8.10), $p + 1$ or O_G/O_H. Initially, it might appear confusing from eqn. (8.7) that the number of elements in the sub-group is in fact multiplied by the number of elements in the group, giving a total number of elements in the factor group of $O_G \times O_H$. This is wrong, however, because one must be careful not to count cosets which are *similar* more than once; indeed, this is the point behind the requirement of an invariant sub-group. Cosets which are merely permutations of one another are considered to be equivalent.

8.2.5 Example of a factor group: $SU(2)/Z_2$

Many group algebras generate groups which are the same except for their maximal Abelian sub-group, called the *centre*. This virtual equivalence is determined by factoring out the centre, leaving only the factor group which has a trivial centre (the identity); thus, factor groups are important in issues of spontaneous symmetry breaking in physics, where one is often interested in the precise group symmetry rather than algebras. As an example of a factor group, consider $SU(2)$. The group elements of $SU(2)$ can be parametrized in terms of $d_G = 3$ parameters, as shown in eqn. (8.131). There is a redundancy in these parameters. For example, one can generate the identity element from each of the matrices $g_1(\theta_1)$, $g_2(\theta_2)$, $g_3(\theta_3)$ by choosing θ_A to be zero.

A non-trivial Abelian sub-group in these generators must come from the diagonal matrix $g_3(\theta_3)$. Indeed, one can show quite easily that g_3 commutes with any of the generators for any $\theta_A \neq 0$, if and only if $\exp(i\frac{1}{2}\theta_3) = \exp(-i\frac{1}{2}\theta_3) = \pm 1$. Thus, there are two possible values of θ_3, arising from one of the generators; these lead to an Abelian sub-group, and the group elements they correspond to are:

$$H = \left\{ \begin{pmatrix} 1 & 0 \\ 0 & 1 \end{pmatrix}, \begin{pmatrix} -1 & 0 \\ 0 & -1 \end{pmatrix} \right\}, \tag{8.20}$$

which form a 2×2 representation of the discrete group Z_2. This sub-group is invariant, because it is Abelian, and we may therefore form the right cosets of H for every other element of the group:

$$H \cdot H = \{ \mathbf{1} , -\mathbf{1} \}$$
$$H \cdot g_1(\theta_1) = \{ g_1(\theta_1) , -g_1(\theta_1) \}$$
$$H \cdot g_1(\theta_1') = \{ g_1(\theta_1') , -g_1(\theta_1') \}$$
$$H \cdot g_1(\theta_1'') = \{ g_1(\theta_1'') , -g_1(\theta_1'') \}$$
$$\vdots$$
$$H \cdot g_2(\theta_2) = \{ g_2(\theta_2) , -g_2(\theta_2) \}$$
$$H \cdot g_2(\theta_2') = \{ g_2(\theta_2') , -g_2(\theta_2') \}$$
$$\vdots$$
$$H \cdot g_3(\theta_3) = \{ g_3(\theta_3) , -g_2(\theta_3) \}$$
$$\vdots \tag{8.21}$$

The last line is assumed to exclude the members of g_3, which generate H, and the elements of g_1 and g_2, which give rise to the identity in Z_2, are also excluded from this list. That is because we are listing distinct group elements rather than the combinations, which are produced by a parametrization of the group.

The two columns on the right hand side of this list are two equivalent copies of the factor group $SU(2)/Z_2$. They are simply mirror images of one another which can be transformed into one another by the action of an element of Z_2. Notice that the full group is divided into two invariant pieces, each of which has half the total number of elements from the full group. The fact that these coset groups are possible is connected with *multiple coverings*. In fact, it turns out that this property is responsible for the double-valued nature of electron spin, or, equivalently, the link between the real rotation group $SO(3)$ ($d_G = 3$) and the complexified rotation group, $SU(2)$ ($d_G = 3$).

8.3 Group representations

A *representation* of a group is a mapping between elements of the group and elements of the general linear group of either real matrices, $GL(n, R)$, or complex matrices, $GL(n, C)$. Put another way, it is a correspondence between the abstract group and matrices such that each group element can be represented in matrix form, and the rule of combination is replaced by matrix multiplication.

8.3.1 Definition of a representation G_R

If each element g of a group G can be assigned a non-singular $d_R \times d_R$ matrix $U_R(g)$, such that matrix multiplication preserves the group combination rule $g_{12} = g_1 \cdot g_2$,

$$U_R(g_{12}) = U_R(g_1 \cdot g_2) = U_R(g_1)\, U_R(g_2), \tag{8.22}$$

then the set of matrices is said to provide a d_R dimensional representation of the group G. The representation is denoted collectively G_R and is composed of matrices U_R. In most cases we shall call group representations U to avoid excessive notation.

8.3.2 Infinitesimal group generators

If one imagines a continuous group geometrically, as a vector space in which every point is a new element of the group, then, using a set of basis vectors, it is possible to describe every element in this space in terms of coefficients to these basis vectors. Matrices too can be the basis of a vector space, which is why matrix representations are possible. The basis matrices which span the vector space of a group are called its generators.

If one identifies the identity element of the group with the origin of this geometrical space, the number of linearly independent vectors required to reach every element in a group, starting from the identity, is the dimension of the space, and is also called the dimension of the group d_G. Note that the number of independent generators, d_G, is unrelated to their size d_R as matrices.

Thus, given that every element of the group lies in this vector space, an arbitrary element can be described by a vector whose components (relative to the generator matrices) uniquely identify that element. For example, consider the group $SU(2)$, which has dimension $d_G = 3$. In the fundamental representation, it has three generators (the Pauli matrices) with $d_R = 2$:

$$T_1 = \frac{1}{2}\begin{pmatrix} 0 & 1 \\ 1 & 0 \end{pmatrix}, \quad T_2 = \frac{1}{2}\begin{pmatrix} 0 & -i \\ i & 0 \end{pmatrix}, \quad T_3 = \frac{1}{2}\begin{pmatrix} 1 & 0 \\ 0 & -1 \end{pmatrix}. \tag{8.23}$$

A general point in group space may thus be labelled by a d_G dimensional vector $(\theta_1, \theta_2, \theta_3)$:

$$\Theta = \theta_1\, T_1 + \theta_2\, T_2 + \theta_3\, T_3. \tag{8.24}$$

A general element of the group is then found by exponentiating this generalized generator:

$$U_R = \exp(i\Theta). \tag{8.25}$$

U_R is then a two-dimensional matrix representation of the group formed from two-dimensional generators. Alternatively, one may exponentiate each generator separately, as in eqn. (8.131) and combine them by matrix multiplication to obtain the same result. This follows from the property that multiplication of exponentials leads to the addition of the arguments.

For continuous groups generally, we can formalize this by writing a Taylor expansion of a group element $U(\theta)$ about the identity $I \equiv U(\mathbf{0})$,

$$U(\theta_A) = \sum_{A=1}^{d_G} \theta_A \left(\frac{\partial U}{\partial \theta_A}\right)\bigg|_{\theta_A=0} + \cdots, \tag{8.26}$$

where d_G is the dimension of the group. We can write this

$$U(\theta) = U(0) + \sum_{A=1}^{d_G} \theta_A T_A + \frac{1}{2!}\theta_A\theta_B T_A T_B + \cdots + O(\theta^3)$$

$$= I + \sum_{A=1}^{d_G} \theta_A T_A + \frac{1}{2!}\theta_A\theta_B T_A T_B + \cdots + O(\theta^3), \tag{8.27}$$

where

$$T_A = \left(\frac{\partial U}{\partial \theta_A}\right)\bigg|_{\theta_A=0}. \tag{8.28}$$

T_A is a matrix generator for the group.

8.3.3 Proper group elements

All infinitesimal group elements can be parametrized in terms of linear combinations of generators T_A; thus, it is normal for group transformations to be discussed in terms of infinitesimal transformations. In terms of the geometrical analogy, infinitesimal group elements are those which are very close to the identity. They are defined by taking only terms to first order in θ in the sum in eqn. (8.27). The coefficients θ_A are assumed to be infinitesimally small, so that all higher powers are negligible. This is expressed by writing

$$U(\delta\theta) = U(0) + \delta\theta_A T_A, \tag{8.29}$$

with an implicit summation over A. With infinitesimal transformations, one does not get very far from the origin; however, the rule of group composition may be used to build (almost) arbitrary elements of the group by repeated application of infinitesimal elements. This is analogous to adding up many infinitesimal vectors to arrive at any point in a vector space.

We can check the consistency of repeatedly adding up N group elements by writing $\delta\theta_A = \theta_A/N$, combining $U(\theta) = U(\delta\theta)^N$ and letting $N \to \infty$. In this limit, we recover the exact result:

$$U(\theta) = \lim_{N\to\infty} \left(I + i\frac{\delta\theta_A}{N} T_A \right) = e^{i\theta_A T_A}, \tag{8.30}$$

which is consistent with the series in eqn. (8.27). Notice that the finite group element is the exponential of the infinitesimal combination of the generators. It is often stated that we obtain a group by *exponentiation* of the generators.

It will prove significant to pay attention to another form of this exponentiation in passing. Eqn. (8.30) may also be written

$$U(\theta) = \exp \left(i\int_0^\theta T_A d\theta'_A \right). \tag{8.31}$$

From this we note that

$$\frac{\partial U(\theta)}{\partial \theta_A} = iT_A \, U(\theta), \tag{8.32}$$

and hence

$$\frac{dU}{U} = iT_A d\theta \equiv \Gamma. \tag{8.33}$$

This quantity, which we shall often label Γ in future, is an infinitesimal linear combination of the generators of the group. Because of the exponential form, it can also be written as a differential change in the group element $U(\theta)$ divided by the value of $U(\theta)$ at that point. This quantity has a special significance in geometry and field theory, and turns up repeatedly in the guise of gauge fields and 'connections'.

Not all elements of a group can necessarily be generated by combining infinitesimal elements of the group. In general, it is only a sub-group known as the *proper group* which can be generated in this way. Some transformations, such as *reflections* in the origin or coordinate reversals with respect to a group parameter are, by nature, discrete and discontinuous. A reflection is an all-or-nothing transformation; it cannot be broken down into smaller pieces. Groups which contain these so-called *large* transformations are expressible as a direct product of a connected, continuous group and a discrete group.

8.3.4 Conjugate representations

Given a set of infinitesimal generators, T^A, one can generate infinitely many more by similarity transformations:

$$T^A \rightarrow \Lambda\, T^A\, \Lambda^{-1}. \tag{8.34}$$

This has the effect of generating an equivalent representation. Any two representations which are related by such a similarity transformation are said to be *conjugate* to one another, or to lie in the same *conjugacy class*. Conjugate representations all have the same dimension d_R.

8.3.5 Congruent representations

Representations of different dimension d_R also fall into classes. Generators which exponentiate to a given group may be classified by *congruency class*. All group generators with different d_R exponentiate to groups which are congruent, modulo their centres, i.e. those which are the same up to some multiple covering. Put another way, the groups formed by exponentiation of generators of different d_R are identical only if one factors out their centres.

A given matrix representation of a group is not necessarily a one-to-one mapping from algebra to group, but might cover all of the elements of a group one, twice, or any integer number of times and still satisfy all of the group properties. Such representations are said to be multiple coverings.

A representation U_R and another representation $U_{R'}$ lie in different *congruence classes* if they cover the elements of the group a different number of times. Congruence is a property of discrete tiling systems and is related to the ability to lay one pattern on top of another such that they match. It is the properties of the generators which are responsible for congruence [124].

8.4 Reducible and irreducible representations

There is an infinite number of ways to represent the properties of a given group on a representation space. A representation space is usually based on some physical criteria; for instance, to represent the symmetry of three quarks, one uses a three-dimensional representation of $SU(3)$, although the group itself is eight-dimensional. It is important to realize that, if one chooses a large enough representation space, the space itself might have more symmetry than the group which one is using to describe a particular transformation. Of the infinity of possible representations, some can be broken down into simpler structures which represent truly invariant properties of the *representation space*.

8.4.1 Invariant sub-spaces

Suppose we have a representation of a group in terms of matrices and vectors; take as an example the two-dimensional rotation group $SO(2)$, with the representation

$$U = \begin{pmatrix} \cos\theta & \sin\theta \\ -\sin\theta & \cos\theta \end{pmatrix}, \tag{8.35}$$

so that the rotation of a vector by an angle θ is accomplished by matrix multiplication:

$$\begin{pmatrix} x_1' \\ x_2' \end{pmatrix} = \begin{pmatrix} \cos\theta & \sin\theta \\ -\sin\theta & \cos\theta \end{pmatrix} \begin{pmatrix} x_1 \\ x_3 \end{pmatrix}. \tag{8.36}$$

It is always possible to find higher-dimensional representations of the same group by simply embedding such a group in a larger space. If we add an extra dimension x_3, then the same rotation is accomplished, since x_1 and x_2 are altered in exactly the same way:

$$\begin{pmatrix} x_1' \\ x_2' \\ x_3' \end{pmatrix} = \begin{pmatrix} \cos\theta_3 & \sin\theta_3 & 0 \\ -\sin\theta_3 & \sin\theta_3 & 0 \\ 0 & 0 & 1 \end{pmatrix} \begin{pmatrix} x_1 \\ x_2 \\ x_3 \end{pmatrix}. \tag{8.37}$$

This makes sense: it is easy to make a two-dimensional rotation in a three-dimensional space, and the same generalization carries through for any number of extra dimensions. The matrix representation of the transformation has zeros and a diagonal 1, indicating that nothing at all happens to the x_3 coordinate. It is irrelevant or ignorable:

$$U = \begin{pmatrix} \cos\theta_3 & \sin\theta_3 & 0 \\ -\sin\theta_3 & \sin\theta_3 & 0 \\ 0 & 0 & 1 \end{pmatrix}. \tag{8.38}$$

A six-dimensional representation would look like this:

$$\begin{pmatrix} x_1' \\ x_2' \\ x_3' \\ x_4' \\ x_5' \\ x_6' \end{pmatrix} = \begin{pmatrix} \cos\theta_3 & \sin\theta_3 & 0 & 0 & 0 & 0 \\ -\sin\theta_3 & \sin\theta_3 & 0 & 0 & 0 & 0 \\ 0 & 0 & 1 & 0 & 0 & 0 \\ 0 & 0 & 0 & 1 & 0 & 0 \\ 0 & 0 & 0 & 0 & 1 & 0 \\ 0 & 0 & 0 & 0 & 0 & 1 \end{pmatrix} \begin{pmatrix} x_1 \\ x_2 \\ x_3 \\ x_4 \\ x_5 \\ x_6 \end{pmatrix}. \tag{8.39}$$

The matrix has a block-diagonal form. These higher-dimensional representations are said to be *reducible*, since they contain *invariant sub-spaces*, or coordinates which remain unaltered by the group. In the six-dimensional case above, the 6×6 matrix factorizes into a direct sum of block-diagonal pieces: a 2×2 piece, which is the actual $SO(2)$ part, and a trivial four-dimensional group composed of only the identity I_4. The direct sum is written

$$SO(2)_6 = SO(2)_2 \oplus I_4. \tag{8.40}$$

When a matrix has the form of eqn. (8.39), or is related to such a form by a similarity transformation

$$\Lambda^{-1} \, U \, \Lambda, \tag{8.41}$$

it is said to be a completely reducible representation of the group. In block-diagonal form, each block is said to be an *irreducible representation* of the group. The smallest representation with all of the properties of the group intact is called the *fundamental representation*. A representation composed of $d_G \times d_G$ matrices, where d_G is the dimension of the group, is called the *adjoint representation*. In the case of $SO(3)$, the fundamental and adjoint representations coincide; usually they do not.

Whilst the above observation might seem rather obvious, it is perhaps less obvious if we turn the argument around. Suppose we start with a 6×6 matrix parametrized in terms of some group variables, θ_A, and we want to know which group it is a representation of. The first guess might be that it is an irreducible representation of $O(6)$, but if we can find a linear transformation Λ which changes that matrix into a block-diagonal form with smaller blocks, and zeros off the diagonal, then it becomes clear that it is really a reducible representation, composed of several sub-spaces, each of which is invariant under a smaller group.

8.4.2 Reducibility

The existence of an invariant sub-space S in the representation space R implies that the matrix representation G_R is reducible. Suppose we have a representation space with a sub-space which is unaffected by the action of the group. By choosing coordinates we can write a group transformation g as

$$\begin{pmatrix} X'_R \\ X'_S \end{pmatrix} = \begin{pmatrix} A(g) & B(g) \\ 0 & C(g) \end{pmatrix} \begin{pmatrix} X_R \\ X_S \end{pmatrix}, \tag{8.42}$$

which shows that the coordinates X_S belonging to the sub-space are independent of the remaining coordinates X_R. Thus no matter how X_R are transformed, X_S will be independent of this. The converse is not necessarily true, but often is. Our representation,

$$U_R(g) = \begin{pmatrix} A(g) & B(g) \\ 0 & C(g) \end{pmatrix}, \tag{8.43}$$

satisfies the group composition law; thus,

$$
U_R(g_1) U_R(g_2) = \begin{pmatrix} A(g_1) & B(g_1) \\ 0 & C(g_1) \end{pmatrix} \begin{pmatrix} A(g_2) & B(g_2) \\ 0 & C(g_2) \end{pmatrix}
$$
$$
= \begin{pmatrix} A(g_1) A(g_2) & A(g_1)B(g_2) + B(g_1)C(g_2) \\ 0 & C(g_1)C(g_2) \end{pmatrix} \tag{8.44}
$$

Comparing this with the form which a true group representation would have:

$$
\begin{pmatrix} A(g_1 \cdot g_2) & B(g_1 \cdot g_2) \\ 0 & C(g_1 \cdot g_2) \end{pmatrix}, \tag{8.45}
$$

one sees that A and C also form representations of the group, of smaller size. B does not, however, and its value is constrained by the condition $B(g_1 \cdot g_2) = A(g_1)B(g_2) + B(g_1)C(g_2)$. A representation of this form is said to be *partially reducible*.

If $B = 0$ in the above, then the two sub-spaces decouple: both are invariant under transformations which affect the other. The representation is then said to be *completely reducible* and takes the block-diagonal form mentioned in the previous section.

8.5 Lie groups and Lie algebras

Groups whose elements do not commute are called *non-Abelian*. The commutativity or non-commutativity of the group elements $U(\theta)$ follows from the commutation properties of the generators T_a, as may be seen by writing the exponentiation operation as a power series. In a non-Abelian group the commutation relations between generators may be written in this form:

$$
[T_a, T_b] = C_{ab}. \tag{8.46}
$$

A special class of groups which is interesting in physics is the *Lie groups*, which satisfy the special algebra,

$$
[T_a, T_b] = -\mathrm{i} f_{ab}{}^c T_c. \tag{8.47}
$$

$f_{ab}{}^c$ is a set of *structure constants*, and all the labels a, b, c run over the group indices from $1, \ldots, d_G$. Eqn. (8.47) is called a *Lie algebra*. It implies that the matrices which generate a Lie group are not arbitrary; they are constrained to satisfy the algebra relation. The matrices satisfy the algebraic Jacobi identity

$$
[T^a, [T^b, T^c]] + [T^b, [T^c, T^a]] + [T^c, [T^a, T^b]] = 0. \tag{8.48}
$$

Many of the issues connected to Lie algebras are analogous to those of the groups they generate. We study them precisely because they provide a deeper level of understanding of groups. One also refers to representations, equivalence classes, conjugacy for algebras.

8.5.1 Normalization of the generators

The structure of the $d_R \times d_R$ dimensional matrices and the constants f_{abc} which make up the algebra relation are determined by the algebra relation, but the normalization is not. If we multiply T_R^a and f_{abc} by any constant factor, the algebra relation will still be true. The normalization of the generators is fixed here by relating the trace of a product of generators to the quadratic Casimir invariant:

$$\mathrm{Tr}\left(T_R^a T_R^b\right) = I_2(G_R)\delta^{ab}, \tag{8.49}$$

where I_2 is called the Dynkin index for the representation G_R. The Dynkin index may also be written as

$$I_2(G_R) = \frac{d_R}{d_G}C_2(G_R) \tag{8.50}$$

where d_R is the dimension (number of rows/columns) of the generators in the representation G_R, and d_G is the dimension of the group. $C_2(G_R)$ is the quadratic Casimir invariant for the group in the representation, G_R: $C_2(G_R)$ and $I_2(G_R)$ are constants which are listed in tables for various representations of Lie groups [96]. d_G is the same as the dimension of the adjoint representation of the algebra G_{adj}, by definition of the adjoint representation. Note, therefore, that $I_2(G_{\mathrm{adj}}) = C_2(G_{\mathrm{adj}})$.

The normalization is not completely fixed by these conditions, since one does not know the value of the Casimir invariant a priori. Moreover, Casimir invariants are often defined with inconsistent normalizations, since their main property of interest is their ability to commute with other generators, rather than their absolute magnitude. The above relations make the Casimir invariants consistent with the generator products. To complete the normalization, it is usual to define the length of the longest roots or eigenvalues of the Lie algebra as 2. This fixes the value of the Casimir invariants and thus fixes the remaining values. For most purposes, the normalization is not very important as long as one is consistent, and most answers can simply be expressed in terms of the arbitrary value of $C_2(G_R)$. Thus, during the course of an analysis, one should not be surprised to find generators and Casimir invariants changing in definition and normalization several times. What is important is that, when comparisons are made between similar things, one uses consistent conventions of normalization and definition.

8.5.2 Adjoint transformations and unitarity

A Lie algebra is formed from the d_G matrices T^a which generate a Lie group. These matrices are $d_R \times d_R$ matrices which act on the vector space, which has been denoted *representation space*. In addition, the d_G generators which fulfil the algebra condition form a basis which spans the group space. Since the group is formed from the algebra by exponentiation, both a Lie algebra A and its group G live on the vector space referred to as group space. In the case of the adjoint representation $G_R = G_{\text{adj}}$, the group and representation spaces coincide ($d_G = d_R$, $a, b, c \leftrightarrow A, B, C$). The adjoint representation is a direct one-to-one mapping of the algebra properties into a set of matrices. It is easy to show that the structure constants themselves form a representation of the group which is adjoint. This follows from the Jacobi identity in eqn. (8.48). Applying the algebra relation (8.47) to eqn. (8.48), we have

$$[T^a, -\mathrm{i} f^{bcd} T^d] + [T^b, -\mathrm{i} f^{cad} T^d] + [T^c, -\mathrm{i} f^{abd} T^d] = 0. \qquad (8.51)$$

Using it again results in

$$\left[-f^{bcd} f^{ade} - f^{cad} f^{bde} - f^{abd} f^{cde} \right] T^e = 0. \qquad (8.52)$$

Then, from the coefficient of T^e, making the identification,

$$\left[T^a \right]_{BC} \equiv \mathrm{i} f^a_{BC} \qquad (8.53)$$

it is straightforward to show that one recovers

$$[T^a, T^b] = -\mathrm{i} f^{abd} T^d. \qquad (8.54)$$

Thus, the components of the structure constants are the components of the matrices in the adjoint representation of the algebra. The representation is uniquely identified as the adjoint since all indices on the structure constants run over the dimension of the group $a, b = 1, \ldots, d_G$.

The group space to which we have been alluding is assumed, in field theory, to be a Hilbert space, or a vector space with a positive definite metric. Representation space does not require a positive definite metric, and indeed, in the case of groups like the Lorentz group of spacetime symmetries, the metric in representation space is indefinite. The link between representation space and group space is made by the adjoint representation, and it will prove essential later to understand what this connection is.

Adjoint transformations can be understood in several ways. Suppose we take a group vector v^a which transforms by the rule

$$v'^a = U_{\text{adj}}{}^a{}_b\, v^b, \qquad (8.55)$$

where

$$U_{\text{adj}} = \exp\left(\mathrm{i} \theta^a\, T^a_{\text{adj}} \right). \qquad (8.56)$$

It is also possible to represent the same transformation using a complete set of arbitrary matrices to form a basis for the group space. For the matrices we shall choose the generators T_R, is an arbitrary representation

$$V_R = v^a\, T^a_R. \qquad (8.57)$$

If we assume that the v^a in eqns. (8.55) and (8.57) are the same components, then it follows that the transformation rule for V_R must be written

$$V_R' = v^{a'} T_R^a = U_R^{-1} V_R U_R, \tag{8.58}$$

where

$$U_R = \exp\left(i\theta^a T_R^a\right). \tag{8.59}$$

This now has the appearance of a similarity transformation on the group space. To prove this, we shall begin with the assumption that the field transforms as in eqn. (8.58). Then, using the matrix identity

$$\exp(A)B\exp(-A) = B + [A, B] + \frac{1}{2!}[A, [A, B]] +$$

$$\frac{1}{3!}[A, [A, [A, B]]] + \cdots, \tag{8.60}$$

it is straightforward to show that

$$V_R' = v^a \left\{ \delta^a_{\ r} - \theta_b f^{ab}_{\ \ r} + \frac{1}{2}\theta_b\theta_c f^{ca}_{\ \ s} f^{bs}_{\ \ r} + \right.$$

$$\left. -\frac{1}{3!}\theta_b\theta_c\theta_d f^{da}_{\ \ q} f^{cq}_{\ \ p} f^{bp}_{\ \ r} + \cdots \right\} T_R^r, \tag{8.61}$$

where the algebra commutation relation has been used. In our notation, the generators of the adjoint representation may written

$$(T_{\text{adj}}^a)^b_{\ c} = i f^{ab}_{\ \ c}, \tag{8.62}$$

and the structure constants are real. Eqn. (8.61) may therefore be identified as

$$V_R' = v^a (U_{\text{adj}})^a_{\ b} T_R^b, \tag{8.63}$$

where

$$U_{\text{adj}} = \exp(i\theta^a T_{\text{adj}}^a). \tag{8.64}$$

If we now define the components of the transformed field by

$$V_R' = v'^a T_R^a, \tag{8.65}$$

in terms of the original generators, then it follows that

$$v'^a = (U_{\text{adj}})^a_{\ b} v^b. \tag{8.66}$$

We can now think of the set of components, v^a and v'^a, as being grouped into d_G-component column *vectors* $\mathbf{v}$ and $\mathbf{v}'$, so that

$$\mathbf{v}' = U_{\text{adj}}\mathbf{v}. \tag{8.67}$$

Thus, we see that the components of a group vector, v^a, always transform according to the adjoint representation, regardless of what type of basis we use to represent them. To understand the significance of this transformation rule, we should compare it with the corresponding tensor transformation rule in representation space. If we use the matrix

$$U_R = [U_R]^A_B \tag{8.68}$$

where $A, B = 1, \ldots, d_R$, as a transformation of some representation space vector ϕ^A or tensor $[V_R]^A_B$, then, by considering the invariant product

$$\phi^\dagger V_R \phi \rightarrow (U\phi)^\dagger U V_R U^{-1} (U\phi), \tag{8.69}$$

we find that the transformation rule is the usual one for tensors:

$$\phi^A = U^A_B \phi^B \tag{8.70a}$$

$$V_{AB} = U^A_C U^B_D V_{CD}. \tag{8.70b}$$

The transformation rule (8.58) agrees with the rule in eqn. (8.70b) provided

$$U^\dagger = U^{-1}. \tag{8.71}$$

This is the unitary property, and it is secured in field theory also by the use of a Hilbert space as the group manifold. Thus, the form of the adjoint transformation represents unitarity in the field theory, regardless of the fact that the indices A, B might have an indefinite metric.

The object V_R, which transforms like $U^{-1} V U$, signifies a change in the disposition of the system. This form is very commonly seen; for example, in dynamical changes:

$$\partial_\mu \phi \rightarrow \partial_\mu (U\phi) = (\partial_\mu U)\phi + U(\partial_\mu \phi)$$
$$= U(\partial_\mu + \Gamma_\mu)\phi \tag{8.72}$$

where

$$\Gamma_\mu = U^{-1}\partial_\mu U. \tag{8.73}$$

This object is usually called a 'connection', but, in this context, it can be viewed as an expression of a change in the dynamical configuration, of the internal constraints on a system. In the following two chapters, we shall see examples of these transformations, when looking at the Lorentz group and gauge symmetries in particular.

8.5.3 Casimir invariants

From the Lie algebra relation in eqn. (8.47), it is straightforward to show that the quadratic sum of the generators commutes with each individual generator:

$$
\begin{aligned}
[T^a, T^b T^b] &= T^a\, T^b T^b - T^b T^b\, T^a \\
&= T^a\, T^b T^b - T^b\, (T^a T^b + \mathrm{i} f^{abc} T^c) \\
&= [T^a, T^b]\, T^b - \mathrm{i} f^{abc}\, T^b T^c \\
&= -\mathrm{i}\, f^{abc}\, [T^c T^b + T^b T^c] \\
&= 0.
\end{aligned}
\tag{8.74}
$$

The last line follows since the bracket is a symmetric matrix, whereas the structure constants are anti-symmetric. In fact, the quadratic sum of the generators is proportional to the identity matrix. This follows also from Schur's lemma:

$$
T^a T^a = \frac{1}{d_G}\, C_2(G_R)\, \mathbf{I}_R,
\tag{8.75}
$$

or

$$
f^a_{\ bc}\, f^{dbc} = -\frac{1}{d_G}\, C_2(G_{\mathrm{adj}}) \delta^{ad}.
\tag{8.76}
$$

8.5.4 Sub-algebra

Just as groups have sub-groups, algebras have sub-algebras. A sub-set, H, of an algebra, A, is called a linear sub-algebra of A if H is a linear sub-space of the group space and is closed with respect to the algebra relation. i.e. for any matrix elements of the sub-algebra h_1, h_2 and h_3, one has

$$
[t_1, t_2] = -\mathrm{i} f_{12}^{\ \ 3} t_3.
\tag{8.77}
$$

This is a non-Abelian sub-algebra. Sub-algebras can also be Abelian:

$$
[h_1, h_2] = 0.
\tag{8.78}
$$

8.5.5 The Cartan sub-algebra

The Cartan sub-algebra is an invariant sub-algebra whose elements generate the *centre* of a Lie group when exponentiated. This sub-algebra has a number of extremely important properties because many properties of the group can be deduced directly from the sub-set of generators which lies in the Cartan sub-algebra.

The generators of the Cartan sub-algebra commute with one another but not necessarily with other generators of the group. Since the Cartan sub-algebra generates the centre of the group (the maximal Abelian sub-group) under exponentiation, Schur's lemma tells us that the group elements found from these are diagonal and proportional to the identity matrix. The Cartan sub-algebra is the sub-set the group generators T^a which are simultaneously diagonalizable in a suitable basis. In other words, if there is a basis in which one of the generators, T^a, is diagonal, then, in general, several of the generators will be diagonal in the same basis. One can begin with a set of generators, T_R^a, in a representation, G_R, and attempt to diagonalize one of them using a similarity transformation:

$$T^{a'} \rightarrow \Lambda T_R^a \Lambda^{-1}. \tag{8.79}$$

The same transformation, Λ, will transform a fixed number of the matrices into diagonal form. This number is always the same, and it is called the *rank* of the group or rank(G). The diagonalizable generators are denoted H^i, where $i = 1, \ldots, \text{rank}(G)$. These form the Cartan sub-algebra. Note that, in the case of the fundamental representation of $SU(2)$, the third Pauli matrix is already diagonal. This matrix is the generator of the Cartan sub-algebra for $SU(2)$ in the $d_R = 2$ representation. Since only one of the generators is diagonal, one concludes that the rank of $SU(2)$ is 1.

8.5.6 Example of diagonalization

The simplest example of a Cartan sub-algebra may be found in the generators of the group $SO(3)$ in the fundamental representation, or identically of $SU(2)$ in the adjoint representation. These matrices are well known as the generators of rotations in three dimensions, and are written:

$$T^1 = \begin{pmatrix} 0 & 0 & 0 \\ 0 & 0 & -i \\ 0 & i & 0 \end{pmatrix}$$

$$T^2 = \begin{pmatrix} 0 & 0 & i \\ 0 & 0 & 0 \\ -i & 0 & 0 \end{pmatrix}$$

$$T^3 = \begin{pmatrix} 0 & -i & 0 \\ i & 0 & 0 \\ 0 & 0 & 0 \end{pmatrix}. \tag{8.80}$$

To find a basis which diagonalizes one of these generators, we pick T^1 to diagonalize, arbitrarily. The self-inverse matrix of eigenvectors for T^1 is easily found. It is given by

$$\Lambda = \begin{pmatrix} -1 & 0 & 0 \\ 0 & \frac{1}{\sqrt{2}} & \frac{-i}{\sqrt{2}} \\ 0 & \frac{i}{\sqrt{2}} & \frac{-1}{\sqrt{2}} \end{pmatrix}. \tag{8.81}$$

Constructing the matrices $\Lambda^{-1} T^a \Lambda$, one finds a new set of generators,

$$T^1 = \begin{pmatrix} 0 & 0 & 0 \\ 0 & 1 & 0 \\ 0 & 0 & -1 \end{pmatrix}$$

$$T^2 = \frac{1}{\sqrt{2}} \begin{pmatrix} 0 & 1 & i \\ 1 & 0 & 0 \\ -i & 0 & 0 \end{pmatrix}$$

$$T^3 = \frac{1}{\sqrt{2}} \begin{pmatrix} 0 & i & 1 \\ -i & 0 & 0 \\ 1 & 0 & 0 \end{pmatrix}. \tag{8.82}$$

Since only one of these is diagonal, rank rank $SU(2) = 1$. Equally, we could have chosen to diagonalize a different generator. This would then have had the same eigenvalues, and it would have been the generator of the Cartan sub-algebra in an alternative basis. None of the generators are specially singled out to generate the sub-algebra. The diagonalizability is an intrinsic property of the algebra.

8.5.7 Roots and weights

The *roots* and *weights* of algebra representations are proportional to eigenvalues of the Cartan sub-algebra generators for different d_R. The roots are denoted α^A and the weights are denoted λ^A. Because the algebra relation ensures exactly d_G independent vectors on the group space, there are d_G independent eigenvalues to be found from the generators.[1] We shall explore the significance of these eigenvalues in the next section.

---

[1] This might seem confusing. If one has rank(G) simultaneously diagonalizable $d_R \times d_R$ matrices, then it seems as though there should be $d_R \times $ rank(G) eigenvalues to discern. The reason why this is not the case is that not all of the generators are independent. They are constrained by the algebra relation. The generators are *linearly independent* but constrained through the quadratic commutator condition

For generators of the Cartan sub-algebra, H^i_R, in a representation G_R, the *weights* are eigenvalues:

$$H^i_R = \begin{bmatrix} \lambda^i_1 & & & \\ & \lambda^i_2 & & \\ & & \lambda^i_3 & \\ & & & \ddots \end{bmatrix}. \tag{8.83}$$

The name *root* is reserved for an eigenvalue of the adjoint representation:

$$H^i_{\text{adj}} = \begin{bmatrix} \alpha^i_1 & & & \\ & \alpha^i_2 & & \\ & & \alpha^i_3 & \\ & & & \ddots \end{bmatrix}. \tag{8.84}$$

The significance of the adjoint representation is that it is a direct one-to-one mapping of intrinsic algebra properties. The roots have a special significance too: the algebra can be defined purely in terms of its roots. The diagonal basis we have referred to above is a step towards showing this, but to see the true significance of the root and weights of an algebra, we need to perform another linear transformation and construct the Cartan–Weyl basis.

8.5.8 The Cartan–Weyl basis

The Cartan–Weyl basis is one of several bases in which the generators of the Cartan sub-algebra are diagonal matrices. To construct this basis we can begin from the diagonal basis, found in the previous section, and form linear combinations of the remaining non-diagonal generators. The motivation for this requires a brief theoretical diversion.

Suppose that Θ and Φ are arbitrary linear combinations of the generators of a Lie algebra. This would be the case if Θ and Φ were non-Abelian gauge fields, for instance

$$\Theta = \theta_a T^a$$
$$\Phi = \phi_a T^a, \tag{8.85}$$

where $a = 1, \ldots, d_G$. Then, consider the commutator eigenvalue equation

$$[\Theta, \Phi] = \alpha \Phi, \tag{8.86}$$

where α is an eigenvalue for the 'eigenvector' Φ. If we write this in component form, using the algebra relation in eqn. (8.47), we have

$$\theta^a \phi^b f_{abc} T^c = \alpha \, \phi_l T^l. \tag{8.87}$$

Now, since the T^a are *linearly* independent we can compare the coefficients of the generators on the left and right hand sides:

$$(\phi^a f_{ab}{}^c - \alpha \delta_b{}^c)\phi^b = 0. \tag{8.88}$$

This equation has non-trivial solutions if the determinant of the bracket vanishes, and thus we require

$$\det |\phi^a f_{ab}{}^c - \alpha \delta_b{}^c| = 0. \tag{8.89}$$

For a d_G dimensional Lie algebra this equation cannot have more than d_G independent roots, α. Cartan showed that if one chooses Θ so that the secular equation has the maximum number of different eigenvalues or roots, then only zero roots $\alpha = 0$ can be degenerate (repeated). If $\alpha = 0$ is r-fold degenerate, then r is the rank of the semi-simple Lie algebra.

The generators associated with zero eigenvalues are denoted H^i, where $i = 1, \ldots, \text{rank}(G)$ and they satisfy

$$[\theta^j H^j, H^i] = 0, \tag{8.90}$$

i.e. they commute with one another. The remaining generators, which they do not commute with are written E_α, for some non-zero α, and they clearly satisfy

$$[\theta^j H^j, E_\alpha] = \alpha E_\alpha. \tag{8.91}$$

We can think of the roots or eigenvalues as vectors living on the invariant sub-space spanned by the generators H^i. The components can be found by allowing the H^i to act on the E_α. Consider

$$[\theta^j H^j, [H_i, E_\alpha]] = [\theta^j H^j, H_i E_\alpha] - [\theta^j H^j, E_\alpha H_i]$$
$$= \alpha[H^i, E_\alpha]. \tag{8.92}$$

This result can be interpreted as follows. If E_α is an 'eigenvector' associated with the eigenvalue α, then there are $\text{rank}(G)$ eigenvectors $[H^i, E_\alpha]$ belonging to the same eigenvalue. The eigenvectors must therefore each be proportional to E_α:

$$[H^i, E_\alpha] = \alpha^i E_\alpha, \tag{8.93}$$

and the components of the vector are defined by

$$\alpha = \alpha^i \theta^i. \tag{8.94}$$

This relation defines the components of a *root vector* on the invariant Cartan sub-space. Comparing eqn. (8.93) with the algebra relation in eqn. (8.47),

$$f_{ia}{}^b = \alpha_i \delta_a{}^b. \tag{8.95}$$

Finally, by looking at the Jacobi identity,

$$[\Theta, [E_\alpha, E_\beta]] + [E_\alpha, [E_\beta, \Theta]] + [E_\beta, [\Theta, E_\alpha]] = 0, \qquad (8.96)$$

we find that

$$[\Theta, [E_\alpha, E_\beta]] = (\alpha + \beta)[E_\alpha, E_\beta]. \qquad (8.97)$$

This means that $[E_\alpha, E_\beta]$ is the eigenvector associated with the root $(\alpha + \beta)$, provided that $\alpha + \beta \neq 0$. If $\alpha + \beta = 0$ then, since the zero eigenvalues are associated with H^i, we must have

$$[E_\alpha, E_{-\alpha}] = f_{\alpha,-\alpha}{}^i H_i$$
$$= \alpha_i H^i. \qquad (8.98)$$

This shows how the E_α act as stepping operators, adding together solutions to the eigenvalue equation. It also implies that if there is a zero root, then there must be *pairs* of roots $\alpha, -\alpha$. In summary,

---

$$[H^i, E_\alpha] = \alpha^i E_\alpha$$

$$[E_\alpha, E_{-\alpha}] = \alpha^i H_i$$

What is the physical meaning of the root vectors? The eigenvalue equation is an equation which tells us how many ways one generator of transformations maps to itself, up to a scalar multiple under the action of the group. The H are invariant sub-spaces of a symmetry group because they only change the magnitude of a symmetry state, not its character. In other words, the Cartan sub-algebra represents the number of simultaneous labels which can be measured or associated with a symmetry constraint. Labels represent physical properties like spin, momentum, energy, etc. The stepping operators for a given representation of the group determine how many independent values of those labels can exist based on symmetry constraints. This is the number of weights in a stepping chain. In the case of rotations, the root/weight eigenvalues represent the spin characteristics of particles. A system with one pair of weights (one property: rotation about a fixed axis) in a $d_R = 2$ representation can only be in a spin up or spin down state because there are only two elements in the stepping chain. A $d_R = 3$ representation has three elements, so the particle can have spin up down or zero etc.

The Chevalley normalization of generators is generally chosen so as to make the magnitude of the longest root vectors equal to $(\alpha, \alpha) = \sqrt{\alpha^a \alpha^a} = 2$.

8.5.9 Group vectors and Dirac notation

In quantum mechanics, Dirac introduced a notation for the eigenvectors of an operator using *bra* and *ket* notation. Dirac's notation was meant to emphasize the role of eigenvectors as projection operators which span a vector space. Dirac's notation is convenient since it is fairly intuitive and is widely used in the physics literature. An eigenvector is characterized by a number of labels, i.e. the eigenvalues of the various operators which share it as an eigenvector. If we label these eigenvalues $\alpha, \beta, \ldots$ and so on, then we may designate the eigenvectors using a field or eigenfunction

$$\psi_{\alpha_i, \beta_j, \ldots} \tag{8.99}$$

or in Dirac notation as a *ket*:

$$|\alpha_i, \beta_j, \ldots\rangle. \tag{8.100}$$

Notice that, in Dirac's notation, the redundant symbol ψ is removed, which helps to focus one's attention on the relevant labels: the eigenvalues themselves. The operators which have these eigenfunctions as simultaneous eigenvectors then produce:

$$A_i \, \psi_{\alpha_i, \beta_j, \ldots} = \alpha_i \, \psi_{\alpha_i, \beta_j, \ldots}$$
$$B_j \, \psi_{\alpha_i, \beta_j, \ldots} = \beta_j \, \psi_{\alpha_i, \beta_j, \ldots} \qquad (i, j \text{ not summed}), \tag{8.101}$$

or, equivalently,

$$A_i \, |\alpha_i, \beta_j, \ldots\rangle = \alpha_i \, |\alpha_i, \beta_j, \ldots\rangle$$
$$B_j \, |\alpha_i, \beta_j, \ldots\rangle = \beta_j \, |\alpha_i, \beta_j, \ldots\rangle \qquad (i, j \text{ not summed}). \tag{8.102}$$

In most physical problems we are interested in group spaces with a positive definite metric, i.e. Hilbert spaces. In that case, the dual vectors are written as a Hermitian conjugate:

$$\psi^{\dagger}_{\alpha_i, \beta_j, \ldots} \tag{8.103}$$

or in Dirac notation as a *bra*:

$$\langle \alpha, \beta, \ldots |. \tag{8.104}$$

The length of a vector is then given by the inner product

$$\langle \alpha_i, \beta_j | \alpha_k, \beta_l \rangle = \psi^{\dagger}_{\alpha_i, \beta_j} \psi_{\alpha_k, \beta_l} = \delta_{ik} \delta_{jl} \times \text{length}. \tag{8.105}$$

The eigenvectors with different eigenvalues are orthogonal and usually normalized to unit length.

The existence of simultaneous eigenvalues depends on the existence of commuting operators. Operators which do not commute, such as x^i, p^j and group generators, T^a, T^b, can be assigned eigenvectors, but they are not all linearly independent; they have a projection which is a particular group element:

$$\langle x | p \rangle = e^{i\, p\, x / \hbar}. \tag{8.106}$$

8.5.10 Example: rotational eigenvalues in three dimensions

In this section, we take a first look at the rotation problem. We shall return to this problem in chapter 11 in connection with angular momentum and spin. The generators of three-dimensional rotations are those of $SO(3)$, or equivalently $su(2)$ in the adjoint representation. The generators are already listed in eqns. (8.80). We define

$$T^2 = T^a T^a$$
$$E_\pm = T_2 \mp iT_3$$
$$H = T_1. \tag{8.107}$$

In this new basis, the generators satisfy the relation

$$[H, E_\pm] = \pm E_\pm. \tag{8.108}$$

The stepping operators are Hermitian conjugates:

$$E_+^\dagger = E_-. \tag{8.109}$$

The generator H labels a *central* generator, or invariant sub-space, and corresponds to the fact that we are considering a special axis of rotation. The eigenvalues of the central generator H are called its *weights* and are labelled Λ_c

$$H |\Lambda_c\rangle = \Lambda_c |\Lambda_c\rangle. \tag{8.110}$$

$|\Lambda_c\rangle$ is an eigenvector of H with eigenvalue Λ_c. The value of the quadratic form, T^2, is also interesting because it commutes with H and therefore has its own eigenvalue when acting on H's eigenfunctions, which is independent of c. It can be evaluated by expressing T^2 in terms of the generators in the new basis:

$$E_+E_- = T_2^2 + T_3^2 - i[T_2, T_3]$$
$$E_-E_+ = T_2^2 + T_3^2 + i[T_2, T_3], \tag{8.111}$$

so that, rearranging and using the algebra relation,

$$T^2 = E_-E_+ + T_1^2 - i[T_2, T_3]$$
$$= E_-E_+ + T_1^2 - i(-iT_1)$$
$$= E_-E_+ + H(H + 1), \tag{8.112}$$

where we have identified $T_1 = H$ in the last line. By the analogous procedure with $\pm$ labels reversed, we also find

$$T^2 = E_+E_- + H(H - 1). \tag{8.113}$$

These forms allow us to evaluate the eigenvalues of T^2 for two of the eigen-functions in the full series. To understand this, we note that the effect of the $E_\pm$ generators is to generate new solutions step-wise, i.e. starting with an arbitrary eigenfunction $|\Lambda_c\rangle$ they generate new eigenfunctions with new eigenvalues. This is easily confirmed from the commutation relation in eqn. (8.108), if we consider the 'new' eigenvector $E_\pm|\Lambda_c\rangle$ from $|\Lambda_c\rangle$ and try to calculate the corresponding eigenvalue:

$$
\begin{aligned}
H\, E_\pm\, |\Lambda_c\rangle &= (E_\pm H + [H, T_\pm])\,|\Lambda_c\rangle \\
&= (E_\pm H \pm E_\pm)\,|\Lambda_c\rangle \\
&= (\Lambda_c \pm 1)\, E_\pm\,|\Lambda_c\rangle.
\end{aligned}
\tag{8.114}
$$

We see that, given any initial eigenfunction of H, the action of $E_\pm$ is to produce a new eigenfunction with a new eigenvalue, which differs by ± 1 from the original, up to a possible normalization constant which would cancel out of this expression:

$$
E_\pm|\Lambda_c\rangle \propto |\Lambda_c \pm 1\rangle.
\tag{8.115}
$$

Now, the number of solutions cannot be infinite because the Schwarz (triangle) inequality tells us that the eigenvalue of T^2 (whose value is not fixed by the eigenvalue of H, since T^2 and T^a commute) must be bigger than any of the individual eigenvalues T^a:

$$
\langle\Lambda_c|E_+E_- + E_-E_+ + H^2|\Lambda_c\rangle > \langle\Lambda_c|H^2|\Lambda_c\rangle,
\tag{8.116}
$$

so the value of H acting on $|\Lambda_c\rangle$ must approach a maximum as it approaches the value of T^2 acting on $|\Lambda_c\rangle$. Physically, the maximum value occurs when all of the rotation is about the $a = 1$ axis corresponding to our chosen Cartan sub-algebra generator, $T_1 = H$.

In other words, there is a highest value, Λ_{max}, and a lowest eigenvalue, Λ_{min}. Now eqns. (8.112) and (8.113) are written in such a way that the first terms contain $E_\pm$, ready to act on any eigenfunction, so, since there is a highest and lowest eigenvalue, we must have

$$
\begin{aligned}
E_+\,|\Lambda_{max}\rangle &= 0 \\
E_-\,|\Lambda_{min}\rangle &= 0.
\end{aligned}
\tag{8.117}
$$

Thus,

$$
T^2|\Lambda_{max}\rangle = \Lambda_{max}(\Lambda_{max} + 1)\,|\Lambda_{max}\rangle,
\tag{8.118}
$$

and

$$
T^2|\Lambda_{min}\rangle = \Lambda_{min}(\Lambda_{min} - 1)\,|\Lambda_{min}\rangle.
\tag{8.119}
$$

From these two points of reference, we deduce that

$$\Lambda_{max}(\Lambda_{max} + 1) = \Lambda_{min}(\Lambda_{min} - 1). \qquad (8.120)$$

This equation has two solutions, $\Lambda_{min} = \Lambda_{max} + 1$ (which cannot exist, since there is no solution higher than Λ_{max} by assumption), and

$$\Lambda_{max} = -\Lambda_{min}, \qquad (8.121)$$

thus

$$T^2 = \Lambda_{max}(\Lambda_{max} + 1)\,\mathbf{I}. \qquad (8.122)$$

The result means that the value T^2 is fixed by the maximum value which H can acquire. Strangely, the value is not Λ_{max}^2 (all rotation about the 1 axis), which one would expect from the behaviour of the rotation *group*. This has important implications for quantum mechanics, since it is the algebra which is important for angular momentum or spin. It means that the total angular momentum can never be all in one fixed direction. As $\Lambda_{max} \to \infty$ the difference becomes negligible.

The constant of proportionality in eqn. (8.115) can now be determined from the Hermitian property of the stepping operators as follows. The squared norm of $E_+|\Lambda_c\rangle$ may be written using eqn. (8.112)

$$\begin{aligned}
|E_+|\Lambda_c\rangle|^2 &= \langle\Lambda_c|E_-E_+|\Lambda_c\rangle \\
&= \langle\Lambda_c|T^2 - H(H + 1)|\Lambda_c\rangle \\
&= \Lambda_{max}(\Lambda_{max} + 1) - \Lambda_c(\Lambda_c + 1) \\
&= (\Lambda_{max} - \Lambda_c)(\Lambda_{max} + \Lambda_c + 1). \qquad (8.123)
\end{aligned}$$

Thus,

$$\begin{aligned}
E_+|\Lambda_c\rangle &= \sqrt{(\Lambda_{max} - \Lambda_c)(\Lambda_{max} + \Lambda_c + 1)}\,|\Lambda_c + 1\rangle \\
E_-|\Lambda_c\rangle &= \sqrt{(\Lambda_{max} + \Lambda_c)(\Lambda_{max} - \Lambda_c + 1)}\,|\Lambda_c - 1\rangle. \qquad (8.124)
\end{aligned}$$

Eqn. (8.121), taken together with eqn. (8.114), implies that the eigenvalues are distributed symmetrically about $\Lambda_c = 0$ and that they are separated by integer steps. This means that the possible values are restricted to

$$\Lambda_c = 0, \pm\frac{1}{2}, \pm1, \pm\frac{3}{2}, \pm2, \ldots, \pm\Lambda_{max}. \qquad (8.125)$$

There are clearly $2\Lambda_{max} + 1$ possible solutions. In the study of angular momentum, Λ_{max}, is called the *spin* up to dimensional factors ($\hbar$). In group theory, this is referred to as the highest weight of the representation. Clearly, this single value characterizes a key property of the representation.

What the above argument does not tell us is the value of $\Lambda_{\max}$. That is determined by the dimension of the irreducible representation which gives rise to rotations. In field theory the value of $\Lambda_{\max}$ depends, in practice, on the number of spacetime indices on field variables. Since the matrices for rotation in three spatial dimensions are fixed by the spacetime dimension itself, the only freedom left in transformation properties under rotations is the number of spacetime indices which can be operated on by a rotational transformation matrix. A scalar (no indices) requires no rotations matrix, a vector (one index) requires one, a rank 2-tensor requires two and so on. The number of independently transforming components in the field becomes essentially blocks of $2\Lambda_{\max} + 1$ and defines the spin of the fields.

8.6 Examples of discrete and continuous groups

Some groups are important because they arise in field theory with predictable regularity; others are important because they demonstrate key principles with a special clarity.

8.6.1 $GL(N, C)$: the general linear group

The group of all complex $N \times N$, non-singular matrices forms a group. This group has many sub-groups which are important in physics. Almost all physical models can be expressed in terms of variables which transform as sub-groups of this group.

(1) Matrix multiplication combines non-singular matrices into new non-singular matrices.

(2) Matrix multiplication is associative.

(3) The identity is the unit matrix

$$I = \begin{pmatrix} 1 & 0 & \dots & 0 \\ 0 & 1 & \dots & 0 \\ 0 & \vdots & 1 & 0 \\ 0 & \dots & 0 & 1 \end{pmatrix}. \tag{8.126}$$

(4) Every non-singular matrix has an inverse, by definition.

The representation space of a collection of matrices is the vector space on which the components of those matrices is defined. Since matrices normally multiply vectors, mapping one vector, v^A, onto another vector, v'^A,

$$v_A \rightarrow v'^A = U_{AB} v^B, \tag{8.127}$$

it is normal to think of these matrices as acting on group vectors. In field theory, these transformations are especially important since the group vectors are multiplets of fields, e.g.

$$\phi(x)_A = \begin{pmatrix} \phi_1(x) \\ \phi_2(x) \\ \vdots \\ \phi_{d_R}(x) \end{pmatrix}, \tag{8.128}$$

where d_R is the dimension of the representation, or the size of the $d_R \times d_R$ matrices. Note: the dimension of a representation (the number of components in a multiplet) is not necessarily the same as the dimension of the group itself. For example: a three-dimensional vector ($d_R = 3$) might be constrained, by some additional considerations, to have only an axial symmetry (group dimension $d_G = 1$, a single angle of rotation); in that case one requires a 3×3 representation of a one-dimensional group, since vectors in three dimensions have three components.

8.6.2 U(N): unitary matrices

$U(N)$ is the set of all unitary matrices of matrix dimension N. An $N \times N$ unitary matrix satisfies

$$U^\dagger U = (U^{\mathrm{T}})^* U = I, \tag{8.129}$$

where I is the $N \times N$ unit matrix, i.e. $U^\dagger = U^{-1}$. When $n = 1$, the matrices are single-component numbers. An $N \times N$ matrix contains N^2 components; however, since the transpose matrix is related to the untransposed matrix by eqn. (8.129), only half of the off-diagonal elements are independent of one another. Moreover, the diagonal elements must be real in order to satisfy the condition. This means that the number of independent *real* elements in a unitary matrix is $(N^2 - N)/2$ complex plus N real means N^2 real numbers. This is called the dimension of the group. $U(N)$ is non-Abelian for $U > 1$.

8.6.3 SU(N): the special unitary group

The special unitary group is the sub-group of $U(N)$ which consists of all unitary matrices with unit determinant. Since the requirement of unit determinant is an extra constraint on the all of the independent elements of the group (i.e. the product of the eigenvalues), this reduces the number of independent elements by one compared with $U(N)$. Thus the dimension of $SU(N)$ is $N^2 - 1$ real components. $SU(N)$ is non-Abelian for $N > 1$. $SU(N)$ has several simple properties:

$$C_2(G_{\text{adj}}) = N$$
$$d_G = N^2 - 1$$
$$d_F = N$$
$$C_2(G_f) = \frac{N^2 - 1}{2N}, \tag{8.130}$$

where $C_2(G)$ is the quadratic Casimir invariant in representation G, d_G is the dimension of the group, and d_F is the dimension of the fundamental representation $R \to F$.

8.6.4 SU(2)

The set of 2×2 unitary matrices with unit determinant has $N^2 - 1 = 3$ elements for $n = 2$. Up to similarity transformations, these may be written in terms of three real parameters $(\theta_1, \theta_2, \theta_2)$:

$$g_1 = \begin{pmatrix} \cos\left(\frac{1}{2}\theta_1\right) & i\sin\left(\frac{1}{2}\theta_1\right) \\ i\sin\left(\frac{1}{2}\theta_1\right) & \cos\left(\frac{1}{2}\theta_1\right) \end{pmatrix} \tag{8.131a}$$

$$g_2 = \begin{pmatrix} \cos\left(\frac{1}{2}\theta_2\right) & \sin\left(\frac{1}{2}\theta_2\right) \\ -\sin\left(\frac{1}{2}\theta_2\right) & \cos\left(\frac{1}{2}\theta_2\right) \end{pmatrix} \tag{8.131b}$$

$$g_3 = \begin{pmatrix} e^{i\frac{1}{2}\theta_3} & 0 \\ 0 & e^{-i\frac{1}{2}\theta_3} \end{pmatrix}. \tag{8.131c}$$

These matrices are the exponentiated Pauli matrices $e^{\frac{i}{2}\sigma_i}$. Using this basis, any element of the group may be written as a product of one or more of these matrices with some θ_i.

8.6.5 U(1): *the set of numbers* $z : |z|^2 = 1$

The set of all complex numbers $U = e^{i\theta}$ with unit modulus forms an Abelian group under multiplication:

(1) $e^{i\theta_1} e^{i\theta_2} = e^{i(\theta_1 + \theta_2)}$.

(2) $(e^{i\theta_1} e^{i\theta_2}) e^{i\theta_3} = e^{i\theta_1} (e^{i\theta_2} e^{i\theta_3})$.

(3) $e^{i\theta} e^{i0} = e^{i\theta}$.

(4) $U^{-1} = U^*$ since $e^{i\theta} e^{-i\theta} = e^{i0} = 1$.

The representation space of this group is the space of complex scalars Φ, with constant modulus:

$$\Phi^*\Phi \to (U\Phi)^*U\Phi = \Phi^* U^*U \Phi = \Phi^*\Phi. \tag{8.132}$$

This group is important in electromagnetism; it is this symmetry group of complex phases which is connected to the existence of a conserved electrical charge.

8.6.6 Z_N: the Nth roots of unity

The Nth roots of unity form a sub-group of $U(1)$. These complex numbers may be written in the form $\exp(2\pi i\frac{p}{N})$, for $p = 0, \ldots, N - 1$. The group Z_N is special because it is not infinite. It has exactly N discrete elements. The group has the topology of a circle, and the elements may be drawn as equi-distant points on the circumference of the unit circle in the complex plane. Z_N is a modulo group. Its elements satisfy modulo N arithmetic by virtue of the multi-valuedness of the complex exponential. The group axioms are thus satisfied as follows:

(1) $\exp\left(2\pi i\frac{p}{N}\right)\exp\left(2\pi i\frac{p'}{N}\right) = \exp\left(2\pi i\frac{p+p'}{N}\right) = \exp\left(2\pi i\left[\frac{p+p'}{N} + m\right]\right)$,
   where N, m, p are integers;

(2) follows trivially from $U(1)$;

(3) follows trivially from $U(1)$;

(4) the inverse exists because of the multi-valued property that

$$\exp\left(-2\pi i\frac{p}{N}\right) = \exp\left(2\pi i\frac{N-p}{N}\right). \tag{8.133}$$

Thus when $p = N$, one arrives back at the identity, equivalent to $p = 0$.

The representation space of this group is undefined. It can represent translations or shifts along a circle for a complex scalar field. Z_2 is sometimes thought of as a reflection symmetry of a scalar field, i.e. $Z_2 = \{1, -1\}$ and $\phi \to -\phi$. An action which depends only on ϕ^2 has this symmetry.

Usually Z_N is discussed as an important sub-group of very many continuous Lie groups. The presence of Z_N as a sub-group of another group usually signifies some multi-valuedness or redundancy in that group. For example, the existence of a Z_2 sub-group in the Lie group $SU(2)$ accounts for the double-valued nature of electron spin.

8.6.7 $O(N)$: the orthogonal group

The orthogonal group consists of all matrices which satisfy

$$U^T U = I \tag{8.134}$$

under normal matrix multiplication. In other words, the transpose of each matrix is the inverse matrix. All such matrices are real, and thus there are $(N^2 - N)/2 +$ $n = N(N + 1)/2$ real components in such a matrix. This is the dimension of the group. The orthogonal group is non-Abelian for $N > 2$ and is trivial for $n = 1$.

The special orthogonal group is the sub-group of $O(N)$ which consists of matrices with unit determinant. This reduces the dimension of the group by one, giving $N(N - 1)/2$ independent components.

8.6.8 $SO(3)$: the three-dimensional rotation group

This non-Abelian group has three independent components corresponding to rotations about three-independent axes in a three-dimensional space. The group elements may be parametrized by the rotation matrices g_i about the given axis i:

$$U_x = \begin{pmatrix} 1 & 0 & 0 \\ 0 & \cos\theta_1 & \sin\theta_1 \\ 0 & -\sin\theta_1 & \cos\theta_1 \end{pmatrix} \tag{8.135}$$

$$U_y = \begin{pmatrix} \cos\theta_2 & 0 & -\sin\theta_2 \\ 0 & 1 & 0 \\ \sin\theta_2 & 0 & \cos\theta_2 \end{pmatrix} \tag{8.136}$$

$$U_z = \begin{pmatrix} \cos\theta_3 & \sin\theta_3 & 0 \\ -\sin\theta_3 & \sin\theta_3 & 0 \\ 0 & 0 & 1 \end{pmatrix}. \tag{8.137}$$

The representation space of this group is a three-dimensional Euclidean space and the transformations rotate three-dimensional vectors about the origin, preserving their lengths but not their directions. Notice that these matrices do not commute; i.e. a rotation about the x axis followed by a rotation about the y axis, is not the same as a rotation about the y axis followed by a rotation about the x axis.

8.6.9 $SO(2)$: the two-dimensional rotation group

This group has only one element, corresponding to rotations about a point in a plane. Any element of $SO(2)$ may be written in the form

$$U = \begin{pmatrix} \cos\theta & \sin\theta \\ -\sin\theta & \cos\theta \end{pmatrix}. \tag{8.138}$$

The representation space of this group is a two-dimensional Euclidean space, and the transformation rotates two-component vectors about the origin. Notice how the matrices parametrizing $SO(3)$ are simply rotations of $SO(2)$ embedded in a three-dimensional framework.

8.7 Universal cover groups and centres

We know that groups can contain other groups, as sub-groups of the whole, and therefore that some are larger than others. The *universal cover group* is defined to be a simply connected group which contains an image of every point in a given Lie group. If we consider an arbitrary Lie group, in general it will have companion groups which are locally the same, but globally different. The best known example of this is the pair $SU(2)$ and $SO(3)$, which are locally isomorphic, but globally different. In fact $SU(2)$ contains two images of $SO(3)$ or covers it twice, or contains two equivalent copies of it. Taking this a step further, if three groups have the same local structure, then they will all be sub-groups of the universal cover groups.

If we begin with a Lie algebra, it is possible to exponentiate the generators of the algebra to form group elements:

$$\Theta = \theta^A T^A \to G = e^{i\Theta}. \tag{8.139}$$

The group formed by this exponentiation is not unique; it depends on the particular representation of the algebra being exponentiated. For instance, the 2×2 representation of $SU(2)$ exponentiates to $SU(2)$, while the 3×3 representation of $SU(2)$ exponentiates to $SO(3)$. Both of these groups are locally isomorphic but differ in their centres. In the case of $SU(2)$ and $SO(3)$, we can relate them by factorizing out the centre of the universal cover group,

$$SU(2)/Z_2 = SO(3). \tag{8.140}$$

From Schur's lemma, we know that the centre of a group is only composed of multiples of the identity matrix, and that, in order to satisfy the rules of group multiplication, they must also have modulus one. It follows from these two facts that any element of the centre of a group can be written

$$g_c = \exp(\pm 2\pi i q/N) \mathbf{I}, \qquad q = 0, \dots, N-1. \tag{8.141}$$

These elements are the Nth roots of unity for some N (in principle infinite, but in practice usually finite). If we start off with some universal cover group then, whose centre is Z_N, there will be many locally isomorphic groups which can be found by factoring out sub-groups of the centre. The largest thing one can divide out is Z_N itself, i.e. the whole centre. The group formed in this way is called the adjoint group, and it is generated by the *adjoint representation*:

$$\frac{\text{group}}{\text{centre of group}} = \text{adjoint group}. \tag{8.142}$$

Table 8.1. Some common Lie algebras and groups.

Algebra	Centre	Cover
A_N	Z_N	$SU(N-1)$
B_N	Z_2	$SO(2N+1)$
C_N	Z_2	$Sp(2N)$
D_N	Z_4 (Nodd)	$SO(2N)$
	$Z_2 \times Z_2$ (Neven)	
E_6	Z_3	E_6
G_2, F_4, E_8	Z_3	

But it is not necessary to factor out the entire centre, one can also factor out a sub-group of the full centre; this will also generate a locally isomorphic group. For example, $SU(8)$ has centre Z_8. We can construct any of the following locally isomorphic groups:

$$SU(8) \qquad SU(8)/Z_8 \qquad SU(8)/Z_4 \qquad SU(8)/Z_2. \qquad (8.143)$$

Some well known Lie groups are summarized in table 8.1.

8.7.1 Centre of $SU(N)$ is Z_N

$SU(N)$ is a simply connected group and functions as its own universal cover group. As the set of $N \times N$ matrices is the fundamental, defining representation, it is easy to calculate the elements of the centre. From Schur's lemma, we know that the centre must be a multiple of the identity:

$$g_c = \alpha \, \mathbf{I}_N. \qquad (8.144)$$

where $\mathbf{I}_N$ is the $N \times N$ identity matrix. Now, $SU(N)$ matrices have unit determinant, so

$$\det I_N = \alpha^N = 1. \qquad (8.145)$$

Thus, the solutions for α are the Nth roots of unity, Z_N.

8.7.2 Congruent algebras: N-ality

Since roots and weights of representations can be drawn as vectors in the Cartan sub-space, different representations produce similar, but not identical, patterns. Elements E_α of the algebra step through chains of solutions, creating a laced lattice-work pattern. Representations which exponentiate to the same group have patterns which are *congruent* to one another [124].

Congruence is a property of discrete sets. The correct terminology is 'congruent to x modulo m'. The property is simplest to illustrate for integers. x is said to be conjugate to y modulo m if $y - x$ is an integer multiple of m:

$$y = x + km, \qquad (8.146)$$

for integer k, m. Congruence modulo m is an equivalence relation, and it sorts numbers into classes or congruent sets. The patterns made by congruent sets can be overlain consistently. The equivalence class, E_x, is the set of all integers which can be found from x by adding integer multiples m to it:

$$
\begin{aligned}
E_x &= \{x + km \mid \text{integer } k\} \\
&= \{\ldots, -2m + x, -m + x, x, x + m, x + 2m, \ldots\}.
\end{aligned} \qquad (8.147)
$$

There are exactly m different congruence classes modulo m, and these partition the integers; e.g. for $m = 4$, we can construct four classes:

$$
\begin{aligned}
E_0 &= \{\ldots, -8, -4, 0, 4, 8, \ldots\} \\
E_1 &= \{\ldots, -7, -3, 1, 5, 9, \ldots\} \\
E_2 &= \{\ldots, -6, -2, 2, 6, 10, \ldots\} \\
E_3 &= \{\ldots, -5, -1, 3, 7, 11, \ldots\}.
\end{aligned} \qquad (8.148)
$$

Lie algebra representations can also be classified into congruence classes. Historically, congruence classes of $SU(N)$ modulo N are referred to as N-ality as a generalization of 'triality' for $SU(3)$. Each congruence class has a label q; $q = 0$ corresponds to no centre, or the adjoint congruence class. The well known algebras contain the following values [56]:

$$q = \sum_{k=1}^{n} \alpha_k \qquad (\text{mod } n + 1) \qquad \text{for } A_n \qquad (8.149)$$

$$q = \alpha_n \qquad (\text{mod } 2) \qquad \text{for } B_n \qquad (8.150)$$

$$q = \alpha_1 + \alpha_3 + \alpha_5 \qquad (\text{mod } 2) \qquad \text{for } C_n \qquad (8.151)$$

$$q = \alpha_1 - \alpha_2 + \alpha_4 - \alpha_5 \qquad (\text{mod } 3) \qquad \text{for } E_6 \qquad (8.152)$$

$$q = \alpha_4 + \alpha_6 + \alpha_7 \qquad (\text{mod } 2) \qquad \text{for } E_7 \qquad (8.153)$$

$$q = 0 \qquad \text{for all representations of } E_7, E_8, F_4, G_2. \qquad (8.154)$$

In the special case of D_n, the congruence classes require classification by a two-component vector:

$$
\begin{aligned}
q_1 &= (\alpha_{n-1} + \alpha_n, 2\alpha_1 + \alpha_3 + \cdots \\
&\quad + 2\alpha_{n-2} + (n-2)\alpha_{n-1} + n\alpha_n + \cdots) \qquad (\text{mod } 2) \qquad \text{odd } n \\
q_2 &= (\alpha_{n-1} + \alpha_n, 2\alpha_1 + 2\alpha_3 + \cdots \\
&\quad + 2\alpha_{n-3} + (n-2)\alpha_{n-1} + n\alpha_n) \qquad (\text{mod } 4) \qquad \text{even } n.
\end{aligned} \qquad (8.155)
$$

The congruence is modulo the order of the centre. The algebra D_n requires a two-dimensional label, since its centre is two-dimensional. E_7, E_8, F_4, and G_2 all have trivial centres, thus they all lie in a single class congruent to the adjoint.

8.7.3 Simple and semi-simple Lie algebras

A Lie algebra is *simple* if it has no proper invariant sub-algebras; i.e if only one element (the identity) commutes with every other in the group. A simple algebra is necessarily semi-simple. A semi-simple Lie algebra can be written in block-diagonal form, as a direct sum of invariant sub-algebras, each of which is a simple Lie algebra

$$A = A_1 \oplus A_2 \oplus A_3 \oplus \cdots A_N, \qquad (8.156)$$

i.e. it factorizes into block-diagonal form with simple blocks. A semi-simple algebra has no Abelian invariant sub-algebras.

8.8 Summary

The existence of a symmetry in a physical system means that it is possible to re-label parameters of a model without changing its form or substance. Identify the symmetries of a physical system and one can distinguish between the freedom a system has to change and the constraints which hold it invariant: symmetries are thus at the heart of dynamics and of perspective.

Symmetries form *groups*, and can therefore be studied with the group theory. Since a symmetry means that some quantity R_ξ does not change, when we vary the action with respect to a parameter ξ, *conservation of R_ξ* is also linked to the existence of the symmetry. All of the familiar conservation laws can be connected to fundamental symmetries.

In the case of electromagnetism, Lorentz covariance was exposed just by looking at the field equations and writing them in terms of $(3 + 1)$ dimensional vectors. The chapters which follow examine the transformations which change the basic variables parametrizing the equations of motion, and the repercussions such transformations have for covariance.

9

Spacetime transformations

An important class of symmetries is that which refers to the geometrical disposition of a system. This includes translational invariance, rotational invariance and boosts. Historically, covariant methods were inspired by the fact that the speed of light in a vacuum is constant for all inertial observers. This follows from Maxwell's equations, and it led Einstein to the special theory of relativity and covariance. The importance of covariance has since been applied to many different areas in theoretical physics.

To discuss coordinate transformations we shall refer to figure 9.1, which shows two coordinate systems moving with a relative velocity $\mathbf{v} = \beta c$. The constancy of the speed of light in any inertial frame tells us that the line element (and the corresponding proper time) must be invariant for all inertial observers. For a real constant Ω, this implies that

$$\mathrm{d}s^2 = \Omega^2 \mathrm{d}s'^2 = \Omega^2(-c^2\mathrm{d}t^2 + \mathbf{dx} \cdot \mathbf{dx}). \tag{9.1}$$

This should not be confused with the non-constancy of the effective speed of light in a material medium; our argument here concerns the vacuum only. This property expresses the constancy, or x-independence, of c. The factor Ω^2 is of little interest here as long as it is constant: one may always re-scale the coordinates to absorb it. Normally one is not interested in re-scaling measuring rods when comparing coordinate systems, since it only make systems harder to compare. However, we shall return to this point in section 9.7.

For particles which travel at the speed of light (massless particles), one has $\mathrm{d}s^2 = 0$ always, or

$$\frac{\mathbf{dx}}{\mathrm{d}t} = c. \tag{9.2}$$

Now, since $\mathrm{d}s^2 = 0$, it is clearly true that $\Omega^2(x)\,\mathrm{d}s^2 = 0$, for any non-singular, non-zero function $\Omega(x)$. Thus the value of c is preserved by a group of

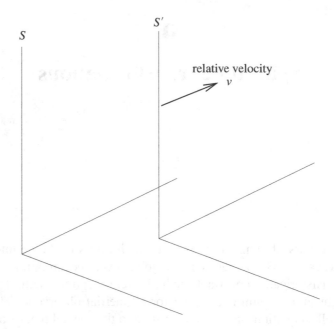

Fig. 9.1. The schematic arrangement for discussing coordinate transformations. Co-ordinate systems $S(x)$ and $S'(x')$ are in relative motion, with speed $\mathbf{v} = \beta c$.

transformations which obey

$$\mathrm{d}s'^2 = \Omega^2(x)\mathrm{d}s^2. \tag{9.3}$$

This set of transformations forms a group called the *conformal* group.

If all particles moved at the speed of light, we would identify this group as being the fundamental symmetry group for spacetime. However, for particles not moving at c, the line element is non-zero and may be characterized by

$$\frac{\mathrm{d}\mathbf{x}}{\mathrm{d}t} = \beta c, \tag{9.4}$$

for some constant $\beta = v/c$. Since we know that, in any frame, a free particle moves in a straight line at constant velocity, we know that β must be a constant and thus

$$\mathrm{d}s'^2 = \mathrm{d}s^2 \neq 0. \tag{9.5}$$

If it were possible for an x-dependence to creep in, then one could transform an inertial frame into a non-inertial frame. The group of transformations which preserve the line element in this way is called the inhomogeneous Lorentz group, or Poincaré group.

In the non-relativistic limit, coordinate invariances are described by the so-called Galilean group. This group is no smaller than the Lorentz group, but space and time are decoupled, and the speed of light does not play a role at all. The non-relativistic limit assumes that $c \to \infty$. Galilean transformations lie closer to our intuition, but they are often more cumbersome since space and time must often be handled separately.

9.1 Parity and time reversal

In an odd number of spatial dimensions ($n = 2l+1$), a parity, or space-reflection transformation $\mathcal{P}$ has the following non-zero tensor components:

$$\mathcal{P}^0{}_0 = 1$$
$$\mathcal{P}^i{}_i = -1, \tag{9.6}$$

where i is not summed in the last line. When this transformation acts on another tensor object, it effects a change of sign on all space components. In other words, each spatial coordinate undergoes $x^i \to -x^i$. The transformation $A \to -A$ is the discrete group $Z_2 = \{1, -1\}$.

In an even number of spatial dimensions ($n = 2l$), this construction does not act as a reflection, since the combination of an even number of reflections is not a reflection at all. In group language, $(Z_2)^{2n} = \{1\}$. It is easy to check that, in two spatial dimensions, reflection in the x_1 axis followed by reflection in the x_2 axis is equivalent to a continuous rotation. To make a true reflection operator in an even number of space dimensions, one of the spatial indices must be left out. For example,

$$\mathcal{P}^0{}_0 = 1$$
$$\mathcal{P}^i{}_i = -1 \ (i = 1, \ldots, n-1)$$
$$\mathcal{P}^i{}_i = +1 \ (i = n). \tag{9.7}$$

The time reversal transformation in any number of dimensions performs the analogous function for time coordinates:

$$T^0{}_0 = -1$$
$$T^i{}_i = 1. \tag{9.8}$$

These transformations belong to the Lorentz group (and others), and are sometimes referred to as large Lorentz transformations since they cannot be formed by integration or repeated combination of infinitesimal transformations.

9.2 Translational invariance

A general translation in space, or in time, is a coordinate shift. A scalar field transforms simply:

$$\phi(x) \to \phi(x + \Delta x). \tag{9.9}$$

The direction of the shift may be specified explicitly, by

$$\phi(t, x^i) \to \phi(t, x^i + \Delta x^i)$$
$$\phi(t, x^i) \to \phi(t + \Delta t, x^i). \tag{9.10}$$

Invariance under such a *constant* shift of a coordinate is almost always a prerequisite in physical problems found in textbooks. Translational invariance is easily characterized by the coordinate dependence of Green functions. Since the Green function is a two-point function, one can write it as a function of x and x' or in terms of variables rotated by 45 degrees, $\frac{1}{\sqrt{2}}(x - x')$ and $\frac{1}{\sqrt{2}}(x + x')$. These are more conveniently defined in terms of a difference and an average (mid-point) position:

$$\tilde{x} = (x - x')$$
$$\overline{x} = \frac{1}{2}(x + x'). \tag{9.11}$$

The first of these is invariant under coordinate translations, since

$$x - x' = (x + a) - (x' + a). \tag{9.12}$$

The second equation is not, however. Thus, in a theory exhibiting translational invariance, the two-point function must depend only on $\tilde{x} = x - x'$.

9.2.1 Group representations on coordinate space

Translations are usually written in an additive way,

$$x^\mu \to x^\mu + a^\mu, \tag{9.13}$$

but, by embedding spacetime in one extra dimension, $d_R = (n + 1) + 1$, one can produce a group vector formulation of the translation group:

$$\begin{pmatrix} x^\mu \\ 1 \end{pmatrix} \to \begin{pmatrix} 1 & a^\mu \\ 0 & 1 \end{pmatrix} \begin{pmatrix} x^{\mu'} \\ 1 \end{pmatrix}. \tag{9.14}$$

This has the form of a group vector multiplication. The final 1 in the column vector is conserved and plays only a formal role. This form is common in computer representations of translation, such as in computer graphics.

A representation of translations which is particularly important in quantum mechanics is the differential coordinate representation. Consider an infinitesimal translation $a_\mu = \epsilon_\mu$. This transformation can be obtained from an exponentiated group element of the form

$$U(\epsilon) = \exp\left(i\theta^A T^A\right) \tag{9.15}$$

by writing

$$U(\epsilon) = \exp\left(i\epsilon_\rho k^\rho\right) \exp\left(i\epsilon_\rho p^\rho / \chi_h\right) = (1 + i\epsilon_\rho p^\rho / \chi_h), \tag{9.16}$$

where

$$p_\mu = \chi_h k_\mu = -i \chi_h \partial_\mu. \tag{9.17}$$

The action of the infinitesimal group element is thus

$$x^\mu \rightarrow U(\epsilon) x^\mu = (1 + \chi_h \epsilon^\rho \partial_\rho x^\mu) = x^\mu + \epsilon^\rho \eta_\rho^{\ \mu} = x^\mu + \epsilon^\mu. \tag{9.18}$$

The reason for writing the generator,

$$T^A \rightarrow p_\mu / \chi_h, \tag{9.19}$$

in this form, is that p_μ is clearly identifiable as a momentum operator which satisfies

$$[x, p] = i \chi_h. \tag{9.20}$$

Thus, it is the momentum divided by a dimensionful scale (i.e. the wavenumber k_μ) which is the generator of translations. In fact, we already know this from Fourier analysis.

The momentum operator closely resembles that from quantum mechanics. The only difference is that the scale χ_h (with dimensions of action), which is required to give p_μ the dimensions of momentum, is not necessarily $\hbar$. It is arbitrary. The fact that $\hbar$ is small is the physical content of quantum mechanics; the remainder is group theory. What makes quantum mechanics special and noticeable is the non-single-valued nature of the exponentiated group element. The physical consequence of a small χ_h is that even a small translation will cause the argument of the exponential to go through many revolutions of 2π. If χ_h is large, then this will not happen. Physically this means that the oscillatory nature of the group elements will be very visible in quantum mechanics, but essentially invisible in classical mechanics. This is why a wavelike nature is important in quantum mechanics.

9.2.2 Bloch's theorem: group representations on field space

Bloch's theorem, well known in solid state physics, is used to make predictions about the form of wavefunctions in systems which have periodic potentials. In metals, for instance, crystal lattices look like periodic arrays of potential wells, in which electrons move. The presence of potentials means that the eigenfunctions are not plane waves of the form

$$e^{ik(x-x')}, \qquad (9.21)$$

for any x, x'. Nevertheless, translational invariance by discrete vector jumps a_i is a property which must be satisfied by the eigenfunctions

$$\phi_k(t, \mathbf{x} + \mathbf{a}) = U(\mathbf{a}) \, \phi_k(t, \mathbf{x}) = e^{i\,\mathbf{k}\cdot\mathbf{a}} \, \phi_k(t, \mathbf{x}). \qquad (9.22)$$

9.2.3 Spatial topology and boundary conditions

Fields which live on spacetimes with non-trivial topologies require boundary conditions which reflect the spacetime topology. The simplest example of this is the case of periodic boundary conditions:

$$\phi(x) = \alpha \, \phi(x + L), \qquad (9.23)$$

for some number α. Periodic boundary conditions are used as a model for homogeneous crystal lattices, where the periodicity is interpreted as translation by a lattice cell; they are also used to simulate infinite systems with finite ones, allowing the limit $L \to \infty$ to be taken in a controlled manner. Periodic boundary conditions are often the simplest to deal with.

The value of the constant α can be specified in a number of ways. Setting it to unity implies a strict periodicity, which is usually over-restrictive. Although it is pragmatic to specify a boundary condition on the field, it should be noted that the field itself is not an observable. Only the probability $P = (\phi, \phi)$ and its associated operator $\hat{P}$ are observables. In Schrödinger theory, for example, $\hat{P} = \psi^*(x)\psi(x)$, and one may have $\psi(x + L) = e^{i\theta(x)}\psi(x)$ and still preserve the periodicity of the probability.

In general, if the field $\phi(x)$ is a complex field or has some multiplet symmetry, then it need only return to its original value up to a gauge transformation; thus $\alpha = U(x)$. For a multiplet, one may write

$$\Phi_A(x + L) = U_A{}^B(x) \, \Phi_B(x). \qquad (9.24)$$

The transformation U is the exponentiated phase factor belonging to the group of symmetry transformations which leaves the action invariant. This is sometimes referred to as a *non-integrable phase*. Note that, for a local gauge transformation, one also has a change in the vector field:

$$A_\mu(x + L) = \beta A_\mu(x). \qquad (9.25)$$

This kind of transformation is required in order to obtain a consistent energy–momentum tensor for gauge symmetric theories (see section 11.5). The value of β depends now on the type of couplings present. From the spacetime symmetry, a real field, A_μ, has only a Z_2 reflection symmetry, i.e. $\beta = \pm 1$, which corresponds heuristically to ferromagnetic and anti-ferromagnetic boundary conditions. Usually $\beta = 1$ to avoid multiple-valuedness.

In condensed matter physics, conduction electrons move in a periodic potential of crystallized valence ions. The potential they experience is thus periodic:

$$V(\mathbf{x}) = V(\mathbf{x} + \mathbf{L}), \tag{9.26}$$

and it follows that, for plane wave eigenfunctions,

$$\phi_k(t, \mathbf{x} + \mathbf{L}) = U(\mathbf{L})\, \phi_k(t, \mathbf{x}) = e^{i\,\mathbf{k}\cdot\mathbf{L}}\, \phi_k(t, \mathbf{x}). \tag{9.27}$$

This is a straightforward application of the scalar translation operator; the result is known as Bloch's theorem.

On toroidal spacetimes, i.e. those which have periodicities in several directions, the symmetries of the boundary conditions are linked in several directions. This leads to boundary conditions called co-cycle conditions [126]. Such conditions are responsible for flux quantization of magnetic fields in the Hall effect [65, 85].

In order to define a self-consistent set of boundary conditions, it is convenient to look at the so-called Wilson loops in the two directions of the torus, since they may be constructed independently of the eigenfunctions of the Hamiltonian. Normally this is presented in such a way that any constant part of the vector potential would cancel out, giving no information about it. This is the co-cycle condition, mentioned below. The Wilson line is defined by

$$W_j(x) = P \exp\left\{ ig \int_{\vec{x}_0}^{\vec{x}} A_j\, dx'_j \right\}, \tag{9.28}$$

j not summed, for some fixed point $\vec{x}_0$. It has an associated Wilson loop $W_j(L'_j)$ around a cycle of length L'_j in the x_j direction by

$$W_j(x_j + L'_j) = W_j(L'_j)\, W_j(x_j). \tag{9.29}$$

The notation here means that the path-dependent Wilson line $W_j(\vec{x})$ returns to the same value multiplied by a phase $W_j(L'_j, \vec{x})$ on translation around a closed curve from x_j to $x_j + L'_j$. The coordinate dependence of the phase usually arises in the context of a uniform magnetic field passing through the torus. In the presence of a constant magnetic field strength, the two directions of the torus are closely linked, and thus one has

$$W_1(u_1 + L_1, u_2) = \exp\left\{ iL_1 u_2 + ic_1 L_1 \right\} W_1(u_1, u_2) \tag{9.30}$$

$$W_2(u_1, u_2 + L_2) = \exp\left\{ ic_2 L_2 \right\} W_2(u_1, u_2). \tag{9.31}$$

At this stage, it is normal to demonstrate the quantization of flux by opening out the torus into a rectangle and integrating around its edges:

$$W_1(u_2 + L_2)W_2(u_1)W_1^{-1}(u_2)W_2^{-1}(u_1 + L_1) = 1. \tag{9.32}$$

This is known as the co-cycle condition, and has the effect of cancelling the contributions to the c's and thus flux quantization is found independently of the values of c_i due to the nature of the path. The most general consistency requirement for the gauge field (Abelian or non-Abelian), which takes into account the phases c_i, has been constructed in ref. [18].

The results above imply that one is not free to choose, say, periodic boundary conditions for bosons and anti-periodic boundary conditions for fermions in the presence of a uniform field strength. All fields must satisfy the same consistency requirements. Moreover, the spectrum may not depend on the constants, c_i, which have no invariant values. One may understand this physically by noting that a magnetic field causes particle excitations to move in circular Landau orbits, around which the line integral of the constant vector potential is null. The constant part of the vector potential has no invariant meaning in the presence of a magnetic field.

In more complex spacetimes, such as spheres and other curved surfaces, boundary conditions are often more restricted. The study of eigenfunctions (spherical harmonics) on spheres shows that general phases are not possible at identified points. Only the eigenvalues ± 1 are consistent with a spherical topology [17].

9.3 Rotational invariance: $SO(n)$

Rotations are clearly of special importance in physics. In n spatial dimensions, the group of rotations is the group which preserves the Riemannian, positive definite, inner product between vectors. In Cartesian coordinates this has the well known form

$$\mathbf{x} \cdot \mathbf{y} = x^i y_i. \tag{9.33}$$

The rotation group is the group of orthogonal matrices with unit determinant $SO(n)$. Rotational invariance implies that the Green function only depends on squared combinations of this type:

$$G(x, x') = G\left((x_1 - x_1')^2 + (x_2 - x_2)^2 + \cdots + (x_n - x_n')^2\right). \tag{9.34}$$

The exception here is the Dirac Green function.

9.3.1 Group representations on coordinate space

Three-dimensional rotations are generated by infinitesimal matrices:

$$T^1 = \begin{pmatrix} 0 & 0 & 0 \\ 0 & 0 & -i \\ 0 & i & 0 \end{pmatrix}$$

$$T^2 = \begin{pmatrix} 0 & 0 & i \\ 0 & 0 & 0 \\ -i & 0 & 0 \end{pmatrix}$$

$$T^3 = \begin{pmatrix} 0 & -i & 0 \\ i & 0 & 0 \\ 0 & 0 & 0 \end{pmatrix} \tag{9.35}$$

which satisfy a Lie algebra

$$[T_i, T_j] = i\epsilon_{ijk} T_k. \tag{9.36}$$

These exponentiate into the matrices for a three-dimensional rotation, parametrized by three Euler angles,

$$R_x \equiv U_x = \begin{pmatrix} 1 & 0 & 0 \\ 0 & \cos\theta_1 & \sin\theta_1 \\ 0 & -\sin\theta_1 & \cos\theta_1 \end{pmatrix} \tag{9.37}$$

$$R_y \equiv U_y = \begin{pmatrix} \cos\theta_2 & 0 & -\sin\theta_2 \\ 0 & 1 & 0 \\ \sin\theta_2 & 0 & \cos\theta_2 \end{pmatrix} \tag{9.38}$$

$$R_z \equiv U_z = \begin{pmatrix} \cos\theta_3 & \sin\theta_3 & 0 \\ -\sin\theta_3 & \sin\theta_3 & 0 \\ 0 & 0 & 1 \end{pmatrix}. \tag{9.39}$$

The rotation group is most often studied in $n = 3$ dimensions, for obvious reasons, though it is worth bearing in mind that its properties differ quite markedly with n. For instance, in two dimensions it is only possible to have rotation about a point. With only one angle of rotation, the resulting rotation group, $SO(2)$, is Abelian and is generated by the matrix

$$T_1 = \begin{pmatrix} 0 & i \\ -i & 0 \end{pmatrix}. \tag{9.40}$$

This exponentiates into the group element

$$U = \begin{pmatrix} \cos\theta & \sin\theta \\ -\sin\theta & \cos\theta \end{pmatrix}. \tag{9.41}$$

A two-dimensional world can also be represented conveniently by adopting complex coordinates on the Argand plane. In this representation, a vector is simply a complex number z, and a rotation about the origin by an angle θ is accomplished by multiplying:

$$z \to e^{i\theta} z. \tag{9.42}$$

9.3.2 Eigenfunctions: circular and spherical harmonics

The eigenfunctions of the rotation operators form a set of basis functions which span representation space. The rotational degrees of freedom in quantum fields can be expanded in terms of these eigenfunctions.

Eigenfunctions in $n = 2$ In two dimensions, there is only a single axis of rotation to consider. Then the action of the rotation operator T_1 has the form

$$-i\partial_\phi |\phi\rangle = \Lambda |\phi\rangle. \tag{9.43}$$

This equation is trivially solved to give

$$|\phi\rangle = e^{i\Lambda\phi}. \tag{9.44}$$

In two spatial dimensions, there are no special restrictions on the value of Λ. Notice that this means that the eigenfunctions are not necessarily single-valued functions: under a complete rotation, they do not have to return to their original value. They may differ by a phase:

$$|\phi + 2\pi\rangle = e^{i\Lambda(\phi+2\pi)} = e^{i\delta} e^{i\Lambda\phi}, \tag{9.45}$$

where $\delta = 2\Lambda\pi$. In higher dimensions δ must be unity because of extra topological restrictions (see below).

Eigenfunctions in $n = 3$ The theory of matrix representations finds all of the irreducible representations of the rotation algebra in $n = 3$ dimensions. These are characterized by their highest weight, or *spin*, with integral and half-integral values. There is another approach, however, which is to use a differential representation of the operators. The advantage of this is that it is then straightforward to find orthonormal basis functions which span the rotational space.

A set of differential operators which satisfies the Lie algebra is easily constructed, and has the form

$$\mathbf{T} = \mathbf{r} \times i\nabla, \tag{9.46}$$

or

$$T_i = i\epsilon_{ijk} x_j \partial_k. \tag{9.47}$$

This has the form of an orbital angular momentum operator $\mathbf{L} = \mathbf{r} \times \mathbf{p}$, and it is no coincidence that it re-surfaces also in chapter 11 in that context with only a factor of $\hbar$ to make the dimensions right. It is conventional to look for the simultaneous eigenfunctions of the operators L_1 and L^2 by writing these operators in spherical polar coordinates (with constant radius):

$$L_1 = i \left(\sin\phi\, \partial_\theta + \cot\theta\, \cos\phi\, \partial_\phi \right)$$
$$L_2 = i \left(-\cos\phi\, \partial_\theta + \cot\theta\, \sin\phi\, \partial_\phi \right)$$
$$L_3 = -i\, \partial_\phi, \tag{9.48}$$

and

$$L^2 = \frac{1}{\sin\theta}\, \partial_\theta \left(\sin\theta\, \partial_\theta \right) + \frac{1}{\sin^2\theta}\, \partial_\phi^2. \tag{9.49}$$

The eigenvectors and eigenvalues involve two angles, and may be defined by

$$L^2 \,|\phi, \theta\rangle = T^2 |\phi, \theta\rangle$$
$$L_3 \,|\phi, \theta\rangle = \Lambda_c |\phi, \theta\rangle. \tag{9.50}$$

The solution to the second equation proceeds as in the two-dimensional case, with only minor modifications due to the presence of the other coordinates. The eigenfunctions are written as a direct product,

$$|\phi, \theta\rangle = \Theta(\theta)\Phi(\phi), \tag{9.51}$$

so that one may identify $\Phi(\phi)$ with the solution to the two-dimensional problem, giving

$$|\phi, \theta\rangle = \Theta(\theta)\, e^{i\Lambda_c \phi}. \tag{9.52}$$

The values of Λ_c are not arbitrary in this case: the solution of the constraints for the θ coordinate imposes extra restrictions, because of the topology of a three-dimensional space. Suppose we consider a rotation through an angle of 2π in the ϕ direction in the positive and negative directions:

$$|\phi + 2\pi\rangle = e^{i\Lambda_c(\phi + 2\pi)} = e^{i\delta}\, e^{i\Lambda_c \phi},$$
$$|\phi - 2\pi\rangle = e^{i\Lambda_c(\phi - 2\pi)} = e^{-i\delta}\, e^{i\Lambda_c \phi}. \tag{9.53}$$

In two spatial dimensions, these two rotations are distinct, but in higher dimensions they are not. This is easily seen by drawing the rotation as a circle with an arrow on it (see figure 9.2). By flipping the circle about an axis in its plane we can continuously deform the positive rotation into the negative one, and vice versa. This is not possible in $n = 2$ dimensions. This means that they are, in fact, different expressions of the same rotation. Thus,

$$e^{i\delta} = e^{-i\delta} = \pm 1. \tag{9.54}$$

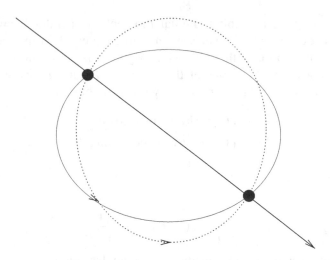

Fig. 9.2. Exchange of particles in two and three spatial dimensions. In the plane, there is only one rotation about the centre of mass which exchanges identical particles. Clockwise and anti-clockwise are inequivalent. In three dimensions or greater, one may rotate this plane around another axis and deform clockwise into anti-clockwise.

These two values are connected with the existence of two types of particle: bosons and fermions, or

$$\Lambda_c = 0, \pm\frac{1}{2}, \pm1, \ldots, \tag{9.55}$$

for integer m. Note that, in older texts, it was normal to demand the single-valuedness of the wavefunction, rather than using the topological argument leading to eqn. (9.54). If one does this, then only integer values of Λ_c are found, and there is an inconsistency with the solution of the group algebra. This illustrates a danger in interpreting results based on coordinate systems indiscriminately. The result here tells us that the eigenfunctions may be either single-valued for integer Λ_c, or double-valued for half-integral Λ_c. In quantum mechanics, it is normal to use the notation

$$T^2 = l(l+1) \tag{9.56}$$

$$\Lambda_c = m. \tag{9.57}$$

If we now use this result in the eigenvalue equation for L^2, we obtain

$$\frac{1}{\sin\theta}\frac{d}{d\theta}\left(\sin\theta\,\frac{d\Theta}{d\theta}\right) + \left(l(l+1) - \frac{m^2}{\sin^2\theta}\right)\Theta = 0. \tag{9.58}$$

Putting $z = \cos\theta$ in this equation turns it into the associated Legendre equation,

$$\frac{d}{dz}\left[(1-z^2)\frac{dP}{dz}\right] + \left[l(l+1) - \frac{m^2}{1-z^2}\right]P = 0, \tag{9.59}$$

where $P = \Theta(\cos\theta)$. The solutions of the associated Legendre equation may be found for integral and half-integral values of Λ_c, though most books ignore the half-integral solutions. They are rather complicated, and their form is not specifically of interest here. They are detailed, for instance, in Gradshteyn and Ryzhik [63]. Since the magnitude of L_3 cannot exceed that of L^2, by virtue of the triangle (Schwartz) inequality,

$$m^2 \leq l(l+1), \tag{9.60}$$

or

$$-l \leq m \leq l. \tag{9.61}$$

The rotational eigenfunctions are

$$|l, m\rangle = N_{lm}\, P_l^m(\cos\theta)\, \mathrm{e}^{\mathrm{i}m\phi}, \tag{9.62}$$

with normalization factor

$$N_{lm} = (-1)^m \sqrt{\left[\frac{2l+1}{4\pi}\frac{(l-m)!}{(l+m)!}\right]}. \tag{9.63}$$

These harmonic eigenfunctions reflect the allowed boundary conditions for systems on spherical spacetimes. They also reflect particle statistics under the interchange of identical particles. The eigenvalues of the spherical harmonics are ± 1 in $3+1$ dimensions, corresponding to (symmetrical) bosons and (anti-symmetrical) fermions; in $2+1$ dimensions, the Abelian rotation group has arbitrary boundary conditions corresponding to the possibility of anyons, or particles with 'any' statistics [83, 89].

9.4 Lorentz invariance

9.4.1 Physical basis

The Lorentz group is a non-compact Lie group which lies at the heart of Einsteinian relativistic invariance. Lorentz transformations are coordinate transformations which preserve the relativistic scalar product

$$x^\mu y_\mu = -x^0 y^0 + x^i y^i, \tag{9.64}$$

and therefore also the line element

$$\mathrm{d}s^2 = g_{\mu\nu}\mathrm{d}x^\mu\mathrm{d}x^\nu. \tag{9.65}$$

Lorentz transformations include, like the Galilean group, translations, rotations and boosts, or changes of relative speed. Under a linear transformation of x^μ, we may write generally

$$x^\mu \to x'^\mu = U^\mu_{\ \nu}x^\nu + a^\mu, \tag{9.66}$$

where a^μ is a constant translation and

$$U^\mu_{\ \nu} = \frac{\partial x'^\mu}{\partial x^\nu} \tag{9.67}$$

is constant.

9.4.2 Lorentz boosts and rotations

A boost is a change of perspective from one observer to another in relative motion to the first. The finite speed of light makes boosts special in Einsteinian relativity. If we refer to figure 9.1 and consider the case of relative motion along the x^1 axis, such that the two frames S and S' coincide at $x^0 = 0$, the Lorentz transformation relating the primed and unprimed coordinates may be written

$$
\begin{aligned}
x'^0 &= \gamma(x^0 - \beta x^1) = x^0 \cosh\alpha - x^1 \sinh\alpha \\
x'^1 &= \gamma(x^1 - \beta x^0) = x^1 \cosh\alpha - x^0 \sinh\alpha \\
x'^2 &= x^2 \\
x'^3 &= x^3,
\end{aligned}
\tag{9.68}
$$

where

$$
\begin{aligned}
\gamma &= 1/\sqrt{1 - \beta^2} \\
\beta^i &= v^i/c \\
\beta &= \sqrt{\beta^i \beta_i} \\
\alpha &= \tanh^{-1}\beta.
\end{aligned}
\tag{9.69}
$$

The appearance of hyperbolic functions here, rather than, say, sines and cosines means that there is no limit to the numerical values of the group elements. The group is said to be *non-compact*. In matrix form, in $(3 + 1)$ dimensional spacetime we may write this:

$$
L(B) = \begin{pmatrix} \gamma & -\gamma\beta & 0 & 0 \\ -\gamma\beta & \gamma & 0 & 0 \\ 0 & 0 & 1 & 0 \\ 0 & 0 & 0 & 1 \end{pmatrix} = \begin{pmatrix} \cosh\alpha & -\sinh\alpha & 0 & 0 \\ -\sinh\alpha & \cosh\alpha & 0 & 0 \\ 0 & 0 & 1 & 0 \\ 0 & 0 & 0 & 1 \end{pmatrix}
\tag{9.70}
$$

where the 'rapidity' $\alpha = \tanh^{-1}\beta$. This may be compared with the explicit form of a rotation about the x^1 axis:

$$
L(R) = \begin{pmatrix} 1 & 0 & 0 & 0 \\ 0 & 1 & 0 & 0 \\ 0 & 0 & \cos\theta & -\sin\theta \\ 0 & 0 & \sin\theta & \cos\theta \end{pmatrix}.
\tag{9.71}
$$

Notice that the non-trivial parts of these matrices do not overlap. This leads to an important result, which we shall derive below, namely that rotations and boosts are independent transformations which can be used to parametrize general transformations.

The form of these matrix representations makes it clear that the n-dimensional group of rotations, $SO(n)$, is a sub-group with irreducible representations

$$L_{\mu\nu}(R) = \begin{pmatrix} 1 & 0 \\ 0 & R_{ij} \end{pmatrix}, \tag{9.72}$$

and similarly that boosts in a single direction also form a sub-group. General boosts in multiple directions do not form a group, however.

The form of a general boost can be derived as a generalization of the formulae in eqns. (9.68) on the basis of general covariance. We can write a general form based on figure 9.1 and eqns. (9.68)

$$dx^{0'} = \gamma(dx^0 - \beta^i dx_i)$$

$$dx^{i'} = \gamma\left(c_1 \delta^i{}_j + c_2 \frac{\beta^i \beta_j}{\beta^2}\right) dx^j - \gamma \beta^i dx^0. \tag{9.73}$$

The unknown coefficients label projection operators for longitudinal and transverse parts with respect to the n-component velocity vector β^i. By squaring the above expressions and using the invariance of the line element

$$ds^2 = -(dx^0)^2 + (dx^i)^2 = -(dx^{0'})^2 + (dx^{i'})^2, \tag{9.74}$$

giving

$$-(dx^{0'})^2 = -\gamma^2\left((dx^0)^2 - 2(\beta^i dx_i)dx^0 + (\beta^i dx_i)^2\right), \tag{9.75}$$

and

$$(dx^{0'})^2 = \left(c_1^2 \delta_{jk} + (2c_1 c_2 + c_2^2)\frac{\beta_j \beta_k}{\beta^2}\right) dx^j dx^k$$
$$+ \gamma^2 \beta^2 (dx^0)^2 - 2\gamma(c_1 + c_2)(\beta^i dx_i)dx^0, \tag{9.76}$$

one compares the coefficients of similar terms with the untransformed ds^2 to obtain

$$c_1 = 1$$
$$c_2 = \gamma - 1. \tag{9.77}$$

Thus, in $1 + n$ block form, a general boost may be written as

$$L_{\mu\nu}(B) = \begin{pmatrix} \gamma & -\gamma\beta^i \\ -\gamma\beta^i & \delta_{ij} + (\gamma - 1)\frac{\beta_i \beta_j}{\beta^2} \end{pmatrix}. \tag{9.78}$$

9.4.3 The homogeneous Lorentz group: $SO(1, n)$

It is convenient to divide the formal discussion of the Lorentz group into two parts. In the first instance, we shall set the inhomogeneous term, a_μ, to zero. A homogeneous coordinate transformation takes the form

$$x^\mu \to x'^\mu = L^\mu_{\ \nu} x^\nu, \tag{9.79}$$

where $L_{\mu\nu}$ is a constant matrix. It does not include translations. After a transformation of the line element, one has

$$\begin{aligned} \mathrm{d}s'^2 &= g'_{\mu\nu}(x')\mathrm{d}x^{\mu'}\mathrm{d}x^{\nu'} \\ &= g_{\mu\nu}(x)L^\mu_{\ \rho}L^\nu_{\ \lambda}\mathrm{d}x^\rho\mathrm{d}x^\lambda. \end{aligned} \tag{9.80}$$

The metric must compensate for this change by transforming like this:

$$g_{\mu\nu}(x) = L^\rho_{\ \mu}L^\lambda_{\ \nu}\, g'_{\rho\lambda}(x'). \tag{9.81}$$

This follows from the above transformation property. We can see this in matrix notation by considering the constant metric tensor $\eta_{\mu\nu} = \mathrm{diag}(-1, 1, 1, 1, \ldots)$, which must be invariant if the scalar product is to be preserved. In a Cartesian basis, we have

$$\begin{aligned} x^\mu y_\mu &= \eta_{\mu\nu}\, x^\mu y^\nu = \eta_{\mu\nu}\, (Lx)^\mu (Ly)^\nu \\ x^\mathrm{T}\eta\, y &= (Lx)^\mathrm{T}\eta\,(Ly) \\ &= x^\mathrm{T}L^\mathrm{T}\eta\, Ly. \end{aligned} \tag{9.82}$$

Comparing the left and right hand sides, we have the matrix form of eqn. (9.81) in a Cartesian basis:

$$\eta = L^\mathrm{T}\eta\, L. \tag{9.83}$$

The matrices L form a group called the homogeneous Lorentz group. We can now check the group properties of the transformation matrices L. The existence of an associative combination rule is automatically satisfied since matrix multiplication has these properties (any representation in terms of matrices automatically belongs to the general linear group $G(n, R)$). Thus we must show the existence of an inverse and thus an identity element. Acting on the left of eqn. (9.83) with the metric

$$\eta\, L^\mathrm{T}\eta\, L = \eta^2 = I = L^{-1}\, L, \tag{9.84}$$

where I is the identity matrix belonging to $GL(n, R)$. Thus, the inverse of L is

$$L^{-1} = \eta\, L^\mathrm{T}\, \eta. \tag{9.85}$$

In components we have

$$(L^{-1})^\mu_{\ \nu} = \eta^{\mu\lambda}L^\rho_{\ \lambda}\eta_{\rho\nu} = L_\nu^{\ \mu}. \tag{9.86}$$

Since the transpose matrix is the inverse, we can write the Lorentz group as $SO(1, 3)$.

Dimension and structure of the group The symmetry in $(n+1)^2$ components of $L_{\mu\nu}$ implies that not all of the components may be chosen independently. The fact that only half of the off-diagonal components are independent means that there are

$$d_G = \frac{(n+1)^2 - (n+1)}{2} \tag{9.87}$$

independent components in $n+1$ dimensions, given by the independent elements of $\tilde{\omega}_{\mu\nu}$ to be defined below. Another way of looking at this is that there are $(n+1)^2$ components in the matrix L^ν_μ, but the number of constraints in eqn. (9.83) limits this number. Eqn. (9.83) tells us that the transpose of the equation is the same, thus the independent components of this equation are the diagonal pieces plus half the off-diagonal pieces. This is turn means that the other half of the off-diagonal equations represent the remaining freedom, or dimensionality of the group. d_G is the dimension of the inhomogeneous Lorentz group. The components of

$$g_{\mu\nu} L^\mu_\alpha L^\nu_\beta = g_{\alpha\beta}$$

may be written out in $1 + n$ form, $\mu = (0, i)$ form as follows:

$$L^0_0 L^0_0 \, g_{00} + L^i_0 L^j_0 \, g_{ij} = g_{00}$$
$$L^0_i L^0_0 \, g_{00} + L^k_i L^l_0 \, g_{kl} = g_{i0} = 0$$
$$L^0_i L^0_j \, g_{00} + L^k_i L^l_j \, g_{ij} = g_{ij}. \tag{9.88}$$

This leads to the extraction of the following equations:

$$(L^0_0)^2 = 1 + L^i_0 L^i_0$$
$$L^0_0 L^0_i = L^k_i L_{k0}$$
$$L^0_i L^{j0} + L_{ki} L^{kj} = \delta_{ij}. \tag{9.89}$$

These may also be presented in a schematic form in terms of a scalar S, a vector $\mathbf{V}$ and an $n \times n$ matrix M:

$$L^\mu_\nu = \begin{pmatrix} S & \mathbf{V}^{\mathrm{T}}_i \\ \mathbf{V}_j & M_{ij} \end{pmatrix}, \tag{9.90}$$

giving

$$S^2 = 1 + \mathbf{V}^i \mathbf{V}_i$$
$$S\mathbf{V}^{\mathrm{T}} = \mathbf{V}^{\mathrm{T}} M$$
$$I = M^{\mathrm{T}} M + \mathbf{V} \mathbf{V}^{\mathrm{T}}. \tag{9.91}$$

It is clear from eqn. (9.90) how the n-dimensional group of rotations, $SO(n)$, is a sub-group of the homogeneous Lorentz group acting on only the spatial components of spacetime vectors:

$$L^\mu_\nu(R) = \begin{pmatrix} 1 & 0 \\ 0 & R_{ij} \end{pmatrix}. \tag{9.92}$$

Notice that it is sufficient to know that $L_0^0 = 1$ to be able to say that a Lorentz transformation is a rotation, since the remaining equations then imply that

$$M^{\mathrm{T}}M = R^{\mathrm{T}}R = I, \tag{9.93}$$

i.e. that the n-dimensional sub-matrix is orthogonal. The discussion of the Lorentz group can, to a large extent, be simplified by breaking it down into the product of a continuous, connected sub-group together with a few discrete transformations. The elements of the group for which $\det L = +1$ form a sub-group which is known as the proper or restricted Lorentz group. From the first line of eqn. (9.89) or (9.91), we have that $L_0^0 \geq 1$ or $L_0^0 \leq -1$. The group elements with $L_0^0 \geq 1$ and $\det L = +1$ form a sub-group called the *proper orthochronous* Lorentz group, or the restricted Lorentz group. This group is continuously connected, but, since there is no continuous change of any parameter that will deform an object with $\det L = +1$ into an object with $\det L = -1$ (since this would involve passing through $\det L = 0$), this sub-group is not connected to group elements with negative determinants. We can map these disconnected sub-groups into one another, however, with the help of the discrete or *large* Lorentz transformations of *parity* (space reflection) and *time reversal*.

Group parametrization and generators The connected part of the homogeneous Lorentz group may be investigated most easily by considering an infinitesimal transformation in a representation which acts directly on spacetime tensors, i.e. a transformation which lies very close to the identity and whose representation indices A, B are spacetime indices μ, ν. This is the form which is usually required, and the only form we have discussed so far, but it is not the only representation of the group, as the discussion in the previous chapter should convince us. We can write such an *infinitesimal* transformation, $L(\epsilon)$, in terms of a symmetric part and an anti-symmetric part, without loss of generality:

$$L(\epsilon) = I + \epsilon(\tilde{\omega} + \overline{\omega}), \tag{9.94}$$

where $\tilde{\omega}$ is an anti-symmetric matrix, and I and $\overline{\omega}$ together form the symmetric part. ϵ is a vanishingly small (infinitesimal) number. Thus we write, with indices,

$$L_\mu^\rho(\epsilon) = \delta_\mu^\rho + \epsilon(\tilde{\omega}_\mu^{\ \rho} + \overline{\omega}_\mu^{\ \rho}). \tag{9.95}$$

Note that, for general utility, the notation commonly appearing in the literature is used here, but beware that the notation is used somewhat confusingly. Some words of explanation are provided below. Substituting this form into eqn. (9.81) gives, to first order in ϵ,

$$g_{\mu\nu}(x)L^{\mu}{}_{\rho}L^{\nu}{}_{\lambda} = g_{\rho\lambda} + \epsilon(\tilde{\omega}_{\rho\lambda} + \overline{\omega}_{\rho\lambda} + \tilde{\omega}_{\lambda\rho} + \overline{\omega}_{\lambda\rho}) + \cdots + O(\epsilon^2).$$
(9.96)

Comparing the left and right hand sides of this equation, we find that

$$\tilde{\omega}_{\mu\nu} = -\tilde{\omega}_{\nu\mu}$$
$$\overline{\omega}_{\mu\nu} = -\overline{\omega}_{\nu\mu} = 0.$$
(9.97)

Thus, the off-diagonal terms in $L(\epsilon)$ are anti-symmetric. This property survives exponentiation and persists in finite group elements with one subtlety, which is associated with the indefinite metric. We may therefore identify the structure of a finite Lorentz transformation, L, in spacetime block form. Note that a Lorentz transformation has one index up and one down, since it must map vectors to vectors of the same type:

$$L^{\nu}{}_{\mu} = \begin{pmatrix} L^0{}_0 & L^i{}_0 \\ L^0{}_j & L^j{}_i \end{pmatrix}.$$
(9.98)

There are two independent (reducible) parts to this matrix representing boosts $\mu, \nu = 0, i$ and rotations $\mu, \nu = i, j$. Although the generator $\tilde{\omega}_{\mu\nu}$ is purely anti-symmetric, the $0, i$ components form a symmetric matrix under transpose since the act of transposition involves use of the metric:

$$\left(L_0{}^i\right)^{\mathrm{T}} = -L^i{}_0 = L_i{}^0.$$
(9.99)

The second, with purely spatial components, is anti-symmetric since the generator is anti-symmetric, and the metric leaves the signs of spatial indices unchanged:

$$\left(L_i{}^j\right)^{\mathrm{T}} = -L^i{}_j.$$
(9.100)

Thus, the summary of these two may be written (with both indices down)

$$L_{\mu\nu} = -L_{\nu\mu}.$$
(9.101)

The matrix generators in a $(3 + 1)$ dimensional representation for the Lorentz group in $(3 + 1)$ spacetime dimensions, $T^{AB} = T^{\mu\nu}$, are given explicitly by

$$T_{3+1}{}^{01} = \begin{pmatrix} 0 & i & 0 & 0 \\ i & 0 & 0 & 0 \\ 0 & 0 & 0 & 0 \\ 0 & 0 & 0 & 0 \end{pmatrix}$$

$$T_{3+1}{}^{02} = \begin{pmatrix} 0 & 0 & i & 0 \\ 0 & 0 & 0 & 0 \\ i & 0 & 0 & 0 \\ 0 & 0 & 0 & 0 \end{pmatrix}$$

$$T_{3+1}{}^{03} = \begin{pmatrix} 0 & 0 & 0 & i \\ 0 & 0 & 0 & 0 \\ 0 & 0 & 0 & 0 \\ i & 0 & 0 & 0 \end{pmatrix}$$

$$T_{3+1}{}^{12} = \begin{pmatrix} 0 & 0 & 0 & 0 \\ 0 & 0 & -i & 0 \\ 0 & i & 0 & 0 \\ 0 & 0 & 0 & 0 \end{pmatrix}$$

$$T_{3+1}{}^{23} = \begin{pmatrix} 0 & 0 & 0 & 0 \\ 0 & 0 & 0 & 0 \\ 0 & 0 & 0 & -i \\ 0 & 0 & i & 0 \end{pmatrix}$$

$$T_{3+1}{}^{31} = \begin{pmatrix} 0 & 0 & 0 & 0 \\ 0 & 0 & 0 & i \\ 0 & 0 & 0 & 0 \\ 0 & -i & 0 & 0 \end{pmatrix}. \tag{9.102}$$

Note that, because of the indefinite metric, only the spatial components of these generators are Hermitian. This will lead us to reparametrize the components in terms of positive definite group indices below. It is now conventional, if not a little confusing, to write a general infinitesimal Lorentz transformation in the form

$$U_R = L_R(\omega) = I_R + \frac{1}{2} i \omega_{\mu\nu} T_R^{\mu\nu}, \tag{9.103}$$

where I_R and T_R are the identity and generator matrices of a given representation G_R. In terms of their components A, B,

$$U^A{}_B = L^A{}_B(\omega) = \delta^A{}_B + \frac{i}{2} \omega_{\rho\sigma} [T_R^{\rho\sigma}]^A{}_B. \tag{9.104}$$

The second term here corresponds to the second term in eqn. (9.95), but the spacetime-specific indices μ in eqn. (9.95) have now been replaced by representation indices A, B, anticipating a generalization to other representations. A general finite element of the group in a representation G_R is obtained by exponentiation,

$$L^A{}_B = \exp\left(\frac{i}{2}\omega_{\rho\sigma}\left[T_R{}^{\rho\sigma}\right]^A{}_B\right) \tag{9.105}$$

Let us take a moment to understand this form, since it appears repeatedly in the literature without satisfactory explanation. The $\omega^{\mu\nu}$ which appears here is not the same as $\epsilon\tilde{\omega}^{\mu\nu}$ a priori (but see the next point). In fact, it plays the role of the group parameters θ^a in the previous chapter. Thus, in the language of the previous chapter, one would write

$$U^A{}_B = L^A{}_B(\epsilon) = \delta^A{}_B + \frac{i}{2}\theta^a[T_R^a]^A{}_B$$

$$L^A{}_B = \exp\left(\frac{i}{2}\theta^a\left[T_R{}^a\right]^A{}_B\right). \tag{9.106}$$

It is easy to see that the use of two indices is redundant notation, since most of the elements of the generators are zeros. It is simply a convenient way to count to the number of non-zero group dimensions d_G in terms of spacetime indices μ, $\nu = 0, \ldots, n+1$ rather than positive definite $a, b = 1, \ldots, d_G$ indices of the group space. The factor of $\frac{1}{2}$ in eqn. (9.105) accounts for the double counting due to the anti-symmetry in the summation over all μ, ν indices. The fact that two indices are used in this summation, rather than the usual one index in T^a, should not lead to confusion. To make contact with the usual notation for generators, we may take the $(3+1)$ dimensional case as an example. In $(3+1)$ dimensions, the homogeneous Lorentz group has $d_G = 6$, and its complement of generators may be written:

$$T^a = \left\{T_{3+1}{}^{10}, T_{3+1}{}^{20}, T_{3+1}{}^{30}, T_{3+1}{}^{12}, T_{3+1}{}^{23}, T_{3+1}{}^{31}\right\}, \tag{9.107}$$

where $a = 1, \ldots, 6$ and the group elements in eqn. (9.105) have the form

$$\exp\left(i\theta^a T^a\right). \tag{9.108}$$

The first three T^a are the generators of boosts (spacetime rotations), while the latter three are the generators of spatial rotations. The anti-symmetric matrix of parameters $\omega_{\mu\nu}$ contains the components of the rapidity α^i from eqn. (9.68) as well as the angles θ^i which characterize rotations. Eqn. (9.105) is general for any representation of the Lorentz group in $n+1$ dimensions with an appropriate set of matrix generators $T_{\mu\nu}$.

Lie algebra in 3 + 1 dimensions The generators above satisfy a Lie algebra relation which can be written in several equivalent forms. In terms of the two-index parametrization, one has

$$[T_R{}^{\mu\nu}, T_R{}^{\rho\sigma}] = \mathrm{i}\, (\eta^{\nu\sigma} T_R{}^{\mu\rho} + \eta^{\mu\sigma} T_R{}^{\nu\rho} - \eta^{\mu\rho} T_R{}^{\nu\sigma} - \eta^{\rho\nu} T_R{}^{\mu\sigma}) \,.$$

$$(9.109)$$

This result applies in any number of dimensions. To see this, it is necessary to tie up a loose end from the discussion of the parameters $\omega_{\mu\nu}$ and $\epsilon\tilde{\omega}_{\mu\nu}$ above. While these two quantities play formally different roles, in the way they are introduced above they are in fact equivalent to one another and can even be defined to be equal. This is not in contradiction with what is stated above, where pains were made to distinguish these two quantities formally. The resolution of this point comes about by distinguishing carefully between which properties are special for a specific representation and which properties are general for all representations. Let us try to unravel this point.

The Lorentz transformation is defined in physics by the effect it has on spacetime reference frames (see figure 9.1). If we take this as a starting point, then we must begin by dealing with a representation in which the transformations act on spacetime vectors and tensors. This is the representation in which $A, B \to \mu\nu$, and we can write an infinitesimal transformation as in eqn. (9.95). The alternative form in eqn. (9.104) applies for any representation. If we compare the two infinitesimal forms, it seems clear that $\tilde{\omega}_{\mu\nu}$ plays the role of a generator T_{AB}, and in fact we can make this identification complete by defining

$$\epsilon\tilde{\omega}^{\mu}{}_{\nu} = \frac{\mathrm{i}}{2} \left[\omega_{\rho\lambda} T_{3+1}^{\rho\lambda} \right]^{\mu}{}_{\nu} \,.$$

$$(9.110)$$

This is made clearer if we make the identification again, showing clearly the representation specific indices:

$$\epsilon\tilde{\omega}^{A}{}_{B} = \frac{\mathrm{i}}{2} \left[\omega_{\rho\lambda} T_{3+1}^{\rho\lambda} \right]^{A}{}_{B} \,.$$

$$(9.111)$$

This equation is easily satisfied by choosing

$$\left[T_{3+1}^{\rho\sigma} \right] \sim \eta^{\rho A} \eta^{\sigma}{}_{B} \,.$$

$$(9.112)$$

However, we must be careful about preserving the anti-symmetry of T_{3+1}, so we have

$$\left[T_{3+1}^{\rho\sigma} \right]^{A}{}_{B} = \frac{2}{\mathrm{i}} \times \frac{1}{2} \left(\eta^{\rho A} \eta^{\sigma}{}_{B} - \eta^{\rho}{}_{B} \eta^{\sigma A} \right) \,.$$

$$(9.113)$$

Clearly, this equation can only be true when A, B representation indices belong to the set of $(3 + 1)$ spacetime indices, so this equation is only true in one representation. Nevertheless, we can use this representation-specific result to determine the algebra relation which is independent of representation as follows. By writing

$$[T_{3+1}^{\mu\nu}]^A{}_B = i\left(\eta^{\mu A}\eta^\nu{}_B - \eta^\mu{}_B\eta^{\nu A}\right)$$
$$[T_{3+1}^{\rho\sigma}]^B{}_C = i\left(\eta^{\rho B}\eta^\sigma{}_C - \eta^\rho{}_C\eta^{\sigma B}\right),$$
(9.114)

it is straightforward to compute the commutator,

$$[T_R^{\mu\nu}, T_R^{\rho\sigma}]^A{}_C,$$
(9.115)

in terms of η tensors. Each contraction over B leaves a new η with only spacetime indices. The remaining η's have mixed A, μ indices and occur in pairs, which can be identified as generators by reversing eqn. (9.113). The result with A, C indices suppressed is given by eqn. (9.109). In fact, the expression is uniform in indices A, C and thus these 'cancel' out of the result; more correctly they may be generalized to any representation.

The representations of the restricted homogeneous Lorentz group are the solutions to eqn. (9.109). The finite-dimensional, irreducible representations can be labelled by two discrete indices which can take values in the positive integers, positive half-integers and zero. This may be seen by writing the generators in a vector form, analogous to the electric and magnetic components of the field strength $F^{\mu\nu}$ in $(3+1)$ dimensions:

$$J^i \equiv T_B^i = \frac{1}{2}\epsilon_{ijk}T^{jk} = (T^{32}, T^{13}, T^{21})$$
$$K^i \equiv T_E^i/c = \qquad T^{0i} = (T^{01}, T^{02}, T^{03}).$$
(9.116)

These satisfy the Lie algebra commutation rules

$$[T_B^i, T_B^j] = i\epsilon^{ijk}T_B^k$$
$$[T_E^i, T_E^j] = -i\epsilon^{ijk}T_E^k/c^2$$
$$[T_E^i, T_B^j] = i\epsilon^{ijk}T_E^k.$$
(9.117)

Also, as with electromagnetism, one can construct the invariants

$$T^a T^a = \frac{1}{2}T_{R\mu\nu}T_R^{\mu\nu} = T_B^2 - T_E^2/c^2$$
$$\frac{1}{8}\epsilon^{\mu\nu\rho\sigma}T_R^{\mu\nu}T_R^{\rho\sigma} = -T_E^i T_{Bi}/c.$$
(9.118)

These quantities are Casimir invariants. They are proportional to the identity element in any representation, and thus their values can be used to label the representations. From this form of the generators we obtain an interesting perspective on electromagnetism: its form is an inevitable expression of the properties of the Lorentz group for vector fields. In other words, the constraints of relativity balanced with the freedom in a vector field determine the form of the action in terms of representations of the restricted group.

The structure of the group can be further unravelled and related to earlier discussions of the Cartan–Weyl basis by forming the new Hermitian operators

$$E_i = \frac{1}{2} \chi_h \left(T_B + i T_E / c \right)$$

$$F_i = \frac{1}{2} \chi_h \left(T_B - i T_E / c \right) \tag{9.119}$$

which satisfy the commutation rules

$$[E_i, E_j] = i \chi_h \, \epsilon_{ijk} E_k$$
$$[F_i, F_j] = i \chi_h \, \epsilon_{ijk} F_k$$
$$[E_i, F_j] = 0. \tag{9.120}$$

The scale factor, χ_h, is included for generality. It is conventional to discuss angular momentum directly in quantum mechanics texts, for which $\chi_h \to \hbar$. For pure rotation, we can take $\chi_h = 1$. As a matter of principle, we choose to write χ_h rather than $\hbar$, since there is no reason to choose a special value for this scale on the basis of group theory alone. The special value $\chi_h = \hbar$ is the value which is measured for quantum mechanical systems. The restricted Lorentz group algebra now has the form of two copies of the rotation algebra $su(2)$ in three spatial dimensions, and the highest weights of the representations of these algebras will be the two labels which characterize the full representation of the Lorentz group representations.

From the commutation rules (and referring to section 8.5.10), we see that the algebra space may be spanned by a set of basis vectors $((2\Lambda_{\max} + 1)(2\Lambda'_{\max} + 1)$ of them). It is usual to use the notation

$$\Lambda_c = \chi_h \, (m_e, m_f)$$
$$\Lambda_{\max} = \chi_h \, (e, f) \tag{9.121}$$

in physics texts, where they are referred to as quantum numbers rather than algebra eigenvalues. Also, the labels j_1, j_2 are often used for e, f, but, in the interest of a consistent and unique notation, it is best not to confuse these with the eigenvalues of the total angular momentum J_i which is slightly different. In terms of these labels, the Lorentz group basis vectors may be written as $|e, m_e; f, m_f\rangle$, where $-e \leq m_e \leq e$, $-f \leq m_f \leq f$, and e, m_e, f, m_f take on integer or half-integer values. The Cartan–Weyl stepping operators are then, by direct transcription from section 8.5.10,

$$E_{\pm}|e, m_e; f, m_f\rangle = (E_1 \pm iE_2)|e, m_e; f, m_f\rangle$$
$$= \chi_h \sqrt{(e \mp m_e)(e \pm m_e + 1)} \, |e, m_e \pm 1; f, m_f\rangle$$
$$E_3|e, m_e; f, m_f\rangle = \chi_h \, m_e|e, m_e; f, m_f\rangle \tag{9.122}$$

and

$$F_{\pm}|e, m_e; f, m_f\rangle = (F_1 \pm iF_2)|e, m_e; f, m_f\rangle$$
$$= \chi_h \sqrt{(f \mp m_f)(f \pm m_f + 1)} \, |e, m_e; f, m_f \pm 1\rangle$$
$$F_3|e, m_j; e, m_e\rangle = \chi_h \, m_f|e, m_e; f, m_f\rangle. \tag{9.123}$$

The algebra has factorized into two $su(2)$ sub-algebras. Each irreducible representation of this algebra may be labelled by a pair (e, f), which corresponds to boosts and rotations, from the factorization of the algebra into E and F parts. The number of independent components in such an irreducible representation is $(2e + 1)(2f + 1)$ since, for every e, f can run over all of its values, and vice versa. The physical significance of these numbers lies in the extent to which they may be used to construct field theories which describe a real physical situations. Let us round off the discussion of representations by indicating how these irreducible labels apply to physical fields.

9.4.4 Different representations of the Lorentz group in 3 + 1 dimensions

The explicit form of the Lorentz group generators given in eqns. (9.102) is called the defining representation. It is also the form which applies to the transformation of a spacetime vector. Using this explicit form, we can compute the Casimir invariants for E_i and F_i to determine the values of e and f which characterize that representation. It is a straightforward exercise to perform the matrix multiplication and show that

$$E^2 = E^i E_i = \frac{1}{4} \chi_h{}^2 (T_B^2 - T_E^2/c^2) = \frac{3}{4} \chi_h{}^2 I_{3+1}, \tag{9.124}$$

where I_{3+1} is the identity matrix for the defining representation. Now, this value can be likened to the general form to determine the highest weight of the representation e:

$$E^2 = \frac{3}{4} \chi_h{}^2 I_{3+1} = e(e + 1) \chi_h{}^2 I_{3+1}, \tag{9.125}$$

whence we deduce that $e = \frac{1}{2}$. The same argument may be applied to F^2, with the same result. Thus, the defining representation is characterized by the pair of numbers $(e, f) = (\frac{1}{2}, \frac{1}{2})$.

The Lorentz transformations have been discussed so far in terms of tensors, but the independent components of a tensor are not always in an obvious form. A vector, for instance, transforms as

$$A^\mu \rightarrow L^\mu{}_\nu A^\nu, \tag{9.126}$$

but a rank 2-tensor transforms with two such Lorentz transformation matrices

$$A^{\mu\nu} \rightarrow L^\mu{}_\rho L^\nu{}_\sigma A^{\rho\sigma}. \tag{9.127}$$

The independent components of a rank 2-tensor might be either diagonal or off-diagonal, and there might be redundant zeros or terms which are identical by symmetry or anti-symmetry, but one could think of re-writing eqn. (9.127) in terms of a single larger matrix acting on a new vector where only the independent components were present, rather than two smaller matrices acting on a tensor. Again, this has to do with a choice of representations. We just pick out the components and re-write the transformations in a way which preserves their content, but changes their form.

Suppose then we do this: we collect all of the independent components of any tensor field into a column vector,

$$A^{\mu\nu\lambda\ldots}{}_{\rho\sigma\ldots} \rightarrow \begin{pmatrix} a_1 \\ a_2 \\ \vdots \\ a_N \end{pmatrix}, \tag{9.128}$$

where N is the total number of independent components in the object being acted upon, and is therefore the dimension of this representation. The array of matrices $L_\mu{}^\nu$ (one for each index) can now be replaced by a single matrix $L_\oplus$ which will have as many independent components as the product of the L's. Often such a single matrix will be reducible into block-diagonal form, i.e. a direct sum of irreducible representations.

The irreducible blocks of any $(3+1)$ spacetime-dimensional Lorentz transformation of arbitrary representation d_R are denoted $D^{(e,f)}(G_R)$. A tensor transformation of rank N might therefore decompose into a number of irreducible blocks in equivalent-vector form:

$$L_\oplus{}^A{}_B = D^{(e_1,f_1)} \oplus D^{(e_2,f_2)} \ldots \oplus D^{(e_N,f_N)}. \tag{9.129}$$

The decomposition of a product of transformations as a series of irreducible representations

$$D^{(A)} \otimes D^{(B)} = \sum_\oplus c_M D^M \tag{9.130}$$

is called the Clebsch–Gordon series. The indices A, B run over $1, \ldots, (2e + 1)(2f + 1)$ for each irreducible block. For each value of e, we may take all the

Table 9.1. Spin/helicity properties of some representations of the Lorentz group in $(3 + 1)$ dimensions.

The number of degrees of freedom (D.F.) $\phi = (2e + 1)(2f + 1)$. Note that the electromagnetic field $F_{\mu\nu}$ lacks the longitudinal mode $m_s = 0$ of the massive vector field A_μ.

Representation (e, f)	'Spin' $m_s = e + f$	D.F. ϕ	Description
$(\frac{1}{2}, 0)$	$\frac{1}{2}$	2	Weyl 2-spinor
$(0, \frac{1}{2})$	$\frac{1}{2}$	2	Weyl 2-spinor
$(0, 0)$	0	1	trivial scalar
$(\frac{1}{2}, 0) \oplus (0, \frac{1}{2})$	$\pm\frac{1}{2}$	4	Dirac 4-spinor
$(\frac{1}{2}, \frac{1}{2})$	$0, \pm 1$	4	4-vector A_μ
$(1, 0) \oplus (0, 1)$	± 1	6	anti-symm. $F_{\mu\nu}$
$(1, 1) \oplus (1, 0) \oplus (0, 1) \oplus (0, 0)$	$0, \pm 1, \pm 2$	16	rank 2-tensor

values of f in turn, and vice versa. So which representation applies to which field? We can look at this in two ways.

- We see that e, f are allowed by the general solution of the Lorentz symmetry. The values are $0, \frac{1}{2}, 1, \ldots$. We then simply construct fields which transform according to these representations and match them with physical phenomena.

- We look at fields which we know about ($\phi, A_\mu, g_{\mu\nu}, \ldots$) and determine what e, f these correspond to.

Some common values of 'spin' are listed in table 9.1. Counting the highest weights of the blocks is not difficult, but to understand the difference between a massless vector field and a massive vector field, for example (both with highest spin weight 1), we must appreciate that these fields have quite different space-time transformation properties. This is explained by the fact that there are two ways in which a spin 1 field can be constructed from irreducible representations of the Lorentz group, and they form inequivalent representations. Since we are dealing with the homogeneous Lorentz group in a given frame, the spin is the same as the total intrinsic angular momentum of the frame, and is defined by a sum of the two vectors

$$S_i \equiv J_i = E_i + F_i, \tag{9.131}$$

with maximum helicity s given by $e + f$; the range of allowed values follows in integer steps from the rules of vector addition (see section 11.7.4). The maximum value is when the vectors are parallel and the minimum value is when they are anti-parallel. Thus

$$s = \pm(e + f), \pm(e + f - 1), \ldots, \pm|e - f|. \tag{9.132}$$

The spin s is just the highest weight of the Lorentz representation. Of all the representations which one might construct for physical models, we can narrow down the possibilities by considering further symmetry properties. Most physical fields do not change their properties under parity transformations or spatial reflection. Under a spatial reflection, the generators E_i, F_i are exchanged:

$$\mathcal{P} E_i \mathcal{P}^{-1} = F_i$$
$$\mathcal{P} F_i \mathcal{P}^{-1} = E_i. \tag{9.133}$$

In order to be consistent with spatial reflections, the representations of parity-invariant fields must be symmetrical in (e, f). This means we can either make irreducible representations of the form

$$(e, e) \tag{9.134}$$

or symmetrized composite representations of the form

$$(e, f) \oplus (f, e), \tag{9.135}$$

such that exchanging $e \leftrightarrow f$ leaves them invariant.

Helicity values for spin 1 For example, a spin 1 field can be made in two ways which correspond to the massless and massive representations of the Poincaré algebra. In the first case, a spin 1 field can be constructed with the irreducible transformational properties of a vector field,

$$\left(\frac{1}{2}, \frac{1}{2} \right). \tag{9.136}$$

A field of this type would exist in nature with spin/helicities $s = 0, \pm 1$. These correspond to: (i) $2s + 1 = 1$, i.e. one longitudinal scalar component A_0, and (ii) $2s + 1 = 3$, a left or right circularly polarized vector field. This characterizes the massive Proca field, A_μ, which describes W and Z vector bosons in the electro-weak theory. However, it is also possible to construct a field which transforms as

$$(1, 0) \oplus (0, 1). \tag{9.137}$$

The weight strings from this representation have only the values $m_s = \pm 1$, the left and right circular polarizations. There is no longitudinal zero component. The values here apply to the photon field, $F_{\mu\nu}$. The symmetrization corresponds to the anti-symmetry of the electromagnetic field strength tensor. The anti-symmetry is also the key to understanding the difference between these two representations.

One reason for looking at this example is that, at first glance, it seems confusing. After all, the photon is also usually represented by a vector potential A_μ, but here we are claiming that a vector formulation is quite different from an anti-symmetric tensor formulation. There is a crucial difference between the massive vector field and the massless vector field, however. The difference can be expressed in several equivalent ways which all knit together to illuminate the theme of representations nicely.

The physical photon field, $F_{\mu\nu}$, transforms like a tensor of rank 2. Because of its anti-symmetry, it can also be written in terms of a *massless* 4-vector potential, which transforms like a *gauge-invariant* vector field. Thus, the massless vector field is associated with the anti-symmetric tensor form. The massive Proca field only transforms like a vector field with no gauge invariance. The gauge invariance is actually a direct manifestation of the difference in transformation properties through a larger invariance group with a deep connection to the Lorentz group. The true equation satisfied by the photon field is

$$\partial_\mu F^{\mu\nu} = (\Box\, \delta^\mu_\nu - \partial^\mu \partial_\nu) A_\mu = 0, \tag{9.138}$$

while the Proca field satisfies

$$(-\Box + m^2) A_\mu = 0. \tag{9.139}$$

This displays the difference between the fields. The photon field has a degree of freedom which the Proca field does not; namely, its vector formulation is invariant under

$$A_\mu \to A_\mu + (\partial_\mu s), \tag{9.140}$$

for any scalar function $s(x)$. The Proca field is not. Because of the gauge symmetry, for the photon, no coordinate transformation is complete without an associated, arbitrary gauge transformation. A general coordinate variation of these fields illustrates this (see section 4.5.2).

$$\text{Photon field} \quad \delta_x A^\mu = \epsilon_\nu F^{\nu\mu}$$
$$\text{Proca field} \quad \delta_x A^\mu = \epsilon_\nu (\partial^\nu A^\mu).$$

The difference between these two results is a gauge term. This has the consequence that the photon's gauge field formulation behaves like an element of the *conformal group*, owing to the spacetime-dependent function $s(x)$. This is very clearly illustrated in section 11.5. The gauge field A_μ must transform like this if the tensor $F_{\mu\nu} = \partial_\mu A_\nu - \partial_\nu A_\mu$ which derives from it is to transform like an element of the Lorentz group. The same is not true of the Proca field, A_μ, which is simply a vector field without complication.

Appearances can therefore be deceptive. The spin 1 vector fields might look the same, but the gauge invariance of the gauge field associates it with an anti-symmetric second-rank tensor. The anti-symmetric property of the photon tensor endows it with a property called *transversality*, which means that the physical excitations of the field E_i, B_i are transverse to the direction of propagation (i.e. to the direction of its momentum or wavenumber) k^i. This is not the case for the Proca field. It has components of its field in the direction of motion, i.e. *longitudinal* components. The extra $s = 0$ mode in the helicity values for the Proca field corresponds to a longitudinal mode.

For a massless field travelling in the x^3 direction, $k_\mu = (k, 0, 0, k)$. Transversality means that

$$k^i F_{i\mu} = \partial^i F_{i\mu} = 0, \tag{9.141}$$

which is guaranteed by Maxwell's equations away from sources. In gauge form,

$$k^i A_i = 0, \tag{9.142}$$

which can always be secured by a gauge transformation. For the massive vector field, the lack of gauge invariance means that this condition cannot be secured.

9.4.5 Other spacetime dimensions

In a different number of spacetime dimensions $n + 1$, the whole of the above $(3 + 1)$ dimensional procedure for finding the irreducible representations must be repeated, and the spin labels must be re-evaluated in the framework of a new set of representations for the Lorentz group. This will not be pursued here.

9.4.6 Factorization of proper Lorentz transformations

From the discussion of the Lie algebra above, one sees that an arbitrary element of the proper or restricted Lorentz group can be expressed as a product of a rotation and a boost. This only applies to the restricted transformations, and is only one possible way of parametrizing such a transformation. The result follows from the fact that a general boost may be written as

$$L(B) = \begin{pmatrix} \gamma & -\gamma\beta^i \\ -\gamma\beta^i & \delta_{ij} + (\gamma - 1)\frac{\beta_i\beta_j}{\beta^2} \end{pmatrix}, \tag{9.143}$$

and a rotation may be written

$$L(R) = \begin{pmatrix} 1 & 0 \\ 0 & R_{ij} \end{pmatrix}. \tag{9.144}$$

The result can be shown starting from a general Lorentz transformation as in eqn. (9.98). Suppose we operate on this group element with an inverse boost (a boost with $\beta^i \to -\beta^i$:

$$L^{-1}(B)L = \begin{pmatrix} \gamma & -\gamma\beta^i \\ -\gamma\beta^i & \delta_{ij} + (\gamma - 1)\frac{\beta_i\beta_j}{\beta^2} \end{pmatrix} \begin{pmatrix} L^0_{\ 0} & L^i_{\ 0} \\ L^0_{\ j} & L^j_{\ i} \end{pmatrix}, \tag{9.145}$$

where we define the velocity to be

$$\beta^i = -\left(\frac{L^i_{\ 0}}{L^0_{\ 0}}\right). \tag{9.146}$$

This makes

$$\gamma = L^0_{\ 0}, \tag{9.147}$$

and it then follows from eqns. (9.89) that this product has the form

$$L^{-1}(B)L = \begin{pmatrix} 1 & 0 \\ 0 & M_i^{\ j} \end{pmatrix} = L(R). \tag{9.148}$$

This result is clearly a pure rotation, meaning that we can rearrange the formula to express the original arbitrary proper Lorentz transformation as a product of a boost and a rotation,

$$L = L(B)L(R). \tag{9.149}$$

9.4.7 The inhomogeneous Lorentz group or Poincaré group in 3 + 1 dimensions

If the inhomogeneous translation term, a_μ, is not set to zero in eqn. (9.66), one is led to a richer and more complex group structure [137]. This is described by the so-called inhomogeneous Lorentz group, or Poincaré group. It is a synthesis of the physics of translations, from earlier in this chapter, and the fixed origin behaviour of the homogeneous Lorentz group. The most general transformation of this group can be written

$$x'^\mu = L^\mu_{\ \nu} x^\nu + a^\mu, \tag{9.150}$$

where a^μ is an x^μ-independent constant translation vector. These transformations cannot be represented by a $d_R = 4$ representation by direct matrix multiplication, but a $d_R = 5$ representation is possible, by analogy with eqn. (9.14), by embedding in one extra dimension:

$$U_{3+1+1}x^\mu = \begin{pmatrix} L^\mu_{\ \nu} & a_\mu \\ 0 & 1 \end{pmatrix} \begin{pmatrix} x^\mu \\ 1 \end{pmatrix} = x^\mu + a^\mu. \qquad (9.151)$$

The generic infinitesimal Poincaré transformation may be written

$$U = 1 + \frac{i}{2}\omega_{\mu\nu}T_R^{\mu\nu} + i\epsilon_\rho\, k_R^\rho, \qquad (9.152)$$

for some scale χ_h with dimensions of action. Inspired by the differential representation for the translation group, we find a differential form for the homogeneous Lorentz group, which might be combined with the translation group in a straightforward way. These are:

$$\begin{aligned} T_{\text{diff}}^{\mu\nu} &= -i(x^\mu\partial^\nu - x^\nu\partial^\mu) \\ J_i &= \frac{1}{2}\epsilon_{ijk}T^{jk} = -\frac{i}{2}\,\chi_h\,\epsilon_{ijk}(x_j\partial_k - x_k\partial_j) \\ K_i &= T_{0i} \\ p_\mu &= \chi_h\,k_\mu = -i\,\chi_h\,\partial_\mu. \end{aligned} \qquad (9.153)$$

An important difference between the inhomogeneous Lorentz group and the homogeneous Lorentz group is that the total angular momentum generator, J_i, is no longer just the intrinsic angular momentum of a field, but it can include orbital angular momentum about a point displaced from the origin. This means that we have to be more careful than before in distinguishing spin s from $j = e + k$ by defining it only in an inertial rest frame with zero momentum. It is easily verified that these representations satisfy the algebra relations. Using these forms, it is a matter of simple algebra to evaluate the full algebraic content of the Poincaré group:

$$[k_\mu, T_{\rho\sigma}] = -i(\eta_{\mu\rho}k_\sigma - \eta_{\mu\sigma}k_\rho), \qquad (9.154)$$

or equivalently

$$\begin{aligned} [k_0, J_i] &= 0 \\ [k_i, J_l] &= -i\,\chi_h\,\epsilon_{ilm}k_m. \end{aligned} \qquad (9.155)$$

These relations are trivial statements about the transformation properties of k_0 (scalar) and k_i (vector) under rotations. Using the definitions above, we also find that

$$\begin{aligned} [k_0, K_i] &= ik_i \\ [k_i, K_j] &= -i\,\chi_h\,\eta_{ij}k_0. \end{aligned} \qquad (9.156)$$

These relations show that a boost K_i affects k_0, k_i, but not k_j for $j \neq i$.

Massive fields It is a curious feature of the Poincaré group, which comes about because it arises in connection with the finite speed of light, that the mass of fields plays a role in their symmetry properties. Physically, massless fields are bound to move at the speed of light so they have no rest frame about which to define intrinsic properties, like spin, which depend on coordinate concepts. It is therefore necessary to find another way to characterize intrinsic rotation. We can expect mass to play a role since it is linked to the momentum, which is the generator of translations.

The Poincaré group leaves invariant the relation

$$\mathbf{p}^2 c^2 + m^2 c^4 = \text{const},\tag{9.157}$$

where $p_\mu = (mc, p_i)$. This is, in fact, a Casimir invariant, $p^\mu p_\mu$, up to dimensional factors. Recall from the discussion of translations that the momentum may be written

$$p_\mu = \chi_h k_\mu,\tag{9.158}$$

where k_μ is the wavenumber or reciprocal lattice vector. As in the case of the other groups, we can label the field by invariant quantities. Here we have the quadratic Casimir invariants

$$J^2 = j(j+1)\,\chi_h{}^2$$
$$p^2 = \mathbf{p}^2 c^2 + m^2 c^4,\tag{9.159}$$

which commute with the group generators and are thus independent of symmetry basis:

$$[p^2, p_\mu] = 0$$
$$[p^2, J_i] = 0$$
$$[p^2, K_i] = 0.\tag{9.160}$$

A covariant rotation operator can be identified which will be useful for discussing intrinsic in chapter 11. It is called the Pauli–Lubanski vector, and it is defined by

$$W_\mu = \frac{1}{2}\,\chi_h\,\epsilon_{\mu\nu\lambda\rho} T^{\nu\lambda} p^\rho.\tag{9.161}$$

The quadratic form, W^2, is Lorentz- and translation-invariant:

$$[W^2, p_\mu] = 0$$
$$[W^2, T_{\mu\nu}] = 0.\tag{9.162}$$

W satisfies

$$W^\mu p_\mu = 0\tag{9.163}$$

and

$$[W_\mu, W_\nu] = i\epsilon_{\mu\nu\rho\sigma} W^\rho p^\sigma \tag{9.164}$$

$$W^2 = -\frac{1}{2} \chi_h{}^2 T^{\mu\nu} T_{\mu\nu} p^2 + \chi_h{}^2 T^{\mu\nu} T_\nu{}^\lambda p_\nu p_\lambda. \tag{9.165}$$

If we consider W_μ in a rest frame where $p_i = 0$, we have

$$W^\mu_{\text{rest}} = -mc(0, J_1, J_2, J_3)_{\text{rest}} = -\frac{1}{2} mc(0, S_1, S_2, S_3), \tag{9.166}$$

where S_i may be thought of as the intrinsic (non-orbital) rotation of the field (called *spin* of the representation), which is defined by

$$S_i = J_i \Big|_{\text{rest}}. \tag{9.167}$$

Thus, W^μ is clearly a 4-vector with the properties of intrinsic rotations in a rest frame. Indeed, evaluating eqn. (9.164) in a rest frame, we find that

$$[W_i, W_j] = -imc\,\epsilon_{ijk} W^k. \tag{9.168}$$

Or setting $W_i = -mc\,J_i$, we recover the rotational algebra

$$[J_i, J_j] = i\,\chi_h\,\epsilon_{ijk} J^k. \tag{9.169}$$

Thus the Poincaré-invariant quadratic form is

$$W^2_{\text{rest}} = m^2 c^2 J^2 = m^2 c^2 j(j+1) \chi_h{}^2 I_R. \tag{9.170}$$

For classifying fields, we are interested in knowing which of the properties of the field can be determined independently (or which simultaneous eigenvalues of the symmetry operators exist). Since the rest mass m is fixed by observation, we need only specify the 3-momentum, p_i, to characterize linear motion. From eqns. (9.155), we find that J_i and p_j do not commute so they are not (non-linearly) independent, but there is a rotation (or angular momentum) which does commute with p_j. It is called the *helicity* and is defined by

$$\lambda \equiv J_i \hat{p}^i, \tag{9.171}$$

where $\hat{p}^i$ is a unit vector in the direction of the spatial 3-momentum. The commutator then becomes

$$[p_i, J_j]p^j = i\,\chi_h\,\epsilon_{ijk} p^k p^j = 0. \tag{9.172}$$

Thus, λ can be used to label the state of a field. A state vector is therefore characterized by the labels ('quantum numbers' in quantum mechanics)

$$|\Theta\rangle \equiv |m, j, p^i, \lambda\rangle, \tag{9.173}$$

i.e. the mass, the linear momentum, the highest weight of the rotational symmetry and the helicity. In a rest frame, the helicity becomes ill defined, so one must choose an arbitrary component of the spin, usually m_j as the limiting value.

We would like to know how these states transform under a given Poincaré transformation. Since the states, as constructed, are manifestly eigenstates of the momentum, a translation simply incurs a phase

$$|\Theta\rangle \rightarrow \exp\left(\mathrm{i}p^\mu a_\mu\right)|\Theta\rangle. \tag{9.174}$$

Homogeneous Lorentz transformations can be used to halt a moving state. The state $|m, j, p^i, \lambda\rangle$ can be obtained from $|m, j, 0, s_i\rangle$ by a rotation into the direction of p_i followed by a boost $\exp(\mathrm{i}\theta^i K_i)$ to set the frame in motion. Thus

$$|m, j, p^i, \lambda\rangle = L\,|m, j, 0, s_i\rangle. \tag{9.175}$$

The sub-group which leaves the momentum p_μ invariant is called the *little group* and can be used to classify the intrinsic rotational properties of a field. For massive fields in $3+1$ dimensions, the little group is covered by $SU(2)$, but this is not the case for massless fields.

Massless fields For massless fields, something special happens as a result of motion at the speed of light in a special direction. It is as though a field is squashed into a plane, and the rotational behaviour becomes two-dimensional and Abelian. The direction of motion decouples from the two orthogonal directions. Consider a state of the field

$$\Theta_\pi\rangle = |m, s, \pi, \lambda\rangle, \tag{9.176}$$

where the momentum $\pi_\mu = \pi(1, 0, 0, 1)$ is in the x^3 direction, and the Lorentz energy condition becomes $p^2 c^2 = 0$ or $p_0 = |p_i|$. This represents a 'particle' travelling in the x^3 direction at the speed of light. The little group, which leaves p_μ invariant, may be found and is generated by

$$\begin{aligned}
\Lambda_1 &= J_1 + K_1 \\
\Lambda_2 &= J_1 - K_1 \\
\Lambda_3 &= J_3.
\end{aligned} \tag{9.177}$$

Clearly, the x^3 direction is privileged. These are the generators of the two-dimensional Euclidean group of translations and rotations called $ISO(2)$ or E_2. It is easily verified from the Poincaré group generators that the little group generators commute with the momentum operator

$$[\Lambda_i, p_\mu]\,|\Theta_\pi\rangle = 0. \tag{9.178}$$

The commutation relations for Λ_i are

$$[\Lambda_3, \Lambda_1] = i\Lambda_2$$
$$[\Lambda_3, \Lambda_2] = -i\Lambda_1$$
$$[\Lambda_1, \Lambda_2] = 0. \tag{9.179}$$

The last line signals the existence of an invariant sub-group. Indeed, one can define a Cartan–Weyl form and identify an invariant sub-algebra H,

$$E_\pm = \Lambda_1 \pm i\Lambda_2$$
$$H = \Lambda_3, \tag{9.180}$$

with Casimir invariant

$$C_2 = \Lambda_1^2 + \Lambda_2^2$$
$$0 = [C_2, \Lambda_i]. \tag{9.181}$$

The stepping operators satisfy

$$[H, E_\pm] = \pm E_\pm, \tag{9.182}$$

i.e. $\Lambda_c = \pm 1$. This looks almost like the algebra for $su(2)$, but there is an important difference, namely the Casimir invariant. Λ_3 does not occur in the Casimir invariant since it would spoil its commutation properties (it has decoupled). This means that the value of $\Lambda_c = m_j$ is not restricted by the Schwarz inequality, as in section 8.5.10, to less than $\pm \Lambda_{max} = \pm j$. The stepping operators still require the solutions for $\Lambda_c = m_j$ to be spaced by integers, but there is no upper or lower limit on the allowed values of the spin eigenvalues. In order to make this agree, at least in notation, with the massive case, we label physical states by Λ_3 only, taking

$$\Lambda_1|\Theta_\pi\rangle = \Lambda_2|\Theta_\pi\rangle = 0. \tag{9.183}$$

Thus, we may take the single value $H = \Lambda_3 = \Lambda_c = m_j = \lambda$ to be the angular momentum in the direction x^3, which is the helicity, since we have taken the momentum to point in this direction. See section 11.7.5 for further discussion on this point.

9.4.8 Curved spacetime: Killing's equation

In a curved spacetime, the result of an infinitesimal translation from a point can depend on the local curvature there, i.e. the translation is position-dependent. Consider an infinitesimal inhomogeneous translation $\epsilon^\mu(x)$, such that

$$x^\mu \rightarrow L^\mu_{\ \nu} x^\nu + \epsilon^\mu(x). \tag{9.184}$$

Then we have

$$\frac{\partial x'^{\mu}}{\partial x^{\nu}} = L^{\mu}_{\ \nu} + (\partial_{\nu}\epsilon^{\mu}), \tag{9.185}$$

and

$$\begin{aligned}
\mathrm{d}s'^{2} &= g_{\mu\nu}\left(L^{\mu}_{\ \rho} + (\partial_{\rho}\epsilon^{\mu})\right)\left(L^{\nu}_{\ \sigma} + (\partial_{\sigma}\epsilon^{\nu})\right)\mathrm{d}x^{\rho}\mathrm{d}x^{\lambda} \\
&= g_{\mu\nu}\left[L^{\mu}_{\ \rho}L^{\nu}_{\ \lambda} + L^{\mu}_{\ \rho}(\partial_{\sigma}\epsilon^{\nu}) + (\partial_{\rho}\epsilon^{\mu})L^{\nu}_{\ \sigma} + \cdots + \mathrm{O}(\epsilon^{2})\right]\mathrm{d}x^{\rho}\mathrm{d}x^{\lambda}.
\end{aligned} \tag{9.186}$$

The first term here vanishes, as above, owing to the anti-symmetry of $\omega_{\mu}^{\ \rho}$. Expanding the second term using eqn. (9.95), and remembering that both $\omega_{\mu\nu}$ and $\epsilon_{\mu}(x)$ are infinitesimal so that $\epsilon^{\mu}\omega_{\rho\sigma}$ is second-order and therefore negligible, we have an additional term, which must vanish if we are to have invariance of the line element:

$$\partial_{\mu}\epsilon_{\nu} + \partial_{\nu}\epsilon_{\mu} = 0. \tag{9.187}$$

The covariant generalization of this is clearly

$$\nabla_{\mu}\epsilon_{\nu} + \nabla_{\nu}\epsilon_{\mu} = 0. \tag{9.188}$$

This equation is known as Killing's equation, and it is a constraint on the allowed transformations, $\epsilon^{\mu}(x)$, which preserve the line element, in a spacetime which is curved. A vector, $\xi^{\mu}(x)$, which satisfies Killing's equation is called a Killing vector of the metric $g_{\mu\nu}$. Since this equation is symmetrical, it has $\frac{1}{2}(n+1)^{2}+(n+1)$ independent components. Since ξ^{μ} has only $n+1$ components, the solution is over-determined. However, there are $\frac{1}{2}(n+1)^{2} - (n+1)$ anti-symmetric components in Killing's equation which are unaffected; thus there must be

$$m = (n+1) + \frac{1}{2}(n+1)^{2} - (n+1) \tag{9.189}$$

free parameters in the Killing vector, in the form:

$$\nabla_{\mu}\xi_{\nu} + \nabla_{\nu}\xi_{\mu} = 0$$
$$\xi_{\mu}(x) = a_{\mu} + \omega_{\mu\nu}x^{\nu}, \tag{9.190}$$

where $\omega_{\mu\nu} = -\omega_{\nu\mu}$. A manifold is said to be 'maximally symmetric' if it has the maximum number of Killing vectors, i.e. if the line element is invariant under the maximal number of transformations.

9.5 Galilean invariance

The relativity group which describes non-Einsteinian physics is the Galilean group. Like the Poincaré group, it contains translations, rotations and boosts.

As a group, it is no smaller, and certainly no less complicated, than the Lorentz group. In fact, it may be derived as the $c \to \infty$ limit of the Poincaré group. But there is one conceptual simplification which makes Galilean transformations closer to our everyday experience: the absence of a cosmic speed limit means that arbitrary boosts of the Galilean transformations commute with one another. This alters the algebra of the generators.

9.5.1 Physical basis

The Galilean group applies physically to objects moving at speeds much less than the speed of light. For this reason, it cannot describe massless fields at all. The care required in distinguishing massless from massive concepts in the Poincaré algebra does not arise here for that simple reason. An infinitesimal Galilean transformation involves spatial and temporal translations, now written separately as

$$x^{i'} = x^i + \delta x^i$$
$$t' = t + \delta t, \tag{9.191}$$

rotations by $\theta^i = \frac{1}{2}\epsilon^{ijk}\omega_{jk}$ and boosts by incremental velocity δv^i

$$x^{i'} = x^i - \delta v^i\, t. \tag{9.192}$$

This may be summarized by the standard infinitesimal transformation form

$$x^{i'} = \left(1 + \frac{\mathrm{i}}{2}\omega_{lm}T^{lm}\right)^i_{\ j} x^j$$
$$x^{i'} = (1 + \mathrm{i}\Theta)^i_{\ j}\, x^j, \tag{9.193}$$

where the matrix

$$\Theta \equiv k_i \delta x^i - \tilde{\omega}\delta t + \theta_i T^i_B + \delta v_i\, T^i_E. \tag{9.194}$$

The exponentiated translational part of this is clearly a plane wave:

$$U \sim \exp\,\mathrm{i}(\mathbf{k} \cdot \delta\mathbf{x} - \tilde{\omega}\delta t). \tag{9.195}$$

Galilean transformations preserve the Euclidean scalar product

$$\mathbf{x} \cdot \mathbf{y} = x^i\, y_i. \tag{9.196}$$

9.5.2 Retardation and boosts

Retardation is the name given to the delay experienced in observing the effect of a phenomenon which happened at a finite distance from the source. The delay is caused by the finite speed of the disturbance. For example, the radiation at great distances from an antenna is retarded by the finite speed of light. A disturbance in a fluid caused at a distant point is only felt later because the disturbance travels at the finite speed of sound in the fluid. The change in momentum felt by a ballistic impulse in a solid or fluid travels at the speed of transport, i.e. the rate of flow of the fluid or the speed of projectiles emanating from the source.

Retardation expresses causality, and it is important in many physical problems. In Galilean physics, it is less important than in Einsteinian physics because cause and effect in a Galilean world (where $v \ll c$) are often assumed to be linked instantaneously. This is the Galilean approximation, which treats the speed of light as effectively infinite. However, retardation transformations become a useful tool in systems where the action is not invariant under boosts. This is because they allow us to derive a covariant form by transforming a non-covariant action. For example, the action for the Navier–Stokes equation can be viewed as a retarded snapshot of a particle field in motion. It is a snapshot because the action is not covariant with respect to boosts. We also derived a retarded view of the electromagnetic field arising from a particle in motion in section 7.3.4.

Retardation can be thought of as the opposite of a boost transformation. A boost transformation is characterized by a change in position due to a finite speed difference between two frames. In a frame x' moving with respect to a frame x we have

$$x^i(t)' = x^i(t) + v^i\, t. \tag{9.197}$$

Rather than changing the position variable, we can change the way we choose to measure time taken for the moving frame to run into an event which happened some distance from it:

$$t_{\text{ret}} = t - \frac{(x' - x)^i}{v^i}. \tag{9.198}$$

Whereas the idea of simultaneity makes this idea more complicated in the Einsteinian theory, here the retarded time is quite straightforward for constant velocity, v^i, between the frames. If we transform a system into a new frame, it is sometimes convenient to parametrize it in terms of a retarded time. To do this, we need to express both coordinates and derivatives in terms of the new quantity. Considering an infinitesimal retardation

$$t_{\text{ret}} = t - \frac{\mathrm{d}x^i}{v^i}, \tag{9.199}$$

it is possible to find the transformation rule for the time derivative, using the requirement that

$$\frac{\mathrm{d}t_{\text{ret}}}{\mathrm{d}t_{\text{ret}}} = 1. \tag{9.200}$$

It may be verified that

$$
\left[\partial_t + v^i\,\partial_i\right]\left[t - \frac{\mathrm{d}x^j}{v^j}\right] = 1. \tag{9.201}
$$

Thus, one identifies

$$
\frac{\mathrm{d}}{\mathrm{d}t_{\mathrm{ret}}} = \partial_t + v^i\,\partial_j. \tag{9.202}
$$

This retarded time derivative is sometimes called the *substantive derivative*. In fluid dynamics books it is written

$$
\frac{\mathrm{D}}{\mathrm{D}t} \equiv \frac{\mathrm{d}}{\mathrm{d}t_{\mathrm{ret}}}. \tag{9.203}
$$

It is simply the retarded-time total derivative. Compare this procedure with the form of the Navier–Stokes equation in section 7.5.1 and the field of a moving charge in section 7.3.4.

9.5.3 Generator algebra

The generators T_B and T_E are essentially the same generators as those which arise in the context of the Lorentz group in eqn. (9.116). The simplest way to derive the Galilean group algebra at this stage is to consider the $c \to \infty$ properties of the Poincaré group. The symbols T_B and T_E help to identify the origins and the role of the generators within the framework of Lorentzian symmetry, but they are cumbersome for more pedestrian work. Symbols for the generators, which are in common usage are

$$
\begin{aligned}
J^i &= T_B^i \\
N^i &= T_E^i.
\end{aligned} \tag{9.204}
$$

These are subtly different from, but clearly related to, the symbols used for rotations and boosts in the Poincaré algebra. The infinitesimal parameters, θ^a, of the group are

$$
\theta^a = \left\{\delta t, \delta x^i, \theta^i, \delta v^i\right\}. \tag{9.205}
$$

In $3 + 1$ dimensions, there are ten such parameters, as there are in the Poincaré group. These are related to the symbols of the Lorentz group by

$$
\begin{aligned}
\delta v_i &= \frac{1}{2}\omega_{0i} \\
\delta x^i &= \epsilon^i \\
\delta t &= \epsilon^0/c,
\end{aligned} \tag{9.206}
$$

and

$$H + mc^2 = cp_0 = \chi_h \, c \, k_0$$
$$H = \chi_h \, \tilde{\omega} \, (= \hbar \tilde{\omega}). \tag{9.207}$$

Note that the zero point is shifted so that the energy H does not include the rest energy mc^2 of the field in the Galilean theory. This is a definition which only changes group elements by a phase and the action by an irrelevant constant. The algebraic properties of the generators are the $c \to \infty$ limit of the Poincaré algebra. They are summarized by the following commutators:

$$[k_i, k_j] = 0$$
$$[N_i, N_j] = 0$$
$$[H, k_i] = 0$$
$$[H, J_i] = 0$$
$$[H, N_i] = i \, \chi_h \, k_i$$
$$[k_i, J_l] = -i \, \chi_h \, \epsilon_{ilm} k_m$$
$$[k_i, N_j] = im \, \chi_h \, \delta_{ij}$$
$$[J_i, N_l] = i\epsilon_{ilm} N_m$$
$$[J_i, J_j] = i\epsilon_{ijk} J_k, \tag{9.208}$$

where $p_0/c \to m$ is the mass, having neglected $H/c = \chi_h \, \tilde{\omega}/c$. The Casimir invariants of the Galilean group are

$$J^i J_i, k^i K_i, N^i N_i. \tag{9.209}$$

The energy condition is now the limit of the Poincaré Casimir invariant, which is singular and asymmetrical:

$$\frac{p^i p_i}{2m} = E \tag{9.210}$$

(see section 13.5).

9.6 Conformal invariance

If we relax the condition that the line element ds^2 must be preserved, and require it only to transform isotropically (which preserves $ds^2 = 0$), then we can allow transformations of the form

$$ds^2 = -dt^2 + dx^2 + dy^2 + dz^2$$
$$\to \Omega^2(x) \left(-dt^2 + dx^2 + dy^2 + dz^2\right), \tag{9.211}$$

where $\Omega(x)$ is a non-singular, non-vanishing function of x^μ. In the action, we combine this with a similar transformation of the fields, e.g. in $n+1$ dimensions,

$$\phi(x) \rightarrow \Omega^{(1-n)/2}\phi(x). \qquad (9.212)$$

This transformation stretches spacetime into a new shape by deforming it with the function $\Omega(x)$ equally in all directions. For this reason, the conformal transformation preserves the angle between any two lines which meet at a vertex, even though it might bend straight lines into curves or vice versa.

Conformal transformations are important in physics for several reasons. They represent a deviation from systems of purely Markov processes. If a translation in spacetime is accompanied by a change in the environment, then the state of the system must depend on the history of changes which occurred in the environment. This occurs, for instance, in the curvature of spacetime, where parallel transport is sensitive to local curvature; it also occurs in gauge theories, where a change in a field's internal variables (gauge transformation) accompanies translations in spacetime, and in non-equilibrium statistical physics where the environment changes alongside dynamical processes, leading to conformal distortions of the phase space. Conformal symmetry has many applications.

Because the conformal transformation is a scaling of the metric tensor, its effect is different for different kinds of fields and their interactions. The number of powers of the metric which occurs in the action (or, loosely speaking, the number of spacetime indices on the fields) makes the invariance properties of the action and the field equations quite different. Amongst all the fields, Maxwell's free equations (a massless vector field in) in $3+1$ dimensions stand out for their general conformal invariance. This leads to several useful properties of Maxwell's equations, which many authors unknowingly take for granted. Scalar fields are somewhat different, and are conformally invariant in $1+1$ dimensions, in the massless case, in the absence of self-interactions. We shall consider these two cases below.

Consider now an infinitesimal change of coordinates, as we did in the case of Lorentz transformations:

$$x^\mu \rightarrow \Lambda^\mu{}_\nu x^\nu + \epsilon^\mu(x). \qquad (9.213)$$

The line element need not be invariant any longer; it may change by

$$ds^{2'} = \Omega^2(x)\, ds^2. \qquad (9.214)$$

Following the same procedure as in eqn. (9.186), we obtain now a condition for eqn. (9.214) to be true. To first order, we have:

$$\Omega^2(x)g_{\mu\nu} = g_{\mu\nu} + \partial_\mu\epsilon_\nu + \partial_\nu\epsilon_\mu. \qquad (9.215)$$

Clearly, ϵ^μ and $\Omega(x)$ must be related in order to satisfy this condition. The relationship is easily obtained by taking the trace of this equation, multiplying through by $g^{\mu\nu}$. This gives, in $n + 1$ dimensions,

$$(\Omega^2 - 1)(n + 1) = 2(\partial_\lambda \epsilon^\lambda). \tag{9.216}$$

Using this to replace $\Omega(x)$ in eqn. (9.215) gives us the equation analogous to eqn. (9.187), but now for the full conformal symmetry:

$$\partial_\mu \epsilon_\nu + \partial_\nu \epsilon_\mu = \frac{2}{(n + 1)} (\partial_\lambda \epsilon^\lambda) g_{\mu\nu}. \tag{9.217}$$

This is the Killing equation for the conformal symmetry. Its general solution in $n + 1$ dimensions, for $n > 1$, is

$$\epsilon^\mu(x) = a^\mu + bx^\mu + \omega^{\mu\nu} x_\nu + 2x^\mu c^\nu x_\nu - c^\mu x^2, \tag{9.218}$$

where $\omega^{\mu\nu} = -\omega^{\nu\mu}$. In $(1 + 1)$ dimensional Minkowski space, eqn. (9.217) reduces to two equations

$$\partial_0 \epsilon_0 = -\partial_1 \epsilon_1$$
$$\partial_0 \epsilon_1 = -\partial_1 \epsilon_0. \tag{9.219}$$

In two-dimensional Euclidean space, i.e. $n = 1$, followed by a Wick rotation to a positive definite metric, this equation reduces simply to the Cauchy–Riemann relations for $\epsilon^\mu(x)$, which is solved by any analytic function in the complex plane. After a Wick rotation, one has

$$\partial_0 \epsilon_0 = \partial_1 \epsilon_1$$
$$\partial_0 \epsilon_1 = -\partial_1 \epsilon_0. \tag{9.220}$$

To see that this is simply the Cauchy–Riemann relations,

$$\frac{d}{dz^*} f(z) = 0, \tag{9.221}$$

we make the identification

$$z = x^0 + ix^1$$
$$f(z) = \epsilon_0 + i\epsilon_1 \tag{9.222}$$

and note that

$$\frac{d}{dz^*} = \frac{1}{2} (\partial_0 + i\partial_1). \tag{9.223}$$

This property of two-dimensional Euclidean space reflects the well known property of analytic functions in the complex plane, namely that they all are conformally invariant and solve Laplace's equation:

$$\nabla^2 f(x^i) = 4 \frac{d}{dz} \frac{d}{dz^*} f(z) = 0. \tag{9.224}$$

It makes two-dimensional, conformal field theory very interesting. In particular it is important for string theory and condensed matter physics of critical phenomena, since the special analyticity allows one to obtain Green functions and conservation laws in the vicinity of so-called fixed points.

9.6.1 Scalar fields in $n + 1$ dimensions

We begin by writing down the action, making the appearance of the metric explicit:

$$S = \int d^{n+1}x \sqrt{g}\frac{1}{c} \left\{ \frac{1}{2}(\partial_\mu\phi)g^{\mu\nu}(\partial_\nu\phi) + V(\phi) - J\phi \right\}. \qquad (9.225)$$

Note the factor of the determinant of the metric in the volume measure: this will also scale in the conformal transformation. We now let

$$g_{\mu\nu} \to \Omega^2(x)\overline{g}_{\mu\nu}$$
$$g \to \Omega^{2(n+1)}(x)\,\overline{g}$$
$$\phi(x) \to \Omega^{(1-n)/2}(x)\overline{\phi}(x)$$
$$J \to \Omega^\alpha(x)\overline{J}, \qquad (9.226)$$

where α is presently unknown. It is also useful to define the 'connection' $\Gamma_\mu = \Omega^{-1}\partial_\mu\Omega$. We now examine the variation of the action under this transformation:

$$\delta S = \int d^{n+1}x \sqrt{\overline{g}}\Omega^{n+1}\frac{1}{c} \left\{ (\partial_\mu\Omega^{(1-n)/2}\delta\overline{\phi})\frac{\overline{g}^{\mu\nu}}{\Omega^2}(\partial_\nu\Omega^{(1-n)/2}\overline{\phi}) \right.$$
$$\left. + \delta V - \Omega^{(1-n)/2+\alpha}\overline{J}\,\delta\overline{\phi} \right\}. \qquad (9.227)$$

Integrating by parts to separate $\delta\overline{\phi}$ gives

$$\delta S = \int d^{n+1}x \sqrt{\overline{g}}\,\Omega^{n+1}\frac{1}{c}$$
$$\left\{ -(1 + n - 2)\Gamma_\mu\,\Omega^{(1-n)/2-2}\delta\overline{\phi}\overline{g}^{\mu\nu}(\partial_\nu\Omega^{(1-n)/2}\overline{\phi}) \right.$$
$$\left. - \Omega^{(1-n)/2-2}\delta\overline{\phi}\overline{g}^{\mu\nu}(\partial_\mu\partial_\nu\Omega^{(1-n)/2}\overline{\phi}) + \delta V \right\}. \qquad (9.228)$$

Notice how the extra terms involving Γ_μ, which arise from derivatives acting on Ω, are proportional to $(1 + n - 2) = n - 1$. These will clearly vanish in $n = 1$ dimensions, and thus we see how $n = 1$ is special for the scalar field. To fully express the action in terms of barred quantities, we now need to commute the factors of Ω through the remaining derivatives and cancel them against the factors in the integration measure. Each time $\Omega^{(1-n)/2}$ passes through a derivative, we pick up a term containing $\frac{1}{2}(1-n)\Gamma_\mu$, thus, provided we have $\alpha = -(n+3)/2$ and $\delta V = 0$, we may write

$$\delta S = \int d^{n+1}x \sqrt{g}\, \frac{1}{c}\left\{-\overline{\square}\,\overline{\phi} - \overline{J}\right\}\delta\overline{\phi} + \text{terms} \times (n-1). \qquad (9.229)$$

Clearly, in $1+1$ dimensions, this equation is conformally invariant, provided the source J transforms in the correct way, and the potential V vanishes. The invariant equation of motion is

$$-\overline{\square}\,\overline{\phi}(x) = \overline{J}. \qquad (9.230)$$

9.6.2 The Maxwell field in $n+1$ dimensions

The conformal properties of the Maxwell action are quite different to those of the scalar field, since the Maxwell action contains two powers of the inverse metric, rather than one. Moreover, the vector source coupling $J^\mu A_\mu$ contains a power of the inverse metric because of the indices on A_μ. Writing the action with metric explicit, we have

$$S = \int d^{n+1}x \sqrt{g}\, \frac{1}{c}\left\{\frac{1}{4}F_{\mu\nu}g^{\mu\rho}g^{\nu\lambda}F_{\rho\lambda} + J_\mu g^{\mu\nu}A_\nu\right\}. \qquad (9.231)$$

We now re-scale, as before, but with slightly different dimensional factors

$$g_{\mu\nu} \rightarrow \Omega^2(x)\overline{g}_{\mu\nu}$$
$$g \rightarrow \Omega^{2(n+1)}(x)\,\overline{g}$$
$$A_\mu(x) \rightarrow \Omega^{(3-n)/2}(x)\overline{A}_\mu(x)$$
$$J_\mu \rightarrow \Omega^\alpha \overline{J}_\mu, \qquad (9.232)$$

and vary the action to find the field equations:

$$\delta S = \int d^{n+1}x \sqrt{\overline{g}}\,\Omega^{n+1}\frac{1}{c}\left\{\partial_\mu(\delta\overline{A}_\nu\Omega^{(3-n)/2})\frac{\overline{g}^{\mu\rho}\overline{g}^{\nu\lambda}}{\Omega^4}F_{\rho\lambda}\right.$$
$$\left. + \overline{J}_\mu\overline{g}^{\mu\nu}\Omega^{(3-n)/2-2+\alpha}\delta\overline{A}_\nu\right\}. \qquad (9.233)$$

Integrating by parts, we obtain

$$\delta S = \int d^{n+1}x \frac{1}{c}\sqrt{\overline{g}}\,\Omega^{(n-3)/2}\,\delta\overline{A}_\nu\left\{(n-3)\Gamma_\mu\overline{g}^{\mu\rho}\overline{g}^{\nu\lambda}F_{\rho\lambda}\right.$$
$$\left. - \partial_\mu F_{\rho\lambda}\overline{g}^{\mu\rho}\overline{g}^{\nu\lambda} + \overline{J}_\mu\overline{g}^{\mu\nu}\Omega^{\alpha-2}\right\}. \qquad (9.234)$$

On commuting the scale factor through the derivatives using

$$\partial_\mu F_{\rho\lambda} = \frac{1}{2}(3-n)\partial_\mu\left[\Gamma_\rho\overline{A}_\lambda - \Gamma_\lambda\overline{A}_\rho\right] + \partial_\mu\overline{F}_{\rho\lambda}, \qquad (9.235)$$

we see that we acquire further terms proportional to $n - 3$. Three dimensions clearly has a special significance for Maxwell's equations, so let us choose $n = 3$ now and use the notation $\overline{\partial}_\mu$ to denote the fact that the derivative is contracted using the transformed metric $\overline{g}_{\mu\nu}$. This gives

$$\delta S = \int \mathrm{d}^{n+1}x \frac{1}{c}\sqrt{\overline{g}}\left\{-\overline{\partial}_\mu \overline{F}^{\mu\nu} + \overline{J}^\nu \Omega^{\alpha-2}\right\}\delta\overline{A}_\nu = 0. \qquad (9.236)$$

Notice that the invariance of the equations of motion, in the presence of a current, depends on how the current itself scales. Suppose we couple to the current arising from a scalar field which has the general form $J_\mu \sim \phi^* \partial_\mu \phi$, then, from the previous section, this would scale by Ω^{n-1}. For $n = 1$, this gives precisely $\alpha = n - 1 = 2$. Note, however, that the matter field itself is not conformally invariant in $n = 3$. As far as the electromagnetic sector is concerned, however, $n = 3$ gives us the conformally invariant equation of motion

$$\overline{\partial}_\mu \overline{F}^{\mu\nu} = \overline{J}^\nu. \qquad (9.237)$$

The above treatment covers only two of the four Maxwell's equations. The others arise from the Bianchi identity,

$$\epsilon^{\mu\nu\lambda\rho}\partial_\mu F_{\lambda\rho} = 0. \qquad (9.238)$$

The important thing to notice about this equation is that it is independent of the metric. All contractions are with the metric-independent, anti-symmetric tensor; the other point is precisely that it is anti-symmetric. Moreover, the field scale factor $\Omega^{3-n}/2$ is simply unity in $n = 3$, thus the remaining Maxwell equations are trivially invariant.

In non-conformal dimensions, the boundary terms are also affected by the scale factor, Ω. The conformal distortion changes the shape of a boundary, which must be compensated for by the other terms. Since the dimension in which gauge fields are invariant is different to the dimension in which matter fields are invariant, no gauge theory can be conformally invariant in flat spacetime. Conformally improved matter theories can be formulated in curved spacetime, however, in any number of dimensions (see section 11.6.3).

9.7 Scale invariance

Conformal invariance is an exacting symmetry. If we relax the x-dependence of $\Omega(x)$ and treat it as a constant, then there are further possibilities for invariance of the action. Consider

$$S = \int (\mathrm{d}x)\left\{\frac{1}{2}(\partial^\mu\phi)(\partial_\mu\phi) + \sum_l \frac{1}{l!}a_l\,\phi^l\right\}. \qquad (9.239)$$

Table 9.2. Scale-invariant potentials.

$n = 1$	$n = 2$	$n = 3$
All	$\frac{1}{6!}g\phi^6$	$\frac{1}{4!}\lambda\phi^4$

Let us scale

$$g_{\mu\nu} \to \overline{g}_{\mu\nu}\,\Omega^2$$
$$\phi(x) \to \overline{\phi}(x)\,\Omega^{-\alpha}, \tag{9.240}$$

where α is to be determined. Since the scale factors now commute with the derivatives, we can secure the invariance of the action for certain l which satisfy,

$$\Omega^{n+1}\,\Omega^{-2-2\alpha} = 1 = \Omega^{-l\alpha}, \tag{9.241}$$

which solves to give $\alpha = \frac{n+1}{2} - 1$, and hence,

$$l = \frac{n+1}{(n+1)/2 - 1}. \tag{9.242}$$

For $n = 3$, $l = 4$ solves this; for $n = 2$, $l = 6$ solves this; and for $n = 1$, it is not solved for any l since the field is dimensionless. We therefore have the globally scale-invariant potentials in table 9.2.

9.8 Breaking spacetime symmetry

The breakdown of a symmetry means that a constraint on the uniformity of a system is lost. This sometimes happens if systems develop structure. For example, if a uniformly homogeneous system suddenly becomes lumpy, perhaps because of a phase transition, then translational symmetry will be lost. If a uniform external magnetic field is applied to a system, rotational invariance is lost. When effects like these occur, one or more symmetry generators are effectively lost, together with the effect on any associated eigenvalues of the symmetry group. In a sense, the loss of a constraint opens up the possibility of more freedom or more variety in the system. In the opposite sense, it restricts the type of transformations which leave the system unchanged. Symmetry breakdown is often associated with the lifting of *degeneracy* of group eigenvalues, or quantum numbers.

There is another sense in which symmetry is said to be broken. Some calculational procedures break symmetries in the sense that they invalidate the assumptions of the original symmetry. For example, the imposition of periodic boundary conditions on a field in a crystal lattice is sometimes said to break Lorentz invariance,

$$\psi(x + L) = \psi(x). \tag{9.243}$$

The existence of a topological property such as periodicity does not itself break the Lorentz symmetry. If there is a loss of homogeneity, then translational invariance would be lost, but eqn. (9.243) does not imply this in any way: it is purely an identification of points in the system at which the wavefunction should have a given value. The field still transforms faithfully as a spacetime scalar. However, the condition in eqn. (9.243) does invalidate the assumptions of Lorentz invariance because the periodicity length L is a constant and we know that a boost in the direction of that periodicity would cause a length contraction. In other words, the fact that the boundary conditions themselves are stated in a way which is not covariant invalidates the underlying symmetry.

Another example is the imposition of a finite temperature scale $\beta = 1/kT$. This is related to the last example because, in the Euclidean representation, a finite temperature system is represented as being periodic in imaginary time (see section 6.1.5). But whether we use imaginary time or not, the idea of a constant temperature is also a non-covariant concept. If we start in a heat bath and perform a boost, the temperature will appear to change because of the Doppler shift. Radiation will be red- and blue-shifted in the direction of travel, and thus it is only meaningful to measure a temperature at right angles to the direction of travel. Again, the assumption of constant temperature does not break any symmetry of spacetime, but the ignorance of the fact that temperature is a function of the motion leads to a contradiction.

These last examples cannot be regarded as a breakdown of symmetry, because they are not properties of the system which are lost, they are only a violation of symmetry by the assumptions of a calculational procedure.

9.9 Example: Navier–Stokes equations

Consider the action for the velocity field:

$$S = \tau \int (\mathrm{d}x) \left\{ \frac{1}{2} \rho v_i (D_t v^i) + \rho v^i v^j (D_{ij}^k v_k) + \frac{\mu}{2} (\partial_i v^i)^2 + J_i v^i \right\}, \tag{9.244}$$

where

$$J_i \equiv F_i + \partial_i P, \tag{9.245}$$

and

$$D_t = \partial_t + \Gamma = \partial_t + \frac{1}{2}\left(\frac{\partial_t \rho}{\rho}\right)$$

$$D_{ij}^k = \delta^l_{\ i}\delta^m_{\ j}\ \partial^k + \Gamma^k_{ij} = \delta^l_{\ i}\delta^m_{\ j}\ \partial^k + \frac{v_i v_j}{v^4}\ \partial_m(v^m v^k), \qquad (9.246)$$

$$\rho\frac{Dv^i}{Dt} + (\partial_i P) - \mu\nabla^2 v^i = F^i, \qquad (9.247)$$

where P is the pressure and F is a generalized force. This might be the effect of gravity or an electric field in the case of a charged fluid.

These connections result from the spacetime dependence of the coordinate transformation. They imply that our transformation belongs to the conformal group rather than the Galilean group, and thus we end up with connection terms

$$\frac{Du^i}{Dt} = (\partial_t + v^j\partial_j)v^i, \qquad (9.248)$$

where

$$\partial_\mu N^\mu = 0 \qquad (9.249)$$

and $N^\mu = (N, Nv^i)$.

10

Kinematical and dynamical transformations

In addition to parameter symmetries, which express geometrical uniformity in spacetime, some symmetries relate to uniformities in the more abstract space of the dynamical variables themselves. These 'internal' symmetries can contain group elements which depend on the spacetime parameters, so that there is a cross-dependence on the internal and external parameters; they are intimately connected to the concept of 'charge' (see also chapter 12).

Internal symmetries are not necessarily divorced from geometrical (parameter) invariances, but they may be formulated independently of them. The link between the two is forged by the spacetime properties of the action principle, through interactions between fields which act as generators for the symmetries (see, for instance, section 11.5).

10.1 Global or rigid symmetries

The simplest symmetries are *global symmetries*, whose properties are independent of spacetime location. For example, the action

$$S = \int (\mathrm{d}x) \left\{ \frac{1}{2}(\partial^\mu \phi)(\partial_\mu \phi) + \frac{1}{2}m^2\phi^2 \right\} \tag{10.1}$$

is invariant under the Z_2 reflection symmetry $\phi(x) \to -\phi(x)$ at all spacetime points. This symmetry would be broken by a term of the form

$$S = \int (\mathrm{d}x) \left\{ \frac{1}{2}(\partial^\mu \phi)(\partial_\mu \phi) + \frac{1}{2}m^2\phi^2 + \frac{1}{3!}\alpha\phi^3 \right\}. \tag{10.2}$$

The next most commonly identified symmetry is the $U(1)$ phase symmetry, which is exhibited by complex fields:

$$\Phi \to e^{i\theta} \Phi. \tag{10.3}$$

256

The action

$$S = \int (\mathrm{d}x) \left\{ \frac{1}{2}(\partial^\mu \Phi^*)(\partial_\mu \Phi) + \frac{1}{2}m^2 \Phi^* \Phi \right\} \tag{10.4}$$

is invariant under this transformation. This symmetry is related to the idea of electric charge. One can say that charge is a book-keeping parameter for this underlying symmetry, or vice versa.

Multi-component fields also possess global symmetries. For instance, the model

$$S = \int (\mathrm{d}x) \left\{ \frac{1}{2}(\partial^\mu \phi_A)(\partial_\mu \phi_A) + \frac{1}{2}m^2 \phi_A \phi_A \right\} \tag{10.5}$$

is invariant under the transformation

$$\phi_A = U_A{}^B \phi_B, \tag{10.6}$$

where

$$U_A{}^B U_B{}^C = \delta_A{}^C, \tag{10.7}$$

or $U^T U = I$. This is the group of orthogonal transformations $O(N)$, where $A, B = 1, \ldots, N$. Multi-level atom bound states can be represented in this way, see, for instance, section 10.6.3. Multi-component symmetries of this kind are form groups which are generally *non-Abelian* (see chapter 23 for further details on the formulation of non-Abelian field theory).

The physical significance of global symmetries is not always clear *a priori*. They represent global correlations of properties over the whole of spacetime simultaneously, which apparently contradicts special relativity. Often the analysis of global symmetries is only a prelude to studying local ones. Even in section 10.6.3, the global symmetry appears only as a special case of a larger local symmetry. One often finds connections between spacetime symmetries and phase symmetries which make local symmetries more natural. This is especially true in curved spacetime or inhomogeneous systems.

In practice, global symmetries are mainly used in non-relativistic, small systems where simultaneity is not an issue, but there is a lingering suspicion that global symmetries are only approximations to more complex local ones.

10.2 Local symmetries

A symmetry is called *local* if it involves transformations which depend on coordinates. Allowing a phase transformation to depend on the coordinates is sometimes referred to as 'gauging the symmetry'. For example, the local version of the complex $U(1)$ symmetry is

$$\Phi \rightarrow e^{i\theta(x)} \phi$$

$$\Gamma_\mu(x) \rightarrow \Gamma_\mu - (\partial_\mu \theta). \tag{10.8}$$

The action now needs to be modified in order to account for the fact that partial derivatives do not commute with these transformations. The partial derivative is exchanged for a covariant one, which includes the connection $\Gamma_\mu(x)$,

$$D_\mu = \partial_\mu + i\Gamma_\mu. \tag{10.9}$$

$$S = \int (dx) \left\{ \frac{1}{2} (D^\mu \Phi^*)(D_\mu \Phi) + \frac{1}{2} m^2 \Phi^* \Phi \right\}. \tag{10.10}$$

The most important way in which abstract field symmetries connect with spacetime properties is through the derivative operator, since this is the generator of dynamical behaviour in continuous, holonomic systems.

10.3 Derivatives with a physical interpretation

Covariance with respect to local symmetries of the action may be made manifest by re-writing the action in terms of an effective derivative. The physical motivation for this procedure is that the ordinary partial derivative does not have an invariant physical interpretation under local symmetry transformations. By adding additional terms, called 'connections', to a partial derivative ∂_μ, one creates an 'effective derivative', D_μ, which does have an invariant meaning. Although the definition of a new object, D_μ, is essentially a notational matter, the notation is important because it assigns a unique interpretation to the new derivative symbol, in any basis. For that reason, D_μ is called a covariant derivative.

There are two related issues in defining derivatives which have a physical interpretation. The first issue has to do with the physical assumption that measurable quantities are associated with Hermitian operators (Hermitian operators have real eigenvalues). The second has to do with form invariance under specific transformations.

10.3.1 Hermiticity

According to the standard interpretation of quantum mechanics, physical quantities are derived from Hermitian operators, since Hermitian operators have real eigenvalues. Hermitian operators are self-adjoint with respect to the scalar product:

$$(\phi|\mathcal{O}|\phi) = (\mathcal{O}^\dagger \phi, \phi) = (\phi, \mathcal{O}\phi), \tag{10.11}$$

or formally

$$\mathcal{O}^\dagger = \mathcal{O}. \tag{10.12}$$

If the operator $\mathcal{O}$ is a derivative operator, it can be moved from the left hand side of the inner product to the right hand side and back by partial integration. This follows from the definition of the inner product. For example, in the case of the Schrödinger field, we have

$$(\psi_1, i\partial_\mu \psi_2) = \int d\sigma \, \psi_1^\dagger \, (-i\partial_\mu \, \psi_2)$$

$$= \int d\sigma \, (-i\partial_\mu \psi_1^\dagger) \, \psi_2$$

$$= -(i\partial_\mu \psi_1, \psi_2). \tag{10.13}$$

Partial integration moves the derivative from ψ_2 to ψ_1 and changes the sign. This sign change means that $i\partial_\mu$ is not a Hermitian operator. In order for a derivative operator to be Hermitian, it must not change sign. Thus, a quadratic derivative, ∂^2, would be Hermitian. For linear derivatives, we should symmetrize the left–right nature of the derivative. Using arrow notation to show the direction in which the derivative acts, we may write

$$i\partial_\mu \rightarrow \frac{i}{2}(\overrightarrow{\partial_\mu} - \overleftarrow{\partial_\mu}) \equiv \frac{i}{2} \overleftrightarrow{\partial_\mu} \, . \tag{10.14}$$

Partial integration preserves the sign of $\overleftrightarrow{\partial_\mu}$.

A second important situation occurs when this straightforward partial integration is obstructed by a multiplying function. This is commonly the situation for actions in curvilinear coordinates where the Jacobian in the volume measure is a function of the coordinates themselves. The same thing occurs in momentum space. To see this, we note that the volume measure in the inner product is

$$d\sigma = |J(x)| d^n x, \tag{10.15}$$

where $J(x)$ is the Jacobian of the coordinates relative to a Cartesian basis. Normally, $J(x) = \sqrt{g_{ij}(x)}$, where $g_{ij}(x)$ is the spatial metric. If we now try to integrate by parts with this volume measure, we pick up an extra term involving the derivative of this function:

$$\int d\sigma \, \psi_1^\dagger \, (-i\partial_\mu \, \psi_2) = \int d\sigma \left(-i\partial_\mu - i\frac{\partial_\mu J}{J} \right) \psi_1^\dagger \, \psi_2. \tag{10.16}$$

This problem affects x derivatives in curved x coordinates and k derivatives in Fourier transform space, on the 'mass shell'. See table 10.1.

The partial derivatives in table 10.1 are clearly not Hermitian. The problem now is the extra term owing to the coordinate-dependent measure. We can solve this problem by introducing an extra term, called a 'connection', which makes the derivative have the right properties under integration by parts. The crux of the matter is to find a linear derivative operator which changes sign under

Table 10.1. Derivatives and measures.

Derivative	Measure
∂_μ	$\sqrt{g_{ij}(x)}\mathrm{d}^n\mathbf{x}$
$\dfrac{\partial}{\partial k_\mu}$	$\dfrac{\mathrm{d}^n\mathbf{k}}{2\omega(k)}$

integration by parts, but does not pick up any new terms. Then we are back to the first example above, and further symmetrization is trivial. Consider the spacetime derivative. The problem will be solved if we define a new derivative by

$$D_\mu = \partial_\mu + \Gamma_\mu, \qquad (10.17)$$

and demand that Γ_μ be determined by requiring that D_μ only change sign under partial integration:

$$\int \mathrm{d}^n\mathbf{x}\,J(x)\,\phi_1(D_\mu\phi_2) = \int \mathrm{d}^n\mathbf{x}\,J(x)(-D_\mu\phi_1)\phi_2. \qquad (10.18)$$

Substituting eqn. (10.17) into eqn. (10.18), we find that Γ_μ must satisfy

$$-(\partial_\mu J) + M\Gamma_\mu = -M\Gamma_\mu, \qquad (10.19)$$

or

$$\Gamma_\mu = \frac{1}{2}\frac{\partial_\mu J}{J}. \qquad (10.20)$$

The new derivative D_μ can be used to construct symmetrical derivatives such as $D^2 = D_\mu D^\mu$ and $\overleftrightarrow{D}_\mu$, by analogy with the partial derivative.

10.3.2 Commutativity with transformations

The problem of additional terms arising due to the presence of functions of the coordinates occurs not just with the integration measure but also with transformations of the fields. Imagine a field theory involving the field variable $\phi(x)$, a simple scalar field satisfying an equation of motion given by

$$-\Box\phi = -\partial_\mu\partial^\mu\phi = 0. \qquad (10.21)$$

We then consider the transformation

$$\phi(x) \rightarrow \phi(x)U(x), \qquad (10.22)$$

where $U(x)$ is an arbitrary function of x. This situation crops up quite often in field theory, when $U(x)$ is a phase transformation. The first thing we notice is that our equation of motion (10.21) is neither covariant nor invariant under this transformation, since

$$\partial_\mu \phi \rightarrow (\partial_\mu \phi(x))U(x) + (\partial_\mu U(x))\phi(x). \tag{10.23}$$

Clearly eqn. (10.21) is only a special case of the equations of motion. Under a transformation we will always pick up new terms, as in eqn. (10.23), since the partial derivative does not commute with an arbitrary function $U(x)$, so $U(x)$ can never be cancelled out of the equations. But, suppose we re-write eqn. (10.23) as

$$(\partial_\mu \phi(x)U(x)) = U(x)\left(\partial_\mu + \frac{\partial_\mu U}{U}\right)\phi(x), \tag{10.24}$$

and define a new derivative

$$D_\mu = (\partial_\mu + \Gamma_\mu), \tag{10.25}$$

where $\Gamma_\mu = U^{-1}(\partial_\mu U) = \partial_\mu \ln U$, then we have

$$\partial_\mu (U(x)\phi(x)) = U(x)D_\mu(\phi(x)). \tag{10.26}$$

We can now try to make eqn. (10.21) covariant. We replace the partial derivative by a covariant one, giving

$$-\partial^2 \phi(x) = -D_\mu D^\mu \phi(x) = 0. \tag{10.27}$$

The covariance can be checked by applying the transformation

$$-D^2(U(x)\phi(x)) = -U(x)\partial^2(\phi(x)) = 0 \tag{10.28}$$

so that the factor of $U(x)$ can now be cancelled from both sides.

At this point, it almost looks as though we have achieved an invariance in the form of the equations, but that is not so. To begin with, the derivative we introduced only works for a specific function $U(x)$, and that function is actually buried in the definition of the new derivative, so all we have done is to re-write the equation in a new notation. If we change the function, we must also change the derivative. Also, if we add a source to the right hand side of the equations, then this argument breaks down. In other words, while the equation is now written in a more elegant way, it is neither covariant nor invariant since the specific values of the terms must still change from case to case.

10.3.3 Form-invariant derivatives

To obtain invariance requires another idea – and this involves a physical assumption. Instead of defining $\Gamma_\mu = U^{-1}(\partial_\mu U)$, we say that Γ_μ is itself a new physical field; in addition, we demand that the transformation law be *extended* to include a transformation of the new field Γ_μ. The new transformation rule is then

$$\phi(x) \rightarrow U(x)\phi(x)$$
$$\Gamma_\mu \rightarrow \Gamma_\mu - \frac{\partial_\mu f}{f}. \qquad (10.29)$$

Γ_μ might be zero in some basis, but not always. Under this new assumption, only the physical fields transform. The covariant derivative is form-invariant, as are the equations of motion, since Γ_μ absorbs the extra term which is picked up by the partial differentiation.

Note how this last step is a physical assumption. Whereas everything leading up to eqn. (10.28) has simply been a mathematical manipulation of the formulation, the assumption that Γ_μ is a new field, which transforms separately, is a physical assumption. This makes symmetries of this type *dynamical symmetries*, rather than coincidental kinematical symmetries, which arise simply as a matter of fortuitous cancellations.

The covariant derivative crops up in several guises – most immediately in connection with the interaction of matter with the electromagnetic field, and the invariance of probabilities under arbitrary choice of quantum phase.

10.4 Charge conjugation

A charge conjugation transformation, for a field with sufficient internal symmetry, is defined to be one which has the following properties on spin 0, $\frac{1}{2}$, and 1 fields:

$$\mathcal{C}\,\phi(x)\,\mathcal{C}^\dagger = \eta_\phi\,\phi^\dagger(x)$$
$$\mathcal{C}\,\psi(x)\,\mathcal{C}^\dagger = \eta_\psi\,\overline{\psi}^{\mathrm{T}}(x)$$
$$\mathcal{C}\,A_\mu(x)\,\mathcal{C}^\dagger = -A_\mu. \qquad (10.30)$$

Under this transformation, the sign of the gauge field (and hence the sign of the charge it represents) is reversed. It is clearly a discrete rather than a continuous transformation. In the complex scalar case, the transformation simply exchanges the conjugate pair of fields. This is easy to see in the formulation of the complex scalar as a pair of real fields (see section 19.7), where the field, A_μ, is accompanied by the anti-symmetric tensor ϵ_{AB}, which clearly changes sign on interchange of scalar field components. In the Dirac spinor case, a more complicated transformation is dictated by the Dirac matrices (see section 20.3.4).

10.5 TCP invariance

The TCP theorem [87, 88, 105, 114] asserts that any local physical Lagrangian must be invariant under the combined action of time reversal (T), parity (P) and charge conjugation (C). More specifically, it claims that the effect of CP should be the same as T. Interactions may be constructed which violate these symmetries separately, but the TCP theorem requires the product of these transformations

$$U_{\text{TCP}} = U_{\text{T}} U_{\text{P}} U_{\text{C}} \tag{10.31}$$

to be conserved:

$$U_{\text{TCP}} \, \phi(x) \, U_{\text{TCP}}^{-1} = \eta_{\text{c}}\eta_{\text{t}}\eta_{\text{p}} \, \phi^{\dagger}(-x)$$
$$U_{\text{TCP}} \, \psi(x) \, U_{\text{TCP}}^{-1} = -\gamma_5\eta_{\text{c}}\eta_{\text{t}}\eta_{\text{p}} \, \psi^{*}(-x)$$
$$U_{\text{TCP}} \, A_{\mu}(x) \, U_{\text{TCP}}^{-1} = -\eta_{\text{c}}\eta_{\text{t}}\eta_{\text{p}} \, A_{\mu}^{\dagger}(-x). \tag{10.32}$$

A choice of phase such that $\eta_{\text{c}}\eta_{\text{t}}\eta_{\text{p}} = 1$ is natural. This transformation has particularly interesting consequence in the case of a spin-$\frac{1}{2}$ field. If one considers a bi-linear term in the action, of the form

$$\Delta\mathcal{L} = \overline{\psi}_1(x) \, \mathcal{O}\psi_2(x), \tag{10.33}$$

then the application of the transformation leads to

$$
\begin{aligned}
U_{\text{TCP}} \, [\overline{\psi}_1(x) \, \mathcal{O}(x)\psi_2(x)] \, U_{\text{TCP}}^{-1} &= U_{\text{TCP}} \, [\psi_1^{\dagger}\gamma^0 \, \mathcal{O}(x)\psi_2(x)] \, U_{\text{TCP}}^{-1} \\
&= [\psi_1(-x)^{\dagger}\gamma_5\gamma^0 \, \mathcal{O}(x)\gamma_5\psi_2(-x)] \\
&= -[\overline{\psi}_1^{\dagger}(-x)\gamma_5 \, \mathcal{O}(x)\gamma_5\psi_2(-x)] \\
&= [\overline{\psi}_1^{\dagger}(-x)\gamma_5 \, \mathcal{O}(x)\gamma_5\psi_2(-x)]^{\dagger}.
\end{aligned}
\tag{10.34}
$$

In the last two lines, a minus sign appears first when commuting γ_5 through γ^0, then a second minus sign must be associated with commuting $\overline{\psi}_1$ and ψ_2. Under the combination of TCP, one also has scalar behaviour

$$\gamma_5 \, \mathcal{O}(x)\gamma_5 = -\mathcal{O}(-x). \tag{10.35}$$

Regardless of what one chooses to view as fundamental, the invariance under TCP and the anti-commutativity of the Dirac field go hand in hand

$$U_{\text{TCP}} \, [\overline{\psi}_1(x) \, \mathcal{O}(x)\psi_2(x)] \, U_{\text{TCP}}^{-1} = [\overline{\psi}_1^{\dagger}(-x)\mathcal{O}(-x)\psi_2(-x)]^{\dagger}. \tag{10.36}$$

What is noteworthy about the TCP theorem is that it relates environmental, spacetime symmetries (space and time reflection) to internal degrees of freedom (charge reflection). This result follows from the locality and Hermiticity of the action, but requires also a new result: the spin-statistics theorem, namely that spin-$\frac{1}{2}$ particles must anti-commute. This means that fermionic variables should be represented by anti-commuting Grassman variables.

10.6 Examples

The following examples show how symmetry requirements and covariance determine the structure of the action under both internal and spacetime symmetries. The link between spacetime and internal symmetry, brought markedly to bear in the TCP theorem, is also reflected through conformal symmetry and transformation connections.

10.6.1 Gauge invariance: electromagnetism

The Schrödinger equation has the form

$$\left(-\frac{\hbar^2}{2m} \partial^i \partial_i + V \right) \psi = i\partial_t \psi. \tag{10.37}$$

The wavefunction $\psi(x)$ is not a direct physical observable of this equation. However, the probability

$$P = |\psi|^2 \tag{10.38}$$

is observable. As the modulus of a complex number, the probability is invariant under phase transformations of the form

$$\psi(x) \to e^{i\theta(x)} \psi(x). \tag{10.39}$$

One expects that the Schrödinger action should be invariant under this symmetry too. It should be clear from the discussion in section 10.3 that this is not the case as long as the phase $\theta(x)$ is x-dependent; to make the Schrödinger equation invariant, we must introduce a new field, A_μ. By appealing to the phenomenology of the Aharonov–Bohm effect, one can identify A_μ with the electromagnetic vector potential.

From eqn. (2.44), one may assume the following form for the covariant derivative:

$$-i\hbar \partial_\mu \to -i\hbar D_\mu = -i\hbar \left(\partial_\mu - i\frac{e}{\hbar} A_\mu \right), \tag{10.40}$$

since it only differs from a completely general expression by some constants c, $\hbar$ and e. In explicit terms, we have chosen $\Gamma_\mu = -i\frac{e}{\hbar c} A_\mu$. The total gauge or phase transformation is now a combination of eqns. (10.37) and (10.39), and to secure invariance of the equation, we must perform both transformations together.

Applying the phase transformation and demanding that D_μ commute with the phase leads to

$$D_\mu(e^{i\theta(x)} \psi(x)) = e^{i\theta(x)} \left(\left(\partial_\mu - i\frac{e}{\hbar}(A_\mu + \partial_\mu s) \right) + i(\partial_\mu \theta) \right) \psi(x),$$

$$= e^{i\theta(x)} D_\mu(\psi(x)) \tag{10.41}$$

where $D_\mu = \partial_\mu - i\frac{e}{\hbar}A_\mu$, and the last line follows provided we take

$$i(\partial_\mu \theta) - i\frac{e}{\hbar}(\partial_\mu s) = 0. \tag{10.42}$$

Both $\theta(x)$ and $s(x)$ are completely arbitrary scalar fields, so this relation merely identifies them to be the same arbitrary quantity. We may therefore write the combined phase and gauge transformations in the final form

$$\psi(x) \rightarrow \psi'(x) = e^{i\frac{e}{\hbar}s(x)}\psi(x)$$
$$A_\mu(x) \rightarrow A'_\mu(x) = A_\mu(x) + (\partial_\mu s(x)), \tag{10.43}$$

and Schrödinger's equation in gauge-invariant form is

$$\left(-\frac{\hbar^2}{2m}D^i D_i + V\right)\psi(x) = i\hbar D_t \psi, \tag{10.44}$$

where $D_t = cD_0$. In terms of the covariant derivative, we can write the field strength tensor as a commutator:

$$[D_\mu, D_\nu] = -2i\frac{e}{\hbar}F_{\mu\nu}. \tag{10.45}$$

This may be compared with eqn. (10.58) in the following section.

10.6.2 Lorentz invariance: gravity

In the presence of a non-trivial metric $g_{\mu\nu}$, i.e. in the curved spacetime of a gravitational field, or in a curvilinear coordinate system, the Lorentz transformation is not merely a passive kinematic transformation, it has the appearance of a dynamical transformation. This change of character is accompanied by the need for a transforming connection, like the ones above, only now using a more complex rule, fit for general tensor fields.

The Lorentz-covariant derivative is usually written ∇_μ, so that covariance is obtained by substituting partial derivatives in the following manner:

$$\partial_\mu \rightarrow \nabla_\mu. \tag{10.46}$$

With Lorentz transformations there is a subtlety, since we are interested in many different representations of the Lorentz group, i.e. in tensors of different rank. For scalar fields, there is no problem for Lorentz transformations. A scalar field does not transform under a Lorentz transformation, so the partial derivative is Hermitian. In other words,

$$\nabla_\mu \phi(x) = \partial_\mu \phi(x). \tag{10.47}$$

For a vector field, however, the story is different. Now the problem is that vectors transform according to the rules of tensor transformations and the partial derivative of a vector field does not commute with Lorentz transformations themselves. To fix this, a connection is required.[1] As before, we look for a connection which makes the derivative commute with the transformation law. Consider the vector field V_μ. Let us transform it from one set of coordinates, ξ^α, ξ^β, to another, x^μ, x^ν. According to the rules of tensor transformation, we have

$$V'_\mu(\xi) = \frac{\partial \xi^\beta}{\partial x^\mu} \, V_\beta(x)$$

$$= (\overset{x}{\partial}_\mu \xi^\beta) \, V_\beta(x)$$

$$= L_\mu^{\;\beta} \, V_\beta(x). \tag{10.48}$$

Let us now introduce a derivative ∇_μ with the property that

$$\nabla(LV) = L(\nabla'V), \tag{10.49}$$

i.e. such that the derivative ∇_μ is form-invariant, but transforms dynamically under a coordinate transformation. Let us write

$$\nabla_\mu = \partial_\mu + \Gamma_{\mu?}, \tag{10.50}$$

where the question mark is to be determined. At this stage, it is not clear just how the indices will be arranged on Γ, since there are several possibilities when acting on a vector field. Let us evaluate

$$\nabla_\mu V'_\nu(x) = \nabla_\mu \left(L_\nu^{\;\beta} V_\beta \right)$$

$$= \nabla_\mu \left((\overset{x}{\partial}_\nu \xi^\beta(x)) V_\beta \right)$$

$$= (\partial_\mu + \Gamma_\mu) \left((\overset{x}{\partial}_\nu \xi^\beta(x)) V_\beta \right)$$

$$= (\partial_\mu \partial_\nu \xi^\beta) V_\beta(x) + (\partial_\nu \xi^\beta)(\partial_\mu V_\beta) + \Gamma_\mu (\partial_\nu \xi^\beta) V_\beta. \tag{10.51}$$

From the assumed form of the transformation, we expect this to be

$$L_\nu^{\;\beta}(\nabla'_\mu V_\beta) = (\overset{x}{\partial}_\mu \xi^\beta)(\partial_\mu + \Gamma'_\mu) V_\beta. \tag{10.52}$$

Comparing eqn. (10.51) and eqn. (10.52), we see that

$$(\overset{x}{\partial}_\mu \xi^\beta)\Gamma_\mu \rightarrow (\overset{x}{\partial}_\mu \xi^\beta)\Gamma'_\mu - (\partial_\mu \partial_\nu \xi^\beta). \tag{10.53}$$

---

[1] There are two ways to derive the connection for Lorentz transformations, one is to look at the Hermitian nature of the derivatives; the other is to demand that the derivative of a tensor always be a tensor. Either way, one arrives at the same answer, for essentially the same reason.

Multiplying through by $(\overset{\xi}{\partial_\alpha} x^\nu)$ and using the chain-rule, we see that the transformation of Γ must be

$$\Gamma \to \Gamma' - (\overset{\xi}{\partial_\alpha} x^\nu)(\overset{x}{\partial_\mu}\overset{x}{\partial_\nu} \xi^\beta). \tag{10.54}$$

This also shows us that there must be three indices on Γ, so that the correct formulation of the vector-covariant derivative is

$$\nabla_\mu V_\nu = \partial_\mu V_\nu - \Gamma^\lambda_{\mu\nu} V_\lambda, \tag{10.55}$$

with transformation rule

$$\Gamma^\beta_{\alpha\mu} \to \Gamma'^\beta_{\alpha\mu} - (\overset{\xi}{\partial_\alpha} x^\nu)(\overset{x}{\partial_\mu}\overset{x}{\partial_\nu} \xi^\beta). \tag{10.56}$$

Thus, demanding commutativity with a dynamical transformation, once again requires the introduction of a corrective term, or connection.

What turns a coordinate transformation into a dynamical transformation is the spacetime dependence of the metric. It makes the coordinate transformation into a spacetime-dependent quantity also, changing its status from a passive kinematical property to an active dynamical one. The non-linearity which is implied by having coordinates which depend on other coordinates is what leads Einstein's theory of gravity to use the concept of intrinsic curvature.

The above procedure can be generalized to any tensor field. Extra terms will be picked up for each index, since there is a coordinate transformation term for each index of a tensor. The sign of the correction depends on whether indices are raised or lowered, because of the mutually reciprocal nature of the transformations in these cases. To summarize, we have spacetime-covariant derivatives defined as follows:

$$\nabla_\mu \phi(x) = \partial_\mu \phi(x)$$
$$\nabla_\mu A_\nu = \partial_\mu A_\nu - \Gamma^\lambda_{\mu\nu} A_\lambda$$
$$\nabla_\mu A^\nu = \partial_\mu A^\nu + \Gamma^\nu_{\mu\lambda} A_\lambda$$
$$\nabla_\mu T^{\mu\sigma}_{\ \ \lambda} = \partial_\mu T^{\mu\sigma}_{\ \ \lambda} + \Gamma^\mu_{\rho\nu} T^{\nu\sigma}_{\ \ \lambda} + \Gamma^\sigma_{\rho\nu} T^{\mu\nu}_{\ \ \lambda} - \Gamma^\kappa T^{\mu\sigma}_{\ \ \kappa}. \tag{10.57}$$

Note that we can express the curvature as a commutator of covariant derivatives:

$$[\nabla_\mu, \nabla_\nu]\xi^\sigma = -R^\lambda_{\ \sigma\mu\nu}\xi_\lambda. \tag{10.58}$$

This may be compared with eqn. (10.45).

10.6.3 The two-level atom in a strong radiation field

It was first realized by Jaynes and Cummings that a semi-classical model of a two-level atom could reproduce the essential features of the quantum theoretical problem [79]. The two-level system has a broad repertoire of applications in physics, from spin models to the micromaser [91]. It is related to a class of Dicke models [37, 57], and, in the so-called rotating wave approximation, it becomes the Jaynes–Cummings model [79] which may be solved exactly. A Hamiltonian analysis of symmetries in this Jaynes–Cummings model is given in ref. [7].

The symmetry techniques and principles of covariant field theory can be applied to the two-level atom to solve the full model and eliminate the need for the so-called rotating wave approximation. Consider the phenomenological two-level system described by the action

$$
S = \int (\mathrm{d}x) \left[-\frac{\hbar^2}{2m} (\partial^i \psi_A)^* (\partial_i \psi_A) - \psi_A^* V_{AB}(t) \psi_B \right.
$$
$$
\left. + \frac{i\hbar}{2} (\psi^* D_t \psi - (D_t \psi)^* \psi) \right], \tag{10.59}
$$

where $A, B = 1, 2$ characterizes the two levels, $i\hbar D_t = i\hbar \partial_t + i\Gamma(t)$ in matrix notation, and $\Gamma = \Gamma_{AB}$ is an off-diagonal anti-symmetrical matrix. At frequencies which are small compared with the light-size of the atom, an atom may be considered electrically neutral. The distribution of charge within the atoms is not required here. In this approximation the leading interaction is a resonant dipole transition. The connection Γ_{AB} plays an analogous role to the electromagnetic vector potential in electrodynamics, but it possesses no dynamics of its own. Rather, it works as a constraint variable, or auxiliary Lagrange multiplier field. There is no electromagnetic vector potential in the action, since the field is electrically neutral in this formulation. Γ_{AB} refers not to the $U(1)$ phase symmetry but to the two-level symmetry. Variation of the action with respect to $\Gamma(t)$ provides us with the conserved current.

$$
\frac{\delta S}{\delta \Gamma_{AB}} = \frac{i}{2} (\psi_A^* \psi_B - \psi_B^* \psi_A), \tag{10.60}
$$

which represents the amplitude for stimulated transition between the levels. The current generated by this connection is conserved only on average, since we are not taking into account any back-reaction. The conservation law corresponds merely to

$$
\partial_t \left(\frac{\delta S}{\delta \Gamma_{AB}} \right) \propto \sin \left(2 \int X(t) \right), \tag{10.61}
$$

where $X(t)$ will be defined later. The potential $V_{AB}(t)$ is time-dependent, and comprises the effect of the level splitting as well as a perturbation mediated by the radiation field. A 'connection' $\Gamma_{21} = -\Gamma_{12}$ is introduced since the diagonalization procedure requires a time-dependent unitary transformation, and thus general covariance demands that this will transform in a different basis. The physics of the model depends on the initial value of this 'connection', and this is the key to the trivial solubility of the Jaynes–Cummings model.

In matrix form we may write the action for the matter fields

$$S = \int (dx) \, \psi_A^* \mathcal{O}_{AB} \psi_B \tag{10.62}$$

where

$$\mathcal{O} = \begin{bmatrix} -\frac{\hbar^2 \nabla^2}{2m} - V_1 - \frac{i\hbar}{2} \hbar D_t & J(t) + i\Gamma_{12} \\ J(t) - i\Gamma_{12} & -\frac{\hbar^2 \nabla^2}{2m} - V_2 - \frac{i\hbar}{2} \overset{\leftrightarrow}{D}_t \end{bmatrix}. \tag{10.63}$$

The level potentials may be regarded as constants in the effective theory. They are given by $V_1 = E_1$ and $V_2 = E_2 - \hbar\Omega_R$ where $\hbar\Omega_R$ is the interaction energy imparted by the photon during the transition, i.e. the continuous radiation pressure on the atom. In the effective theory, we must add this by hand, since we have separated the levels into independent fields which are electrically neutral; it would follow automatically in a complete microscopic theory. The quantum content of this model is now that this recoil energy is a quantized unit of $\hbar\Omega$, the energy of a photon at the frequency of the source. Also, the amplitude of the source, J, would be quantized and proportional to the number of photons on the field. If one switches off the source (which models the photon's electric field), this radiation energy does not automatically go to zero, so this form is applicable mainly to continuous operation (stimulation). The origin of the recoil is clear, however: it is the electromagnetic force's interaction with the electron, transmitted to the nucleus by binding forces. What we are approximating is clearly a $J^\mu A_\mu$ term for the electron, with neutralizing background charge.

It is now desirable to perform a unitary transformation on the action $\psi \to U\psi$, $\mathcal{O} \to U\mathcal{O}U^{-1}$, which diagonalizes the operator $\mathcal{O}$. Clearly, the connection Γ_{AB} will transform under this procedure by

$$\Gamma \to \Gamma + \frac{i\hbar}{2} \left(U(\partial_t U^{-1}) - (\partial_t U)U^{-1} \right) \tag{10.64}$$

since a time-dependent transformation is required to effect the diagonalization. For notational simplicity we define $\hat{L} = -\frac{\hbar^2 \nabla^2}{2m} - \frac{i}{2}\hbar \overset{\leftrightarrow}{D}_t$, so that the secular equation for the action is:

$$(\hat{L} - E_1 - \lambda)(\hat{L} - E_2 + \hbar\Omega - \lambda) - (J^2 + \Gamma_{12}^2) = 0. \tag{10.65}$$

Note that since $J \overset{\leftrightarrow}{\partial}_t J = 0$ there are no operator difficulties with this equation. The eigenvalues are thus

$$\lambda_\pm = \hat{L} - \overline{E}_{12} + \hbar\Omega \pm \sqrt{\frac{1}{4}(\tilde{E}_{21} - \hbar\Omega)^2 + J^2 + \Gamma_{12}^2} \tag{10.66}$$

$$\equiv \hat{L} - \overline{E}_{12} + \hbar\Omega \pm \sqrt{\hbar^2\tilde{\omega}^2 + J^2 + \Gamma_{12}^2} \qquad (10.67)$$

$$\equiv \hat{L} - \overline{E}_{12} + \hbar\Omega \pm \hbar\omega_R, \qquad (10.68)$$

where $\overline{E}_{12} = \frac{1}{2}(E_1 + E_2)$ and $\tilde{E}_{21} = (E_2 - E_1)$. For notational simplicity we define $\tilde{\omega}$ and ω_R. One may now confirm this procedure by looking for the eigenvectors and constructing U^{-1} as the matrix of these eigenvectors. This may be written in the form

$$U^{-1} = \begin{pmatrix} \cos\theta & -\sin\theta \\ \sin\theta & \cos\theta \end{pmatrix}, \qquad (10.69)$$

where

$$\cos\theta = \frac{\hbar(\tilde{\omega} + \omega_R)}{\sqrt{\hbar^2(\tilde{\omega} + \omega_R)^2 + J^2 + \Gamma_{12}^2}} \qquad (10.70)$$

$$\sin\theta = \frac{\sqrt{J^2 + \Gamma_{12}^2}}{\sqrt{\hbar^2(\tilde{\omega} + \omega_R)^2 + J^2 + \Gamma_{12}^2}}. \qquad (10.71)$$

The change in the connection $\Gamma(t)$ is thus off-diagonal and anti-symmetric, as required by the gauge symmetry conservation law:

$$U\partial_t U^{-1} = \begin{pmatrix} 0 & \partial_t\theta \\ -\partial_t\theta & 0 \end{pmatrix}. \qquad (10.72)$$

The time derivative of $\theta(t)$ may be written in one of two forms, which must agree

$$(\partial_t\theta) = \frac{\partial_t\cos\theta}{-\sin\theta} = \frac{\partial_t\sin\theta}{\cos\theta}. \qquad (10.73)$$

This provides a consistency condition, which may be verified, and leads to the proof of the identities

$$\omega_R\partial_t\omega_R = J\,\partial_t\,J + \Gamma\,\partial_t\,\Gamma \qquad (10.74)$$

and

$$\sqrt{J^2 + \Gamma^2}(\partial_t + \Lambda)\sqrt{J^2 + \Gamma^2} + (\tilde{\omega} + \omega_R)(\partial_t + \Lambda)(\tilde{\omega} + \omega_R) = 0 \qquad (10.75)$$

for arbitrary $J(t)$ and $\Gamma(t)$, where

$$\Lambda = -\frac{1}{2}\frac{\partial_t\left((\tilde{\omega} + \omega_R)^2 + J^2 + \Gamma^2\right)}{(\tilde{\omega} + \omega_R)^2 + J^2 + \Gamma^2}. \qquad (10.76)$$

These relations are suggestive of a conformal nature to the transformation and, with a little manipulation using the identities, one evaluates

$$\Gamma_{12}/\hbar = (\partial_t \theta) = \frac{(J \, \partial_t \, J + \Gamma \, \partial_t \, \Gamma)}{\omega_R \sqrt{J^2 + \Gamma^2}} \left[1 - \frac{(\tilde{\omega} + \omega_R)(\tilde{\omega} + 2\omega_R)}{(\tilde{\omega} + \omega_R)^2 + J^2 + \Gamma^2} \right].$$

(10.77)

This quantity vanishes when $J^2 + \Gamma^2$ is constant with respect to time. Owing to the various identities, the result presented here can be expressed in many equivalent forms. In particular, it is zero when $\tilde{\omega} = 0$. The equations of motion for the transformed fields are now

$$\left[\begin{array}{cc} \hat{L} - \overline{E}_{12} + \hbar\omega_R & i\partial_t\theta \\ -i\partial_t\theta & \hat{L} - \overline{E}_{12} - \hbar\omega_R \end{array} \right] \left(\begin{array}{c} \psi_+ \\ \psi_- \end{array} \right) = 0.$$

(10.78)

In this basis, the centre of mass motion of the neutral atoms factorizes from the wavefunction, since a neutral atom in an electromagnetic field is free on average. The two equations in the matrix above may therefore be unravelled by introducing a 'gauge transformation', or 'integrating factor',

$$\psi_\pm(x) = e^{\pm i \int_0^t X(t')dt'} \, \overline{\psi}(x),$$

(10.79)

where the free wavefunction in $n = 3$ dimensions is

$$\overline{\psi}(x) = \int \frac{d\omega}{(2\pi)} \frac{d^n k}{(2\pi)^n} \, e^{i(\mathbf{k}\cdot\mathbf{x} - \omega t)} \delta(\chi)$$

(10.80)

is a general linear combination of plane waves satisfying the dispersion relation for centre of mass motion

$$\chi = \frac{\hbar^2 \mathbf{k}^2}{2m} + \hbar(\Omega - \omega) - \overline{E}_{12} = 0.$$

(10.81)

The latter is enforced by the delta function. This curious mixture of continuous (ω) and discontinuous (Ω) belies the effective nature of the model and the fact that its validity is only for a continuous operation (an eternally sinusoidal radiation source which never starts or stops). The relevance of the model is thus limited by this. Substituting this form, we identify $X(t)$ as the integrating factor for the uncoupled differential equations. The complete solution is therefore

$$\psi_\pm(x) = e^{\mp i \int_0^t (\omega_R + i\partial_t\theta)dt'} \, \overline{\psi}(x).$$

(10.82)

Notice that this result is an exact solution, in the sense of being in a closed form. In the language of a gauge theory this result is gauge-dependent. This is because our original theory was not invariant under time-dependent transformations. The covariant procedure we have applied is simply a method to transform the equations into an appealing form; it does not imply invariance of the results under a wide class of sources.

That this system undergoes transitions in time may be seen by constructing wavefunctions which satisfy the boundary conditions where the probability of being in one definite state of the system is zero at $t = 0$. To this end we write $\Psi_1 = \frac{1}{2}(\psi_+ + \psi_-)$ and $\Psi_0 = \frac{1}{2i}(\psi_+ - \psi_-)$. In order to proceed beyond this point, it becomes necessary to specify the initial value of Γ_{12}. This choice carries with it physical consequences; the model is not invariant under this choice. The obvious first choice is to set this to zero. This would correspond to not making the rotating wave approximation in the usual two-level atom, with a cosine perturbation. Focusing on the state Ψ_0 which was unoccupied at $t = 0$ for $\Gamma_{12} = 0$,

$$
\Psi_0 = \sin \left(\int_0^t dt' \left[\sqrt{\tilde{\omega}^2 + \hbar^{-2} J_0^2 \cos^2(\Omega t')} \right. \right.
$$
$$
\left. \left. - i\tilde{\omega} \frac{J_0 \Omega \sin(\Omega t')}{2\hbar \omega_R} \left[\tilde{\omega} + \frac{J_0^2 \cos^2(\Omega t')}{\hbar^2(\tilde{\omega} + \omega_R)} \right]^{-1} \right] \right) \overline{\psi}(x). \qquad (10.83)
$$

We are interested in the period, and the amplitude of this quantity, whose squared norm may be interpreted as the probability of finding the system in the prepared state, given that it was not there at $t = 0$. Although the integral is then difficult to perform exactly, it is possible to express it in terms of Jacobian elliptic integrals, logarithms and trig functions. Nevertheless it is clear that $\tilde{\omega} = \frac{1}{2}(\tilde{E}_{21}/\hbar - \Omega)$ is the decisive variable. When $\hbar\tilde{\omega} \ll J_0$ is small, the first term is $J_0 \cos(\Omega t)$ and the second term is small. This is resonance, although the form of the solution is perhaps unexpected. The form of the wavefunction guarantees a normalized result which is regular at $\tilde{\omega} = 0$, and one has $\Psi_0 \sim \sin \left(\int_0^t dt' \frac{J_0}{\hbar} \cos(\Omega t') \right)$, which may be compared with the standard result of the Jaynes–Cummings model $\Psi_0 \sim \sin(J_0 t/\hbar)$. In the quantum case the amplitude of the radiation source, J_0, is quantized as an integral number, N_Ω, of photons of frequency Ω. Here we see modulation of the rate of oscillation by the photon frequency (or equivalently the level spacing). In a typical system, the photon frequency is several tens of orders of magnitude larger than the coupling strength $J_0 \ll \hbar\Omega \sim \tilde{E}_{12}$ and thus there is an extremely rapid modulation of the wavefunction. This results in an almost chaotic collapse–revival behaviour with no discernible pattern, far from the calm sinusoidal Rabi oscillations of the Jaynes–Cummings model. If $\hbar\tilde{\omega} \sim J_0$, the second term is of order unity, and then, defining the normalized resonant amplitude

$$
A = \frac{J_0}{\sqrt{\hbar^2 \tilde{\omega}^2 + J_0^2}}, \qquad (10.84)
$$

one has

$$\Psi_0 \sim \sin\left(\frac{J_0\Omega}{A} \, E\left(\Omega t, A\right) - A \int d(\Omega t) \, \frac{\sin(\Omega t)}{\sqrt{1 - A^2 \sin^2(\Omega t)}}\right) \overline{\psi}(x).$$
(10.85)

The Jacobian elliptical integral $E(\alpha, \beta)$ is a doubly periodic function, so one could expect qualitatively different behaviour away from resonance. On the other hand, far from resonance, $\hbar\tilde{\omega} \gg J_0$, the leading term of the connection becomes $\Psi_0 \sim \sin(\tilde{\omega}t)\,\overline{\psi}(x) \sim \sin(\Omega t)\,\overline{\psi}(x)$, and the effect of the level spacing is washed out.

One can also consider other values for the connection. Comparing Γ_{12} to the off-diagonal sources $\gamma^\mu D_\mu$, predicted on the basis of unitarity in effective non-equilibrium field theory [13], one obtains an indication that, if the initial connection is in phase with the time derivative of the perturbation, then one can effectively 're-sum' the decay processes using the connection. This is a back-reaction effect of the time-dependent perturbation, or a renormalization in the language of ref. [13]. If one chooses $\Gamma_{12} = J_0 \sin(\Omega t)$, this has the effect of making the off-diagonal terms in the action not merely cosines but a complex conjugate pair $J_0 \exp(\pm i\Omega t)$. This corresponds to the result one obtains from making the rotating wave approximation near resonance. This initial configuration is extremely special. With this choice, one has exactly

$$\Psi_0 = \sin\left(\int_0^t dt' \left[\sqrt{\tilde{\omega}^2 + \hbar^{-2} J_0^2}\right]\right) \overline{\psi}(x).$$
(10.86)

The stability of the solution is noteworthy, and the diagonalizing transformation is rendered trivial. The connection $\partial_t \theta$ is now zero under the diagonalizing transformation. Thus, the above result is exact, and it is the standard result of the approximated Jaynes–Cummings model. This indicates that the validity of the Jaynes–Cummings model does not depend directly on its approximation, but rather on the implicit choice of a connection.

10.7 Global symmetry breaking[2]

The dynamical properties of certain interacting fields lead to solution surfaces whose stable minima favour field configurations, which are ordered, over random ones. Such fields are said to display the phenomenon of spontaneous *ordering*, or spontaneous *symmetry breaking*. This is a phenomenon in which the average behaviour of the field, in spite of all its fluctuations, is locked into a sub-set of its potential behaviour, with less symmetry. A classic example of

---

[2] $\hbar = c = \mu_0 = \epsilon_0 = 1$ in this section.

this is the alignment of spin in ferromagnetism, in which rotational symmetry is broken into a linear alignment.

Spontaneous symmetry breaking can be discussed entirely within the framework of classical field theory, but it should be noted that its dependence on interactions raises the problem of negative energies and probabilities, which is only fully resolved in the quantum theory of fields.

When a *continuous* global symmetry is broken (i.e. when its average state does not express the full global symmetry), one sees the appearance of massless modes associated with each suppressed symmetric degree of freedom. These massless modes are called Nambu–Goldstone bosons [59, 60, 99, 100]. To see how they arise, consider the action

$$S = \int (dx) \left\{ \frac{1}{2}(\partial^\mu \phi_A)(\partial_\mu \phi_A) + \frac{1}{2}m^2 \phi_A \phi_A + \frac{\lambda}{4!}(\phi_A \phi_A)^2 \right\}. \quad (10.87)$$

The interaction potential $V(\phi) = \frac{1}{2}m^2\phi^2 + \frac{\lambda}{4!}\phi^4$ has a minimum at

$$i e \hbar c^2 \frac{\partial V(\phi)}{\partial \phi_A} = m^2 \phi_A + \frac{\lambda}{6}\phi_A(\phi_B \phi_B) = 0. \quad (10.88)$$

This would therefore be the equilibrium value for the average field. Note that a non-zero value for $\langle \phi \rangle$, within a bounded potential $\lambda > 0$, is possible only if $m^2 < 0$. Suppose one now considers the effect of fluctuations, or virtual processes, in the field. Following the procedure of chapter 6, one may split the field into an average (constant) part $\langle \phi \rangle$ and a fluctuating (quickly varying) part φ,

$$\phi_A = \langle \phi \rangle_A + \varphi_A. \quad (10.89)$$

Expressed in terms of these parts, the terms of the action become:

$$(\partial^\mu \phi_A)(\partial_\mu \phi_A) \to (\partial^\mu \varphi)(\partial_\mu \varphi)$$
$$\frac{1}{2}m^2 \phi_A \phi_A \to \frac{1}{2}(\langle \phi \rangle_A \langle \phi \rangle_A + 2\langle \phi \rangle_A \varphi_A + \varphi_A \varphi_A)$$
$$(\phi_A \phi_A)^2 \to (\langle \phi \rangle_A \langle \phi \rangle_A) + 4(\langle \phi \rangle_A \langle \phi \rangle_A)(\langle \phi \rangle_B \varphi_B)$$
$$+ 2(\varphi_A \varphi_A)(\langle \phi \rangle_B \langle \phi \rangle_B) + 4(\varphi_A \langle \phi \rangle_A)(\varphi_B \langle \phi \rangle_B)$$
$$+ 4(\langle \phi \rangle_A \varphi_A)(\varphi_B \varphi_B) + (\varphi_A \varphi_A)^2. \quad (10.90)$$

To quadratic order, the action therefore takes the form

$$S = \int (dx) \left\{ \frac{1}{2}(\partial^\mu \varphi)(\partial_\mu \varphi) + \frac{1}{2}\varphi_A \left(m^2 + \frac{\lambda}{6}\langle \phi \rangle^2 \right) \varphi_A \right.$$
$$\left. + \frac{\lambda}{6}\varphi_A(\langle \phi \rangle_A \langle \phi \rangle_B)\varphi_B + \cdots \right\}. \quad (10.91)$$

If the action is evaluated at the minimum of the potential, substituting for the minimum $\langle\phi\rangle_A$, the quadratic masslike terms do not vanish, nor is any asymmetry created. The action is still invariant under rotations in A, B space, with a different mass matrix $\lambda/3\langle\phi\rangle_A\langle\phi\rangle_B$. However, if one postulates that it is favourable to select a particular combination for $\langle\phi\rangle_A$, e.g. let $A, B = 1, 2$ and $\langle\phi\rangle_1 = 0$, $\langle\phi\rangle_2 = \langle\phi\rangle$, thus breaking the symmetry between degenerate choices, then the quadratic terms become:

$$\frac{1}{2}\varphi_1\left(m^2 + \frac{\lambda}{6}\langle\phi\rangle^2\right)\varphi_1 + \frac{1}{2}\varphi_2\left(m^2 + \frac{\lambda}{2}\langle\phi\rangle^2\right)\varphi_2. \tag{10.92}$$

The first of these terms, evaluated at the minimum, vanishes, meaning that φ_1 is a massless excitation at the equilibrium solution. It is a Nambu–Goldstone boson, which results from the selection of a special direction. The rotational A, B symmetry of the fluctuating field φ_A is still present, but the direction of the average field is now chosen at all points.

In this two-dimensional rotational example, the special direction was chosen by hand, using the *ad hoc* assumption that the scalar field would have an energetically favoured ordered state. Clearly, one could have chosen any direction (linear combination of ϕ_A from the rotational invariance), and the result would be the same, due to the original symmetry. Since these are all equivalent, it takes only the energetic selection of any one of them to lead to an ordering, and thus spontaneous symmetry breaking. In the parametrization

$$\Phi = \frac{1}{\sqrt{2}}\rho\,e^{i\theta} \tag{10.93}$$

the symmetry properties of the action become even more transparent. The action is now:

$$S = \int (dx)\left[\frac{1}{2}(\partial^\mu\rho)(\partial_\mu\rho) + \frac{1}{2}m^2\rho^2 + \frac{\lambda}{4!}\rho^4\right]. \tag{10.94}$$

This, assuming a stable average state $\rho \to \langle\rho\rangle + \rho$, gives, to quadratic order:

$$S = \int (dx)\left\{\rho\left[-\Box + m^2 + \frac{\lambda}{2}\langle\rho\rangle^2\right]\rho + \langle\rho\rangle^2\,\theta(-\Box)\theta + \cdots\right\} \tag{10.95}$$

The radial θ excitation is clearly massless. This parametrization has presented several technical challenges in the quantum theory however, so we shall not pursue it in detail.

The foregoing argument can be generalized to any continuous global group, either Abelian or non-Abelian. Suppose that the action

$$S = \int (dx)\left\{T(\partial_\mu\phi_A) - V(\phi_A)\right\} \tag{10.96}$$

is invariant under a symmetry group G, of dimension d_G; then, if it is energetically favourable for the field to develop a stable average $\langle\phi\rangle_A$ with restricted behaviour, such that

$$\phi_i \to \langle\phi\rangle_i + \varphi_i \tag{10.97}$$

for a sub-set of the components $i \in A$, there must a minimum in the potential, such that

$$\left.\frac{\partial V}{\partial \phi_i}\right|_{\phi=\langle\phi\rangle} = 0. \tag{10.98}$$

The field there splits into two parts:

$$\phi_A \to \begin{cases} \langle\phi\rangle_i + \varphi_i & \in H \\ \phi_i & \in G/H \end{cases}. \tag{10.99}$$

The first part has a stable average and small fluctuations around this value. The remainder of the components are unconstrained fluctuations, which are orthogonal in the group theoretical sense from the others. For the components with non-zero averages, one may expand the potential around the minimum:

$$V(\phi_A) = V(\phi_A)\Big|_{\phi_A=\langle\phi\rangle_A} + \frac{\partial^2 V}{\partial\varphi_A \partial\varphi_B}\Big|_{\phi_A=\langle\phi\rangle_A} \varphi_A \varphi_B + \cdots. \tag{10.100}$$

The form and value of the potential are unchanged by a group transformation G, since the action is invariant under G. Moreover, by assumption of a minimum, one must have

$$M_{AB} = \frac{\partial^2 V}{\partial\varphi_i \partial\varphi_j}\Big|_{\phi_A=\langle\phi\rangle_A} \geq 0. \tag{10.101}$$

To determine whether any of the components of this have to be zero, one uses the assumption that the average state is invariant under the sub-group H. Invariance under H means that

$$V(U_H\langle\phi\rangle) = V(\langle\phi\rangle) + \frac{\partial^2 V(\langle\phi\rangle)}{\partial\varphi_i \partial\varphi_j} \delta_H\langle\phi\rangle_i \delta_H\langle\phi\rangle_j + \cdots; \tag{10.102}$$

thus, $\delta_H\langle\phi\rangle_i = 0$ and M_{ij}^2 is arbitrary, since the transformation itself is null-potent at $\langle\phi\rangle$. However, if one transforms the average state by an element which does not belong to the restricted group H, then $\delta_G\langle\phi\rangle \neq 0$, and

$$V(U_G\langle\phi\rangle) = V(\langle\phi\rangle) + \frac{\partial^2 V(\langle\phi\rangle)}{\partial\varphi_A \partial\varphi_B} \delta_G\langle\phi\rangle_A \delta_G\langle\phi\rangle_B + \cdots. \tag{10.103}$$

Thus, for any A, B which do not belong to i, j, the mass terms $M^2_{AB} = 0$ for invariance of the potential. These are the massless modes. There are clearly $\dim G/H = d_G - d_H$ of these massless elements, which correspond to all of the fluctuations which are not constrained by the average state.

This argument does not depend on whether the group is Abelian or non-Abelian (except that the coset dimension G/H does not apply to groups like $U(1)$), only on the fact that a stable average emerges, picking out a special direction in group space. Since even a single group generator, corresponding to a single component of the field, generates a sub-group, the average field lies in a group of its own (the factor group). If the group H is an Abelian sub-group, such as Z_N, (generated by the Cartan sub-algebra of the full Lie algebra), then the resulting factor group shares the same algebra as the full group, only the centre of the group is broken. This changes the dimension of the representation, but does not change the universal cover group for the symmetry. If H is not an Abelian sub-group, then the basic algebra of the symmetry must also change.

The Nambu–Goldstone mechanism is a relative suppression of certain fluctuations, rather than a breakdown of fundamental symmetry. For example, in a crystal, with an R^n symmetry, the crystal lattice breaks up translations into R^n/Z_N, leading to massless vector fields, which are phonons.

It is not clear from the above that the choice of symmetry breaking potential is actually feasible: it has not been shown that the fluctuations around the average state are small enough to sustain the average value that was assumed. This requires a more lengthy calculation, using the generating functionals of chapter 6. Moreover, unless the result of the calculation can be determined entirely by quadratic terms, one is forced to use quantum field theory to calculate the expectation values, since there are questions of negative energies and probabilities which are only resolved by operator ordering in the second quantization. General theorems exist which prohibit the existence of Goldstone bosons, due to infra-red divergences, and thus global symmetry breaking in less than three spatial dimensions cannot occur by this mechanism [27, 97].

The occurrence of *spontaneous* symmetry breaking assumes that it will be possible to find a system in which the effective mass squared in the action is less than zero. Clearly no such fundamental fields exist: they would be tachyonic. However, composite systems, or systems influenced by external forces, can have effective mass-squared terms which have this property. This is exploited in heuristic studies of phase transitions, where one often writes the mass term as:

$$m^2(T) = \left(\frac{T - T_c}{T_c}\right) m_0^2 \tag{10.104}$$

which gives rise to a second-order phase transition at critical temperature T_c ($n > 2$), i.e. a change from an ordered average state at low temperature to a disordered state above the critical temperature.

10.8 Local symmetry breaking[3]

The predominance of gauge theories in real physical models leads one to ask whether symmetry breaking phenomena could occur in local gauge theories. Here one finds a subtly different mechanism, originally pointed out by Anderson [3], inspired by an observation of Schwinger [117], and rediscovered in the context of non-Abelian field theory by Higgs [68, 69, 70]. It is called the Anderson–Higgs mechanism, or simply the Higgs mechanism.

The action for this model is that of a complex scalar field coupled to the electromagnetic field. It is sometimes used as a simple Landau–Ginsburg model of super-conductivity (see section 12.6). It is also referred to as scalar electrodynamics. A straightforward non-Abelian generalization is used in connection with the Standard Model; this is discussed in many other references [136]. The action in complex form is written

$$S = \int (\mathrm{d}x) \left\{ (D^\mu \Phi)^\dagger (D_\mu \Phi) + m^2 \Phi^\dagger \Phi + \frac{\lambda}{3!} (\Phi^\dagger \Phi)^2 + \cdots \right.$$
$$\left. + \frac{1}{4} F^{\mu\nu} F_{\mu\nu} \right\}. \quad (10.105)$$

Here we have only written a Φ^4 interaction explicitly, with coupling constant λ. Other interactions are also possible depending on the criteria for the model. In the quantum theory, restrictions about renormalizability exclude higher powers of the field in $3 + 1$ dimensions. In $2 + 1$ dimensions one may add a term $\frac{8g}{6!}(\Phi^\dagger \Phi)^3$. Odd powers of the fields are precluded by the fact that the action must be real. The covariant derivative is usually written $D_\mu = \partial_\mu + ieA_\mu$. The conserved current generated by the gauge field A_μ is therefore

$$\frac{\delta S_\phi}{\delta A^\mu} = J_\mu = ie(\Phi^\dagger (D_\mu \Phi) - (D_\mu \Phi)^\dagger \Phi). \quad (10.106)$$

The action clearly has a basic $U(1)$ symmetry. An alternative form of the action is obtained by re-writing the complex field in terms of two real component fields ϕ_A, where $A = 1, 2$, as follows:

$$\Phi(x) = \frac{1}{\sqrt{2}} (\phi_1 + i\phi_2). \quad (10.107)$$

The covariant derivative acting on the fields can then be expanded in real and imaginary parts to give

$$D_\mu \phi_A = \partial_\mu \phi_A - e\epsilon_{AB} \phi_B A_\mu. \quad (10.108)$$

---

[3] $\hbar = c = \mu_0 = \epsilon_0 = 1$ in this section.

The action then takes the more complicated form

$$S = \int (dx) \left\{ \frac{1}{2}(\partial^\mu \phi_A)(\partial_\mu \phi_A) - e(\partial^\mu \phi_A)\epsilon_{AB} A_\mu \phi_B \right.$$
$$\left. + \frac{1}{2}e^2 \epsilon_{AB}\epsilon_{AC}\phi_B\phi_C A^\mu A_\mu + \frac{\lambda}{4!}(\phi_A\phi_A)^2 + \frac{1}{4}F^{\mu\nu}F_{\mu\nu} \right\}.$$

(10.109)

Expressed in this language, the conserved current becomes

$$J_\mu = e\,\epsilon_{AB}\,(\phi_A D_\mu \phi_B).$$

(10.110)

This shows the anti-symmetry of the current with respect to the field components in this $O(2)$ formulation.

Suppose, as before, that one component of the scalar field develops a constant non-zero expectation value $\phi_1 \to \langle\phi\rangle + \varphi_1$; the action can be expanded around this solution. Once again, this must be justified by an energy calculation to show that such a configuration is energetically favourable; is non-trivial and will not be discussed here. It is interesting to compare what happens in the presence of the Maxwell field with the case in the previous section. The part of the action, which is quadratic in φ_1, ϕ_2, A_μ is the dynamical part of the fluctuations. It is given by

$$S^{(2)} = \int (dx) \left\{ \frac{1}{2}\varphi_1 \left[-\Box + m^2 + \frac{\lambda}{2}\langle\phi\rangle^2 \right] \varphi_1 \right.$$
$$+ \frac{1}{2}\phi_2 \left[-\Box + m^2 + \frac{\lambda}{6}\langle\phi\rangle^2 + e^2\langle\phi\rangle^2 \right] \phi_2$$
$$\left. + 2e\varphi_1 A^\mu (\partial_\mu \langle\phi\rangle) + \frac{1}{2}A_\mu \left[-\Box + e^2\langle\phi\rangle^2 \right] A^\mu \right\}$$

(10.111)

This may be diagonalized with the help of the procedure analogous to eqn. (A.11) in Appendix A. The identity

$$\frac{1}{2}\phi_2 A\phi_2 + B\phi_2 = \frac{1}{2}(\phi + BA^{-1})A(\phi_2 + A^{-1}B) - \frac{1}{2}BA^{-1}B$$

(10.112)

with

$$A = \left[-\Box + m^2 + \frac{\lambda}{6}\langle\phi\rangle^2 + e^2\langle\phi\rangle^2 \right]$$
$$B = -2e\varphi_1 A^\mu (\partial_\mu \langle\phi\rangle)$$

(10.113)

results in an action of the form

$$
S^{(2)} = \int (dx) \left\{ \frac{1}{2} \varphi_1 \left[-\Box + m^2 + \frac{\lambda}{2} \langle \phi \rangle^2 \right] \varphi_1 \right.
$$

$$
\left. + \frac{1}{2} A_\mu \left[(-\Box + e^2 \langle \phi \rangle^2) g^{\mu\nu} + G \partial^\mu \partial^\nu \right] A_\nu \right\} \tag{10.114}
$$

where G is a gauge-dependent term. The details of this action are less interesting than its general characteristics. Unlike the case of the global symmetry, there is only one remaining scalar field component. The component which corresponds to the Goldstone boson, disappears in the variable transformations and re-appears as a mass for the vector field. The lack of a Goldstone boson is also interesting, since it circumvents the problems associated with Goldstone bosons in lower dimensions $n < 3$ [27, 97]. Although it is only an idealized effective theory, this local symmetry breaking mechanism indicates that symmetry breaking is indeed possible when one relaxes the rigidity of a global group.

The transmutation of the massless scalar excitation into a mass for the vector field can be seen even more transparently in the *unitary gauge*. The unitary gauge is effected by the parametrization

$$
\Phi = \frac{1}{\sqrt{2}} \rho \, e^{i\theta} \tag{10.115}
$$

$$
B_\mu = A_\mu + \frac{1}{e} \partial_\mu \theta \tag{10.116}
$$

so that the action becomes

$$
S = \int dV \left\{ \frac{1}{4} F^{\mu\nu} F_{\mu\nu} + \frac{1}{2} (\partial^\mu \rho)(\partial_\mu \rho) + \frac{1}{2} e^2 \rho^2 B^\mu B_\mu \right.
$$

$$
\left. + \frac{1}{2} m^2 \rho^2 + \frac{\lambda}{4!} \rho^4 \right\} \tag{10.117}
$$

What looks like a gauge transformation by a phase θ is now a dynamical absorption of the Goldstone boson. This is sometimes stated by saying that the Goldstone boson is 'eaten up' by the gauge field, as if the photon were some elementary particular Pacman. A more field theoretical description is to say that the Goldstone mode modulates the fluctuations of the electromagnetic field, making them move in a wavefront. This wavefront impedes the fluctuations by an amount that depends upon the gauge coupling constant e. The result is an effective mass for the gauge fluctuations, or a gap in their spectrum of excitations. However one states it, the Goldstone field ceases to be a separate excitation due to the coupling: its modulation of the vector field's zero point energy breaks the gauge invariance of the fluctuations and it re-appears, with a new status, as the extra mode of the vector field.

It cannot be emphasized enough that the assumption that there exists a stable average state of lower symmetry than the fluctuations of the theory is *ad hoc*, and its consistency has to be proven. Even today, this remains one of the toughest challenges for quantum field theory.

10.9 Dynamical symmetry breaking mechanisms

The Nambu–Goldstone or Anderson–Higgs models of symmetry breaking cannot be fundamental theories, because they do not explain how the mass-squared terms, in their Lagrangians, can become negative. As such, they must be regarded as effective actions for deeper theories. Moreover, their apparent reliance on the existence of an arbitrary scalar field has been controversial, since, in spite of the best efforts of particle physicists, no one has to date observed a Higgs scalar particle. The introduction of a scalar field is not the only way in which gauge symmetries can be broken, however. At least two other possibilities exist. Both rely on quantum dynamical calculations, but can be mentioned here.

One such mechanism was suggested in connection with field theories on topologically non-trivial spacetimes (e.g. the torus), based on an idea by Ford [52], that non-trivial average states, such as vortices could occur around topological singularities in spacetime. The main idea is that a gauge field $A_\mu \to \langle A_\mu \rangle + A_\mu$ (either Abelian or non-Abelian) can acquire a non-zero expectation value around a hole in spacetime. In simply connected spacetimes (without holes), such constant vector field configurations are gauge-equivalent to zero and thus have no invariant meaning. However, around a topological singularity, such transformations are restricted by the cohomology of the manifold. One example is that of a periodic crystal, which has the same boundary conditions as the surface of a torus, and is therefore relevant in solid state physics.

In the Abelian theory, the phenomenon is a purely classical, statistical effect, though for non-Abelian symmetries the non-linearity makes it the domain of quantum field theory. It is equivalent to there being a constant magnetic flux through the centre of the hole. In some theories, such expectation values might occur spontaneously, by the dynamics of the model (without having to assume a negative mass squared *ad hoc*). In the Abelian case, this results only in a phase. However, it was later explored in the context of non-Abelian symmetries by Hosotani [72] and Toms [129] and developed further in refs. [17, 19, 20, 21, 32, 33]. Such models are of particular interest in connection with grand unified theories, such as Kaluza–Klein and string theory, where extra dimensions are involved. Topological singularities also occur in lower dimensions in the form of vortices and the Aharonov–Bohm effect.

The second mechanism is the Coleman and Weinberg mechanism [28], which is a purely quantum effect for massless fields, whereby a non-trivial average state can be created truly spontaneously, by the non-linearities of massless scalar electrodynamics. Quantum fluctuations themselves lead to the attainment of an ordered state. It is believed that this mechanism leads to a first-order phase transition [66, 86], rather than the second-order transitions from the Goldstone and Higgs models.

11

Position and momentum

Field theory is ripe with objects referred to colloquially as coordinates and momenta. These conjugate pairs play a special role in the dynamical formulation but do not necessarily imply any dimensional relationship to actual positions or momenta.

11.1 Position, energy and momentum

In classical particle mechanics, point particles have a definite position in space at a particular time described by a dynamical trajectory $\mathbf{x}(t)$. The momentum $\mathbf{p(t)} = m \frac{d\mathbf{x}(t)}{dt}$. In addition, one has the energy of the particle, $\frac{p^2}{2m} + V$, as a book-keeping parameter for the history of the particle's momentum transactions.

In the theory of fields, there is no *a priori* notion of particles: no variable in the theory represents discrete objects with deterministic trajectories; instead there is a continuous field covering the whole of space and changing in time. The position x is a coordinate parameter, not a dynamical variable. As Schwinger puts it, the coordinates in field theory play the role of an abstract measurement apparatus [119], a ruler or measuring rod which labels the stage on which the field evolves. Table 11.1 summarizes the correspondence.

The quantum theory is constructed by replacing the classical measures of position, momentum and energy with operators satisfying certain commutation relations:

$$[\mathbf{x}, \mathbf{p}] = i\hbar \tag{11.1}$$

and

$$[t, E] = -i\hbar. \tag{11.2}$$

These operators have to act on something, and indeed they act on the fields, but the momentum and energy are represented by the operators themselves

Table 11.1. Dynamical variables.

Canonical position	Particle mechanics	Field theory
Parameter space	t	$\mathbf{x}, t$
Dynamical variable	$\mathbf{x}(t)$	$\phi(\mathbf{x}, t)$

independently of the nature of the fields. Let us see why this must be so. The obvious solution to the commutators above is to represent t and $\mathbf{x}$ by algebraic variables and E and $\mathbf{p}$ as differential operators:

$$p_i = -i\hbar\partial_i$$
$$E = i\hbar\partial_t. \tag{11.3}$$

If we check the dimensions of these operator expressions, we find that $\hbar\partial_i$ has the dimensions of momentum and that $\hbar\partial_t$ has the dimensions of energy. In other words, even though these operators have no meaning until they act on some field, like this

$$p_i\psi = -i\hbar\partial_i\psi$$
$$E\psi = i\hbar\partial_t\psi, \tag{11.4}$$

it is the operator, or its eigenvalues, which represent the momentum and energy. The field itself is merely a carrier of the information, which the operator extracts. In this way, it is possible for the classical analogues of energy and momentum, by assumption, to be represented by the same operators for all the fields. Thus the dimensions of these quantities are correct regardless of the dimensions of the field.

The expectation values of these operators are related to the components of the energy–momentum tensor (see section 11.3),

$$\overline{p}_i c = -\int d\sigma^0 \theta_{0i} = \langle p_i c \rangle$$
$$E_p = \int d\sigma^0 \theta_{00} = \langle H_D \rangle. \tag{11.5}$$

H_D is the differential Hamiltonian operator, which through the equations of motion is related to $i\hbar\partial_t$. The relationship does not work for the Klein–Gordon field, because it is quadratic in time derivatives. Because of their relationship with classical concepts of energy and momentum, E_p and P_i may also be considered as mechanical energy and momenta.

Table 11.2.   Canonical pairs for the fields.

Field	'X'	'P'
Klein–Gordon	ϕ	$\hbar^2 c^2 \partial_0 \phi$
Dirac	ψ	$\psi^\dagger$
Schrödinger	ψ	$i\hbar\psi^*$
Maxwell	A_μ	D_{0i}

Separate from these manifestations of mechanical transport are a number of other conjugate pairs. The field q itself is a basic variable in field theory, whose canonical conjugate $\partial_0 q$ is often referred to as a conjugate momentum; see table 11.2.   That these quantities do not have the dimensions of position and momentum should be obvious from these expressions; thus, it should be clear that they are in no way connected with the mechanical quantities known from the classical theory. In classical electrodynamics there is also a notion of 'hidden' momentum which results from self-interactions [71] in the field.

11.2 Particles and position

The word particle is dogged by semantic confusion in the quantum theory of matter. The classical meaning of a particle, namely a localized pointlike object with mass and definite position, no longer has a primary significance in many problems. The quantum theory of fields is often credited with re-discovering the particle concept, since it identifies countable, discrete objects with a number operator in Fock space. The objects which are counted by this operator are really *quanta*, not particles in the classical sense. They are free, delocalized, plane wave objects with infinite extent. This is no problem for physics. In fact, it is possible to speak of momentum and energy transfer, without discussing the nature of the objects which carry these labels. However, it is sometimes important to discuss localizability.

In spite of their conceptual demotion, it is clear that pointlike particle events are measured by detectors on a regular basis and thus have a practical significance. Accordingly, one is interested in determining how sharply it is possible to localize a particle in space, i.e. how sharp a peak can the wavefunction, and hence the probability, develop? Does this depend on the nature of the field, for instance, the other quantum numbers, such as mass and spin? This question was asked originally by Wigner and collaborators in the 1940s and answered for general mass and spin [6, 101].

The localizability of different types of particle depends on the existence of a Hermitian position operator which can measure it. This is related to the issue of physical derivatives in section 10.3. Finding such an operator is simple in the case of the non-relativistic Schrödinger field, but is less trivial for relativistic fields. In particular, massless fields, such as the photon, which travel at the speed of light, seem unlikely candidates for localization since they can never be halted in one place.

11.2.1 Schrödinger field

The Schrödinger field has a scalar product

$$
\begin{aligned}
(\psi, \psi) &= \int d^n\mathbf{x}\, \psi^*(x)\psi(x) \\
&= \int \frac{d^n\mathbf{k}}{(2\pi)^n}\, \psi^*(k)\psi(k).
\end{aligned}
\tag{11.6}
$$

Its wavefunctions automatically have positive energy, and thus the position operator may be written

$$
\begin{aligned}
(\psi, \hat{\mathbf{x}}\psi) &= \int d^n\mathbf{x}\, \psi^*(x)\hat{\mathbf{x}}\psi(x) \\
&= \int \frac{d^n\mathbf{k}}{(2\pi)^n}\, \psi^*(k)\left(i\frac{\partial}{\partial\mathbf{k}}\right)\psi(k).
\end{aligned}
\tag{11.7}
$$

This is manifestly Hermitian. If one translates one of these wavefunctions a distance **a** from the other, then, using

$$
\psi(a) = e^{i\mathbf{k}\cdot\mathbf{a}}\psi(0),
\tag{11.8}
$$

one has

$$
\begin{aligned}
(\psi(a), \psi(0)) &= \int d^n\mathbf{x}\, \psi^*(0)\psi(0) \equiv \delta(a) \\
&= \int d^n\mathbf{x}\, e^{i\mathbf{k}\cdot\mathbf{a}}.
\end{aligned}
\tag{11.9}
$$

This is an identity. It shows that the Schrödinger wavefunction can be localized with delta-function precision. Point particles exist.

11.2.2 Klein–Gordon field

The Klein–Gordon field does not automatically have only positive energy solutions, so we must restrict the discussion to the set of solutions which have strictly positive energy. The scalar product on this positive energy manifold is

$$(\phi^{(+)}, \phi^{(+)}) = \int d^n\mathbf{x} \, (\phi^{(+)*} \overleftrightarrow{\partial_0} \phi^{(+)}),$$

$$= \int (dk) \, \phi^{(+)*}(k)\phi^{(+)}(k) \, \theta(-k_0)\delta(p^2c^2 + m^2c^4)$$

$$= \int \frac{(dk)}{2|p_0|} e^{-ik\cdot a} |\phi_0^{(+)}|^2. \tag{11.10}$$

A translation by a such that $\phi^{(+)}(a) = e^{ik\cdot a}\phi_0(k)$ makes the states orthogonal;

$$(\phi^{(+)}(a), \phi^{(+)}(0)) = \delta^n(\mathbf{a})$$

$$= \int (dk) e^{-ik\cdot a}$$

$$= \int \frac{(dk)}{2|p_0|} e^{-ik\cdot a} |\phi_0^{(+)}|^2. \tag{11.11}$$

For the last two lines to agree, we must have

$$\phi_0^{(+)}(k) = \sqrt{2|p_0|}, \tag{11.12}$$

and thus the extent of the field about the point $\mathbf{a}$ is given by

$$\phi^{(+)}(\mathbf{x} - \mathbf{a}) = \int \frac{(dk)}{\sqrt{2|p_0|}} e^{-ik\cdot(\mathbf{x}-\mathbf{a})}, \tag{11.13}$$

which is not a delta function, and thus the Klein–Gordon particles do not exist in the same sense that Schrödinger particles do. There exist only approximately localizable concentrations of the field. The result of this integral in n dimensions can be expressed in terms of Bessel functions. For instance, in $n = 3$,

$$\phi^{(+)}(a) \sim \left(\frac{m}{r}\right)^{\frac{5}{4}} H_{\frac{5}{4}}^{(1)}(imr) \tag{11.14}$$

where $r = |\mathbf{x} - \mathbf{a}|$. This lack of sharpness is reflected in the nature of the position operator $\hat{\mathbf{x}}$ acting on these states:

$$(\phi^{(+)}(a), \hat{\mathbf{x}}\phi^{(+)}(a)) = \int \frac{(dk)}{2|p_0|} \phi^*(k) \, \hat{\mathbf{x}} \, \phi(k). \tag{11.15}$$

Clearly, the partial derivative $\frac{\partial}{\partial \mathbf{k}}$ is not a Hermitian operator owing to the factors of p_0 in the measure. It is easy to show (see section 10.3) that the addition of the connection term,

$$\hat{\mathbf{x}} = i\frac{\partial}{\partial \mathbf{k}} + \frac{i}{2}\frac{\mathbf{k}}{p_0^2}, \tag{11.16}$$

is what is required to make this operator Hermitian.

11.2.3 Dirac field

The Dirac field also has both positive and negative energy states, and particle wavefunctions must be restricted to positive energies. It shares with the Klein–Gordon field the inability to produce sharp delta-function-like configurations of the field. The expression for the position operator is extremely complicated for the spin-$\frac{1}{2}$ particles, owing to the constraints imposed by the γ-matrices. Although the procedure is the same, in principle, as for the Klein–Gordon field, the details are aggravated by the complexity of the field equations for the Dirac field.

The scalar product for localizable solutions is now, by analogy with eqn. (11.11),

$$(\psi^{(+)}, \psi^{(+)}) = \int \frac{(\mathrm{d}\mathbf{k})}{(2p_0)^2} |\psi|^2, \tag{11.17}$$

since there is no time derivative in the scalar product. Restricting to positive energies is also more complex, owing to the matrix nature of the equation. The normalized positive energy solutions include factors of

$$N = \sqrt{\frac{E}{E + mc^2}} = \sqrt{\frac{-p_0}{(-p_0 + mc)}}, \tag{11.18}$$

giving

$$(\psi^{(+)}, \hat{\mathbf{x}} \psi^{(+)}) = \int \frac{(\mathrm{d}\mathbf{k})}{(2p_0)^2} u^\dagger N \, \hat{\mathbf{x}} \, N u. \tag{11.19}$$

A suitable Hermitian operator for the position

$$\hat{\mathbf{x}} = N \left(-\mathrm{i}\frac{\partial}{\partial \mathbf{k}} + \Gamma \right) N \tag{11.20}$$

must now take into account all of these factors of the momentum.

11.2.4 Spin s fields in 3 + 1 dimensions

The generalization to any half-integral and integral massive spin fields can be accomplished using Dirac's construction for spin $\frac{1}{2}$. It is only sketched here. A spin-s field may be written as a direct product of $2s$ spin-$\frac{1}{2}$ blocks. Following Wigner *et al.* [6, 101], the wavefunction may be written in momentum space as

$$\psi(k)_\alpha \tag{11.21}$$

where $\alpha = 1, \ldots, 2s$ represents the components of $2s$ four-component spin blocks (in total $2s \times 4$ components). The sub-spinors satisfy block-diagonal equations of motion:

$$(\gamma_\alpha^\mu p_\mu + mc)\psi_\alpha = 0. \tag{11.22}$$

The γ-matrices all satisfy the Clifford algebra relation (see chapter 20),

$$\{\gamma_\alpha^\mu, \gamma_\alpha^\nu\} = -2g^{\mu\nu}. \tag{11.23}$$

The scalar product for localizable positive energy solutions may thus be found by analogy with eqn. (11.17):

$$
\begin{aligned}
(\psi_1, \psi_2) &= \int (\mathrm{d}p)\overline{\psi}_1\, \gamma_1^0 \dots \gamma_{2s}^0\, \psi_2 \\
&= \int (\mathbf{dp}) \left(\frac{|mc|}{p_0}\right)^{2s+1} \gamma_1^\dagger \gamma_2, \tag{11.24}
\end{aligned}
$$

since, in the product over blocks, each normalization factor is multiplied in turn. Wigner *et al.* drop the factors of the mass arbitrarily in their definitions, since these contribute only dimensional factors. It is the factors of p_0 which affect the localizability of the fields. The localizable wavefunction is thus of the form

$$|\psi|^2 \sim p_0^{2s+1}. \tag{11.25}$$

The normalization of the positive energy spinors is

$$\sum_\xi |u|^2 = \left(\frac{p_0 + mc}{2p_0}\right)^{2s}. \tag{11.26}$$

Combining the factors of momentum, one arrives at a normalization factor of

$$N = \left(\frac{p_0}{p_0 + mc}\right)^s \times \sqrt{p_0^{2s+1}} \tag{11.27}$$

and a Hermitian position operator of the form

$$(\psi, \hat{\mathbf{x}}\psi) = \int \frac{(\mathbf{dp})}{2p_0^{2s+1}} \left(u\, N \left(-\mathrm{i}\frac{\partial}{\partial\mathbf{k}} + \Gamma\right) N\, u\right). \tag{11.28}$$

Notice that the extra factors of the momentum lead to a greater de-localization. This expression contains the expressions for spin 0 and spin $\frac{1}{2}$ as special cases. For massless fields, the above expressions hold for spin 0 and spin $\frac{1}{2}$, but break down for spin 1, i.e. the photon.

11.3 The energy–momentum tensor $\theta_{\mu\nu}$

Translational invariance of the action implies the conservation of momentum. Time-translation invariance implies the conservation of energy. Generally, invariance of one variable implies the conservation of its conjugate variable. In this section, we see how symmetry under translations of coordinates leads to the definition of energy, momentum and shear stress in a mechanical system of fields.

In looking at dynamical variations of the action, we have been considering changes in the function $\phi(x)$. Now consider variations in the field which occur because we choose to translate or transform the coordinates x^μ, i.e.

$$\delta_x \phi(x) = (\partial_\mu \phi(x)) \delta x^\mu, \tag{11.29}$$

where we use δ_x to distinguish a coordinate variation and

$$\delta x^\mu = x'^\mu - x^\mu. \tag{11.30}$$

The variation of the action under such a change is given by

$$\delta S = \int (dx') \mathcal{L}(x') - \int (dx) \mathcal{L}(x), \tag{11.31}$$

which is manifestly zero, in the absence of boundaries, since the first term is simply a re-labelling of the second. We shall consider the action of an infinitesimal change δx^μ and investigate what this tells us about the system. Since we are not making a dynamical variation, we can expect to find quantities which are constant with respect to dynamics.

To calculate eqn. (11.31), we expand the first term formally about x:

$$\begin{aligned} \mathcal{L}(x') &= \mathcal{L}(x) + \delta \mathcal{L}^{(1)} + \cdots \\ &= \mathcal{L}(x) + (\partial_\mu \mathcal{L}) \delta x^\mu + O((\delta x)^2). \end{aligned} \tag{11.32}$$

The volume element transforms with the Jacobian

$$(dx') = \det \left(\frac{\partial x'^\mu}{\partial x^\nu} \right) (dx), \tag{11.33}$$

thus, we require the determinant of

$$\overset{x}{\partial}_\nu x'^\mu = \delta^\mu_{\ \nu} + (\partial_\nu \delta x^\nu). \tag{11.34}$$

This would be quite difficult to compute generally, but fortunately we only require the result to first order in δx^μ. Writing out the infinite-dimensional matrix explicitly, it is easy to see that all the terms which can contribute to first order lie on the diagonal:

$$\begin{pmatrix} 1 + \partial_1 \delta x^1 & \partial_1 \delta x^2 & \cdots \\ \partial_2 \delta x^1 & 1 + \partial_2 \delta x^2 & \cdots \\ \vdots & \vdots & \end{pmatrix}. \tag{11.35}$$

Now, the determinant is the product of all the terms along the diagonal, plus some other terms involving off-diagonal elements which do not contribute to first order; thus, it is easy to see that we must have

$$\det(\overset{x}{\partial_\nu} x'^\mu) = 1 + \partial_\mu \delta x^\mu + O((\delta x)^2). \tag{11.36}$$

Using this result in eqn. (11.34), we obtain, to first order,

$$\delta S = \int (\mathrm{d}x) \left\{ \delta \mathcal{L}^{(1)} + (\partial_\mu \delta x^\mu) \mathcal{L} \right\}. \tag{11.37}$$

Let us now use this result to consider the total variation of the action under a combined dynamical and coordinate variation. In principle, we should proceed from here for each Lagrangian we encounter. To make things more concrete, let us make the canonical assumption that we have a Lagrangian density which depends on some generic field $q(x)$ and its derivative $\partial_\mu q(x)$. This assumption leads to correct results in nearly all cases of interest – it fails for gauge theories, because the definition of the velocity is not gauge-covariant, but we can return to that problem later. We take

$$\mathcal{L} = \mathcal{L}\left(q(x), (\partial_\mu q(x)), x^\mu\right). \tag{11.38}$$

Normally, in a conservative system, x^μ does not appear explicitly, but we can include this for generality. Let us denote a functional variation by δq as previously, and the total variation of $q(x)$ by

$$\delta_\mathrm{T} q = \delta q + (\partial_\mu q) \delta x^\mu. \tag{11.39}$$

The total variation of the action is now

$$\delta_\mathrm{T} S = \int (\mathrm{d}x) \left\{ \frac{\delta \mathcal{L}}{\delta q} \delta q + \frac{\delta \mathcal{L}}{\delta(\partial_\mu q)} \delta(\partial_\mu q) + (\partial_\mu \mathcal{L}) \delta x^\mu + (\partial_\mu \delta x^\mu) \mathcal{L} \right\},$$
$$\tag{11.40}$$

where the first two terms originate from the functional variation in eqn. (4.21) and the second two arise from the coordinate change in eqn. (11.32). We now make the usual observation that the δ variation commutes with the partial derivative (see eqn. (4.19)), and thus we may integrate by parts in the second and fourth terms of this expression to give

$$\delta_\mathrm{T} S = \int (\mathrm{d}x) \left\{ \left(\frac{\delta \mathcal{L}}{\delta q} - \partial_\mu \frac{\delta \mathcal{L}}{\delta(\partial_\mu q)} \right) \delta q \right\}$$
$$+ \int (\mathrm{d}x) \left\{ \partial_\mu \left[\frac{\delta \mathcal{L}}{\delta(\partial_\mu q)} \delta q + \mathcal{L} \delta x^\mu \right] \right\}. \tag{11.41}$$

One identifies the first line as being that which gives rise to the Euler–Lagrange field equations. This term vanishes by virtue of the field equations, for any classically acceptable path. The remaining surface term can be compared with eqn. (4.62) and represents a generator for the combined transformation. We recognize the canonical momentum Π_μ from eqn. (4.66). To display this term in its full glory, let us add and subtract

$$\frac{\delta \mathcal{L}}{\delta(\partial_\mu q)}(\partial_\nu q)\delta x^\nu \qquad (11.42)$$

to the surface term, giving

$$\delta_T S = \frac{1}{c}\int d\sigma^\mu \left\{ \Pi_\mu(\delta q + (\partial_\nu q)\delta x^\nu) - \theta_{\mu\nu}\delta x^\nu \right\}$$

$$= \frac{1}{c}\int d\sigma^\mu \left\{ \Pi_\mu \delta_T q - \theta_{\mu\nu}\delta x^\nu \right\}, \qquad (11.43)$$

where we have defined

$$\theta_{\mu\nu} = \frac{\delta \mathcal{L}}{\delta(\partial^\mu q)}(\partial_\nu q) - \mathcal{L}g_{\mu\nu}. \qquad (11.44)$$

This quantity is called the *energy–momentum tensor*. Its $\mu, \nu = 0, 0$ component is the total energy density or Hamiltonian density of the system. Its $\mu, \nu = 0, i$ components are the momentum components. In fact, if we expand out the surface term in eqn. (11.43) we have terms of the form

$$\Pi \delta q - H \delta t + \mathbf{p}\delta \mathbf{x} + \cdots . \qquad (11.45)$$

This shows how elegantly the action principle generates all of the dynamical entities of our covariant system and their respective conjugates (the delta objects can be thought of as the conjugates to each of the dynamical generators). Another way of expressing this is to say

- Π is the generator of q translations,

- H is the generator of t translations,

- $\mathbf{p}$ is the generator of $\mathbf{x}$ translations,

and so on. That these differential operators are the generators of causal changes can be understood from method 2 of the example in section 7.1. A single partial derivative has a complementary Green function which satisfies

$$\partial_x G(x, x') = \delta(x, x'). \qquad (11.46)$$

This Green function is simply the Heaviside step function $\theta(t - t')$ from Appendix A, eqn. (A.2). What this is saying is that a derivative picks out a direction for causal change in the system. In other words, the response of the system to a source is channelled into a change in the coordinates and vice versa.

11.3.1 Example: classical particle mechanics

To illustrate the energy–momentum tensor in the simplest of cases, we return to the classical system, with the Lagrangian given by eqn. (4.5). This Lagrangian has no $\mu\nu$ indices, so our dogged Lorentz-covariant formalism is strictly wasted, but we may take μ to stand for the time t or position i and use the general expression. Recognizing that the metric for classical particles is $\delta_{\mu\nu}$ rather than $g_{\mu\nu}$, we have

$$
\begin{aligned}
\theta_{tt} &= \frac{\partial L}{\partial \dot{q}_i}\dot{q}_i - L\delta_{tt} \\
&= p_i \dot{q}^i - L \\
&= \frac{1}{2}m\dot{q}^2 + V(q) \\
&= H.
\end{aligned}
\tag{11.47}
$$

The off-diagonal spacetime components give the momentum,

$$
\theta_{ti} = \frac{\partial L}{\partial \dot{q}_j}\frac{\partial q_j}{\partial q_i} = p_j \delta^j_{\ i} = p_i = m\dot{q}_i,
\tag{11.48}
$$

and

$$
\theta_{ii} = -L,
\tag{11.49}
$$

which has no special interpretation. The off-diagonal ij components vanish in this case.

The analogous analysis can be carried out for relativistic point particles. Using the action in eqn. (4.32), one finds that

$$
\begin{aligned}
\theta_{\tau\tau} &= \frac{\partial L}{\partial^t \mathbf{x}}(\partial_t \mathbf{x}) + L \\
&= \frac{\partial L}{\partial^\tau \mathbf{x}}(\partial_\tau \mathbf{x}) + L \\
&= m\mathbf{u}^2 - \frac{1}{2}m\mathbf{u}^2 + V' \\
&= \frac{1}{2}m\mathbf{u}^2 + V,
\end{aligned}
\tag{11.50}
$$

where $\mathbf{u} = d\mathbf{x}/d\tau$ is the velocity, or

$$
\theta_{tt} = \frac{1}{2}m\mathbf{v}^2 + V.
\tag{11.51}
$$

11.3.2 Example: the complex scalar field

The application of eqn. (11.44) for the action

$$
S = \int (dx)\left\{\hbar^2 c^2 (\partial^\mu \phi_A)^*(\partial_\mu \phi_A) + m^2 c^4 \phi_A^* \phi_A + V(\phi)\right\},
\tag{11.52}
$$

gives us the following components for the energy–momentum tensor:

$$\theta_{00} = \frac{\partial \mathcal{L}}{\partial(\partial^0 \phi_A)}(\partial_0 \phi_A) + \frac{\partial \mathcal{L}}{\partial(\partial^0 \phi_A^*)}(\partial_0 \phi_A^*) - \mathcal{L}g_{00}$$
$$= \hbar^2 c^2 \left[(\partial_0 \phi_A^*)(\partial_0 \phi_A) + (\partial_i \phi_A^*)(\partial_i \phi_A)\right] + m^2 c^4 + V(\phi).$$
$$(11.53)$$

Thus, the last line defines the Hamiltonian density $\mathcal{H}$, and the Hamiltonian is given by

$$H = \int d\sigma \, \mathcal{H}. \qquad (11.54)$$

The off-diagonal spacetime components define a momentum:

$$\theta_{0i} = \theta_{i0} = \frac{\partial \mathcal{L}}{\partial(\partial^0 \phi)_A}(\partial_i \phi)_A + \frac{\partial \mathcal{L}}{\partial(\partial^0 \phi_A^*)}(\partial_i \phi_A^*)$$
$$= \hbar^2 c^2 \left\{(\partial_0 \phi_A^*)(\partial_i \phi_A) + (\partial_0 \phi_A)(\partial_i \phi_A^*)\right\}. \qquad (11.55)$$

Taking the integral over all space enables us to integrate by parts and write this in a form which turns out to have the interpretation of the expectation value (inner product) of the field momentum (see chapter 9):

$$\int d\sigma \, \theta_{0i} = -\hbar^2 c^2 \int d\sigma \, \left(\phi^* \partial_i \partial_0 \phi - (\partial_0 \phi^*) \partial_i \phi\right)$$
$$= -(\phi, \, p_i c \phi), \qquad (11.56)$$

where $p = -i\hbar \partial_i$. The diagonal space components are given by

$$\theta_{ii} = \frac{\partial \mathcal{L}}{\partial(\partial^i \phi_A)}(\partial_i \phi_A) + \frac{\partial \mathcal{L}}{\partial(\partial^i \phi_A^*)}(\partial_i \phi_A^*) - \mathcal{L}$$
$$= 2\hbar^2 c(\partial_i \phi^*)(\partial_i \phi) - \mathcal{L}, \qquad (11.57)$$

where i is not summed. Similarly, the off-diagonal 'stress' components are given by

$$\theta_{ij} = \frac{\partial \mathcal{L}}{\partial(\partial^i \phi_A)}(\partial_j \phi_A) + \frac{\partial \mathcal{L}}{\partial(\partial^i \phi_A)}(\partial_j \phi_A)$$
$$= \hbar^2 c^2 \left\{(\partial_i \phi_A^*)(\partial_j \phi_A) + (\partial_j \phi_A^*)(\partial_i \phi_A)\right\}$$
$$= \hbar^{-1} c(\phi_A, \, p_i p_j \phi_A). \qquad (11.58)$$

From eqn. (11.57), we see that the trace over spatial components in $n + 1$ dimensions is

$$\sum_i \theta_{ii} = \mathcal{H} - 2m^2 c^4 \phi_A^2 - 2V(\phi) + (n - 1)\mathcal{L}, \qquad (11.59)$$

so that the full trace gives

$$\theta^{\mu}_{\ \mu} = g^{\mu\nu}\theta_{\nu\mu} = -2m^2c^4\phi_A^2 - 2V(\phi) + (n-1)\mathcal{L}. \tag{11.60}$$

Note that this vanishes in $1 + 1$ dimensions for zero mass and potential.

11.3.3 Example: conservation

We can also verify the energy–momentum conservation law, when the fields satisfy the equations of motion. We return to this issue in section 11.8.1. For the simplest example of a scalar field with action,

$$S = \int (\mathrm{d}x) \left\{ \frac{1}{2}(\partial^{\mu}\phi)(\partial_{\mu}\phi) + \frac{1}{2}m^2\phi^2 \right\}. \tag{11.61}$$

Using eqn. (11.44), we obtain the energy–momentum tensor

$$\theta_{\mu\nu} = \frac{1}{2}(\partial_{\mu}\phi)(\partial_{\nu}\phi) - \frac{1}{2}m\phi^2. \tag{11.62}$$

The spacetime divergence of this is

$$\partial^{\mu}\theta_{\mu\nu} = -(-\Box\,\phi + m^2\phi)(\partial_{\nu}\phi) = 0. \tag{11.63}$$

The right hand side vanishes as a result of the equations of motion, and thus the conservation law is upheld.

It is interesting to consider what happens if we add a potential $V(x)$ to the action. This procedure is standard practice in quantum mechanics, for instance. This can be done by shifting the mass in the action by $m^2 \rightarrow m^2 + V(x)$. The result of this is the following expression:

$$\begin{aligned}
\partial^{\mu}\theta_{\mu\nu} &= (\Box\,\phi - (m^2 + V(x))\phi)(\partial_{\nu}\phi) + (\partial_{\nu}V)\phi^2 \\
&= (\partial_{\nu}V(x))\phi^2. \tag{11.64}
\end{aligned}$$

The first term vanishes again by virtue of the equations of motion. The spacetime-dependent potential does not vanish, however. Conservation of energy is only assured if there are no spacetime-dependent potentials. This illustrates an important point, which is discussed more generally in section 11.8.1.

The reason that the conservation of energy is violated here is that a static potential of this kind is not physical. All real potentials change in response to an interaction with another field. By making a potential static, we are claiming that the form of $V(x)$ remains unchanged no matter what we scatter off it. It is an immovable barrier. Conservation is violated because, in a physical system, we would take into account terms in the action which allow $V(x)$ to change in response to the momentum imparted by ϕ. See also exercise 1, at the end of this chapter.

11.4 Spacetime invariance and symmetry on indices

For reasons which should become apparent in section 11.6.1, the energy–momentum tensor, properly defined under maximal symmetry, is symmetrical under interchange of its indices. This reflects the symmetry of the metric tensor under interchange of indices. If the Lorentz symmetry is broken, however (for instance, in the non-relativistic limit), then this property ceases to apply. In a relativistic field theory, a non-symmetrical tensor may be considered simply incorrect; in the non-relativistic limit, only the spatial part of the tensor is symmetrical.

11.5 $\theta_{\mu\nu}$ for gauge theories

Consider the Maxwell action

$$S = \int (\mathrm{d}x) \left\{ \frac{1}{4\mu_0} F^{\mu\nu} F_{\mu\nu} - J^\mu A_\mu \right\}. \tag{11.65}$$

A direct application of the formula in eqn. (11.44) gives an energy–momentum tensor which is not gauge-invariant:

$$\theta_{\mu\nu} = \frac{\partial \mathcal{L}}{\partial (\partial^\mu A^\alpha)} (\partial_\nu A^\alpha) - \frac{1}{4\mu_0} F^{\rho\sigma} F_{\rho\sigma} g_{\mu\nu}. \tag{11.66}$$

The explicit appearance of A_μ in this result shows that this definition cannot be physical for the Maxwell field. The reason for this lack of gauge invariance can be traced to an inaccurate assumption about the nature of a translation, or conformal transformation of the gauge field [44, 76]; it is related to the gauge invariance of the theory. The expression for $\theta_{\mu\nu}$ in eqn. (11.44) relies on the assumption in eqn. (11.29) that the expression for the variation in the field by change of coordinates is given by

$$\delta_x A_\mu = (\partial_\alpha A_\mu)\delta x^\alpha. \tag{11.67}$$

It is clear that this translation is not invariant with respect to gauge transformations, but this seems to be wrong. After all, potential differences are observable as electric and magnetic fields between two points, and observable quantities should be gauge-invariant. In terms of this quantity, the energy–momentum tensor can be written as

$$\theta_{\mu\nu}\delta x^\nu = \frac{\partial \mathcal{L}}{\partial (\partial^\mu A^\alpha)} (\delta_x A^\alpha) - \frac{1}{4\mu_0} F^{\rho\sigma} F_{\rho\sigma} g_{\mu\nu}\delta x^\nu. \tag{11.68}$$

Suppose now that we use this as a more fundamental definition of $\theta_{\mu\nu}$. Our problem is then to find a more appropriate definition of $\delta_x A_\mu$, which leads to a gauge-invariant answer. The source of the problem is the implicit assumption that the field at one point in spacetime should have the same phase as the field at another point. In other words, under a translation of coordinates, we should expect the field to transform like a vector only up to a gauge transformation. Generalizing the transformation rule for the vector potential to account for this simple observation cures the problem entirely. The correct definition of this variation was derived in section 4.5.2.

The correct (gauge-invariant) transformation is now found by noting that we may write

$$\delta_x A_\mu = (\partial_\nu A'_\mu(x))\epsilon^\nu + (\overset{x'}{\partial_\mu} \epsilon^\nu)A_\nu$$
$$= \epsilon_\nu F^\nu_\mu + \partial_\mu(\epsilon_\nu A^\nu). \tag{11.69}$$

This last term has the form of a gauge-invariant translation plus a term which can be interpreted as a gauge transformation $\partial^\mu s$ (where $s = \epsilon_\nu A^\nu$). Thus we may now re-define the variation $\delta_x A^\mu$ to include a simultaneous gauge transformation, leading to the gauge-invariant expression

$$\delta_x A^\mu(x) \equiv \delta_x A^\mu - \partial^\mu s = \epsilon_\nu F^{\nu\mu}, \tag{11.70}$$

where $\epsilon^\mu = \delta x^\mu$. The most general description of the translation ϵ^μ, in $3 + 1$ dimensions is a 15-parameter solution to Killing's equation for the conformal symmetry [76],

$$\partial_\mu\epsilon_\nu + \partial_\nu\epsilon_\mu - \frac{1}{2}g_{\mu\nu}\partial_\gamma\epsilon^\gamma = 0, \tag{11.71}$$

with solution

$$\epsilon^\mu(x) = a^\mu + bx^\mu + \omega^{\mu\nu}x_\nu + 2x^\mu c^\nu x_\nu - c^\mu x^2, \tag{11.72}$$

where $\omega^{\mu\nu} = -\omega^{\nu\mu}$. This explains why the conformal variation in the tensor $T_{\mu\nu}$ gives the correct result for gauge theories: the extra freedom can accommodate x-dependent scalings of the fields, or gauge transformations.

The anti-symmetry of $F_{\mu\nu}$ will now guarantee the gauge invariance of $\theta_{\mu\nu}$. Using this expression in eqn. (11.43) for the energy–momentum tensor (recalling $\epsilon^\mu = \delta x^\mu$) gives

$$\theta'_{\mu\nu} = \frac{\delta\mathcal{L}}{\delta(\partial^\mu A^\alpha)}F_\nu{}^\alpha - \mathcal{L}g_{\mu\nu}$$
$$= 2\frac{\delta\mathcal{L}}{\delta F^{\mu\alpha}}F_\nu{}^\alpha - \mathcal{L}g_{\mu\nu}$$
$$= \mu_0^{-1}F_{\mu\alpha}F_\nu{}^\alpha - \frac{1}{4\mu_0}F^{\rho\sigma}F_{\rho\sigma}g_{\mu\nu}. \tag{11.73}$$

This result is manifestly gauge-invariant and can be checked against the traditional expressions obtained from Maxwell's equations for the energy density and the momentum flux. It also agrees with the Einstein energy–momentum tensor $T_{\mu\nu}$.

The components in $3 + 1$ dimensions evaluate to:

$$
\begin{aligned}
\theta_{00} &= \mu_0^{-1} \left(F_{0i} F_0{}^i - \mathcal{L} g_{00} \right) \\
&= \frac{E_i E_i}{c^2 \mu_0} + \frac{1}{2\mu_0} \left(B_i B_i - \frac{E_i E_i}{c^2} \right) \\
&= \frac{1}{2\mu_0} \left(\frac{\mathbf{E}^2}{c^2} + \mathbf{B}^2 \right) \\
&= \frac{1}{2} (\mathbf{E} \cdot \mathbf{D} + \mathbf{B} \cdot \mathbf{H}),
\end{aligned}
\tag{11.74}
$$

which has the interpretation of an energy or Hamiltonian density. The spacetime off-diagonal components are given by

$$
\begin{aligned}
\theta_{0j} = \theta_{j0} &= \mu_0^{-1} F_{0i} F_j{}^i \\
&= \mu_0^{-1} \epsilon_{ijk} E_i B_k / c \\
&= -\frac{(\mathbf{E} \times \mathbf{H})_k}{c},
\end{aligned}
\tag{11.75}
$$

which has the interpretation of a 'momentum' density for the field. This quantity is also known as Poynting's vector divided by the speed of light. The conservation law is

$$
\partial^\mu \theta_{\mu 0} = -\frac{1}{c} \partial_t \mathcal{H} + \partial_i (\mathbf{H} \times \mathbf{E})^i = \frac{1}{c} \partial_\mu S^\mu = 0,
\tag{11.76}
$$

which may be compared with eqns. (2.70) and (2.73). Notice finally that

$$
\frac{\delta S}{\delta x^0} = -\int d\sigma \, \theta_{00},
\tag{11.77}
$$

and thus that

$$
\frac{\delta S}{\delta t} = -H,
\tag{11.78}
$$

which is the energy density or Hamiltonian. We shall have use for this relation in chapter 14.

11.6 Another energy–momentum tensor $T_{\mu\nu}$

11.6.1 Variational definition

Using the action principle and the Lorentz invariance of the action, we have viewed the energy–momentum tensor $\theta_{\mu\nu}$ as a generator for translations in space and time. There is another quantity which we can construct which behaves as an energy–momentum tensor: it arises naturally in Einstein's field equations of general relativity as a source term for matter. This tensor is defined by the variation of the action with respect to the metric tensor:

$$T_{\mu\nu} = \frac{2}{\sqrt{g}} \frac{\delta S}{\delta g^{\mu\nu}}. \tag{11.79}$$

Clearly, this definition assumes that the action is covariant with respect to the metric $g_{\mu\nu}$, so we should not expect this to work infallibly for non-relativistic actions.

The connection between $T_{\mu\nu}$ and $\theta_{\mu\nu}$ is rather subtle and has to do with conformal transformations. Conformal transformations (see section 9.6) are related to re-scalings of the metric tensor, and they form a super-group, which contains and extends the Lorentz transformation group; thus $T_{\mu\nu}$ admits more freedom than $\theta_{\mu\nu}$. As it turns out, this extra freedom enables it to be covariant even for local gauge theories, where fields are re-defined by spacetime-dependent functions. The naive application of Lorentz invariance for scalar fields in section 11.3 does not automatically lead to invariance in this way; but it can be fixed, as we shall see in the next section. The upshot of this is that, with the exception of the Maxwell field and the Yang–Mills field, these two tensors are the same.

To evaluate eqn. (11.79), we write the action with the metric made explicit, and write the variation:

$$\delta S = \int d^{n+1}x \sqrt{g} \left(\frac{1}{\sqrt{g}} \frac{\delta g}{\delta g^{\mu\nu}} \mathcal{L} + \frac{\delta \mathcal{L}}{\delta g^{\mu\nu}} \right), \tag{11.80}$$

where we recall that $g = -\det g_{\mu\nu}$. To evaluate the first term, we note that

$$\frac{\delta g}{\delta g^{\mu\nu}} = -\frac{\delta \det g_{\mu\nu}}{\delta g^{\mu\nu}}, \tag{11.81}$$

and use the identity

$$\ln \det g_{\mu\nu} = \operatorname{Tr} \ln g_{\mu\nu}. \tag{11.82}$$

Varying this latter result gives

$$\delta \ln(\det g_{\mu\nu}) = \operatorname{Tr} \delta \ln g_{\mu\nu}, \tag{11.83}$$

or

$$\frac{\delta(\det g_{\mu\nu})}{\det g_{\mu\nu}} = \frac{\delta g_{\mu\nu}}{g^{\mu\nu}}. \tag{11.84}$$

Using this result, together with eqn. (11.81), in eqn. (11.80), we obtain

$$T_{\mu\nu} = 2 \frac{\partial \mathcal{L}}{\partial g^{\mu\nu}} - g_{\mu\nu}\mathcal{L}. \tag{11.85}$$

This definition is tantalizingly close to that for the Lorentz symmetry variation, except for the replacement of the first term. In many cases, the two definitions give the same result, but this is not the case for the gauge field, where $T_{\mu\nu}$ gives the correct answer, but a naive application of $\theta_{\mu\nu}$ does not. The clue as to their relationship is to consider how the metric transforms under a change of coordinates (see chapter 25). Relating a general action $g_{\mu\nu}$ to a locally inertial frame $\eta_{\mu\nu}$, one has

$$g_{\mu\nu} = V_\mu{}^\alpha V_\nu{}^\beta \, \eta_{\alpha\beta}, \tag{11.86}$$

where the *vielbein* $V_\mu^\alpha = \partial'_\mu x^\alpha$, so that

$$g^{\mu\nu}(\partial_\mu \phi)(\partial_\nu \phi) = \eta^{\alpha\beta} V^\mu_\alpha V^\nu_\beta (\partial_\mu \phi)(\partial_\nu \phi). \tag{11.87}$$

In terms of these quantities, one has

$$T_{\mu\nu} = \frac{2}{\sqrt{g}} \frac{\delta S}{\delta g^{\mu\nu}} = \frac{V_{\alpha\mu}}{\det V} \frac{\delta S}{\delta V_\alpha{}^\mu}. \tag{11.88}$$

Thus, one sees that variation with respect to a vector, as in the case of $\theta_{\mu\nu}$ will only work if the vector transforms fully covariantly under every symmetry. Given that the maximal required symmetry is the conformal symmetry, one may regard $T_{\mu\nu}$ as the correct definition of the energy–momentum tensor.

11.6.2 The trace of the energy–momentum tensor $T_{\mu\nu}$

The conformal invariance of the field equations is reflected in the trace of the energy–momentum tensor $T_{\mu\nu}$, which we shall meet in the next chapter. Its trace vanishes for actions which are conformally invariant. To see this, we note that, in a conformally invariant theory,

$$\frac{\delta S}{\delta \Omega} = 0. \tag{11.89}$$

If we express this in terms of the individual partial transformations, we have

$$\frac{\delta S}{\delta \Omega} = \frac{\delta S}{\delta g^{\mu\nu}} \frac{\delta g^{\mu\nu}}{\delta \Omega} + \frac{\delta S}{\delta \phi} \frac{\delta \phi}{\delta \Omega} = 0. \tag{11.90}$$

Assuming that the transformation is invertible, and that the field equations are satisfied,

$$\frac{\delta S}{\delta \phi} = 0, \tag{11.91}$$

we have

$$\frac{1}{2} \sqrt{g} \, T_{\mu\nu} \frac{\delta g^{\mu\nu}}{\delta \Omega} = 0. \tag{11.92}$$

Since $\frac{\delta g^{\mu\nu}}{\delta \Omega}$ must be proportional to $g^{\mu\nu}$, we have simply that

$$T_{\mu\nu}g^{\mu\nu} = \text{Tr } T_{\mu\nu} = 0. \tag{11.93}$$

A similar argument applies to the tensor $\theta_{\mu\nu}$, since the two tensors (when defined correctly) agree. In the absence of conformal invariance, one may expand the trace in the following way:

$$T^{\mu}_{\mu} = \beta_i \mathcal{L}^i, \tag{11.94}$$

where $\mathcal{L}^i$ are terms in the Lagrangian of ith order in the fields. β^i is then called the beta function for this term. It occurs in renormalization group and scaling theory.

11.6.3 The conformally improved $T_{\mu\nu}$

The uncertainty in the definition of the energy–momentum tensors $\theta_{\mu\nu}$ and $T_{\mu\nu}$ is usually understood as the freedom to change boundary conditions by adding total derivatives, i.e. surface terms, to the action. However, another explanation is forthcoming: such boundary terms are generators of symmetries, and one would therefore be justified in suspecting that symmetry covariance plays a role in the correctness of the definition. It has emerged that covariance, with respect to the conformal symmetry, frequently plays a role in elucidating a sensible definition of this tensor. While this symmetry might seem excessive in many physical systems, where one would not expect to see such a symmetry, its structure encompasses a generality which ensures that all possible terms are generated, before any limit is taken.

In the case of the energy–momentum tensor, the conformal symmetry motivates improvements not only for gauge theories, but also with regard to scaling anomalies. The tracelessness of the energy–momentum tensor for a massless field is only guaranteed in the presence of conformal symmetry, but such a symmetry usually demands a specific spacetime dimensionality. What is interesting is that a fully covariant, curved spacetime formulation of $T_{\mu\nu}$ leads to an invariant definition, which ensures a vanishing T^{μ}_{μ} in the massless limit [23, 26, 119].

The freedom to add total derivatives means that one may write

$$T_{\mu\nu} \to T_{\mu\nu} + \nabla^{\rho}\nabla^{\sigma} m_{\mu\nu\rho\sigma}, \tag{11.95}$$

where $m_{\mu\nu\rho\sigma}$ is a function of the metric tensor, and is symmetrical on μ, ν and ρ, σ indices; additionally it satisfies:

$$m_{\mu\nu\rho\sigma} + m_{\rho\nu\sigma\mu} + m_{\sigma\nu\mu\rho} = 0. \tag{11.96}$$

These are also the symmetry properties of the Riemann tensor (see eqn. (25.24)). This combination ensures that the additional terms are conserved:

$$\nabla^\mu \Delta T_{\mu\nu} = \nabla^\mu \nabla^\rho \nabla^\sigma m_{\mu\nu\rho\sigma} = 0. \tag{11.97}$$

The properties of the Riemann tensor imply that the following additional invariant term may be added to the action:

$$\Delta S = \int (\mathrm{d}x) \, \xi \, m^{\mu\nu\rho\sigma} R_{\mu\nu\rho\sigma}. \tag{11.98}$$

For spin-0 fields, the only invariant combination of correct dimension is

$$m^{\mu\nu\rho\sigma} = \left(g^{\mu\nu} g^{\rho\sigma} - \frac{1}{2} g^{\rho\nu} g^{\mu\sigma} - \frac{1}{2} g^{\rho\mu} g^{\nu\sigma} \right) \phi^2, \tag{11.99}$$

which gives the term

$$\Delta S = \int \frac{1}{2} \xi R \phi^2, \tag{11.100}$$

where R is the scalar curvature (see chapter 25). Thus, the modified action, which must be temporarily interpreted in curved spacetime, is

$$S = \int (\mathrm{d}x) \left\{ \frac{1}{2} (\nabla^\mu \phi)(\nabla_\mu \phi) + \frac{1}{2} (m^2 + \xi R) \phi^2 \right\}, \tag{11.101}$$

where $(\mathrm{d}x) = \sqrt{g} \mathrm{d}^{n+1} x$. Varying this action with respect to the metric leads to

$$T_{\mu\nu} = (\nabla_\mu \phi)(\nabla_\nu \phi) - \frac{1}{2} g_{\mu\nu} \left[(\nabla^\lambda \phi)(\nabla_\lambda \phi) + m^2 \phi^2 \right]$$
$$+ \xi (\nabla_\mu \nabla_\nu - g_{\mu\nu} \square) \phi^2. \tag{11.102}$$

Notice that the terms proportional to ξ do not vanish, even in the limit $R \to 0$, i.e. $\nabla_\mu \to \partial_\mu$. The resulting additional piece is a classic $(n+1)$ dimensional, transverse (conserved) vector displacement. Indeed, it has the conformally invariant form of the Maxwell action, stripped of its fields. The trace of this tensor may now be computed, giving:

$$T^\mu_\mu = \left[\frac{1-n}{2} + 2\xi n \right] (\nabla_\mu \phi)(\nabla_\nu \phi) - \frac{1}{2} (n+1) m^2 \phi^2. \tag{11.103}$$

One now sees that it is possible to choose ξ such that it vanishes in the massless limit; i.e.

$$T^\mu_\mu = -\frac{1}{2} (n+1) m^2 \phi^2, \tag{11.104}$$

where

$$\xi = \frac{n-1}{4n}. \tag{11.105}$$

This value of ξ is referred to as conformal coupling. In $3+1$ dimensions, it has the value of $\frac{1}{6}$, which is often assumed explicitly.

11.7 Angular momentum and spin[1]

The topic of angular momentum in quantum mechanics is one of the classic demonstrations of the direct relevance of group theory to the nature of microscopic observables. Whereas linear momentum more closely resembles its Abelian classical limit, the microscopic behaviour of rotation at the level of particles within a field is quite unexpected. The existence of intrinsic, half-integral spin **S**, readily predicted by representation theory of the rotation group in $3 + 1$ dimensions, has no analogue in a single-valued differential representation of the orbital angular momentum **L**.

11.7.1 Algebra of orbital motion in $3 + 1$ dimensions

The dynamical commutation relations of quantum mechanics fix the algebra of angular momentum operators. It is perhaps unsurprising, at this stage, that the canonical commutation relations for position and momentum actually correspond to the Lie algebra for the rotation group. The orbital angular momentum of a body is defined by

$$\mathbf{L} = \mathbf{r} \times \mathbf{p}. \tag{11.106}$$

In component notation in n-dimensional Euclidean space, one writes

$$L_i = \epsilon_{ijk} x^j p^k. \tag{11.107}$$

The commutation relations for position and momentum

$$[x^i, p^j] = \mathrm{i}\, \chi_h\, \delta^{ij} \tag{11.108}$$

then imply that (see section 11.9)

$$[L_i, L_j] = \mathrm{i}\, \chi_h\, \epsilon_{ijk} L_k. \tag{11.109}$$

This is a Lie algebra. Comparing it with eqn. (8.47) we see the correspondence between the generators and the angular momentum components,

$$T^a \leftrightarrow L^a / \chi_h$$
$$f_{abc} = -\epsilon_{abc}, \tag{11.110}$$

with the group space $a, b, c \leftrightarrow i, j, k$ corresponding to the Euclidean spatial basis vectors. What this shows, however, is that the group theoretical description of rotation translates directly into the operators of the dynamical theory, with a

---

[1] A full understanding of this section requires a familiarity with Lorentz and Poincaré symmetry from section 9.4.

dimensionful scale χ_h, which in quantum mechanics is $\chi_h = \hbar$. This happens, as discussed in section 8.1.3, because we are representing the dynamical variables (fields or wavefunctions) as tensors which live on the representation space of the group (spacetime) by a mapping which is adjoint (the group space and representation space are the same).

11.7.2 The nature of angular momentum in $n + 1$ dimensions

In spite of its commonality, the nature of rotation is surprisingly non-intuitive, perhaps because many of its everyday features are taken for granted. The freedom for rotation is intimately linked to the dimension of spacetime. This much is clear from intuition, but, as we have seen, the physics of dynamical systems depends on the group properties of the transformations, which result in rotations. Thus, to gain a true intuition for rotation, one must look to the properties of the rotation group in $n + 1$ dimensions.

In one dimension, there are not enough degrees of freedom to admit rotations. In $2 + 1$ dimensions, there is only room for one axis of rotation. Then we have an Abelian group $U(1)$ with continuous eigenvalues $\exp(i\theta)$. These 'circular harmonics' or eigenfunctions span this continuum. The topology of this space gives boundary conditions which can lead to any statistics under rotation. i.e. anyons.

In $3 + 1$ dimensions, the rank 2-tensor components of the symmetry group generators behave like two separate 3-vectors, those arising in the timelike components T^{0i} and those arising in the spacelike components $\frac{1}{2}\epsilon^{ijk}T_{ij}$; indeed, the electric and magnetic components of the electromagnetic field are related to the electric and magnetic components of the Lorentz group generators. Physically, we know that rotations and coils are associated with magnetic fields, so this ought not be surprising. The rotation group in $3 + 1$ dimensions is the non-Abelian $SO(3)$, and the maximal Abelian sub-group (the centre) has eigenvalues ± 1. These form a Z_2 sub-group and reflect the topology of the group, giving rise to two possible behaviours under rotation: symmetrical and anti-symmetrical boundary conditions corresponding in turn to Bose–Einstein and Fermi–Dirac statistics.

In higher dimensions, angular momentum has a tensor character and is characterized by n-dimensional spherical harmonics [130].

11.7.3 Covariant description in $3 + 1$ dimensions

The angular momentum of a body at position $\mathbf{r}$, about an origin, with momentum $\mathbf{p}$, is defined by

$$\mathbf{J} = \mathbf{L} + \mathbf{S} = (\mathbf{r} \times \mathbf{p}) + \mathbf{S}. \tag{11.111}$$

The first term, constructed from the cross-product of the position and linear momentum, is the contribution to the orbital angular momentum. The second term, **S**, is the spin, or intrinsic angular momentum, of the body. The total angular momentum is a conserved quantity and may be derived from the energy–momentum tensor in the following way. Suppose we have a conserved energy–momentum tensor $\theta_{\mu\nu}$, which is symmetrical in its indices (Lorentz-invariant), then

$$\partial_\mu \theta^{\mu\nu} = 0. \tag{11.112}$$

We can construct the new axial tensor,

$$L^{\mu\nu\lambda} = x^\nu \theta^{\lambda\mu} - x^\lambda \theta^{\nu\mu}, \tag{11.113}$$

which is also conserved, since

$$\partial_\mu L^{\mu\nu\lambda} = \theta^{\lambda\nu} - \theta^{\nu\lambda} = 0. \tag{11.114}$$

Comparing eqn. (11.113) with eqn. (11.111), we see that $L^{\mu\nu\lambda}$ is a generalized vector product, since the components of $\mathbf{r} \times \mathbf{p}$ are of the form $L_1 = r_2 p_3 - r_3 p_2$, or $L_i = \epsilon_{ijk} r_j p_k$. We may then identify the angular momentum 2-tensor as the anti-symmetrical matrix

$$J^{\mu\nu} = \int d\sigma \; L^{0\mu\nu} = -J^{\nu\mu}, \tag{11.115}$$

which is related to the generators of homogeneous Lorentz transformations (generalized rotations on spacetime) by

$$J^{\mu\nu}\Big|_{p_i=0} = \chi_h \, T_{3+1}^{\mu\nu}; \tag{11.116}$$

see eqn. (9.95). The ij components of $J^{\mu\nu}$ are simply the components of $\mathbf{r} \times \mathbf{p}$. The $i0$ components are related to boosts. Clearly, this matrix is conserved,

$$\partial_\mu J^{\mu\nu} = 0. \tag{11.117}$$

Since the coordinates x^μ appear explicitly in the definition of $J^{\mu\nu}$, it is not invariant under translations of the origin. Under the translation $x^\mu \to x^\mu + a^\mu$, the components transform into

$$J^{\mu\nu} \to J^{\mu\nu} + (a^\mu p^\nu + a^\mu p^\mu). \tag{11.118}$$

(see eqn. (11.5)). This can be compared with the properties of eqn. (9.153). To isolate the part of $T_{\mu\nu}$ which is intrinsic to the field (i.e. is independent of position), we may either evaluate in a rest frame $p_i = 0$ or define, in $3 + 1$ dimensions, the *dual* tensor

$$S_{\mu\nu} = \frac{1}{2}\epsilon_{\mu\nu\rho\lambda} J^{\lambda\rho} = S_{\mu\nu}^*. \tag{11.119}$$

The anti-symmetry of the Levi-Cevita tensor ensures that the extra terms in eqn. (11.118) cancel. We may therefore think of this as being the generator of the intrinsic angular momentum of the field or spin. This dual tensor is rather formal though and not very useful in practice. Rather, we consider the Pauli–Lubanski vector as introduced in eqn. (9.161). We define a spin 4-vector by

$$-\frac{1}{2}mc\, S_\mu \equiv \chi_h\, W_\mu = \frac{1}{2}\epsilon_{\mu\nu\rho\lambda} J^{\nu\rho} p^\lambda, \qquad (11.120)$$

so that, in a rest frame,

$$\chi_h\, W^\mu_{\text{rest}} = -\frac{1}{2}mc(0, S^i), \qquad (11.121)$$

where S^i is the intrinsic spin angular momentum, which is defined by

$$S^i = J^i\Big|_{p^i=0} = \chi_h\, T_{B\,i} = \frac{1}{2}\chi_h\, \epsilon_{ijk} T^{jk}_{\text{R}}, \qquad (11.122)$$

with eigenvalues $s(s+1)\,\chi_h{}^2$ and $m_s\,\chi_h$, where $s = e + f$.

11.7.4 Intrinsic spin of tensor fields in 3 + 1 dimensions

Tensor fields are classified by their intrinsic spin in $3 + 1$ dimensions. We speak of fields with intrinsic spin $0, \frac{1}{2}, 1, \frac{3}{2}, 2, \ldots$. These labels usually refer to $3 + 1$ dimensions, and may differ in other number of dimensions since they involve counting the number of independent components in the tensors, which differs since the representation space is spacetime for the Lorentz symmetry. The number depends on the dimension and transformation properties of the matrix representation, which defines a rotation of the field. The homogeneous (translation independent) Lorentz group classifies these properties of the field in $3 + 1$ dimensions,

Field	Spin
$\phi(x)$	0
$\psi_\alpha(x)$	$\frac{1}{2}$
A_μ	1
Ψ^μ_α	$\frac{3}{2}$
$g_{\mu\nu}$	2

where $\mu, \nu = 0, 1, 2, 3$. Although fields are classified by their spin properties, this is not enough to be able to determine the rotational modes of the field. The mass also plays a role. This is perhaps most noticeable for the spin-1 field A_μ. In the massless case, it has helicities $\lambda = \pm 1$, whereas in the massive case it can take on the additional value of zero. The reason for the difference follows from a difference in the true spacetime symmetry of the field in the two cases. We shall explore this below.

From section 9.4.3 we recall that the irreducible representations of the Lorentz group determine the highest weight or spin $s \equiv e + f$ of a field. If we set the generators of boosts to zero by taking $\omega_{0i} T^{0i} = 0$ in eqn. (9.95), then we obtain the pure spatial rotations of section 8.5.10. Then the generators of the Lorentz group E_i and F_i become identical, and we may define the *spin* of a representation by the operator

$$S_i = E_i + F_i = \chi_h T_{B\,i}. \tag{11.123}$$

The Casimir operator for the defining (vector field) representation is then

$$S^2 = \chi_h^2 T_B^2 = \chi_h^2 \begin{pmatrix} 0 & 0 & 0 & 0 \\ 0 & 2 & 0 & 0 \\ 0 & 0 & 2 & 0 \\ 0 & 0 & 0 & 2 \end{pmatrix}. \tag{11.124}$$

This shows that the rotational 3-vector part of the defining representation forms an irreducible module, leaving an empty scalar component in the time direction. One might expect this; after all, spatial rotations ought not to involve timelike components. If we ignore the time component, then we easily identify the spin of the vector field as follows. From section 8.5.10 we know that in representation G_R, the Casimir operator is proportional to the identity matrix with value

$$S^2 = S^i S_i = s(s+1)\chi_h^2 I_R, \tag{11.125}$$

and $s = e + f$. Comparing this with eqn. (11.124) we have $s(s+1) = 2$, thus $s = 1$ for the vector field. We say that a vector field has spin 1.

Although the vector transformation leads us to a value for the highest weight spin, this does not necessarily tell us about the intermediate values, because there are two ways to put together a spin-1 representation. One of these applies to the massless (transverse) field and the other to the massive Proca field, which was discussed in section 9.4.4. As another example, we take a rank 2-tensor field. This transforms like

$$G_{\mu\nu} \to L_\mu^{\ \rho} L_\nu^{\ \sigma} G_{\rho\sigma}. \tag{11.126}$$

In other words, two vector transformations are required to transform this, one for each index. The product of two such matrices has an equivalent vector form with irreducible blocks:

$$\underbrace{(1, 1)}_{\substack{\text{traceless} \\ \text{symmetric}}} \quad \oplus \quad \underbrace{(1, 0) \oplus (0, 1)}_{\text{anti-symmetric}} \quad \oplus \quad \underbrace{(0, 0)}_{\text{trace}}.$$

This is another way of writing the result which was found in section 3.76 using more pedestrian arguments. The first has $(2e + 1)(2f + 1) = 9$ $(e = f = 1)$ spin $e + f = 2$ components; the second two blocks are six spin-1 parts; and the last term is a single scalar component, giving 16 components in all, which is the number of components in the second-rank tensor.

Another way to look at this is to compare the number of spatial components in fields with $2s + 1$. For scalar fields (spin 0), $2s + 1$ gives one component. A 4-vector field has one scalar component and $2s + 1 = 3$ spatial components (spin 1). A spin-2 field has nine spatial components: one scalar (spin-0) component, three vector (spin-1) components and $2s + 1 = 5$ remaining spin-2 components. This is reflected in the way that the representations of the Lorentz transformation matrices reduce into diagonal blocks for spins 0, 1 and 2. See ref. [132] for a discussion of spin-2 fields and covariance.

It is coincidental for $3 + 1$ dimensions that spin-0 particles have no Lorentz indices, spin-1 particles have one Lorentz index and spin-2 particles have two Lorentz indices.

What is the physical meaning of the spin label? The spin is the highest weight of the representation which characterizes rotational invariance of the system. Since the string of values produced by the stepping operators moves in integer steps, it tells us how many distinct ways, $m + m'$, a system can spin in an 'equivalent' fashion. In this case, equivalent means about the same axis.

11.7.5 Helicity versus spin

Helicity is defined by

$$\lambda = J_i \, \hat{p}^i. \tag{11.127}$$

Spin s and helicity λ are clearly related quite closely, but they are subtly different. It is not uncommon to refer loosely to helicity as spin in the literature since that is often the relevant quantity to consider. The differences in rotation algebras, as applied to physical states are summarized in table 11.3. Because the value of the helicity is not determined by an upper limit on the total angular momentum, it is conventional to use the component of the spin of the irreducible representation for the Lorentz group which lies along the direction of the direction of travel. Clearly these two definitions are not the same thing. In the massless case, the labels for the helicity are the same as those which would occur for m_j in the rest frame of the massive case.

From eqn. (11.127) we see that the helicity is rotationally invariant for massive fields and generally Lorentz-invariant for massless $p_0 = 0$ fields.

Table 11.3. Spin and helicity.

	Casimir	$\Lambda_c = m_j$
Massive	$j(j+1)$	$0, \pm\frac{1}{2}, \pm 1, \ldots, \pm j$
Massless	0	$0, \pm\frac{1}{2}, \pm 1, \ldots, \infty$

It transforms like a pseudo-scalar, since J_i is a pseudo-vector. Thus, the sign of helicity changes under parity transformations, and a massless particle which takes part in parity conserving interactions must have both helicity states $\pm\lambda$, i.e. we must represent it by a (reducible) symmetrized pair of irreducible representations:

$$\begin{pmatrix} + & 0 \\ 0 & - \end{pmatrix} \quad \text{or} \quad \begin{pmatrix} 0 & + \\ - & 0 \end{pmatrix}. \tag{11.128}$$

The former is the case for the massless Dirac field ($\lambda = \pm\frac{1}{2}$), while the latter is true for the photon field $F^{\mu\nu}$ ($\lambda = \pm 1$), where the states correspond to left and right circularly polarized radiation. Note that, whereas a massive particle could have $\lambda = 0, \pm 1$, representing left transverse, right transverse and longitudinal angular momentum, a massless (purely transverse) field cannot have a longitudinal mode, so $\lambda = 0$ is absent. This can be derived more rigorously from representation theory.

In refs. [45, 55], the authors study massless fields with general spin and show that higher spins do not necessarily have to be strictly conserved; only the Dirac-traceless part of the divergence has to vanish.

11.7.6 Fractional spin in $2+1$ dimensions

The Poincaré group in $2 + 1$ dimensions shares many features of the group in $3 + 1$ dimensions, but also conceals many subtleties [9, 58, 77]. These have specific implications for angular momentum and spin. In two spatial dimensions, rotations form an Abelian group $SO(2) \sim U(1)$, whose generators can, in principle, take on eigenvalues which are unrestricted by the constraints of spherical harmonics. This leads to continuous phases [89, 138], particle statistics and the concept of fractional spin. It turns out, however, that there is a close relationship between vector (gauge) fields and spin in $2 + 1$ dimensions, and that fractional values of spin can only be realized in the context of a gauge field coupling. This is an involved topic, with a considerable literature, which we shall not delve into here.

11.8 Work, force and transport in open systems

The notion of interaction and force in field theory is unlike the classical picture of particles bumping into one another and transferring momentum. Two fields interact in the manner of two waves passing through one another: by interference, or *amplitude modulation*. Two fields are said to interact if there is a term in the action in which some power of one field multiplies some power of another. For example,

$$S_{\text{int}} = \int (\mathrm{d}x) \left\{ \phi^2 A_\mu A^\mu \right\}. \tag{11.129}$$

Since the fields multiply, they modulate one another's behaviour or perturb one another. There is no explicit notion of a force here, and precisely what momentum is transferred is rather unclear in the classical picture; nevertheless, there is an interaction. This can lead to scattering of one field off another, for instance.

The source terms in the previous section have the form of an interaction, in which the coupling is linear, and thus they exert what is referred to as a *generalized force* on the field concerned. The word generalized is used because J does not have the dimensions of force – what is important is that the source has an *influence* on the behaviour of the field.

Moreover, if we place all such interaction terms on the right hand side of the equations of motion, it is clear that interactions also behave as sources for the fields (or currents, if you prefer that name). In eqn. (11.129), the coupling between ϕ and A_μ will lead to a term in the equations of motion for ϕ and for A_μ, thus it acts as a source for both fields.

We can express this in other words: *an interaction can be thought of as a source which transfers some 'current' from one field to another.* But be wary that what we are calling heuristically 'current' might be different in each case and have different dimensions.

A term in which a field multiplies itself, ϕ^n, is called a self-interaction. In this case the field is its own source. Self-interactions lead to the scattering of a field off itself. The classical notion of a force was described in terms of the energy–momentum tensor in section 11.3.

11.8.1 The generalized force $F_\nu = \partial_\mu T^{\mu\nu}$

There is a simple proof which shows that the tensor $T_{\mu\nu}$ is conserved, provided one has Lorentz invariance and the classical equations of motion are satisfied. Consider the total dynamical variation of the action

$$\delta S = \int \frac{\delta S}{\delta g^{\mu\nu}} \delta g^{\mu\nu} + \int \frac{\delta S}{\delta q} \delta q = 0. \tag{11.130}$$

Since the equations of motion are satisfied, the second term vanishes identically, leaving

$$\delta S = \frac{1}{2}\sqrt{g}\int (\mathrm{d}x)T_{\mu\nu}\delta g^{\mu\nu}. \qquad (11.131)$$

For simplicity, we shall assume that the metric $g_{\mu\nu}$ is independent of x, so that the variation may be written (see eqn. (4.88))

$$\delta S = \int (\mathrm{d}x)T_{\mu\nu}\left[g_{\mu\lambda}(\partial_\nu\epsilon^\lambda) + g_{\lambda\nu}(\partial_\mu\epsilon^\lambda)\right] = 0. \qquad (11.132)$$

Integrating by parts, we obtain

$$\delta S = \int (\mathrm{d}x)\left[-2\partial_\mu T^{\mu\nu}\right]\epsilon_\nu = 0. \qquad (11.133)$$

Since $\epsilon^\mu(x)$ is arbitrary, this implies that

$$\partial_\mu T^{\mu\nu} = 0, \qquad (11.134)$$

and hence $T^{\mu\nu}$ is conserved. From this argument, it would seem that $T^{\mu\nu}$ must always be conserved in every physical system, and yet one could imagine constructing a physical model in which energy was allowed to leak away. The assumption of Lorentz invariance and the use of the equations of motion provide a catch, however. While it is true that the energy–momentum tensor is conserved in any *complete* physical system, it does not follow that energy or momentum is conserved in every part of a system individually. If we imagine taking two partial systems and coupling them together, then those two systems can exchange energy. In fact, energy will only be conserved if the systems are in perfect balance: if, on the other hand, one system does work on the other, then energy flows from one system to the other. No energy escapes the total system, however.

Physical systems which are coupled to other systems, about which we have no knowledge, are called *open systems*. This is a matter of definition. Given any closed system, we can make an open system by isolating a piece of it and ignoring the rest. Clearly a description of a piece of a system is an incomplete description of the total system, so it appears that energy is not conserved in the small piece. In order to see conservation, we need to know about the whole system. This situation has a direct analogue in field theory. Systems are placed in contact with one another by interactions, often through currents or sources. For instance, Dirac matter and radiation couple through a term which looks like $J^\mu A_\mu$. If we look at only the Dirac field, the energy–momentum tensor is not conserved. If we look at only the radiation field, the energy–momentum tensor is not conserved, but the sum of the two parts is. The reason is that we have to be 'on shell' – i.e., we have to satisfy the equations of motion.

Consider the following example. The (incomplete) action for the interaction between the Dirac field and the Maxwell field is

$$S = \int (dx) \left\{ \frac{1}{4\mu_0} F^{\mu\nu} F_{\mu\nu} - J^{\mu} A_{\mu} \right\}, \tag{11.135}$$

where $J^{\mu} = \overline{\psi} \gamma^{\mu} \psi$. Now, computing the energy–momentum tensor for this action, we obtain

$$\partial_{\mu} T^{\mu\nu} = F^{\mu\nu} J_{\mu}. \tag{11.136}$$

This is not zero because we are assuming that the current J^{μ} is not zero. But this is not a consistent assumption in the action, because we have not added any dynamics for the Dirac field, only the coupling $J^{\mu} A_{\mu}$. Consider the field equation for ψ from eqn. (11.135). Varying with respect to $\overline{\psi}$,

$$\frac{\delta S}{\delta \overline{\psi}} = i e \gamma^{\mu} A_{\mu} \psi = 0. \tag{11.137}$$

This means that either $A_{\mu} = 0$ or $\psi = 0$, but both of these assumptions make the right hand side of eqn. (11.136) zero! So, in fact, the energy–momentum tensor is conserved, as long as we obey the equations of motion given by the variation of the action.

The 'paradox' here is that we did not include a piece in the action for the Dirac field, but that we were sort of just assuming that it was there. This is a classic example of writing down an incomplete (open) system. The full action,

$$S = \int (dx) \left\{ \frac{1}{4\mu_0} F^{\mu\nu} F_{\mu\nu} - J^{\mu} A_{\mu} + \overline{\psi} (\gamma^{\mu} \partial_{\mu} + m) \psi \right\}, \tag{11.138}$$

has a conserved energy–momentum tensor, for more interesting solutions than $\psi = 0$.

From this discussion, we can imagine the imbalance of energy–momentum on a partial system as resulting in an external force on this system, just as in Newton's second law. Suppose we define the generalized external force by

$$F^{\nu} = \int d\sigma \ \partial_{\mu} T^{\mu\nu}. \tag{11.139}$$

The spatial components are

$$F^{i} = \int d\sigma \ \partial_{0} T^{0i} = \partial_{t} P^{i} = \frac{d\mathbf{p}}{dt}, \tag{11.140}$$

which is just Newton's second law. Compare the above discussion with eqn. (2.73) for the Poynting vector.

An important lesson to learn from this is that a source is not only a generator for the field (see section 14.2) but also a model for what we do not know about an external system. This is part of the essence of source theory as proposed by Schwinger. For another manifestation of this, see section 11.3.3.

11.8.2 Work and power

In chapter 5 we related the imaginary part of the Feynman Green function to the instantaneous rate at which work is done by the field. We now return to this problem and use the energy–momentum tensor to provide a new perspective on the problem.

In section 6.1.4 we assumed that the variation of the action with time, evaluated at the equations of motion, was the energy of the system. It is now possible to justify this; in fact, it should already be clear from eqn. (11.78). We can go one step further, however, and relate the power loss to the notion of an open system. If a system is open (if it is coupled to sources), it does work, w. The rate at which it does work is given by

$$\frac{dw}{dt} = \int d\sigma \; \partial_\mu T^{\mu 0}. \tag{11.141}$$

This has the dimensions of energy per unit time. It is clearly related to the variation of the action itself, evaluated at value of the field which satisfies the field equations, since

$$\Delta w = -\int d\sigma dt \; \partial_\mu T^{\mu 0} = -\frac{\delta S}{\delta t}\bigg|_{\text{field eqns}}. \tag{11.142}$$

The electromagnetic field is the proto-typical example here. If we consider the open part of the action (the source coupling),

$$S_J = \int (dx) \; J^\mu A_\mu, \tag{11.143}$$

then, using

$$A_\mu = \int (dx) \; D_{\mu\nu}(x, x') J^\nu(x'), \tag{11.144}$$

we have

$$\delta S[A_J] = \delta \int (dx) \; J^\mu \delta A_\mu$$

$$= \int (dx)(dx') J^\mu(x) D_{\mu\nu}(x, x') \delta J^\nu(x')$$

$$= \int (dx)(\partial_\mu T^{\mu 0}) \delta t$$

$$= \Delta w \delta t. \tag{11.145}$$

The Green function we choose here plays an important role in the discussion, as noted in section 6.1.4. There are two Green functions which can be used in eqn. (11.144) as the inverse of the Maxwell operator: the retarded Green function and the Feynman Green function. The key expression here is

$$W = \frac{1}{2} \int (\mathrm{d}x)(\mathrm{d}x') J^{\mu}(x) D_{\mu\nu}(x, x') J^{\nu}(x'). \qquad (11.146)$$

Since the integral is spacetime symmetrical, only the symmetrical part of the Green function contributes to the integral. This immediately excludes the retarded Green function

11.8.3 Hydrodynamic flow and entropy

Hydrodynamics is not usually regarded as field theory, but it is from hydrodynamics (fluid mechanics) that we derive notions of macroscopic transport. All transport phenomena and thermodynamic properties are based on the idea of *flow*. The equations of hydrodynamics are the Navier–Stokes equations. These are non-linear vector equations with highly complex properties, and their complete treatment is outside the scope of this book. In their linearized form, however, they may be solved in the usual way of a classical field theory, using the methods of this book. We study hydrodynamics here in order to forge a link between field theory and thermodynamics. This is an important connection, which is crying out to be a part of the treatment of the energy–momentum tensor. We should be clear, however, that this is a phenomenological addition to the field theory for statistically large systems.

A fluid is represented as a velocity field, $U^{\mu}(x)$, such that each point in a system is moving with a specified velocity. The considerations in this section do not depend on the specific nature of the field, only that the field is composed of matter which is flowing with the velocity vector U^{μ}. Our discussion of flow will be partly inspired by the treatment in ref. [134], and it applies even to relativistic flows. As we shall see, the result differs from the non-relativistic case only by a single term. A stationary field (fluid) with maximal spherical symmetry, in flat spacetime, has an energy–momentum tensor given by

$$T_{00} = \mathcal{H}$$
$$T_{0i} = T_{i0} = 0$$
$$T_{ij} = P\delta_{ij}. \qquad (11.147)$$

In order to make this system flow, we may perform a position-dependent boost which places the observer in relative motion with the fluid. Following a boost, the energy–momentum tensor has the form

$$T^{\mu\nu} = Pg^{\mu\nu} + (P + \mathcal{H})U^{\mu}U^{\nu}/c^2. \qquad (11.148)$$

The terms have the dimensions of energy density. P is the pressure exerted by the fluid (clearly a thermodynamical average variable, which summarizes the microscopic thermal motion of the field). $\mathcal{H}$ is the internal energy density of the field. Let us consider the generalized thermodynamic force $F^\mu = \partial_\nu T^{\mu\nu}$. In a closed thermodynamic system, we know that the energy–momentum tensor is conserved:

$$F^\mu = \partial_\nu T^{\mu\nu} = 0, \tag{11.149}$$

and that the matter density $N(x)$ in the field is conserved,

$$\partial_\mu N^\mu = 0, \tag{11.150}$$

where $N_\mu = N(x)U_\mu$. If we think of the field as a plasma of particles, then $N(x)$ is the number of particles per unit volume, or number density. Due to its special form, we may write

$$\partial_\mu N^\mu = (\partial_\mu N)U^\mu + (\partial_\mu U^\mu), \tag{11.151}$$

which provides a hint that the velocity boost acts like a local scaling or conformal transformation on space

$$-c^2 dt^2 + dx_i dx^i \rightarrow -c^2 dt^2 + \Omega^2(U)dx_i dx^i. \tag{11.152}$$

The average rate of work done by the field is zero in an ideal, closed system:

$$\begin{aligned}
\frac{dw}{dt} &= \int d\sigma \; U_\nu F^\nu \\
&= \int d\sigma \; \left[U^\mu \partial_\mu P - \partial_\mu \left((P + \mathcal{H})U^\mu \right) \right] \\
&= 0.
\end{aligned} \tag{11.153}$$

Now, noting the identity

$$N\partial_\mu \left(\frac{P + \mathcal{H}}{N} \right) = \partial(P + \mathcal{H}) - \left(\frac{\partial_\mu N}{N} \right)(P + \mathcal{H}), \tag{11.154}$$

we may write

$$\frac{dw}{dt} = \int d\sigma \; U^\mu \left[\partial_\mu P - N \left(\frac{P + \mathcal{H}}{N} \right) \right]. \tag{11.155}$$

Then, integrating by parts, assuming that U^μ is zero on the boundary of the system, and using the identity in eqn. (11.151)

$$\begin{aligned}
\frac{dw}{dt} &= -\int d\sigma \; NU^\mu \left[P\partial_\mu \left(\frac{1}{N} \right) + \partial_\mu \left(\frac{\mathcal{H}}{N} \right) \right] \\
&= -\int d\sigma \; NU^\mu \left[P\partial_\mu V + \partial_\mu H \right], \tag{11.156}
\end{aligned}$$

where V is the volume per particle and H is the internal energy. This expression can be compared with

$$T\,dS = P\,dV + dH. \tag{11.157}$$

Eqn. (11.156) may be interpreted as a rate of entropy production due to the hydrodynamic flow of the field, i.e. it is the rate at which energy becomes unavailable to do work, as a result of energy diffusing out over the system uniformly or as a result of internal losses. We are presently assuming this to be zero, in virtue of the conservation law, but this can change if the system contains hidden degrees of freedom (sources/sinks), such as friction or viscosity, which convert mechanical energy into heat in a non-useful form. Combining eqn. (11.156) and eqn. (11.157) we have

$$-\int d\sigma\, N U^\mu (\partial_\mu S) T = \int d\sigma\, U_\nu \partial_\mu T^{\mu\nu} = 0. \tag{11.158}$$

From this, it is useful to define a covariant entropy density vector $\mathcal{S}^\mu$, which symbolizes the rate of loss of energy in the hydrodynamic flow. In order to express the right hand side of eqn. (11.158) in terms of gradients of the field and the temperature, we integrate by parts and define. Let

$$c(\partial_\mu \mathcal{S}^\mu) = \partial_\mu \left(\frac{U_\nu}{T} \right) T^{\mu\nu}, \tag{11.159}$$

where

$$\mathcal{S}^\mu = N S U^\mu - \frac{U_\nu T^{\mu\nu}}{T}. \tag{11.160}$$

The zeroth component, $c\mathcal{S}^0 = NS$, is the entropy density, so we may interpret $\mathcal{S}^\mu$ as a spacetime entropy vector. Let us now assume that hidden losses can cause the conservation law to be violated. Then we have the rate of entropy generation given by

$$c(\partial_\mu \mathcal{S}^\mu) = \left[-\frac{1}{T}(\partial_\nu U_\mu) + \frac{1}{T^2}(\partial_\nu T)U_\mu \right] T^{\mu\nu}. \tag{11.161}$$

We shall assume that the temperature is independent of time, since the simple arguments used to address statistical issues at the classical level do not take into account time-dependent changes properly: the fluctuation model introduced in section 6.1.5 gives rise only to instantaneous changes or steady state flows. If we return to the co-moving frame in which the fluid is stationary, we have

$$U_i = \partial_\mu U^0 = \partial_t T = 0, \tag{11.162}$$

and thus

$$(\partial_\mu S^\mu) = -\left[-\frac{1}{c^2 T}(\partial_t U_i) + \frac{1}{T^2}(\partial_i T)\right]T^{0i}$$
$$+ \frac{1}{2T}(\partial_i U_j + \partial_j U_i)T^{ij}. \tag{11.163}$$

Note the single term which vanishes in the non-relativistic limit $c \to \infty$. This is the only sign of the Lorentz covariance of our formulation. Also, we have used the symmetry of T^{ij} to write $\partial_i U_j$ in an ij-symmetric form.

So far, these equations admit no losses: the conservation law cannot be violated: energy cannot be dissipated. To introduce, phenomenologically, an expression of dissipation, we need so-called *constitutive relations* which represent average 'frictional forces' in the system. These relations provide a linear relationship between gradients of the field and temperature and the rate of entropy generation, or energy stirring. The following well known forms are used in elementary thermodynamics to define the thermal conductivity κ in terms of the heat flux Q_i and the temperature gradient; similarly the viscosity η in terms of the pressure P:

$$Q_i = -\kappa \frac{dT}{dx^i}$$
$$P_{ij} = -\eta \frac{\partial U_i}{\partial x^j}. \tag{11.164}$$

The relations we choose to implement these must make the rate of entropy generation non-negative if they are to make thermodynamical sense. It may be checked that the following definitions fulfil this requirement in n spatial dimensions:

$$T^{0i} = -\kappa\left(\partial_i T + T\partial_t U_i/c^2\right)$$
$$T^{ij} = -\eta\left(\partial_i U_j + \partial_j U_i - \frac{2}{n}(\partial_k U^k)\delta_{ij}\right) - \zeta(\partial_k U^k)\delta_{ij}, \tag{11.165}$$

where κ is the thermal conductivity, η is the shear viscosity and ζ is the bulk viscosity. The first term in this last equation may be compared with eqn. (9.217). This makes use of the definition of *shear* σ_{ij} for a vector field V_i as a conformal deformation

$$\Delta_{ij} = \partial_i V_j + \partial_j V_i - \frac{2}{n}(\partial_k V^k)\delta_{ij}. \tag{11.166}$$

This is a measure of the non-invariance of the system to conformal, or shearing transformations. Substituting these constitutive equations into eqn. (11.163), one obtains

$$
\begin{aligned}
c\partial_\mu \mathcal{S}^\mu &= \frac{\kappa}{T^2}(\partial_i T + T\partial_0 U_i/c)(\partial^i T + T\partial_0 U^i/c) \\
&= \frac{\eta}{2T}\left(\partial_i U_j + \partial_j U_i\right)\left(\partial^i U^j + \partial^j U^i\right) \\
&= \left(\zeta + \frac{4}{n}\zeta\right)\frac{1}{T}(\partial_k U^k)^2.
\end{aligned}
\tag{11.167}
$$

11.8.4 Thermodynamical energy conservation

The thermodynamical energy equations supplement the conservation laws for mechanical energy, but they are of a different character. These energy equations are average properties for bulk materials. They summarize collective microscopic conservation on a macroscopic scale.

$$
\partial_\mu T^{\mu\nu} = H + T\mathrm{d}S + P\mathrm{d}V + \mathrm{d}F
\tag{11.168}
$$

$$
S = k\ln\Omega
\tag{11.169}
$$

$$
T\mathrm{d}S = kT\frac{\mathrm{d}\Omega}{\Omega} = \frac{1}{\beta}\frac{\mathrm{d}\Omega}{\Omega}.
\tag{11.170}
$$

11.8.5 Kubo formulae for transport coefficients

In section 6.1.6, a general scheme for computing transport coefficients was presented, but only the conductivity tensor was given as an example. Armed with a knowledge of the energy–momentum tensor, entropy and the dissipative processes leading to viscosity, we are now in a position to catalogue the most important expressions for these transport coefficients. The construction of the coefficients is based on the general scheme outlined in section 6.1.6. In order to compute these coefficients, we make use of the assumption of linear dissipation, which means that we consider only first-order gradients of thermodynamic averages. This assumes a slow rate of dissipation, or a linear relation of the form

$$
\langle\text{variable}\rangle = k\nabla_\mu\langle\text{source}\rangle,
\tag{11.171}
$$

where ∇_μ represents some spacetime gradient. This is the so-called constitutive relation. The expectation values of the variables may be derived from the generating functional W in eqn. (6.7) by adding source terms, or variables conjugate to the ones we wish to find correlations between. The precise meaning of the sources is not important in the linear theory we are using, since the source

Table 11.4. Conductivity tensor.

Component	Response	Measure
σ_{00}/c^2	induced density	charge compressibility
σ_{0i}/c	density current	–
σ_{ii}	induced current	linear conductivity
σ_{ij}	induced current	transverse (Hall) conductivity

cancels out of the transport formulae completely (see eqn. (6.66)). Also, there is a symmetry between the variables and their conjugates. If we add source terms

$$S \rightarrow S + \int (\mathrm{d}x)(J \cdot A + J^\mu A_\mu + J^{\mu\nu} A_{\mu\nu}), \qquad (11.172)$$

then the J's are sources for the A's, but conversely the A's are also sources for the J's.

We begin therefore by looking at the constitutive relations for the transport coefficients, in turn. The generalization of the conductivity derived in eqn. (6.75) for the spacetime current is

$$J_\mu = \sigma^{\mu\nu} \partial_t A^\nu. \qquad (11.173)$$

Although derivable directly from Ohm's law, this expresses a general dissipative relationship between any current J^μ and source A^μ, so we would expect this construction to work equally well for any kind of current, be it charged or not. From eqn. (11.171) and eqn. (6.66) we have the Fourier space expression for the spacetime conductivity tensor in terms of the Feynman correlation functions

$$\sigma_{\mu\nu}(\omega) = \lim_{\mathbf{k}\to 0} \frac{\mathrm{i}}{\hbar\omega} \int (\mathrm{d}x) \mathrm{e}^{-\mathrm{i}k(x-x')} \langle J_\mu(x) J_\nu(x') \rangle, \qquad (11.174)$$

or in terms of the retarded functions. In general the products with Feynman boundary conditions are often easier to calculate, since there are theorems for their factorization.

$$\sigma_{\mu\nu}(\omega)\Big|_\beta \equiv \lim_{\mathbf{k}\to 0} \frac{(1 - \mathrm{e}^{-\hbar\beta\omega})}{\hbar\omega} \int (\mathrm{d}x) \mathrm{e}^{-\mathrm{i}k(x-x')} \langle J_\mu(x) J_\nu(x') \rangle. \qquad (11.175)$$

The D.C. conductivity is given by the $\omega \to 0$ limit of this expression. The components of this tensor are shown in table 11.4: The constitutive relations for the viscosities are given in eqn. (11.165). From eqn. (6.67) we have

$$\langle T_{\mu\nu}(x) \rangle = \frac{\delta W}{\delta J^{\mu\nu}(x)} \qquad (11.176)$$

and

$$\frac{\delta \langle T_{\mu\nu}(x) \rangle}{\delta J^{\rho\sigma}(x')} = \frac{i}{\hbar} \langle T_{\mu\nu}(x) T_{\rho\sigma}(x') \rangle$$

$$= \frac{i}{\hbar} \langle T_{\mu\rho}(x) T_{\nu\sigma}(x') \rangle, \tag{11.177}$$

where the last line is a consequence of the connectivity of Feynman averaging. Note that this relation does not depend on our ability to express $W[J^{\mu\nu}]$ in a quadratic form analogous to eqn. (6.35). The product on the right hand side can be evaluated by expressing $T_{\mu\nu}$ in terms of the field. The symmetry of the energy–momentum tensor implies that

$$J^{\mu\nu} = J^{\nu\mu}, \tag{11.178}$$

and, if the source coupling is to have dimensions of action, $J_{\mu\nu}$ must be dimensionless. The only object one can construct is therefore

$$J^{\mu\nu} = g^{\mu\nu}. \tag{11.179}$$

Thus, the source term is the trace of the energy–momentum tensor, which vanishes when the action is conformally invariant. To express eqn. (11.165) in momentum space, we note that Fourier transform of the velocity is the phase velocity of the waves,

$$U^i(x) = \gamma \frac{\mathrm{d}x^i}{\mathrm{d}t} = \gamma \int \frac{\mathrm{d}^n k}{(2\pi)^n} e^{ik^\mu x_\mu} \frac{\omega}{k_i}$$

$$= \gamma \int \frac{\mathrm{d}^n k}{(2\pi)^n} e^{ik^\mu x_\mu} \frac{\omega k^i}{\mathbf{k}^2}. \tag{11.180}$$

The derivative is given by

$$\partial_j U^i = i\gamma \int \frac{\mathrm{d}^n k}{(2\pi)^n} e^{ik^\mu x_\mu} \frac{\omega k^i k_j}{\mathbf{k}^2}. \tag{11.181}$$

Thus, eqn. (11.165) becomes

$$\langle T_{ij} \rangle = -\left(\zeta + \frac{4}{n}\eta\right) i\gamma \omega g_{ij} - \eta\, i\gamma \omega \frac{k_i k_j}{\mathbf{k}^2}. \tag{11.182}$$

Comparing this with eqn. (11.176), we have, for the spatial components,

$$-\left(\zeta + \frac{4}{n}\eta\right) g_{ij} g_{lm} - \eta \frac{k_i k_j}{\mathbf{k}^2} g_{lm} =$$

$$\frac{1}{\hbar\omega} \int (\mathrm{d}x)\, e^{-ik(x-x')} \langle T_{il}(x) T^i_{\ m}(x') \rangle. \tag{11.183}$$

Contracting both sides with $g^{ij} g^{lm}$ leaves

$$
\left(\zeta(\omega) + \frac{4-n}{n} \, \eta(\omega) \right) =
$$

$$
\lim_{\mathbf{k} \to 0} -\frac{1}{n^2 \hbar \omega} \int (\mathrm{d}x) e^{-ik(x-x')} \langle T_{ij}(x) T^{ij}(x') \rangle. \tag{11.184}
$$

The two viscosities cannot be separated in this relation, but η can be related to the diffusion coefficient, which can be calculated separately. Assuming causal (retarded relation between field and source), at finite temperature we may use eqn. (6.74) to write

$$
\left(\zeta(\omega) + \frac{4-n}{n} \, \eta(\omega) \right) \Big|_{\beta} \equiv
$$

$$
\lim_{\mathbf{k} \to 0} -\frac{(1 - e^{-\hbar \omega \beta})}{n^2 \hbar \omega} \int (\mathrm{d}x) e^{-ik(x-x')} \langle T_{ij}(x) T^{ij}(x') \rangle. \tag{11.185}
$$

The temperature conduction coefficient κ is obtained from eqn. (11.165). Following the same procedure as before, we obtain

$$
\frac{i}{\hbar} \int (\mathrm{d}x) e^{-ik^\mu (x-x')_\mu} \langle T^{0i}(x) T^{0j}(x')
$$

$$
= -g^{0j} \kappa (\partial^i T + T \partial_t U^i / c^2) \rangle
$$

$$
= -i g^{0j} \kappa (k^i T - T \gamma \omega^2 k^i / \mathbf{k}^2). \tag{11.186}
$$

Rearranging, we get

$$
\kappa(\omega) = \lim_{\mathbf{k} \to 0} -\frac{g_{0j} k_i (1 - e^{-\hbar \omega \beta})}{\hbar \left(\mathbf{k}^2 - \gamma \omega^2 / c^2 \right)} \int (\mathrm{d}x) e^{-ik^\mu (x-x')_\mu} \langle T^{0i}(x) T^{0j}(x') \rangle. \tag{11.187}
$$

To summarize, we note a list of properties with their relevant fluctuations and conjugate sources. See table 11.5.

11.9 Example: Radiation pressure

The fact that the radiation field carries momentum means that light striking a material surface will exert a pressure equal to the change in momentum of the light there. For a perfectly absorbative surface, the pressure will simply be equal to the momentum striking the surface. At a perfectly reflective (elastic) surface, the change in momentum is twice the momentum of the incident radiation in that the light undergoes a complete change of direction. Standard expressions for the radiation pressure are for reflective surfaces.

Table 11.5.   Fluctuation generators.

Property	Fluctuation	Source
Electromagnetic radiation	A_μ	J^μ
Electric current	J_i	A^i
Compressibility	N_0	A^0
Temperature current	T	T^{0i} (heat Q)

The pressure (kinetic energy density) in a relativistic field is thus

$$P_i = -2T_{0i} = p_i c/\sigma \qquad (11.188)$$

with the factor of two coming from a total reversal in momentum, and σ being the volume of the uniform system outside the surface. Using the arguments of kinetic theory, where the kinetic energy density of a gas with average velocity $\langle v \rangle$ is isotropic in all directions,

$$\frac{1}{2}m\langle v^2 \rangle = \frac{1}{2}m(v_x^2 + v_y^2 + v_z^2) \sim \frac{3}{2}mv_x^2, \qquad (11.189)$$

we write

$$P_i \sim \frac{1}{3}\langle P \rangle. \qquad (11.190)$$

Thus, the pressure of diffuse radiation on a planar reflective surface is

$$P_i = -\frac{2}{3}T_{0i}. \qquad (11.191)$$

Using eqn. (7.88), we may evaluate this, giving:

$$P_i = -\frac{2}{3}\frac{(\mathbf{E} \times \mathbf{H})_i}{c} = \frac{2}{3}\epsilon_0 E^2. \qquad (11.192)$$

Exercises

Although this is not primarily a study book, it is helpful to phrase a few outstanding points as problems, to be demonstrated by the reader.

(1) In the action in eqn. (11.61), add a kinetic term for the potential $V(x)$

$$\Delta S = \int (\mathrm{d}x)\, \frac{1}{2}(\partial^\mu V)(\partial_\mu V). \qquad (11.193)$$

Vary the total action with respect to $\phi(x)$ to obtain the equation of motion for the field. Then vary the action with respect to $V(x)$ and show that this leads to the equations

$$-\Box\phi + (m^2 + V)\phi = 0$$

$$-\Box V + \frac{1}{2}\phi^2 = 0.$$

Next show that the addition of this extra field leads to an extra term in the energy–momentum tensor, so that

$$\theta_{\mu\nu} = \frac{1}{2}(\partial_\mu\phi)(\partial_\nu\phi) + \frac{1}{2}(\partial_\mu V)(\partial_\nu V) - \frac{1}{2}(m^2 + V)\phi^2. \quad (11.194)$$

Using the two equations of motion derived above, show that

$$\partial^\mu\theta_{\mu\nu} = 0 \quad (11.195)$$

so that energy conservation is now restored. This problem demonstrates that energy conservation can always be restored if one considers all of the dynamical pieces in a physical system. It also serves as a reminder that fixed potentials such as $V(x)$ are only a convenient approximation to real physics.

(2) Using the explicit form of a Lorentz boost transformation, show that a fluid velocity field has an energy–momentum tensor of the form,

$$T^{\mu\nu} = Pg^{\mu\nu} + (P + \mathcal{H})U^\mu U^\nu/c^2. \quad (11.196)$$

Start with the following expressions for a spherically symmetrical fluid at rest:

$$T_{00} = \mathcal{H}$$
$$T_{0i} = T_{i0} = 0$$
$$T_{ij} = P\delta_{ij}. \quad (11.197)$$

(3) Consider a matter current $N^\mu = (N, N\mathbf{v}) = N(x)U^\mu(x)$. Show that the conservation equation $\partial_\mu N^\mu = 0$ may be written

$$\partial_\mu N^\mu = [\partial_t + \pounds_v], \quad (11.198)$$

where $\pounds_D = U^i\partial_i + (\partial_i U^i)$. This is called the Lie derivative. Compare this with the derivatives found in section 10.3 and the discussion found in section 9.6. See also ref. [111] for more details of this interpretation.

(4) By writing the orbital angular momentum operator in the form $L_i = \epsilon_{ijk}x^j p^k$ and the quantum mechanical commutation relations $[x^i, p^j] = i\hbar\delta^{ij}$ in the form $\epsilon_{ijk}x^j p^k = i\hbar$, show that

$$L_i\epsilon_{ilm} = [x_l, p_m] = i\hbar\,\delta_{lm}, \qquad (11.199)$$

and thence

$$\epsilon_{ilm}L_i L_l = i\hbar L_m. \qquad (11.200)$$

Hence show that the angular momentum components satisfy the algebra relation

$$[L_i, L_j] = i\hbar\,\epsilon_{ijk}\,L_k. \qquad (11.201)$$

Show that this is the Lie algebra for $so(3)$ and determine the dimensionless generators T^a and structure constants f_{abc} in terms of L_i and $\hbar$.

12

Charge and current

The idea of charge intuitively relates to that of fields and forces. Charge is that quality or attribute of matter which determines how it will respond to a particular kind of force. It is thus a label which distinguishes forces from one another. The familiar charges are: electric charge, which occurs in Maxwell's equations; the mass, which occurs both in the laws of gravitation and inertia; the colour charge, which attaches to the strong force; and a variety of other labels, such as strangeness, charm, intrinsic spin, chirality, and so on. These attributes are referred to collectively as 'quantum numbers', though a better name might be 'group numbers'.

Charge plays the role of a quantity conjugate to the forces which it labels. Like all variables which are conjugate to a parameter (energy, momentum etc.) charge is a book-keeping parameter which keeps track of a closure or conservation principle. It is a currency for the property it represents. This indicates that the existence of charge ought to be related to a symmetry or conservation law, and indeed this turns out to be the case. An important application of symmetry transformations is the identification of conserved 'charges', and vice versa.

12.1 Conserved current and Noether's theorem

As seen in section 11.3, the spacetime variation of the action reveals a structure which leads to conservation equations in a closed system. The conservation equations have the generic form

$$\partial_t \rho + \vec{\nabla} \cdot \mathbf{J} = \partial_\mu J^\mu = 0, \tag{12.1}$$

for some 'current' J^μ. These are continuity conditions, which follow from the action principle (section 2.2.1). One can derive several different, but equally valid, continuity equations from the action principle by varying the action with respect to appropriate parameters. This is the essence of what is known as *Noether's theorem*.

In practice, one identifies the conservation law $\partial_\mu J^\mu = 0$ for current J_μ by varying the action with respect to a parameter, conjugate to its charge. This leads to two terms upon integration by parts: a main term, which vanishes (either with the help of the field equations, or by straightforward cancellation), and a surface term, which must vanish independently for stationary action $\delta S = 0$. The surface term can be written in the form

$$\delta S = \int (dx)(\partial_\mu J^\mu)\delta\lambda = 0 \tag{12.2}$$

for some J_μ; then we say that we have discovered a conservation law for the current J_μ and parameter λ.

This is most easily illustrated with the aid of examples. As a first example, we shall use this method to prove that the electric current is conserved for a scalar field. We shall set $c = \hbar = 1$ for simplicity here. The gauged action for a complex scalar field is

$$S = \int (dx) \left\{ \hbar^2 c^2 (D^\mu \phi)^* (D_\mu \phi) + m^2 c^4 \phi^* \phi \right\}. \tag{12.3}$$

Consider now a gauge transformation in which $\phi \to e^{ies}\phi$, and vary the action with respect to $\delta s(x)$:

$$\delta S = \int (dx) \hbar^2 c^2 \Big\{ (D^\mu(-ie\delta s)e^{-ies}\phi)^*(D_\mu e^{ies}\phi)$$
$$+ (D^\mu e^{-ies}\phi)(D^\mu(ie\delta s)e^{ies}\phi) \Big\}. \tag{12.4}$$

Now using the property (10.41) of the gauge-covariant derivative that the phase commutes through it, we have

$$\delta S = \int (dx) \left\{ (D^\mu(-ie\delta s)\phi)^*(D_\mu\phi) + (D^\mu\phi)(D^\mu(ie\delta s)\phi) \right\}. \tag{12.5}$$

We now integrate by parts to remove the derivative from δs and use the equations of motion $(-D^2 + m^2)\phi = 0$ and $-(D^{*2} + m^2)\phi^* = 0$, which leaves only the surface (total derivative) term

$$\delta S = \int (dx)\delta s(\partial_\mu J^\nu), \tag{12.6}$$

where

$$J^\mu = ie\hbar^2 c^2(\phi^*(D^\mu\phi) - (D^\mu\phi)^*\phi). \tag{12.7}$$

Eqn. (12.2) can be written

$$\frac{1}{c}\int d\sigma^\mu J_\mu = \text{const.} \tag{12.8}$$

In other words, this quantity is a constant of the motion. Choosing the canonical spacelike hyper-surface for σ, eqn. (12.8) has the interpretation

$$\frac{1}{c} \int d\sigma \, J_0 = \int dx^1 \ldots dx^n \rho = Q, \tag{12.9}$$

where ρ is the charge density and Q is therefore the total charge. In other words, Noether's theorem tells us that the total charge is conserved by the dynamical evolution of the field.

As a second example, let us consider dynamical variations of the field $\delta\phi$. Anticipating the discussion of the energy–momentum tensor, we can write eqn. (11.43) in the form

$$\delta S = \int (dx)(\partial_\mu J^\mu) = 0, \tag{12.10}$$

where we have defined the 'current' as

$$J_\mu \delta\lambda \sim \Pi_\mu \delta q - \theta_{\mu\nu} \delta x^\nu. \tag{12.11}$$

This is composed of a piece expressed in terms of the canonical field variables, implying that canonical momentum is conserved for field dynamics,

$$\partial^\mu \Pi_\mu = 0, \tag{12.12}$$

and there is another piece for the mechanical energy–momentum tensor, the parameter is the spacetime displacement δx^μ. This argument is usually used to infer that the canonical momentum and the energy–momentum tensor,

$$\partial^\mu \theta_{\mu\nu} = 0, \tag{12.13}$$

are conserved; i.e. the conservation of mechanical energy and momentum.

If the action is complete, each variation of the action leads to a form which can be interpreted as a conservation law. If the action is incomplete, so that conservation cannot be maintained with the number of degrees of freedom given, then this equation appears as a constraint which restricts the system. In a conservative system, the meaning of this equation is that 'what goes in goes out' of any region of space. Put another way, in a conservative system, the essence of the field cannot simply disappear, it must move around by flowing from one place to another.

Given a conservation law, we can interpret it as a law of conservation of an abstract charge. Integrating the conservation law over spacetime,

$$\int (dx) \, \partial_\mu J^\mu = \int d\sigma^\mu \, J_\mu = \text{const.} \tag{12.14}$$

Table 12.1.   Conjugates and generators.

	Q	v
Translation	p_i	x^i
Time development	$-H$	t
Electric phase	e	$\theta = \int A_\mu dx^\mu$
Non-Abelian phase	gT^a	$\theta^a = \int A_\mu^a dx^\mu$

If we choose $d\sigma^\mu$, i.e. $\mu = 0$, to be a spacelike hyper-surface (i.e. a surface of covariantly constant time), then this defines the total charge of the system:

$$Q(t) = \int d\sigma \rho(x) = \int d^n\mathbf{x}\ \rho(x). \tag{12.15}$$

Combining eqns. (12.14) and (12.15), we can write

$$\int d^n\mathbf{x}\ \partial_\mu J^\mu = -\partial_t \int d\sigma \rho + \int d\sigma^i J_i = 0. \tag{12.16}$$

The integral over J_i vanishes since the system is closed, i.e. no current flows in or out of the total system. Thus we have (actually by assumption of closure)

$$\frac{dQ(t)}{dt} = 0. \tag{12.17}$$

This equation is well known in many forms. For the conservation of electric charge, it expresses the basic assumption of electromagnetism that charge is conserved. In mechanics, we have the equation for conservation of momentum

$$\frac{dp^i}{dt} = \frac{d}{dt} \int d\sigma\ \theta_0^i = 0. \tag{12.18}$$

The conserved charge is formally the generator of the symmetry which leads to the conservation rule, i.e. it is the conjugate variable in the group transformation. In a group transformation, we always have an object of the form:

$$e^{iQv}, \tag{12.19}$$

where Q is the generator of the symmetry and v is the conjugate variable which parametrizes the symmetry (see table 12.1). Noether's theorem is an expression of symmetry. It tells us that – if there is a symmetry under variations of a parameter in the action – then there is a divergenceless current associated with that symmetry and a corresponding conserved charge. The formal statement is:

> The invariance of the Lagrangian under a one-parameter family
> of transformations implies the existence of a divergenceless current
> and associated conserved 'charge'.

Noether's theorem is not the only approach to finding conserved currents, but it is the most well known and widely used [2]. The physical importance of conservation laws for dynamics is that

> A local excess of a conserved quantity cannot simply
> disappear –it can only relax by spreading slowly
> over the entire system.

12.2 Electric current J_μ for point charges

Electric current is the rate of flow of charge

$$I = \frac{dQ}{dt}. \tag{12.20}$$

Current density (current per unit area, in three spatial dimensions) is a vector, proportional to the velocity $\mathbf{v}$ of charges and their density ρ:

$$J^i = \rho_e v^i. \tag{12.21}$$

By adding a zeroth component $J^0 = \rho c$, we may write the spacetime-covariant form of the current as

$$J^\mu = \rho_e \beta^\mu, \tag{12.22}$$

where $\beta^\mu = (c, \mathbf{v})$. For a point particle at position $x_0(t)$, we may write the charge density using a delta function. The n-dimensional spatial delta function has the dimensions of density and the charge of the particle is q. The current per unit area J^i is simply q multiplied by the velocity of the charge:

$$J^0/c = \rho(x) = q\, \delta^n(\mathbf{x} - \mathbf{x}_p(t))$$
$$J^i = \rho(x) \frac{dx^i(t)}{dt}. \tag{12.23}$$

Relativistically, it is useful to express the current in terms of the velocity vectors β^μ and U^μ. For a general charge distribution the expressions are

$$J^\mu(x) = \rho c \beta^\mu$$
$$= \rho c \gamma^{-1} U^\mu. \tag{12.24}$$

Table 12.2.   Currents for various fields.

Field	Current
Point charges, velocity $\mathbf{v}$	$J^0 = e\rho c$
	$\mathbf{J} = e\rho\mathbf{v}$
Schrödinger field	$J^0 = e\psi^*\psi$
	$\mathbf{J} = \mathrm{i}\frac{e\hbar}{2m}(\psi^*(\mathbf{D}\psi) - (\mathbf{D}\psi)^*\psi)$
Klein–Gordon field	$J_\mu = \mathrm{i}e\hbar c^2(\phi^*(D_\mu\phi) - (D_\mu)^*\phi)$
Dirac field	$J_\mu = \mathrm{i}ec\overline{\psi}\gamma_\mu\psi$

Thus, for a point particle,

$$
\begin{aligned}
J^\mu &= qc\beta^\mu\,\delta^n(\mathbf{x} - \mathbf{x}_p(t))\\
&= qc\int \mathrm{d}t\,\delta^{n+1}(x - x_p(t))\beta^\mu\\
&= q\int \mathrm{d}\tau\,\delta^{n+1}(x - x_p(\tau))U^\mu.
\end{aligned}
\tag{12.25}
$$

12.3 Electric current for fields

The form of the electric current in terms of field variables is different for each of the field types, but in each case we may define the current by

$$
J^\mu = \frac{\delta S_{\mathrm{M}}}{\delta A_\mu}
\tag{12.26}
$$

where S_M is the action for matter fields, including their gauge-invariant coupling to the Maxwell field A_μ, but not including the Maxwell action (eqn. (21.1)) itself. The action must be one consisting of complex fields, since the gauge symmetry demands invariance under arbitrary complex phase transformations. A single-component, non-complex field does not give rise to an electric current. The current density for quanta with charge e may be summarized in terms of the major fields as seen in table 12.2. The action principle displays the form of these currents in a straightforward way and also clarifies the interpretation of the source as a current. For example, consider the complex Klein–Gordon field in the presence of a source:

$$
S = S_{\mathrm{M}} + S_J = \int (\mathrm{d}x)\left\{\hbar^2 c^2 (D^\mu\phi)^*(D_\mu\phi) - J^\mu A_\mu\right\},
\tag{12.27}
$$

where terms independent of A_μ have been omitted for simplicity. Using eqn. (12.26), and assuming that J_μ is independent of A_μ, one obtains

$$\frac{\delta S_M}{\delta A_\mu} = ie\hbar c^2 \left[\phi^*(D^\mu\phi) - (D^\mu\phi)^*\phi \right] = J^\mu. \tag{12.28}$$

Note carefully here: although the left and right hand sides are numerically equal, they are not formally identical, since J_μ was assumed to be independent of A_μ under the variation, whereas the left hand side is explicitly dependent on A_μ through the covariant derivative. Sometimes these are confused in the literature leading to the following error.

It is often stated that the coupling for the electromagnetic field to matter can be expressed in the form:

$$S_M = S_M[A_\mu = 0] + \int (dx) J^\mu A_\mu. \tag{12.29}$$

In other words, the total action can be written as a sum of a matter action (omitting A_μ, or with partial derivatives instead of covariant derivatives), plus a linear source term (which is supposed to make up for the gauge parts in the covariant derivatives) plus the Maxwell action. This is incorrect because, for any matter action which has quadratic derivatives (all fields except the Dirac field), one cannot write the original action as the current multiplying the current, just as

$$x^2 \neq \left(\frac{d}{dx} x^2 \right) x. \tag{12.30}$$

In our case,

$$\frac{\delta S}{\delta A_\mu} A^\mu \neq S. \tag{12.31}$$

The Dirac field does not suffer from this problem. Given the action plus source term,

$$S = S_M + S_J = \int (dx) \left\{ -\frac{1}{2} i\hbar c \overline{\psi} (\gamma^\mu \overrightarrow{D}_\mu - \gamma^\mu \overleftarrow{D}_\mu^\dagger)\psi \right\}, \tag{12.32}$$

the variation of the action equals

$$\frac{\delta S_M}{\delta A_\mu} = iqc\overline{\psi}\gamma^\mu\psi = J^\mu. \tag{12.33}$$

In this unique instance the source and current are formally and numerically identical, and we may write

$$S_M = S_M[A_\mu = 0] + J^\mu A_\mu. \tag{12.34}$$

12.4 Requirements for a conserved probability

According to quantum theory, the probability of finding a particle at a position
x at time t is derived from an invariant product of the fields. Probabilities must
be conserved if we are to have a particle theory which makes sense. For the
Schrödinger wavefunction, this is simply $\psi^*\psi$, but this is only true because this
combination happens to be a conserved density $N(x)$ for the Schrödinger action.

In order to establish a probability interpretation for other fields, one may use
Noether's theorem. In fact, we have already done this. A conserved current is
known from the previous section: namely the electric current, but there seems to
be no good reason to require the existence of electric charge in order to be able to
speak of probabilities. We would therefore like to abstract the invariant structure
of the conserved quantity without referring specifically to electric charge – after
all, particles may have several charges, nuclear, electromagnetic etc – any one
of these should do for counting particle probabilities.

Rather than looking at local gauge transformations, we therefore turn to
global phase transformations[1] and remove the reference in the argument of the
phase exponential to the electric charge. Consider first the Schrödinger field,
described by the action

$$
S = \int d\sigma\, dt \left\{ -\frac{\hbar^2}{2m}(\partial^i\psi)^\dagger(\partial_i\psi) - V\psi^*\psi + \frac{i}{2}(\psi^*\partial_t\psi - \psi\partial_t\psi^*) \right\}.
$$

$$(12.35)$$

The variation of the action with respect to constant δs under a phase transformation $\psi \to e^{is}\psi$ is given by

$$
\delta S = \int (dx) \left\{ -\frac{\hbar^2}{2m}\left[-i\delta s(\partial^i\psi^*)(\partial_i\psi) + (\partial^i\psi^*)i\delta s(\partial_i\psi) \right] \right.
$$

$$
\left. + i\left[-i\delta s\,\psi^*\partial_t\psi + i\delta s\,\psi^*\partial_t\psi \right] \right\}.
$$

$$(12.36)$$

Note that the variation δs need not vanish simply because it is independent of x,
(see comment at end of section). Integrating by parts and using the equation of
motion,

$$
-\frac{\hbar^2}{2m}\nabla^2\psi + V = i\frac{\partial\psi}{\partial t},
$$

$$(12.37)$$

we obtain the expression for the continuity equation:

$$
\delta S = \int (dx)\delta s \left(\partial_t J^t + \partial_i J^i \right) = 0,
$$

$$(12.38)$$

---

[1] Global gauge transformations are also called *rigid* since they are fixed over all space.

where

$$J^t = \psi^*\psi = \rho$$

$$J^i = \frac{i\hbar^2}{2m}\left[\psi^*(\partial^i\psi) - (\partial^i\psi^*)\psi\right], \tag{12.39}$$

which can be compared to the current conservation equation eqn. (12.1). ρ is the probability density and J^i is the probability current. The conserved probability, by Noether's theorem, is therefore

$$P = \int d\sigma\,\psi^*(x)\psi(x), \tag{12.40}$$

and this can be used to define the notion of an inner product between two wavefunctions, given by the overlap integral

$$(\psi_1, \psi_2) = \int d\sigma\,\psi_1^*(x)\psi_2(x). \tag{12.41}$$

Thus we see how the notion of an invariance of the action leads to the identification of a conserved probability for the Schrödinger field.

Consider next the Klein–Gordon field. Here we are effectively doing the same thing as before in eqn. (12.4), but keeping s independent of x and setting $D_\mu \to \partial_\mu$ and $e \to 1$:

$$S = \int (dx)\hbar^2 c^2 \left\{(\partial^\mu e^{-is}\phi^*)(\partial_\mu e^{is}\phi)\right\}$$

$$\delta S = \int (dx)\hbar^2 c^2 \left[(\partial^\mu\phi^*(-i\delta s)e^{-is})(\partial_\mu\phi e^{is}) + c.c.\right]$$

$$= \int (dx)\delta s(\partial_\mu J^\mu), \tag{12.42}$$

where

$$J^\mu = -i\hbar^2 c^2(\phi^*\partial^\mu\phi - \phi\partial^\mu\phi^*). \tag{12.43}$$

The conserved 'charge' of this symmetry can now be used as the definition of the inner product between fields:

$$(\phi_1, \phi_2) = i\hbar c \int d\sigma^\sigma (\phi_1^*\partial_\sigma\phi_2 - (\partial_\sigma\phi_1)^*\phi_2), \tag{12.44}$$

or, in non-covariant form,

$$(\phi_1, \phi_2) = i\hbar c \int d\sigma (\phi_1^*\partial_0\phi_2 - (\partial_0\phi_1)^*\phi_2). \tag{12.45}$$

This is now our notion of probability.

Here we have shown that a conserved probability can be attributed to any complex field as a result of symmetry under rigid (global) phase transformations. One should be somewhat wary of the physical meaning of rigid gauge transformations, since this implies a notion of correlation over arbitrary distances and times (a fact which apparently contradicts the finite speed of communication imposed by relativity). Global transformations should probably be regarded as an idealized case. In general, one requires the notion of a charge and associated gauge field, but not necessarily the electromagnetic gauge field. An additional point is: does it make physical sense to vary an object which does not depend on any dynamical variables $\mathbf{x}$, t? How should it vary without any explicit freedom to do so? These points could make one view rigid (global) gauge transformations with a certain skepticism.

12.5 Real fields

A cursory glance at the expressions for the electric current show that J_μ vanishes for real fields. Formally this is because the gauge (phase) symmetry cannot exist for real fields, since the phase is always fixed at zero. Consequently, there is no conserved current for real fields (though the energy–momentum tensor is still conserved). In the second-quantized theory of real fields (which includes the photon field), this has the additional effect that the number of particles with a given momentum is not conserved.

The problem is usually resolved in the second-quantized theory by distinguishing between excitations of the field (particles) with positive energy and those with negative energy. Since the relativistic energy equation $E^2 = p^2 c^2 + m^2 c^4$ admits both possibilities. We do this by writing the real field as a sum of two parts:

$$\phi = \phi^{(+)} + \phi^{(-)}, \tag{12.46}$$

where $\phi^{(+)*} = \phi^{(-)}$. $\phi^{(+)}$ is a complex quantity, but the sum $\phi^{(+)} + \phi^{(-)}$ is clearly real. What this means is that it is possible to define a conserved current and therefore an inner product on the manifold of positive energy solutions $\phi^{(+)}$,

$$(\phi_1^{(+)}, \phi_2^{(+)}) = i\hbar c \int d\sigma^\mu (\phi_1^{(+)*} \partial_\mu \phi_2^{(+)} - (\partial_\mu \phi_1^{(+)})^* \phi_2^{(+)}), \tag{12.47}$$

and another on the manifold of negative energy solutions $\phi^{(-)}$. Thus there is local conservation of probability (though charge still does not make any sense) of particles and anti-particles separately.

12.6 Super-conductivity

Consider a charged particle in a uniform electric field E_i. The force on the particle leads to an acceleration:

$$q E^i = m\ddot{x}^i. \tag{12.48}$$

Assuming that the particle starts initially from rest, and is free of other influences, at time t it has the velocity

$$\dot{x}^i(t) = \frac{q}{m} \int_0^t E^i \, dt'. \tag{12.49}$$

This movement of charge represents a current (charge multiplied by velocity). If one considers N such identical charges, then the current is

$$J^i(t) = Nq\dot{x}^i = \frac{Nq^2}{m} \int_0^t E^i \, dt'. \tag{12.50}$$

Assuming, for simplicity, that the electric field is constant, at time t one has

$$J^i(t) = \frac{Nq^2 t}{m} E^i$$

$$\equiv \sigma E^i. \tag{12.51}$$

The last line is Ohm's law, $V = IR$, re-written in terms of the current density J^i and the reciprocal resistance, or conductivity $\sigma = 1/R$. This shows that a free charge has an ohmic conductivity which is proportional to time. It tends to infinity. Free charges are super-conducting.

The classical theory of ohmic resistance assumes that charges are scattered by the lattice through which they pass. After a mean free time of τ, which is a constant for a given material under a given set of thermodynamical conditions, the conductivity is $\sigma = Nq^2\tau/m$. This relation assumes hidden dissipation, and thus can never emerge naturally from a fundamental formulation, without modelling the effect of collisions as a transport problem. Fundamentally, all charges super-conduct, unless they are scattered by some impedance. The methods of linear response theory may be used for this.

If one chooses a gauge in which the electric field may be written

$$E^i = -\partial_t A^i, \tag{12.52}$$

then substitution into eqn. (12.50) gives

$$J^i = -\Lambda A^i, \tag{12.53}$$

where $\Lambda = Nq^2/m$. This is known as London's equation, and was originally written down as a phenomenological description of super-conductivity.

The classical model of super-conductivity seems naive in a modern, quantum age. However, the quantum version is scarcely more sophisticated. As noted in ref. [135], the appearance of super-conductivity is a result only of symmetry properties of super-conducting materials at low temperature, not of the detailed mechanism which gives rise to those symmetry properties.

Super-conductivity arises because of an ordered state of the field in which the inhomogeneities of scattering centres of the super-conducting material become invisible to the average state. Consider such a state in a scalar field. The super-conducting state is one of great uniformity, characterized by

$$\partial_\mu \langle \phi(x) \rangle = \langle A_0 \rangle = 0. \tag{12.54}$$

The average value of the field is thus locked in a special gauge. In this state, the average value of the current is given by

$$\langle J_i \rangle = \langle ie\hbar c^2 (\phi^*(D_\mu \phi) - (D_\mu)^* \phi) \rangle. \tag{12.55}$$

The time derivative of this is:

$$\begin{aligned} \partial_t \langle J_i \rangle &= -e^2 c^2 \partial_t \langle A_i \rangle \\ &= e^2 c^2 \langle E_i \rangle. \end{aligned} \tag{12.56}$$

This is the same equation found for the classical case above. For constant external electric field, it leads to a current which increases linearly with time, i.e. it becomes infinite for infinite time. This corresponds to infinite conductivity. Observe that the result applies to statistical averages of the fields, in the same way that spontaneous symmetry breaking applies to statistical averages of the field, not individual fluctuations (see section 10.7). The individual fluctuations about the ground state continue to probe all aspects of the theory, but these are only jitterings about an energetically favourable super-conducting mean field. The details of how the uniform state becomes energetically favourable require, of course, a microscopic theory for their explanation. This is given by the BCS theory of super-conductivity [5] for conventional super-conductors. More recently, unusual materials have given rise to super-conductivity at unusually high temperatures, where an alternative explanation is required.

12.7 Duality, point charges and monopoles

In covariant notation, Maxwell's equations are written in the form

$$\begin{aligned} \partial_\mu F^{\mu\nu} &= -\mu_0 J^\nu \\ \epsilon_{\mu\nu\rho\sigma} \partial^\nu F^{\rho\sigma} &= 0. \end{aligned} \tag{12.57}$$

If one defines the dual F^* of a tensor F by one-half its product with the anti-symmetric tensor, one may write

$$F^*_{\mu\nu} = \frac{1}{2} \epsilon_{\mu\nu\rho\sigma} F^{\rho\sigma}, \tag{12.58}$$

and Maxwell's equations become

$$\partial_\mu F^{\mu\nu} = -\mu_0 J^\nu$$
$$\partial_\mu F^{*\mu\nu} = 0. \tag{12.59}$$

The similarity between these two equations has prompted some to speculate as to whether a dual current, J_m^μ, could not exist:

$$\partial_\mu F^{\mu\nu} = -\mu_0 J^\nu$$
$$\partial_\mu F^{*\mu\nu} = -\mu_0 J_m^\nu. \tag{12.60}$$

This would imply an equation of the form

$$\nabla \cdot \mathbf{B} = (\partial_i B^i) = \mu_0 \rho_m \tag{12.61}$$

and the existence of magnetic monopoles. The right hand side of these equations is usually thought of as a source term, or forcing term, for the differential terms on the left hand side. The existence of pointlike singularities is an interesting issue, since it touches the limits of the smooth differential formalism used to express the theory of electromagnetism and drives home the reasoning behind the model of pointlike charges which physicists have adopted.

Consider a Coulomb field surrounding a point. Up to a factor of $4\pi\epsilon_0$, the electric field has the vectorial form

$$E_i = \frac{x_i}{|\mathbf{x}|^m}, \tag{12.62}$$

in n dimensions. When $n = 3$ we have $m = 3$ for the Coulomb field, i.e. a $1/r^2$ force law. The derivative of this field is

$$\partial_i E_j = \partial_i \left(\frac{x_j}{\sqrt{(x^k x_k)}} \right)$$
$$= \left(\delta_{ij} - m \frac{x_i x_j}{x^k x_k} \right). \tag{12.63}$$

From this, we have that

$$\vec{\nabla} \cdot \mathbf{E} = \partial^i E_i = \frac{(n - m)}{|\mathbf{x}|^m},$$
$$(\vec{\nabla} \times \mathbf{E})_k = \epsilon_{ijk} \partial_i E_j = 0. \tag{12.64}$$

The last result follows entirely from the symmetry on the indices: the product of a symmetric matrix and an anti-symmetric matrix is zero. What we see is that, in $n > 2$ dimensions, we can find a solution $n = m$ where the field satisfies the equation of motion identically, except at the singularity $x_i = 0$, where the solution does not exist. In other words, a field can exist without a source, everywhere except at the singular point.

In fact, this is an illusion of the differential formulation of Maxwell's equations; it highlights a conceptual difficulty. The core difficulty is that the equations are really non-local, in the sense that they relate a field at one point to a source at another. This requires an integration over the intermediate points to be well defined differentially. The differential form of Maxwell's equations is really a shorthand for the integral procedure.

At the singular point, the derivative does not exist, and Maxwell's equation becomes meaningless. We can assign a formal meaning to the differential form and do slightly better, as it turns out, by using the potential A_μ, since this can be regularized choosing variables in which the singularity disappears. In that way we can assign a formal meaning to the field around a point and justify the introduction of a source for the field surrounding the singularity using an integral formulation. The formulation we are looking for is in terms of Green functions. Green functions are, in a sense, a regularization scheme for defining the meaning of an ambiguous, irregular (infinite) expression. This is also the first in a long litany of cases where it is necessary to *regularize*, or re-formulate infinite, badly defined expressions in the physics of fields, which result from assumptions about pointlike structure and Green functions.

In terms of the vector potential A_μ, choosing the so-called Coulomb gauge $\partial_i A^i = 0$, we have

$$E_i/c = -\partial_0 A_i - \partial_i A_0, \qquad (12.65)$$

so that the divergence of the electric field is

$$\partial_i E^i = -\nabla^2 \phi = \rho. \qquad (12.66)$$

Note that we set $\epsilon_0 = 1$ for the purpose of this schematic. The charge density for a point particle with charge q at the origin is written as

$$\begin{aligned} \rho &= q\delta^3(x) \\ &= q\delta(x)\delta(y)\delta(z) \\ &= \frac{q}{4\pi r^2}\delta(r). \end{aligned} \qquad (12.67)$$

Thus, in polar coordinates, about the origin,

$$-\nabla^2 \phi(r) = \frac{q\delta(r)}{4\pi \epsilon_0 r}. \qquad (12.68)$$

The Green function $G(x, x')$ is defined as the object which satisfies the equation

$$-\nabla^2 G(x, x') = \delta(x - x'). \qquad (12.69)$$

If we compare this definition to the Poisson equation for the potential $\phi(x)$ in eqn. (12.68), we see that $G(x, x')$ can interpreted as the scalar potential for a delta-function source at $x = x'$, with unit charge. Without repeating the content of chapter 5, we can simply note the steps in understanding the singularity at the origin. In the case of the Coulomb potential in three dimensions, the answer is well known:

$$\phi(r) = \frac{1}{4\pi r}. \tag{12.70}$$

We can use this to verify the consistency of the Green function definition of the field, in lieu of a more proper treatment later. By multiplying the Poisson equation by the Green function, one has

$$\int d^3\mathbf{x}' \, (-\nabla^2 \phi(x)) G(x, x') = \int d^3\mathbf{x}' \, \rho(x') G(x, x'). \tag{12.71}$$

Integrating by parts, and using the definition of $G(x, x')$,

$$\phi(x) = \int d^3\mathbf{x}' \, \rho(x') G(x, x'). \tag{12.72}$$

Substituting the polar coordinate forms for $\phi(r)$ and using the fact that $G(r, r')$ is just $\phi(r - r')$ in this instance, we have

$$\phi(r) = \frac{1}{4\pi r} = \int \frac{1}{4\pi(r - r')} \frac{\delta(r')}{4\pi r'^2} \, 4\pi r'^2 dr. \tag{12.73}$$

This equation is self-consistent and avoids the singular nature of the r' integration by virtue of cancellations with the integration measure $\int d^3\mathbf{x}' = 4\pi r'^2 dr$. We note that both the potential and the field are still singular at the origin. What we have achieved here, however, is to show that the singularity is related to a delta-function source (well defined under integration). Without the delta-function source ρ, the only consistent solution is $\phi = $ const. in the equation above. Thus we do, in fact, need the source to explain the central Coulomb field.

In fact, the singular structure noted here is a general feature of *central fields*, or conservative fields, whose curl vanishes. A non-vanishing curl, incidentally, indicates the presence of a magnetic field, and thus requires a source for the magnetism, or a magnetic monopole.

The argument for magnetic monopoles is based on the symmetry of the differential formulation of Maxwell's equations. We should pay attention to the singular nature of pointlike sources when considering this point. If we view everything in terms of singularities, then a magnetic monopole exists trivially: it is the Lorentz boost of a point charge, i.e. a string of current. The existence of other monopoles can be inferred from other *topological* singularities in the spacetime occupied by the field.

13
The non-relativistic limit

In some branches of physics, such as condensed matter and quantum optics, one deals exclusively with non-relativistic models. However, there are occasionally advantages to using a relativistic formulation in quantum theory; by embedding a theory in a larger framework, one often obtains new insights. It is therefore useful to be able to take the non-relativistic limit of generally covariant theories, both as an indication of how large or small relativistic effects are and as a cultural bridge between covariant physics and non-relativistic quantum theory.

13.1 Particles and anti-particles

There is no unified theory of particles and anti-particles in the non-relativistic field theory. Formally there are two separate theories. When we take the non-relativistic limit of a relativistic theory, it splits into two disjoint theories: one for particles, with only positive definite energies, and one for anti-particles, with only negative definite energies. Thus, a non-relativistic theory cannot describe the interaction between matter and anti-matter.

The Green functions and fields reflect this feature. The positive frequency Wightman function goes into the positive energy particle theory, while the negative frequency Wightman function goes into the negative energy anti-particle theory. The objects which one then refers to as the Wightman functions of the non-relativistic field theory are asymmetrical. In normal Schrödinger field theory for matter, one says that the zero temperature negative frequency Wightman function is zero.[1]

---

[1] At finite temperature it must have a contribution from the heat bath for consistency.

13.2 Klein–Gordon field

13.2.1 The free scalar field

We begin by considering the Klein–Gordon action for a real scalar field, since this is the simplest of the cases and can be treated at the level of the action. It also reveals several subtleties in the way quantities are defined and the names various quantities go by. In particular, we must recall that relativistic theories have an indefinite metric, while non-relativistic theories can be thought of as having a Euclidean, definite metric. Since one is often interested in the non-relativistic limit in connection with atomic systems, we illustrate the emergence of atomic levels by taking a two-component scalar field, in which the components have different potential energy in the centre of mass frame of the field. This is incorporated by adopting an effective mass $m_A = m + E_A/c^2$.

Consider the action:

$$S = \int (dx) \left\{ \frac{1}{2} \hbar^2 c^2 (\partial^\mu \phi_A)(\partial_\mu \phi_A) + \frac{1}{2} m_A^2 c^4 \phi_A \phi_A \right\}. \tag{13.1}$$

The variation of our action, with respect to the atomic variables, leads to

$$\delta S = \int (dx) \delta \phi_A (-\hbar^2 c^2 \Box + m_A^2 c^4) \phi_A + \hbar^2 c \int d\sigma_x^\mu (\phi_A \partial_\mu \phi_A). \tag{13.2}$$

The vanishing of the first term leads to the field equation

$$\hbar^2 c^2 \left(-\Box + \frac{m_A^2 c^2}{\hbar^2} \right) \phi_A(x) = 0. \tag{13.3}$$

The second (surface) term in this expression shows that any conserved probability must transform like an object of the form $\phi_A \partial_\mu \phi_A$. In fact, the real scalar field has no conserved current from which to derive a notion of locally conserved probability, but we may note the following. Any *complex* scalar field φ has a conserved current, which allows one to define the inner product

$$(\varphi_A, \varphi_B) = i\hbar c \int d\sigma_x^\mu (\varphi_A^* \partial_\mu \varphi_B - (\partial_\mu \varphi_A^*) \varphi_B), \tag{13.4}$$

where $d\sigma_x^\mu$ is the volume element on a spacelike hyper-surface through spacetime. This result is central even to the real scalar field, since a real scalar field does have a well defined probability density in the non-relativistic limit. To see this, we observe that the real scalar field $\phi(x)$ may be decomposed into positive and negative frequency parts:

$$\phi(x) = \phi^{(+)}(x) + \phi^{(-)}(x), \tag{13.5}$$

where $\phi^{(+)}(x)$ is the positive frequency part of the field, $\phi^{(-)}(x)$ is the negative frequency part of the field and $\phi^{(+)}(x) = (\phi^{(-)}(x))^*$. Since the Schrödinger equation has no physical negative energy solutions, one must discard the negative frequency half of the spectrum when reducing the Klein–Gordon field to a Schrödinger field. This leads to well known expressions for the probability density. Starting with the probability

$$p = 2i\hbar c \int d\sigma_x (\phi \partial_0 \phi) \tag{13.6}$$

and letting

$$\phi^{(+)}(x) = \frac{\psi(x)}{\sqrt{2mc^2}} \ , \qquad \phi^{(-)}(x) = \frac{\psi^*(x)}{\sqrt{2mc^2}}, \tag{13.7}$$

one obtains

$$p = \frac{i\hbar}{2mc} \int d\sigma_x (\psi + \psi^*) \partial_0 (\psi + \psi^*). \tag{13.8}$$

Assuming only that $\phi(x)$ may be expanded in a complete set of plane waves $\exp(i\mathbf{k} \cdot \mathbf{x} - \omega t)$, satisfying the free wave equation $\hbar^2 \omega^2 = \hbar^2 \mathbf{k}^2 c^2 + m^2 c^4$, then in the non-relativistic limit $\hbar^2 \mathbf{k}^2 \ll m^2 c^4$, we may make the effective replacement $i\hbar \partial_0 \to mc$ to lowest order. Thus we have

$$p = \int d\sigma_x \psi^*(x) \psi(x), \tag{13.9}$$

which is the familiar result for non-relativistic particles. It is easy to check that p is a dimensionless quantity using our conventions.

This observation prompts us to define the invariant inner product of two fields ϕ_A and ϕ_B by

$$(\phi_A, \phi_B) = i\hbar c \int d\sigma_x \frac{1}{2} (\phi_A^* \partial_0 \phi_B - (\partial_0 \phi_A^*) \phi_B). \tag{13.10}$$

The complex conjugate symbol is only a reminder here of how to take the non-relativistic limit, since ϕ_A is real. This product vanishes unless $A \neq B$, thus it must represent an amplitude to make a transition from ϕ_1 to ϕ_2 or vice versa. The non-relativistic limit of this expression is

$$(\phi_A, \phi_B) \to \frac{1}{2} \int d\sigma_x \left[\psi_A^* \psi_B + \psi_B^* \psi_A \right]. \tag{13.11}$$

Since $\psi(x)$ is the field theoretical destruction operator and $\psi^*(x)$ is the creation operator, this is now manifestly a transition matrix, annihilating a lower state and creating an upper state or vice versa. The apparent A, B symmetry of eqn. (13.11) is a feature only of the lowest order term. Higher order corrections to this expression are proportional to $E_1 - E_2$, the energy difference between the two field levels, owing to the presence of ∂_0.

Using the interaction term $\overline{P}$, we may compute the non-relativistic limit of the action in eqn. (13.1). This procedure is unambiguous only up to re-definitions of the origin for the arbitrary energy scale. Equivalently, we are free to define the mass used to scale the fields in any convenient way. The simplest procedure is to re-scale the fields by the true atomic mass, as in eqn. (13.7). In addition, we note that the non-relativistic energy operator $i\hbar\partial_t$ is related to the non-relativistic energy operator $i\hbar\tilde{\partial}_t$ by a shift with respect to the rest energy of particles:

$$i\hbar\partial_t = mc^2 + i\hbar\tilde{\partial}_t. \tag{13.12}$$

This is because the non-relativistic Hamiltonian does not include the rest energy of particles, its zero point is shifted so as to begin just about the rest energy. Integrating the kinetic term by parts so that $(\partial_\mu\phi)^2 \to \phi(-\Box)\phi$ and substituting eqn. (13.7) into eqn. (13.1) gives

$$S = \int d\sigma_x dt \frac{1}{2}(\psi + \psi^*)_A \left\{ \frac{\hbar^2\tilde{\partial}_t^2}{2mc^2} - i\hbar\tilde{\partial}_t + \frac{E_A^2}{2mc^2} \right.$$

$$\left. + E_A - \frac{\hbar^2}{2m}\nabla^2 \right\} (\psi + \psi^*)_A. \tag{13.13}$$

If we use the fact that $\psi_A(x)$ is composed of only positive plane wave frequencies, it follows that terms involving ψ^2 or $(\psi^*)^2$ vanish since they involve delta functions imposing a non-satisfiable condition on the energy $\delta(mc^2+\hbar\tilde{\omega})$, where both m and $\tilde{\omega}$ are greater than zero. This assumption ceases to be true only if there is an explicit time dependence in the action, indicating a non-equilibrium scenario, or if the mass of the atoms goes to zero (in which case the NR limit is unphysical). We are therefore left with

$$S_{\text{NR}} = \lim_{c\to\infty} \int d\sigma_x dt \left\{ \frac{i}{2} \left(\psi_A^*(\tilde{\partial}_t\psi_A) - (\tilde{\partial}_t\psi_A^*)\psi_A \right) - \psi_A^* H\psi_A \right\}, \tag{13.14}$$

where the differential operator H_A is defined by

$$H_A = -\frac{\nabla^2}{2m} + E_A + \frac{1}{2mc^2}(E_A^2 + \tilde{\partial}_t^2), \tag{13.15}$$

and we have re-defined the action by a sign in passing to a Euclideanized non-relativistic metric. It is now clear that, in the NR limit $c \to \infty$, the final two terms in H_A become negligible, leading to the field equation

$$H_A\psi_A(x) = i\hbar\tilde{\partial}_t\psi_A(x), \tag{13.16}$$

which is the Schrödinger equation of a particle of mass m moving in a constant potential of energy E_A with a dipole interaction. The fact that it is possible to identify what is manifestly the Hamiltonian H in such an easy way is a special property of theories which are linear in the time derivative.

The direct use of the action (a non-physical quantity) in this way requires some care, so it is useful to confirm the above derivation with an approach based on the field equations, which are physical. As an additional spice, we also choose to scale the two components of the field by a factor involving the effective mass m_A rather than the true atomic mass m. The two fields are then scaled differently. This illustrates another viewpoint, namely of the particles as two species with a truly different mass, as would be natural in particle physics. We show that the resulting field equations have the same form in the non-relativistic limit, up to a shift in the arbitrary zero point energy.

Starting from eqn. (13.3), we define new pseudo-canonical variables by

$$P_A = \sqrt{\frac{\omega_A}{2}} \left(\phi_A + \frac{i}{\omega_A} \dot{\phi}_A \right)$$

$$Q_A = \frac{1}{\sqrt{2\omega_A}} \left(\phi_A - \frac{i}{\omega_A} \dot{\phi}_A \right), \tag{13.17}$$

where $\hbar\omega_A \to m_A c^2$ in the non-relativistic limit, and the time dependence of the fields is of the form of a plane wave $\exp(-i\omega_A t)$, for $\omega_A > 0$. This is the same assumption that was made earlier. We note that, owing to this assumption, the field $P_A(x)$ becomes large compared with $Q_A(x)$ in this limit. Substituting this transformation into the field equation (13.3) and neglecting Q, one obtains

$$i\hbar \partial_t P_A = -\frac{\hbar^2}{2m_A} \nabla^2 P_A + \frac{1}{2} m_A c^2 P_A. \tag{13.18}$$

These terms have a natural physical interpretation: the first term on the right hand side is the particle kinetic term for the excited and unexcited atoms in our system. The second term is the energy offset of the two levels in the atomic system.

Our new point of view now leads to a free particle kinetic term with a mass m_A, rather than the true atomic mass m. There is no contradiction here, since E_A is small compared to mc^2, so we can always expand the reciprocal mass to first order. Expanding these reciprocal masses m_A we obtain

$$m_A^{-1} = m^{-1} + O\left(\frac{E_A}{m^2 c^2} \to 0 \right) \tag{13.19}$$

showing that a consistent NR limit requires us to drop the A-dependent pieces.

Eqn. (13.18) may then be compared with eqn. (13.16). It differs only by a shift in the energy. A shift by the average energy level $\frac{1}{2}(E_1 + E_2)$ makes these equations identical.

13.2.2 Non-relativistic limit of $G_F(x, x')$

As we have already indicated, the non-relativistic theory contains only positive energy solutions. We also noted in section 5.5 that the Schrödinger Green function $G_{NR}(x, x')$ satisfied purely retarded boundary conditions. There was no Feynman Green function for the non-relativistic field. Formally, this is a direct result of the lack of negative energy solutions to the Schrödinger equation (or anti-particles, in the language of quantum field theory). We shall now show that object, which we refer to as the Feynman Green function, becomes the non-relativistic retarded Green function in the limit $c \to \infty$. The same argument applies to the relativistic retarded function, and it is clear from eqn. (5.74) that the reason is the vanishing of the negative frequency Wightman function in the non-relativistic limit.

We begin with eqn. (5.95) and reinstate c and $\hbar$:

$$G_F(x, x') = c \int \frac{\mathrm{d}^{n+1}k}{(2\pi)^{n+1}} \frac{c}{2\hbar\omega_k} \frac{e^{ik\Delta x}}{\hbar^2 c}$$
$$\left[\frac{1}{(c\hbar k_0 + \hbar\omega_k - i\epsilon)} - \frac{1}{(c\hbar k_0 - \hbar\omega_k + i\epsilon)} \right]. \qquad (13.20)$$

In order to compare the relativistic and non-relativistic Green functions, we have to re-scale the relativistic function by the rest energy, as in eqn. (13.7), since the two objects have different dimensions. Let

$$2mc^2 \, G_F(x, x') \to G_{F,NR}, \qquad (13.21)$$

so that the dimensions of $G_{F,NR}$ are the same as those for G_{NR}:

$$\left(-\frac{\hbar^2}{2m}\Box + \frac{1}{2}mc^2 \right) G_{F,NR} = \delta(\mathbf{x}, \mathbf{x}')\delta(t, t') = c\delta(x, x');$$
$$\left(-\frac{\hbar^2}{2m}\nabla^2 - i\hbar\partial_t \right) G_{NR} = \delta(\mathbf{x}, \mathbf{x}')\delta(t, t'). \qquad (13.22)$$

Next, we must express the relativistic energy $\hbar\omega$ in terms of the non-relativistic energy $\hbar\tilde{\omega}$ and examine the definition of ω_k with c reinstated,

$$ck_0 = -\omega = -\left(\tilde{\omega} + \frac{mc^2}{\hbar} \right)$$
$$\hbar\omega_k = \sqrt{\hbar^2 c^2 \mathbf{k}^2 + m^2 c^4}. \qquad (13.23)$$

The change of $k_0 \to -\omega/c$, both in the integral limits and the measure, means that we effectively replace $\mathrm{d}k_0 \to \mathrm{d}\tilde{\omega}/c$. In the non-relativistic limit of large c, the square-root in the preceding equation can be expanded using the binomial theorem,

$$\hbar\omega_k = mc^2 + \frac{\hbar^2\mathbf{k}^2}{2m} + \mathrm{O}\left(\frac{1}{c^2}\right). \qquad (13.24)$$

Substituting these results into eqn. (13.20), we have for the partial fractions

$$\frac{1}{c\hbar k_0 + \hbar \omega_k - i\epsilon} = \frac{1}{\frac{\hbar^2 \mathbf{k}^2}{2m} - \hbar \tilde{\omega} - i\epsilon}$$

$$\frac{1}{c\hbar k_0 - \hbar \omega_k + i\epsilon} = \frac{1}{-\frac{\hbar^2 \mathbf{k}^2}{2m} - \hbar \tilde{\omega} - 2mc^2 + i\epsilon}, \tag{13.25}$$

while the pre-factor becomes

$$d\tilde{\omega} \frac{2mc^2}{2\hbar \omega_k} = \left(1 + \frac{\hbar^2 \mathbf{k}^2}{2m^2 c^2} + O\left(\frac{1}{c^4}\right) \right)^{-1}. \tag{13.26}$$

Taking the limit $c \to \infty$ in these expressions causes the second partial fraction in eqn. (13.25) to vanish. This is what removes the negative energy solutions from the non-relativistic theory. The remainder may now be written as

$$G_{\text{F,NR}}(x, x') = \int \frac{d^n \mathbf{k}}{(2\pi)^n} \frac{d\tilde{\omega}}{2\pi} \left(\frac{\hbar^2 \mathbf{k}^2}{2m} - \tilde{\omega} - i\epsilon \right)^{-1}. \tag{13.27}$$

We see that this is precisely the expression obtained in eqn. (5.140). It has poles in the lower half-plane for positive frequencies. It is therefore a retarded Green function and satisfies a Kramers–Kronig relation.

13.3 Dirac field

The non-relativistic limit of the Dirac equation is more subtle than that for scalar particles since the fields are spinors and the γ-matrices imply a constraint on the components of the spinors. There are several derivations of this limit in the literature, all of them at the level of the field equations. Here we base our approach, as usual, on the action and avoid introducing specific solutions or making assumptions about their normalization.

13.3.1 The free Dirac field

The Dirac action may be written

$$S_{\text{D}} = \int (\mathrm{d}x) \overline{\psi} \left(-\frac{1}{2} i \hbar c (\gamma^\mu \overrightarrow{\partial_\mu} - \gamma^\mu \overleftarrow{\partial}_\mu^\dagger) + mc^2 \right) \psi. \tag{13.28}$$

We begin by re-writing this in terms of the two-component spinors χ (see chapter 20) and with non-symmetrical derivatives for simplicity. The latter choice is of no consequence and only aids notational simplicity:

$$S_D = \int (dx)\psi^\dagger \gamma^0 (-i\hbar c\gamma^\mu \partial_\mu + mc^2)\psi$$

$$= \int (dx)(\chi_1^\dagger \chi_2^\dagger) \begin{pmatrix} -i\hbar\partial_t - mc^2 & -i\hbar c\sigma^i\partial_i \\ -i\hbar c\sigma^i\partial_i & -i\hbar\partial_t + mc^2 \end{pmatrix} \begin{pmatrix} \chi_1 \\ \chi_2 \end{pmatrix}.$$

$$(13.29)$$

This block matrix can be diagonalized by a unitary transformation. The eigenvalue equation is

$$(-i\hbar\partial_t - mc^2 - \lambda)(-i\hbar\partial_t + mc^2 - \lambda) + \hbar^2 c^2 \sigma^i \sigma^j \partial_i \partial_j = 0. \quad (13.30)$$

Noting that

$$\sigma^i \sigma^j \partial_i \partial_j = \partial^i \partial_i + i\epsilon^{ijk}\partial_i \partial_j \sigma_k, \quad (13.31)$$

the eigenvalues may be written as

$$\lambda_\pm = -i\hbar\partial_t \pm \sqrt{m^2 c^4 - \hbar^2 c^2 (\partial^i \partial_i + i\epsilon^{ijk}\partial_i \partial_j \sigma_k)}. \quad (13.32)$$

Thus, the action takes on a block-diagonal form

$$S_D = \int (dx)\overline{\psi}^\dagger \gamma^0 (-i\hbar c\gamma^\mu \partial_\mu + mc^2)\psi$$

$$= \int (dx)(\chi_1^\dagger \chi_2^\dagger) \begin{pmatrix} \lambda_+ & 0 \\ 0 & \lambda_- \end{pmatrix} \begin{pmatrix} \chi_1 \\ \chi_2 \end{pmatrix}. \quad (13.33)$$

In the non-relativistic limit, $c \to \infty$, we may expand the square-root in the eigenvalues

$$\lambda_\pm = -i\hbar\partial_t \pm mc^2 \left(1 - \frac{\hbar^2(\partial^i\partial_i + i\epsilon^{ijk}\partial_i\partial_j\sigma_k)}{2m^2 c^2} + O(c^{-4}) + \cdots \right).$$

$$(13.34)$$

The final step is to re-define the energy operator by the rest energy of the field, for consistency with the non-relativistic definitions:

$$\lambda_\pm = -i\hbar\tilde{\partial}_t - mc^2 \pm mc^2 \left(1 - \frac{\hbar^2 \nabla^2}{2m^2 c^2} + O(c^{-4}) + \cdots \right). \quad (13.35)$$

Thus, in the limit, $c \to \infty$, the two eigenvalues, corresponding to positive and negative energy, give

$$\lambda_+ = -i\hbar\tilde{\partial}_t - \frac{\hbar^2 \nabla^2}{2m}$$

$$\lambda_- = \infty. \quad (13.36)$$

Apart from an infinite contribution to the zero point energy which may be re-defined (renormalized) away, and making an overall change of sign as in the Klein–Gordon case, the non-relativistic action is

$$S_{\mathrm{D}} \rightarrow \int (\mathrm{d}x) \left\{ \chi^{\dagger} \left(i\hbar \tilde{\partial}_t + \frac{\hbar^2 \nabla^2}{2m} \right) \chi \right\}. \tag{13.37}$$

13.3.2 The Dirac Green function

The non-relativistic limit of the Dirac Green function may be inferred quite straightforwardly from the Green function for the scalar field. The Dirac Green function $S(x, x')$ satisfies the relation

$$(-i\hbar c \gamma^\mu \partial_\mu + mc^2) S(x, x') = c\delta(x, x'). \tag{13.38}$$

We also know that the squared operator in this equation leads to a Klein–Gordon operator, thus

$$(i\hbar c \gamma^\mu \partial_\mu + mc^2) S(x, x') = G(x, x'), \tag{13.39}$$

so operating on eqn. (13.38) with this conjugate operator leaves us with

$$(-\hbar^2 c^2 \Box + m^2 c^4) G(x, x') = c\delta(x, x'). \tag{13.40}$$

Both sides of this equation are proportional to a spinor identity matrix, which therefore cancels, leaving a scalar equation. Since we know the limiting properties of $G(x, x')$ from section 13.2.2, we may take the limit by introducing unity in the form $2mc^2/2mc^2$, such that $2mc^2 G(x, x') = G_{\mathrm{NR}}(x, x')$ and the operator in front is divided by $2mc^2$. After re-defining the energy operator, as in eqn. (13.12), the limit of $c \rightarrow \infty$ causes the quadratic time derivative to vanish, leaving

$$\left(-\frac{\hbar^2}{2m} \nabla^2 - i\hbar \tilde{\partial}_t \right) G_{\mathrm{NR}}(x, x') = \delta(\mathbf{x}, \mathbf{x}') \delta(t, t'). \tag{13.41}$$

This is the scalar Schrödinger Green function relation. To get the Green function for the two-component spinors found in the preceding section, it may be multiplied by a two-component identity matrix.

13.3.3 Spinor electrodynamics

The interaction between electrons and radiation complicates the simple procedure outlined in the previous section. The minimal coupling to radiation via the gauge potential $A_\mu(x)$ involves x-dependence, which means that the derivatives do not automatically commute with the diagonalization procedure. We must therefore modify the discussion to account for this, in particular taking more care with time reversal invariance. In addition, we must consider the reaction of the electronic matter to the presence of an electromagnetic field. This leads to a polarization of the field, or effective refractive index (see section 21.2 for a simple discussion of classical polarization). The action for electrodynamics is thus

$$S_{\text{QED}} = \int (dx) \left\{ \overline{\psi} \left(-\frac{1}{2} i\hbar c (\gamma^\mu \overrightarrow{D}_\mu - \gamma^\mu \overleftarrow{D}_\mu^\dagger) + mc^2 \right) \psi + \frac{1}{4\mu_0} F^{\mu\nu} G_{\mu\nu} \right\},$$

(13.42)

where $G_{\mu\nu}$ is the covariant displacement field, defined in eqn. (21.62). We proceed once again by re-writing this in terms of the two-component spinors χ. We consider the matter and radiation terms separately. The matter action is given by

$$S_{\text{D}} = \int (dx) \psi^\dagger \gamma^0 (-i\hbar c \gamma^\mu D_\mu + mc^2) \psi$$

$$= \int (dx) (\chi_1^\dagger \chi_2^\dagger) \begin{pmatrix} -i\frac{\hbar}{2} \overleftrightarrow{D}_t - mc^2 & -i\hbar c \sigma^i D_i \\ -i\hbar c \sigma^i D_i & -i\frac{\hbar}{2} \overleftrightarrow{D}_t + mc^2 \end{pmatrix} \begin{pmatrix} \chi_1 \\ \chi_2 \end{pmatrix}.$$

(13.43)

In electrodynamics, the covariant derivative is $D_\mu = \partial_\mu + i\frac{e}{\hbar} A_\mu$, from which it follows that

$$[D_\mu, D_\nu] = i\frac{e}{\hbar} F_{\mu\nu}.$$

(13.44)

The block matrix in eqn. (13.43) can be diagonalized by a unitary transformation. The symmetrized eigenvalue equation is

$$\left(-i\frac{\hbar}{2} \overleftrightarrow{D}_t - mc^2 - \lambda \right) \left(-i\frac{\hbar}{2} \overleftrightarrow{D}_t + mc^2 - \lambda \right) + \hbar^2 c^2 \sigma^i \sigma^j D_i D_j = 0,$$

(13.45)

or

$$\lambda^2 + 2i\hbar \lambda D_t + \hbar^2 c^2 \sigma^i \sigma^j D_i D_j - \hbar^2 D_t^2 - m^2 c^4 - i\frac{\hbar}{2} (\overleftrightarrow{\partial_t \lambda}) = 0,$$

(13.46)

where the last term arises from the fact that the eigenvalues themselves depend on x due to the gauge field. It is important that this eigenvalue equation be time-symmetrical, as indicated by the arrows. We may write this in the form

$$\lambda = -i\frac{\hbar}{2} \overleftrightarrow{D}_t \pm \sqrt{m^2 c^4 - \hbar^2 c^2 \sigma^i \sigma^j D_i D_j + i\frac{\hbar}{2} (\overleftrightarrow{\partial_t \lambda})}$$

(13.47)

and we now have an implicit equation for the positive and negative energy roots of the operator λ. The fact that the derivative term $\partial_t \lambda$ is a factor of c^2 smaller than the other terms in the square-root means that this contribution will always be smaller than the others. In the strict non-relativistic limit $c \to \infty$ it is completely negligible. Since the square-root contains operators, we represent it by its binomial expansion

$$(1+x)^n = 1 + nx + \frac{n(n-1)}{2}x^2 + \cdots, \tag{13.48}$$

after extracting an overall factor of mc^2, thus:

$$\lambda = -\mathrm{i}\frac{\hbar}{2}\overleftrightarrow{D}_t \pm \left[mc^2 - \frac{\hbar^2 \sigma^i \sigma^j D_i D_j}{2m} - \frac{\hbar^4 \left(\sigma^i \sigma^j D_i D_j\right)^2}{8m^3 c^2} \right.$$
$$\left. +\mathrm{i}\frac{\hbar}{4mc^2}(\overleftrightarrow{\partial_t \lambda}) + \cdots \right]. \tag{13.49}$$

The final term, ∂_t, can be evaluated to first order by iterating this expression. Symmetrizing over time derivatives, the first order derivative of eqn. (13.49) is

$$(\overleftrightarrow{\partial_t \lambda})^{(1)} = \mp \frac{\hbar^2}{2m}\sigma^i \sigma^j \, (D_i \overleftrightarrow{\partial_t} D_j)$$
$$= \mp \frac{\mathrm{i}e\hbar}{2m}\sigma^i \sigma^j \, (D_i E_j - E_i D_j) \tag{13.50}$$

since we may add and subtract $\partial_i A_t$ with impunity. To go to next order, we must substitute this result back into eqn. (13.49) and take the time derivative again. This gives a further correction

$$(\overleftrightarrow{\partial_t \lambda})^{(2)} = \mp \mathrm{i}\frac{\hbar}{4mc^2}\partial_t \left[\frac{\mathrm{i}\hbar}{4mc^2}\left(\frac{\mathrm{i}e\hbar}{2m}\sigma^i \sigma^j \, (D_i E_j - E_i D_j) \right) \right] \tag{13.51}$$

Noting the energy shift $-\mathrm{i}\hbar\partial_t \to -\mathrm{i}\hbar\tilde{\partial}_t - mc^2$ and taking the positive square-root, we obtain the non-relativistic limit for the positive half of the solutions:

$$S_\mathrm{D} \to \int (\mathrm{d}x)\left\{ \chi^\dagger \left(\mathrm{i}\hbar\tilde{D}_t + \frac{\hbar^2 D^i D_i}{2m} - \frac{e\hbar B^i \sigma_i}{2m} \right.\right.$$
$$-\frac{e\hbar^2}{8m^2 c^2}\sigma^i \sigma^j \, (D_i E_j - E_i D_j)$$
$$-\frac{\hbar^4}{8m^3 c^2}\left((D^i D_i) - \frac{e}{\hbar}(\sigma^i B_i) \right)^2$$
$$\left.\left. -\mathrm{i}\frac{e\hbar^3}{32m^3 c^4}\sigma^i \sigma^j \, \overleftrightarrow{\partial_t} \, (D_i E_j - E_i D_j) + \cdots \right)\chi \right\}, \tag{13.52}$$

where $B_k = \frac{1}{2}\epsilon_{ijk}F_{jk}$ and the overall sign has been changed to conform to the usual conventions. The negative root in eqn. (13.47) gives a similar result for anti-particles. The fifth term contains $\partial^i E_i$, which is called the Darwin term. It corresponds to a correction to the point charge interaction due to the fact that Dirac 'particles' are spread out over a region with radius of the order of the Compton wavelength $\hbar/mc$. In older texts, this is referred to as the *Zitterbewegung*, since, if one insists on a particle interpretation, it is necessary to imagine the particles jittering around their average position in a kind of random walk. The $\sigma^i B_i$ term is a Zeeman splitting term due to the interaction of the magnetic field with particle trajectories.

Note that our diagonalization of the Dirac action leads to no coupling between the positive and negative energy solutions. One might expect that interactions with A_μ which couple indiscriminately with both positive and negative energy parts of the field would lead to an implicit coupling between positive and negative energy parts. This is not the case classically, however, since the vector potential A_μ leads to no non-linearities with respect to ψ.

Radiative corrections (fluctuation corrections) in the relativistic fields give rise to back-reaction terms both in the fermion sector and in the electromagnetic sector. The effect of photon $D_{\mu\nu}$ exchange leads to an effective quartic interaction

$$S_\Pi = \int (\mathrm{d}x)(\mathrm{d}x')\,(\overline{\psi}(x')\gamma^\mu\psi(x'))\,D_{\mu\nu}(x,x')\,(\overline{\psi}(x)\gamma^\nu\psi(x)). \quad (13.53)$$

The photon propagator is clearly a non-local and gauge-dependent quantity. Non-locality is a feature of the full theory, and reflects the fact that the finite speed of light disallows an instantaneous response in the field during collisions (there is an intrinsic non-elasticity in relativistic particle scattering). Working to a limited order in $1/c$ makes the effective Lagrangian effectively local, however, since the non-local derivative expansion is truncated. The gauge dependence of the Lagrangian is more subtle. In order to obtain a physically meaningful result, one requires an effective Lagrangian which produces gauge-fixing independent results. This does not necessarily mean that the Lagrangian needs to be gauge-independent, however. The reason is that the Lagrangian is no longer covariant with respect to the necessary symmetries to make this apparent.

Gauge invariance is related to a conformal/Lorentz symmetry of the relativistic gauge field, so one would expect a loss of Lorentz invariance to result in a breakdown of invariance under choice of gauge-fixing condition. In fact, a non-relativistic effective Lagrangian is not unique: its form is indeed gauge-dependent. Physical results cannot be gauge-dependent, however, provided one works to consistent order in the expansion of the original covariant theory. Thus, the gauge condition independence of the theory will be secured by working to consistent order in the smallness parameters, regardless of the actual gauge chosen. Propagators and Lagrangian are gauge-dependent, but in just the right way to provide total gauge independence.

Turning to the photon sector, we seek to account for the effects of vacuum and medium polarization in leading order kinetic terms for the photon. To obtain the non-relativistic limit of the radiation terms, it is advantageous to have a model of the dielectric medium in which photons propagate. Nevertheless, some progress can be made on the basis of a generic linear response. We therefore use linear response theory and assume a constitutive relation for the polarization of the form

$$G^{\mu\nu} = F^{\mu\nu} + \int (\mathrm{d}x) \, \chi(x, x') F^{\mu\nu}. \tag{13.54}$$

The second term is a correction to the local field, which is proportional to the field itself. Perhaps surprisingly, this relation plays a role even in the vacuum, since quantum field theory predicts that the field ψ may be polarized by the back-reaction of field fluctuations in A_μ. Since the susceptibility $\chi(x, x')$ depends on the dynamics of the non-relativistic matter field, one expects this polarization to break the Lorentz invariance of the radiation term. This occurs because, at non-relativistic speeds, the interaction between matter and radiation splits into electric and magnetic parts which behave quite differently. From classical polarization theory, we find that the momentum space expression for the susceptibility takes the general form

$$\chi(\omega) \sim \frac{Ne^2\omega^2/\epsilon_0 m}{\omega_0^2 - i\gamma\omega + \omega^2}. \tag{13.55}$$

In an electron plasma, where there are no atoms which introduce interactions over and above the ones we are considering above, the natural frequency of oscillations can only be $\omega_0 \sim mc^2/\hbar$. These are the only scales from which to construct a frequency. The significance of this value arises from the correlations of the fields on the order of the Compton wavelength which lead to an elastic property of the field. This is related to the Casimir effect and to the Zitterbewegung mentioned earlier. It is sufficient to note that such a system has an ultra-violet resonance, where $\omega_0 \gg \omega$ in the non-relativistic limit. This means that $\chi(\omega)$ can be expanded in powers of ω/ω_0. From the equations of motion, $\hbar\omega \sim \hbar^2\mathbf{k}^2/2m$; thus, the expansion is in powers of the quantity

$$\frac{\omega}{\omega_0} \sim \frac{\hbar\mathbf{k}^2/2m}{\hbar(mc^2/\hbar)} = \frac{\hbar\mathbf{k}^2}{m^2c^2}. \tag{13.56}$$

It follows that the action for the radiation may be written in the generic form

$$S_\mathrm{M} = \int (\mathrm{d}x) \left\{ \frac{C_\mathrm{E}}{2} A_0 \left[\nabla^2 \left(\frac{-\hbar\nabla^2}{m^2c^2} + \cdots \right) A_0 \right] \right.$$
$$\left. + \frac{C_B}{2} A_i \left[(-\nabla^2 g_{ij} + \partial_i\partial_j) \left(\frac{-\hbar\nabla^2}{m^2c^2} + \cdots \right) \right] A_j \right\}. \tag{13.57}$$

This form expresses only the symmetries of the field and dimensional scales of the system. In order to evaluate the constants C_E and C_B in this expression, it would necessary to be much more specific about the nature of the polarization. For a plasma in a vacuum, the constants are equal to unity in the classical approximation. The same form for the action would apply in the case of electrons in an ambient polarizable medium, below resonance. Again, to determine the constants in that case, one would have to introduce a model for the ambient matter or input effective values for the constants by hand.

13.4 Thermal and Euclidean Green functions

There are two common formulations of thermal Green functions. At thermal equilibrium, where time is an irrelevant variable on average, one can rotate to a Euclidean, imaginary time formulation, as in eqn. (6.46), where the imaginary part of time places the role of an inverse temperature β. Alternatively one can use a real-time formulation as in eqn. (6.61).

The non-relativistic limit of Euclideanized field theory is essentially no different from the limit in Minkowski spacetime, except that there is no direct concept of retarded or advanced boundary conditions in terms of poles in the propagator. There is nevertheless still a duplicity in the solutions with positive and negative, imaginary energy. This duplicity disappears in the non-relativistic limit, as before, since half of the spectrum is suppressed. The relativistic, Euclidean Green function, closely related to the zero-temperature Feynman Green function, is given by

$$G_\beta(x, x') = \int \frac{d\omega}{2\pi} \frac{d^n \mathbf{k}}{(2\pi)^n} \frac{e^{ik(x-x')}}{p_\beta^2 c^2 + m^2 c^4}, \tag{13.58}$$

where the zeroth component of the momentum is given by the Matsubara frequencies $p_\beta^0 = 2n\pi/\beta\hbar c$:

$$2mc^2 G_\beta(x, x') = \int \frac{d\omega}{2\pi} \frac{d^n \mathbf{k}}{(2\pi)^n} \frac{e^{ik(x-x')}}{\frac{p_\beta^2}{2m} + \frac{1}{2}mc^2}. \tag{13.59}$$

Shifting the energy $ip_\beta^0 \to mc^2 + i\tilde{p}_\beta^0$ leaves us with

$$G_{NR\beta}(x, x') = \int \frac{d\omega}{2\pi} \frac{d^n \mathbf{k}}{(2\pi)^n} \frac{e^{ik(x-x')}}{\frac{\mathbf{p}^2}{2m} - i\hbar\tilde{\omega}}, \tag{13.60}$$

which is the Green function for the Euclidean action

$$S = \int (dx)\chi^\dagger \left[\frac{\hbar^2 \nabla^2}{2m} + \hbar\tilde{\partial}_\tau \right] \chi. \tag{13.61}$$

In the real-time formulation, in which we retain the auxiliary time dependence, the thermal character of the Green functions is secured through the momentum space boundary condition in eqn. (6.54), known in quantum theory as the Kubo–Martin–Schwinger relation. Considering the boundary terms in eqn. (6.61) and following the procedure in section 13.2.2, one has

$$2mc^2\, 2\pi i\, f(k)\theta(k_0)\delta(p^2c^2 + m^2c^4) \rightarrow$$
$$2mc^2\, 2\pi i\, f(k)\theta(k_0)\frac{\delta(p_0 c - \hbar\tilde{\omega}_k)}{2\hbar\omega_k}. \tag{13.62}$$

In the large c limit, $\hbar\omega_k \rightarrow mc^2$, thus the $c \rightarrow \infty$ limit of this term is simply

$$2\pi i f(\tilde{\omega}_k), \tag{13.63}$$

where $\hbar\omega_k = mc^2 + \hbar\tilde{\omega}_k$.

Intimately connected to this form is the Kubo–Martin–Schwinger (KMS) relation. We looked at this relation in section 6.1.5, and used it to derive the form of the relativistic Green functions. Notice that the zero-temperature, negative frequency parts of the Wightman functions do not contribute to the derivation of this relation in eqn. (6.56). For this reason, the form of the relationship in eqn. (6.54) is unchanged,

$$-G^{(+)}(\tilde{\omega}) = e^{\beta\tilde{\omega}}G^{(-)}(\tilde{\omega}). \tag{13.64}$$

This use of the non-relativistic energy in both the relativistic and non-relativistic cases is important and leads to a subtlety in the Euclidean formulation. From the simplistic viewpoint of a Euclidean imaginary-time theory, the meaning of a thermal distribution is different in the relativistic and non-relativistic cases. The Boltzmann factor changes from

$$e^{-\beta(\hbar\tilde{\omega}+mc^2)} \rightarrow e^{-\beta\hbar\tilde{\omega}}. \tag{13.65}$$

This change is reflected also in a change in the time dependence of wave modes,

$$e^{+i(\tilde{\omega}+mc^2/\hbar)\tau} \rightarrow e^{+i\tilde{\omega}\tau}. \tag{13.66}$$

The shift is necessary to reflect the change in dynamical constraints posed by the equations of motion. However, the Boltzmann condition applies (by convention) to the non-relativistic energy. It is this energy scale which defines the temperature we know.

Another way of looking at the change in the Boltzmann distribution is from the viewpoint of fluctuations. Thermal fluctuations give rise to the Boltzmann factor, and these must have a special causal symmetry: emission followed by absorption. These processes are mediated by the Green functions, which reflect the equations of motion and are therefore unambiguously defined. As we take the non-relativistic limit, the meaning of the thermal distribution changes character subtly. The positive frequency/energy condition changes from being $\theta(\omega) = \theta(\tilde{\omega} + mc^2/\hbar)$ to $\theta(\tilde{\omega})$ owing to the re-definition of the zero point energy. Looking back at eqn. (6.56), we derived the Bose–Einstein distribution using the fact that

$$\theta(-\omega)e^{\beta\omega} = 0. \qquad (13.67)$$

But one could equally choose the zero point energy elsewhere and write

$$\theta(-(\omega + \Delta\omega))e^{\beta'(\omega+\Delta\omega)} = 0. \qquad (13.68)$$

As long as the Green functions are free of interactions which couple the energy scale to a third party, we can re-label the energy freely by shifting the variable of integration in eqn. (5.64). In an interacting theory, the meaning of such a re-labelling is less clear.

In a system which is already in thermal equilibrium, one might argue that the interactions are not relevant. Interactions are only important in the approach to equilibrium and to the final temperature. With a new definition of the energy, a temperature has the same role as before, but the temperature scale β' is modified.

This might seem slightly paradoxical, but the meaning it clear. The KMS condition expressed by eqn. (6.54) simply indicates that the fluctuations mediated by given Green functions should be in thermal balance. The same condition may be applied to any virtual process, based on any equilibrium value or zero point energy. If we change the Green functions, we change the condition and the physics underpinning it. In each case, one obtains an equilibrium distribution of the same general form, but the meaning depends on the original Green functions. In order to end up with equivalent temperature scales, one must use equivalent energy scales. Relativistic energies and non-relativistic energies are not equivalent, and neither are the thermal distributions obtained from these. In the non-relativistic case, thermal fluctuations comprise kinetic fluctuations in particle motion. In the relativistic case, the energy of the particles themselves is included.

Two thermal distributions

$$e^{\hbar\beta\omega} = e^{\hbar(\beta+\Delta\beta)(\omega+\Delta\beta)} \qquad (13.69)$$

are equivalent if

$$\frac{\beta + \Delta\beta}{\beta} = \frac{\omega}{\omega + \Delta\omega}. \qquad (13.70)$$

These two viewpoints are related by a renormalization of the energy or chemical potential; the reason why such a renormalization is required is precisely because of the change in energy conventions which affects the Euclidean formulation.

13.5 Energy conservation

The speed of light is built into the covariant notation of the conservation law

$$\partial_\mu \theta^{\mu\nu} = 0. \tag{13.71}$$

We must therefore ascertain whether the conservation law is altered by the limit $c \to \infty$. From eqns. (11.5) and (11.44), one may write

$$\partial_\mu \theta^{\mu\nu} = \frac{1}{c}\partial_t \theta^{0\mu} + \partial_i \theta^{i\mu}$$
$$= \partial_t \theta^{t\mu} + \partial_i \theta^{i\mu}. \tag{13.72}$$

It is apparent from eqn. (11.5) that, as $c \to \infty$,

$$\theta_{0i} \to \infty$$
$$\theta_{i0} \to 0. \tag{13.73}$$

Splitting μ into space and time components, we have, for the time component,

$$\partial_\mu \theta^{\mu 0} = \frac{1}{c}\partial_\mu \theta^{\mu t}$$
$$= \frac{1}{c}\left[\partial_t \theta^{tt} + \partial_i \theta^{it}\right]$$
$$= \frac{1}{c}[\partial_t H] = 0. \tag{13.74}$$

Because of the limit, this equation is ambiguous, but the result is sensible if we interpret the contents of the brackets as being zero. For the space components one has

$$\partial_t \theta^{ti} + \partial_j \theta^{ji} = 0$$
$$\partial_t \overline{p} + \partial_j \sigma^{ji} = 0, \tag{13.75}$$

where σ_{ij} is the stress tensor. Thus, energy conservation is preserved but it becomes divided into two separate statements, one about the time independence of the total Hamiltonian, and another expressing Newton's law that the rate of change of momentum is equal to the applied force.

13.6 Residual curvature and constraints

The non-relativistic limit does not always commute with the limit of zero curvature, nor with that of dimensional reduction, such as projection in order to determine the effective dynamics on a lower-dimensional constraint surface [15]. Such a reduction is performed by derivative expansion, in which every derivative seeks out orders of the curvature of the embedded surface. Since the non-relativistic limit is also an expansion in terms of small derivatives, there is an obvious connection between these. In particular, the shape of a constraint surface can have specific implications for the consistency of the non-relativistic limit [29, 30, 80, 94, 95]. Caution should be always exercised in taking limits, to avoid premature loss of information.

14

Unified kinematics and dynamics

This chapter sews together some of the threads from the earlier chapters to show the relationships between apparently disparate dynamical descriptions of physics.

14.1 Classical Hamiltonian particle dynamics

The traditional formulation Schrödinger quantum mechanics builds on a Hamiltonian formulation of dynamical systems, in which the dynamics describe not only particle coordinates q but also their momenta p. The interesting feature of the Hamiltonian formulation, in classical mechanics, is that one deals only with quantities which have a direct physical interpretation. The disadvantage with the Hamiltonian approach in field theory is its lack of manifest covariance under Lorentz transformations: time is singled out explicitly in the formulation.[1] Some important features of the Hamiltonian formulation are summarized here in order to provide an alternative view to dynamics with some different insights.

The Hamiltonian formulation begins with the definition of the momentum p_i conjugate to the particle coordinate q_i. This quantity is introduced since it is expected to have a particular physical importance. Ironically, one begins with the Lagrangian, which is unphysical and is to be eliminated from the discussion. The Lagrangian is generally considered to be a function of the particle coordinates q_i and their time derivatives or velocities $\dot{q}^i$. The momentum is then conveniently defined from the Lagrangian,

$$p_i = \frac{\partial L}{\partial \dot{q}_i}. \tag{14.1}$$

---

[1] Actually, time is singled out in a special way even in the fully covariant Lagrangian formulation, since time plays a fundamentally different role from space as far as the dynamics are concerned. The main objection which is often raised against the Hamiltonian formulation is the fact that the derivation of covariant results is somewhat clumsy.

This is not the only way in which one could define a momentum, but it is convenient to use a definition which refers only to the abstract quantities L and $\dot{q}_i$ in cases where the Lagrangian and its basic variables are known, but other physical quantities are harder to identify. This extends the use of the formalism to encompass objects which one would not normally think of as positions and momenta. The total time derivative of the Lagrangian is

$$\frac{\mathrm{d}L(q, \dot{q}, t)}{\mathrm{d}t} = \frac{\partial L}{\partial q_i} \dot{q}_i + \frac{\partial L}{\partial \dot{q}_i} \ddot{q} + \frac{\partial L}{\partial t}, \tag{14.2}$$

which may be written

$$\frac{\mathrm{d}}{\mathrm{d}t} \left\{ \dot{q}_i \frac{\partial L}{\partial \dot{q}_i} - L \right\}. \tag{14.3}$$

Now, if the Lagrangian is not explicitly time-dependent, $\frac{\partial L}{\partial t} = 0$, then the quantity in the curly braces must be constant with respect to time, so, using eqn. (14.1), we may define the Hamiltonian H by

$$H = \text{const.} = p\dot{q} - L. \tag{14.4}$$

Notice that this definition involves time derivatives. When we consider the relativistic case, timelike quantities are often accompanied by a sign from the metric tensor, so the form of the Hamiltonian above should not be treated as sacred.

14.1.1 Hamilton's equations of motion

The equations of motion in terms of the new variables may be obtained in the usual way from the action principle, but now treating q_i and p_i as independent variables. Using the Lagrangian directly to obtain the action gives us

$$S = \int \mathrm{d}t \, \{p\dot{q} - L\}. \tag{14.5}$$

However, from earlier discussions about symmetrical derivatives, we know that the correct action is symmetrized about the derivatives. Thus, the action is given by

$$S = \int \mathrm{d}t \, \left\{ \frac{1}{2}(p\dot{q} - q\dot{p}) - L \right\}. \tag{14.6}$$

Varying this action with fixed end-points, one obtains (integrating the $p\dot{q}$ term by parts)

$$\frac{\delta S}{\delta q(t)} = -\dot{p} - \frac{\partial H}{\partial q} = 0$$

$$\frac{\delta S}{\delta p(t)} = \dot{q} - \frac{\partial H}{\partial p}. \tag{14.7}$$

Hence, Hamilton's two equations of motion result:

$$\dot{p} = -\frac{\partial H}{\partial q} \tag{14.8}$$

$$\dot{q} = \frac{\partial H}{\partial p}. \tag{14.9}$$

Notice that this is a pair of equations. This is a result of our insistence on introducing an extra variable (the momentum) into the formulation.

14.1.2 Symmetry and conservation

One nice feature of the Hamiltonian formulation is that invariances of the equations of motion are all identifiable as a generalized translational invariance. If the action is independent of a given coordinate

$$\frac{\partial L}{\partial q_n} = 0, \tag{14.10}$$

then

$$\frac{\partial L}{\partial \dot{q}_n} = p_n = \text{const.}; \tag{14.11}$$

i.e. the momentum associated with that coordinate is constant, or is *conserved*. The coordinate q_n is then called an *ignorable* coordinate.

14.1.3 Symplectic transformations

We started originally with an action principle, which treated only $q(t)$ as a dynamical variable, and later introduced (artificially) the independent momentum variable p.[2] The fact that we now have twice the number of dynamical variables seems unnecessary. This intuition is further borne out by the observation that, if we make the substitution

$$q \rightarrow -p \tag{14.12}$$

$$p \rightarrow q \tag{14.13}$$

in eqn. (14.9), then we end up with an identical set of equations, with only the roles of the two equations switched. This transformation represents an

---

[2] In many textbooks, the Lagrangian formulation is presented as a function of coordinates q and velocities $\dot{q}$. Here we have bypassed this discussion by working directly with variations of the action, where it is possible to integrate by parts and perform functional variations. This makes the usual classical Lagrangian formalism redundant.

invariance of the Hamilton equations of motion, and hints that the positions and momenta are really just two sides of the same coin.

Based on the above, one is motivated to look for a more general linear transformation in which p and q are interchanged. In doing so, one must be a little cautious, since positions and momenta clearly have different engineering dimensions. Let us therefore introduce the quantities $\hat{p}$ and $\hat{q}$, which are re-scaled by a constant Ω with dimensions of mass per unit time in order that they have the same dimensions:

$$\hat{p} = p/\sqrt{\Omega}$$
$$\hat{q} = q\sqrt{\Omega}. \tag{14.14}$$

The product of $\hat{q}$ and $\hat{p}$ is independent of this scale, and this implies that the form of the equations of motion is unchanged:

$$\dot{\hat{p}} = -\frac{\partial H}{\partial \hat{q}} \tag{14.15a}$$

$$\dot{\hat{q}} = \frac{\partial H}{\partial \hat{p}}. \tag{14.15b}$$

Let us consider, then, general linear combinations of q and p and look for all those combinations which leave the equations of motion invariant. In matrix form, we may write such a transformation as

$$\begin{pmatrix} \hat{q}' \\ \hat{p}' \end{pmatrix} = \begin{pmatrix} a & b \\ c & d \end{pmatrix} \begin{pmatrix} \hat{q} \\ \hat{p} \end{pmatrix}. \tag{14.16}$$

The derivatives associated with the new coordinates are

$$\frac{\partial}{\partial \hat{q}'} = \frac{1}{2}\left(\frac{1}{a}\frac{\partial}{\partial \hat{q}} + \frac{1}{b}\frac{\partial}{\partial \hat{p}} \right)$$

$$\frac{\partial}{\partial \hat{p}'} = \frac{1}{2}\left(\frac{1}{c}\frac{\partial}{\partial \hat{q}} + \frac{1}{d}\frac{\partial}{\partial \hat{p}} \right). \tag{14.17}$$

We may now substitute these transformed coordinates into the Hamilton equations of motion (14.15) and determine the values of a, b, c, d for which the equations of motion are preserved. From eqn. (14.15b), one obtains

$$a\dot{\hat{q}} + b\dot{\hat{p}} = \frac{1}{2}\left(\frac{1}{c}\frac{\partial H}{\partial q} + \frac{1}{d}\frac{\partial H}{\partial p} \right). \tag{14.18}$$

This equation is a linear combination of the original equation in (14.15) provided that we identify

$$2ad = 1$$
$$2bc = -1. \tag{14.19}$$

Substitution into eqn. (14.15a) confirms this. The determinant of the transformation matrix is therefore

$$ad - bc = 1 \tag{14.20}$$

and we may write, in full generality:

$$U = \begin{pmatrix} a & b \\ c & d \end{pmatrix} = \frac{1}{\sqrt{2}} \begin{pmatrix} e^{i\theta} & ie^{i\phi} \\ ie^{-i\phi} & e^{-i\theta}. \end{pmatrix} \tag{14.21}$$

This is the most general transformation of $\hat{p}, \hat{q}$ pairs which leaves the equations of motion invariant. The set of transformations, which leaves the Poisson bracket invariant, forms a group known as the symplectic group $sp(2, C)$. If we generalize the above discussion by adding indices to the coordinates and momenta $i = 1, \ldots, n$, then the group becomes $sp(2n, C)$.

Since we have now shown that p and q play virtually identical roles in the dynamical equations, there is no longer any real need to distinguish them with separate symbols. In *symplectic notation*, many authors write both coordinates and momenta as Q_i, where $i = 1, \ldots, 2n$ grouping both together as *generalized coordinates*.

14.1.4 Poisson brackets

Having identified a symmetry group of the equations of motion which is general (i.e. which follows entirely from the definition of the conjugate momentum in terms of the Lagrangian), the next step is to ask which quantities are invariant under this symmetry group. A quantity of particular interest is the so-called *Poisson bracket*.

If we apply the group transformation to the derivative operators,

$$\begin{pmatrix} \hat{D}_+ \\ \hat{D}_- \end{pmatrix} \equiv U(\theta, \phi) \begin{pmatrix} \frac{\partial}{\partial \hat{q}} \\ \frac{\partial}{\partial \hat{p}} \end{pmatrix}, \tag{14.22}$$

then it is a straightforward algebraic matter to show that, for any two functions of the dynamical variables A, B, the Poisson bracket, defined by

$$(D_+X)(D_-Y) - (D_-X)(D_+Y) \equiv [X, Y]_{\hat{p}\hat{q}}, \tag{14.23}$$

is independent of θ and ϕ and is given in all bases by

$$[X, Y]_{pq} = \frac{\partial X}{\partial q} \frac{\partial Y}{\partial p} - \frac{\partial Y}{\partial q} \frac{\partial X}{\partial p}. \tag{14.24}$$

Notice, in particular that, owing to the product of pq in the denominators, this bracket is even independent of the re-scaling by Ω in eqn. (14.14).

We shall return to the Poisson bracket to demonstrate its importance to the variational formalism and dynamics after a closer look at symmetry transformations.

14.1.5 General canonical transformations

The linear combinations of p, q described in the previous section form a symmetry which has its origins in the linear formulation of the Hamiltonian method. Symplectic symmetry is not the only symmetry which might leave the equations of motion invariant, however. More generally, we might expect the coordinates and momenta to be changed into quite different functions of the dynamical variables:

$$q \to q'(p, q, t)$$
$$p \to p'(p, q, t). \tag{14.25}$$

Changes of variable fit this general description, as does the time development of p and q. We might, for example, wish to change from a Cartesian description of physics to a polar coordinate basis, which better reflects the symmetries of the problem. Any such change which preserves the form of the field equations is called a canonical transformation.

It turns out that one can effect general infinitesimal transformations of coordinates by simply adding total derivatives to the Lagrangian. This is closely related to the discussion of continuity in section 4.1.4. Consider the following addition

$$L \to L + \frac{dF}{dt}, \tag{14.26}$$

for some arbitrary function $F(q, p, t)$. Normally, one ignores total derivatives in the Lagrangian, for the reasons mentioned in section 4.4.2. This is because the action is varied, with the end-points of the variation fixed. However, if one relaxes this requirement and allows the end-points to vary about *dynamical variables which obey the equations of motion*, then these total derivatives (often referred to as surface terms in field theory), have a special and profound significance. Our programme and its notation are the following.

- We add the total time derivative of a function $F(q, p, t)$ to the Lagrangian so that

$$S \to S + \int dt \, \frac{dF}{dt}$$
$$= S + F \Big|_{t_1}^{t_2}. \tag{14.27}$$

- We vary the action and the additional term and define the quantity G_ξ, which will play a central role in transformation theory, by

$$\delta_\xi F \equiv G_\xi, \tag{14.28}$$

so that

$$\delta S \to \delta S + G. \tag{14.29}$$

- We may optionally absorb the change in the variation of the action G into the generalized coordinates by making a transformation, which we write formally as

$$q \rightarrow q + \delta q = q + R_\xi \, \delta\xi. \tag{14.30}$$

R_ξ is called the auxiliary function of the transformation, and is related to G_ξ, which is called the generator of the transformation. This transforms the coordinates and momenta by an infinitesimal amount.

Let us now illustrate the procedure in practice. The variation of the action may be written

$$\delta S = \delta \int dt \, (p\dot{q} - H) + \int dt \, \delta\dot{F},$$
$$= \int dt \left(\left(-\dot{p} - \frac{\partial H}{\partial q} \right) \delta q + \frac{\partial \dot{G}}{\partial q} \delta q + \frac{\partial \dot{G}}{\partial q} \delta q \right). \tag{14.31}$$

In the last line we have expanded the infinitesimal change $\delta F = G$ in terms of its components along q and p. This can always be done, regardless of what general change G represents. We can now invoke the modified action principle and obtain the equations of motion:

$$\frac{\delta S}{\delta q} = 0 = -\dot{p} - \frac{\delta H}{\delta q} + \frac{\delta \dot{G}}{\delta q} = -(\dot{p} + \delta\dot{p}) - \frac{\delta H}{\delta q}$$

$$\frac{\delta S}{\delta p} = 0 = \dot{q} - \frac{\delta H}{\delta p} + \frac{\delta \dot{G}}{\delta p} = (\dot{q} + \delta\dot{q}) - \frac{\delta H}{\delta p}, \tag{14.32}$$

where we have identified

$$\delta\dot{p} = -\frac{\partial \dot{G}}{\partial q}$$

$$\delta\dot{q} = \frac{\partial \dot{G}}{\partial p}, \tag{14.33}$$

or, on integrating,

$$\delta p = -\frac{\partial G}{\partial q} = R_a^p \delta\xi^a$$

$$\delta q = \frac{\partial G}{\partial p} = R_a^q \delta\xi^a. \tag{14.34}$$

Notice that G is infinitesimal, by definition, so we may always write it in terms of a set of infinitesimal parameters $\delta\xi$, but ξ need not include q, p now since the q, p dependence was already removed to infinitesimal order in eqn. (14.31).[3] It is now possible to see why G is referred to as the generator of infinitesimal canonical transformations.

---

[3] The expressions are not incorrect for p, q variations, but they become trivial cases.

14.1.6 Variations of dynamical variables and Poisson brackets

One of the most important observations about variational dynamics, as far as the extension to quantum field theory is concerned, is that variational changes in any dynamical variable can be expressed in terms of the invariant Poisson bracket between that variable and the generator of the variation:

$$\delta X(p, q, t) = [X, G_\xi]_{pq} \tag{14.35}$$

To see this, it is sufficient to use eqns. (14.34) in the differential expansion of the function:

$$\delta X = \frac{\partial X}{\partial q_i}\delta q_i + \frac{\partial X}{\partial p_i}\delta p_i. \tag{14.36}$$

Substituting for δq_i and δp_i gives eqn. (14.35). These relations are exemplified as follows.

- Generator of time translations: $G_t = -H\delta t$;

$$\delta X = [X, H]\delta t. \tag{14.37}$$

Noting that the change in X is the dynamical evolution of the function, but that the numerical value of t is unaltered owing to linearity and the infinitesimal nature of the change, we have that

$$\delta X = -\left(\frac{\mathrm{d}X}{\mathrm{d}t} - \frac{\partial X}{\partial t}\right)\mathrm{d}t = [X, H]\delta t. \tag{14.38}$$

Thus, we arrive at the equation of motion for the dynamical variable X:

$$\frac{\mathrm{d}X}{\mathrm{d}t} = [X, H] + \frac{\partial X}{\partial t}. \tag{14.39}$$

This result has the following corollaries:

$$\dot{q} = [q, H]$$
$$\dot{p} = [p, H]$$
$$1 = [H, t]. \tag{14.40}$$

The first two equations are simply a thinly concealed version of the Hamilton equations (14.9). The third, which is most easily obtained from eqn. (14.37), is an expression of the time independence of the Hamiltonian. An object which commutes with the Hamiltonian is conserved.

- Generator of coordinate translations: $G_q = p\delta q$.

  It is interesting to note that, if we consider the variation in the coordinate q with respect to the generator for q itself, the result is an identity which summarizes the completeness of our dynamical system:

  $$\delta q = [q, G_q]_{pq}$$
  $$\delta q = [q, p]_{pq}\delta q$$
  $$\Rightarrow 1 = [q, p]_{pq}. \qquad (14.41)$$

  In Lorentz-covariant notation, one may write

  $$[x^\mu, p_\nu] = \delta^\mu{}_\nu, \qquad (14.42)$$

  where $p_\mu = (-H/c, \mathbf{p})$. This result pervades almost all of dynamics arising from Lagrangian/Hamiltonian systems. In the quantum theory it is supplanted by commutation relations, which have the same significance as the Poisson bracket, though they are not directly related.

14.1.7 Derivation of generators from the action

Starting from the correctly symmetrized action in eqn. (14.6), the generator of infinitesimal canonical transformations for a variable ξ is obtained from the surface contribution to the variation, with respect to ξ.[4] For example,

$$\delta_q S = \int dt \left(-\dot{p} - \frac{\partial H}{\partial q} \right) \delta q + \frac{1}{2} p\delta q$$
$$= 0 + \frac{1}{2} G_q, \qquad (14.43)$$

where we have used the field equation to set the value of the integral in the first line to zero, and we identify

$$G_q = p\delta q. \qquad (14.44)$$

Similarly,

$$\delta_p S = \int dt \left(\dot{q} - \frac{\partial H}{\partial p} \right) \delta p - \frac{1}{2} q\delta p$$
$$= \frac{1}{2} G_p, \qquad (14.45)$$

---

[4] The constants of proportionality are rather inconsistent in this Hamiltonian formulation. If one begins with the action defined in terms of the Lagrangian, the general rule is: for actions which are linear in the time derivatives, the surface contribution is one-half the corresponding generator; for actions which are quadratic in the time derivatives, the generator is all of the surface contribution.

hence

$$G_p = -q\delta p. \qquad (14.46)$$

For time variations,

$$\delta_t S = -H\delta t$$
$$= G_t. \qquad (14.47)$$

The generators are identified, with numerical values determined by convention. The factors of one-half are introduced here in order to eliminate irrelevant constants from the Poisson bracket. This is of no consequence, as mentioned in the next section; it is mainly for aesthetic purposes.

Suppose we write the action in the form

$$S = \int \{pdq - Hdt\}, \qquad (14.48)$$

where we have cancelled an infinitesimal time differential in the first term. It is now straightforward to see that

$$\frac{\partial S}{\partial t} + H = 0. \qquad (14.49)$$

This is the Hamilton–Jacobi equation of classical mechanics. From the action principle, one may see that this results from boundary activity, by applying a general boundary disturbance F:

$$S \to S + \int (dx)\, \partial_\mu F. \qquad (14.50)$$

$\delta F = G$ is the generator of infinitesimal canonical transformations, and

$$\frac{\partial G}{\partial q} = \int d\sigma^\mu\, R^a_\mu \delta\xi_a. \qquad (14.51)$$

Notice from eqn. (11.43) that

$$\int d\sigma^\mu G_\mu = \int d\sigma^\mu (\Pi_\mu \delta q - \theta_{\mu\nu}\delta x^\nu), \qquad (14.52)$$

which is to be compared with

$$\delta S = p\delta q - Hdt. \qquad (14.53)$$

Moreover, from this we have the Hamilton–Jacobi equation (see eqn. (11.78)),

$$\frac{\delta S}{\delta x^0} = -\frac{1}{c}\int d\sigma^\mu\, \theta_{\mu 0} = -\frac{H}{c} \qquad (14.54)$$

or

$$\frac{\delta S}{\delta t} + H = 0. \qquad (14.55)$$

14.1.8 Conjugate variables and dynamical completeness

The commutator functions we have evaluated above are summarized by

$$[q_A, p_B]_{pq} = \delta_{AB}$$
$$[t, H]_{pq} = 1. \tag{14.56}$$

These equations are a formal expression of the completeness of the set of variables we have chosen to parametrize the dynamical equations. Not every variational equation necessarily has coordinates and momenta, but every set of conservative dynamical equations has pairs of conjugate variables which play the roles of p and q. If one is in possession of a complete set of such variables (i.e. a set which spans all of phase space), then an arbitrary state of the dynamical system can be represented in terms of those variables, and it can be characterized as being canonical.

Ignorable coordinates imply that the dimension of phase space is effectively reduced, so there is no contradiction in the presence of symmetries.

Given the definition of the Poisson bracket in eqn. (14.24), the value of $[q, p]_{pq} = 1$ is unique. But we could easily have defined the derivative differently up to a constant, so that we had obtained

$$[q_A, p_B]'_{pq} = \alpha \, \delta_{AB}. \tag{14.57}$$

What is important is not the value of the right hand side of this expression, but the fact that it is constant for all conjugate pairs. In any closed, conservative system, the Hamiltonian time relation is also constant, but again the normalization could easily be altered by an arbitrary constant. These are features which are basic to the geometry of phase space, and they carry over to the quantum theory for commutators. There it is natural to choose a different value for the constant and a different but equivalent definition of completeness.

14.1.9 The Jacobi identity and group algebra

The anti-symmetrical property of the Poisson bracket alone is responsible for the canonical group structure which it generates and the completeness argument above. This may be seen from an algebraic identity known as the Jacobi identity. Suppose that we use the bracket $[A, B]$ to represent *any* object which has the property

$$[A, B] = -[B, A]. \tag{14.58}$$

The Poisson bracket and the commutator both have this property. It may be seen, by writing out the combinations explicitly, that

$$[A, [B, C]] + [B, [C, A]] + [C, [A, B]] = 0. \tag{14.59}$$

This result does not depend on whether A, B, C commute. This equation is known as the Jacobi identity. It is closely related to the Bianchi identity in eqn. (2.27).

Any objects which satisfy this identity also satisfy a Lie algebra. This is easily seen if we identify a symbol

$$T_A(B) \equiv [A, B]. \qquad (14.60)$$

Then, re-writing eqn. (14.59) so that all the C elements are to the right,

$$[A, [B, C]] - [B, [A, C]] - [[A, B], C] = 0, \qquad (14.61)$$

we have

$$T_A T_B(C) - T_B T_A(C) = T_{[A,B]}(C), \qquad (14.62)$$

or

$$[T_A, T_B] = T_{[A,B]}(C). \qquad (14.63)$$

14.2 Classical Lagrangian field dynamics

14.2.1 Spacetime continuum

In the traditional classical mechanics, one parametrizes a system by the coordinates and momenta of pointlike particles. In order to discuss continuous matter or elementary systems, we move to field theory and allow a smooth dependence on the **x** coordinate.

In field theory, one no longer speaks of discrete objects with positions or trajectories (world-lines) $q(t)$ or $x(\tau)$. Rather x, t take on the roles of a ruler or measuring rod, which is positioned and oriented by the elements of the Galilean or Lorentz symmetry groups. Schwinger expresses this by saying that space and time play the role of an abstract measurement apparatus [119], which means that **x** is no longer $q(t)$, the position of an existing particle. It is simply a point along some ruler, or coordinate system, which is used to measure position. The position might be occupied by a particle or by something else; then again, it might not be.

14.2.2 Poisson brackets of fields

The Poisson bracket is not really usable in field theory, but it is instructive to examine its definition as an invariant object. We begin with the relativistic scalar field as the prototype.

The Poisson bracket of two functions X and Y may be written in one of two ways. Since the dynamical variables in continuum field theory are now $\phi_A(x)$

and $\Pi_A(x)$, one obvious definition is the direct transcription of the classical particle result for the canonical field variables, with only an additional integral over all space owing to the functional dependence on **x**. By analogy with Poisson brackets in particle mechanics, the bracketed quantities are evaluated at equal times.

$$[X, Y]_{\phi\Pi} = \int d\sigma_x \left(\frac{\partial X}{\partial \phi_A(x)} \frac{\partial Y}{\partial \Pi_A(x)} - \frac{\partial Y}{\partial \phi_A(x)} \frac{\partial X}{\partial \Pi_A(x)} \right). \quad (14.64)$$

With this definition we have

$$[\phi(x), \Pi(x')]_{\phi\Pi}\bigg|_{t=t'} = \delta(\mathbf{x}, \mathbf{x'})$$

$$\int d\sigma \, [\phi(x), \Pi(x')]_{\phi\Pi}\bigg|_{t=t'} = 1; \quad (14.65)$$

thus, the familiar structure is reproduced. It should be noted, however, that the interpretation of these results is totally different to that for classical particle mechanics. Classically, $q_A(t)$ is the position of the Ath particle as a function of time. $\phi_A(x)$ on the other hand refers to the Ath species of scalar field (representing some unknown particle symmetry, or different discrete states, but there is no inference about localized particles at a definite position and particular time). To think of $\phi(x)$, $\Pi(x)$ as an infinite-dimensional phase space (independent variables at every new value of **x**) is not a directly useful concept. The above form conceals a number of additional subtleties, which are best resolved by abandoning the Hamiltonian variables in favour of a pure description in terms of the field and its Green functions.

It is now possible to define the Poisson bracket using the fields and Green functions, ignoring the Hamiltonian idea of conjugate momentum. In this language, we may write the invariant Poisson bracket in terms of a directional functional derivative, for any two functions X and Y.

$$[X, Y]_\phi \equiv \mathcal{D}_X Y - \mathcal{D}_Y X, \quad (14.66)$$

where

$$\mathcal{D}_X Y \equiv \int (dx) \frac{\delta Y}{\delta A(x)} \lim_{\xi \to 0} \delta_X \phi^A(x), \quad (14.67)$$

and

$$\delta_X \phi_A(x) = \int (dx') G^r_{AB}(x, x') \delta X_B(x'). \quad (14.68)$$

Since we are looking at Lorentz-invariant quantities, there are several possible choices of causal boundary conditions, and we must define the causal nature of the variations. The natural approach is to use a retarded variation by introducing the retarded Green function explicitly in order to connect the source δX_B to the response $\delta_X \phi$. In terms of the small parameter ξ, we may write $\delta X_B = X_{,B} \xi$, or

$$\delta_X \phi_A(x) = \int (dx') G^r_{AB}(x, x') X_{,B} \xi. \tag{14.69}$$

Using this in eqn. (14.66), we obtain, in condensed notation,

$$[X, Y]_\phi = X_{,A} \, G^{AB}_r \, Y_{,B} - Y_{,A} \, G^{AB}_r \, X_{,B}, \tag{14.70}$$

or, in uncondensed notation,

$$[X, Y]_\phi = \int (dx)(dx') \left(\frac{\delta X}{\delta A(x)} \, G^{AB}_r(x, x') \, \frac{\delta Y}{\delta B(x')} \right.$$
$$\left. - \frac{\delta Y}{\delta A(x)} \, G^{AB}_r(x', x) \, \frac{\delta X}{\delta B(x')} \right). \tag{14.71}$$

Now, using eqns. (5.74) and (5.71), we note that

$$2 \tilde{G}_{AB}(x, x') = G^{AB}_r(x, x') - G^{BA}_r(x', x), \tag{14.72}$$

so, re-labelling indices in the second term of eqn. (14.71), we have (condensed)

$$[X, Y]_\phi = 2Y_{,A} \tilde{G}^{AB} X_{,B} \tag{14.73}$$

or (uncondensed)

$$[X, Y]_\phi = 2 \int (dx)(dx') \frac{\delta Y}{\delta \phi_A(x)} (x) \tilde{G}^{AB}(x, x') \frac{\delta X}{\delta \phi_B(x')}. \tag{14.74}$$

The connection between this expression and the operational definition in terms of Hamiltonian variables in eqn. (14.64) is not entirely obvious from this expression, but we hand-wave the reasonableness of the new expression by stretching formalism. From eqn. (5.73), one can write formally

$$\tilde{G}_{AB}(x, x') \Big|_{t=t'} = \delta_{AB} \delta(\mathbf{x}, \mathbf{x}') \frac{1}{\partial_0} \tag{14.75}$$

and thus, hand-wavingly, at equal times,

$$\frac{\delta}{\delta \phi_A} \tilde{G}_{AB} \frac{\delta}{\delta \phi_A} \sim \frac{\delta}{\delta \phi_A} \frac{\delta}{\delta (\partial_0 \phi_A)} \sim \frac{\delta}{\delta \phi_A} \frac{\delta}{\delta \Pi_A}. \tag{14.76}$$

Although we have diverged from a covariant expression in eqn. (14.74) by singling out a spacelike hyper-surface in eqn. (14.76), this occurs in a natural way as a result of the retarded boundary conditions implicit in the causal variations. Manifest covariance of notation cannot alter the fact that time plays a special role in dynamical systems. Clearly, one has

$$
\begin{aligned}
[\phi(x), \Pi(x')]_\phi\Big|_{t=t'} &= \int (\mathrm{d}y)(\mathrm{d}y') \, \frac{\delta\phi(x)}{\delta\phi_A(y)} \tilde{G}_{AB}(y, y') \frac{\delta(\partial_0\phi(x'))}{\delta\phi_B(y')} \\
&= \int (\mathrm{d}y)(\mathrm{d}y') \, \frac{\delta\phi(x)}{\delta\phi_A(y)} - \tilde{\partial}_0 G_{AB}(y, y') \frac{\delta\phi(x')}{\delta\phi_B(y')} \\
&= \delta(\mathbf{x}, \mathbf{x}).
\end{aligned}
\tag{14.77}
$$

The Poisson bracket is only unique if the variables are observable, i.e. if they are invariant quantities.

14.3 Classical statistical mechanics

Statistical mechanics provides a natural point of departure from particle mechanics. Although tethered to classical particle notions in the form of canonical Hamiltonian relations, it seeks to take the limit $N \to \infty$ of infinite numbers of discrete particles. It thereby moves towards a continuum representation of matter, which is a step towards field theory. To understand field theory fully, it is necessary to acknowledge a few of its roots in statistical mechanics. By definition, statistical mechanics is about many-particle systems.

14.3.1 Ensembles and ergodicity

An ensemble is formally a collection of 'identical' systems. The systems in an ensemble are identical in the sense that they contain the same dynamical variables and properties, not in the sense that each system is an exact image of every other with identical values for all its variables (that would be a useless concept). The concept of ensembles is useful for discussing the random or (more correctly) unpredictable development of systems under sufficiently similar conditions. An ensemble is a model for the possible ways in which one system might develop, taking into account a random or unpredictable element. If one takes a snapshot of systems in an ensemble at any time, the outcome could have happened in any of the systems, and may indeed happen in the future in any or all of them if they were allowed to run for a sufficient period of time. Ensembles are used to discuss the process of averaging over possible outcomes.

The ergodic hypothesis claims that the time average of a system is the same as an ensemble average in the limit of large times and large numbers of ensembles. In the limit of infinite time and ensembles, this hypothesis can be proven. The implication is that it does not matter how we choose to define the average properties of a complex (statistical) system, the same results will always be obtained. The ergodic hypothesis is therefore compatible with the continuum hypothesis, but can be expected to fail when one deals with measurably finite times or countably finite ensembles. Much of statistical mechanics and much of quantum theory assumes the truth of this hypothesis.

14.3.2 Expectation values and correlations

Macroscopic observables are the expectation values of variables, averaged over time, or over many similar particle systems (an ensemble). The expectation value of a dynamical variable $X(q, p)$ is defined by the ensemble average. For N particles in a fixed volume V, one has

$$\langle X \rangle_{pq} = \overline{X}(t) = \frac{\int d^N q \, d^N p \, \rho(q, p, t) \, X(q, p, t)}{\int d^N q \, d^N p \, \rho(q, p, t)}, \tag{14.78}$$

where ρ is the density of states in phase space in the fixed volume V. This is sometimes written

$$\langle X \rangle_{pq} = \text{Tr}(\rho X) \tag{14.79}$$

The integral in eqn. (14.78) is interpreted as an ensemble average because it integrates over every possible position and momentum for the particles. All possible outcomes are taken into account because the integral averages over all possible outcomes for the system, which is like averaging over a number of systems in which one (by the rules of chance) expects every possibility to occur randomly.

Suppose one defines the generating or partition functional $Z_{pq}[J(t)]$ by

$$Z_{pq}[J(t)] = \int d^N q \, d^N p \, \rho(q, p, t) \, e^{-\int J_X \, X \, dt'}, \tag{14.80}$$

and the 'transformation function' by

$$W_{pq}[J(t)] = -\ln Z_{pq}[X(t)], \tag{14.81}$$

then the average value of X can be expressed as a functional derivative in the following way:

$$\langle X(t) \rangle = -\frac{\delta W[J(t)]}{\delta J(t)}. \tag{14.82}$$

Similarly, the correlation function is

$$\langle X(t) X(t') \rangle = \frac{\delta^2 W[J(t)]}{\delta J(t) \delta J(t')}. \tag{14.83}$$

Notice how this is essentially the Feynman Green function, providing a link between statistical physics and mechanics through this symmetrical Green function.

14.3.3 Liouville's theorem

An important theorem in statistical mechanics clarifies the definition of time evolution in many-particle systems, where it is impractical to follow the trajectory of every particle. This theorem applies to closed, conservative systems.

A given point in phase space represents a specific state of a many-particle system. The density ρ of points in phase space can itself be thought of as a dynamical variable which deforms with time in such a way that the number of points in an infinitesimal volume element is constant. The overall density of points is constant in time since the number of particles is constant and the volume is a constant, by assumption:

$$\frac{d\rho}{dt} = 0, \tag{14.84}$$

or, equivalently,

$$\frac{\partial \rho}{\partial t} + [\rho, H]_{pq} = 0. \tag{14.85}$$

This last form is an expression of how the local density at a fixed point (q, p) in phase space (a fixed state) varies in time. When a dynamical system is in static equilibrium, the density of states at any point must be a constant, thus

$$[\rho, H]_{pq} = 0. \tag{14.86}$$

In a classical Hamiltonian time development, regions of phase space tend to spread out, distributing themselves over the whole of phase space (this is the essence of ergodicity); Liouville's theorem tells us that they do so in such a way as to occupy the same total volume when the system is in statistical equilibrium.

Another way of looking at this is in terms of the distribution function for the field. If the number of states does not change, as is the case for a free field, then

$$\frac{d}{dt} f(p, x) = 0. \tag{14.87}$$

By the chain-rule we may write

$$\left[\frac{\partial}{\partial t} + \left(\frac{\partial x^i}{\partial t}\right) \partial_i + \left(\frac{\partial p^i}{\partial t}\right) \frac{\partial}{\partial p^i}\right] f(p, x) = 0. \tag{14.88}$$

The rate of change of momentum is just the force. In a charged particle field (plasma) this is the Lorentz force $F_i = q E_i + \epsilon_{ijk} v^j B^k$.

14.3.4 Averaged dynamical variations

Since the expectation value is a simple product-weighted average, Liouville's theorem tells us that the time variation of expectation values is simply the expectation value of the time variation, i.e. these two operations commute because the time derivative of ρ is zero:

$$
\begin{aligned}
\frac{d}{dt} \langle X \rangle_{pq} &= \frac{d}{dt} \text{Tr}(\rho\, X) \\
&= \text{Tr} \frac{d\rho}{dt} X + \text{Tr}\rho \frac{dX}{dt} \\
&= \text{Tr}\rho \frac{dX}{dt} \\
&= \left\langle \frac{dX}{dt} \right\rangle_{pq}.
\end{aligned}
\tag{14.89}
$$

This may also be written as

$$
\frac{d}{dt} \langle X \rangle_{pq} = \left\langle \frac{\partial X}{\partial t} + [X, H] \right\rangle_{pq},
\tag{14.90}
$$

or, more generally for variations, as

$$
\langle \delta_\xi X \rangle_{pq} = \langle [X, G_\xi] \rangle_{pq}.
\tag{14.91}
$$

Again, the similarity to the mechanical theory is striking.

14.4 Quantum mechanics

The discovery of de Broglie waves in electron diffraction experiments by Davisson and Germer (1927) and Thomson (1928) effectively undermined the status of particle coordinates as a fundamental dynamical variable in the quantum theory. The wavelike nature of light and matter cannot be reconciled with discrete labels $q_A(t)$ at the microscopic level. A probabilistic element was necessary to explain quantum mechanics. This is true even of single particles; it is not merely a continuum feature in the limit of large numbers of particles, such as one encounters in statistical mechanics. Instead it was necessary to find a new, more fundamental, description of matter in which both the wavelike properties and impulsive particle properties could be unified. Such a description is only possible by a more careful study of the role played by invariance groups.

Because of the cumbersome nature of the Poisson bracket for continuum theory, continuum theories are not generally described with Poisson algebras. Instead, an equivalent algebra arises naturally from the de Broglie relation $p_\mu = \hbar k_\mu$: namely commutator algebras. The important properties one wishes to preserve are the anti-symmetry of the conjugate pair algebra, which leads to the canonical invariances.

In classical mechanics, $q(t)$ does not transform like a vector under the action of symmetry groups, dynamical or otherwise. A more direct route to the development of the system is obtained by introducing an eigenvector basis in group space which does transform like a vector and which employs operators to extract the dynamical information.

14.4.1 Non-relativistic quantum mechanics in terms of groups and operators

Schrödinger's formulation of quantum mechanics is postulated by starting with the Galilean energy conservation relation

$$E = \frac{\mathbf{p}^2}{2m} + V \tag{14.92}$$

and making the operator replacements $E \rightarrow i\hbar\partial_t$ and $\mathbf{p} \rightarrow -i\hbar\nabla$. The solution of this equation, together with the interpretation of the wavefunction and a specification of boundary conditions, is quantum mechanics. It is interesting nonetheless to examine quantum mechanics as a dynamical system in order to identify its relationship to classical mechanics. The main physical assumptions of quantum mechanics are the de Broglie relation $p_\mu = \hbar k_\mu$ and the interpretation of the wavefunction as a probability amplitude for a given measurement. The remainder is a purely group theoretical framework for exploiting the underlying symmetries and structure.

From a group theoretical viewpoint, quantum mechanics is simpler than classical mechanics, and has a more natural formulation. The use of Poisson brackets to describe field theory is not practical, however. Such a formulation would require intimate knowledge of Green functions and boundary conditions, and would involve complicated functional equations. To some degree, this is the territory of quantum field theory, which is beyond the scope of this work. In the usual approach, canonical invariances are made possible by the introduction of a vector space description of the dynamics. It is based upon the algebra of observables and the method of eigenvalues.

The wavefunction or field Since a particle position cannot be a wave (a particle is by definition a localized object), and neither can its momentum, it is postulated that the wavelike nature of quantum propagation is embodied in a *function of state* for the particle system called the *wavefunction* and that all physical properties (called *observables*) can be extracted from this function by Hermitian operators. The wavefunction $\psi(\mathbf{x}, t)$ is postulated to be a vector in an abstract multi-dimensional Hilbert space, whose magnitude and direction contains all the information about the particle, in much the same way that phase space plays the same role for classical particle trajectories.

The fact that the wavefunction is a vector is very convenient from the point of view of the dynamics (see section 8.1.3), since it means that the generators of invariance groups can operate directly on them by multiplication. This leads to a

Table 14.1. Dynamical formulations.

Classical	Schrödinger	Heisenberg
$q(t)$	$\hat{x}\psi(\mathbf{x}, t)$	$\hat{x}(t)\psi(\mathbf{x})$
$p(t)$	$\hat{p}\psi(\mathbf{x}, t)$	$\hat{p}(t)\psi(\mathbf{x})$

closer connection with the group theory that explains so much of the dynamics. It means that any change in the system, characterized by a group transformation U, can be expressed in the operational form

$$\psi' = U\,\psi. \tag{14.93}$$

This is much simpler than the pair of equations in (14.34). It is, in fact, more closely related to eqn. (14.35) because of the group structure in eqn. (14.63), as we shall see below.

Operator-valued position and momentum $q(t)$ and $p(t)$ may be effectively supplanted as the dynamical variables by the wavefunction. To represent the position and momentum of particles, one makes a correspondence with operators according to one of two equivalent prescriptions (table 14.1). The choice depends on whether one wishes to place the time development of the system in the definition of the operators, or whether it should be placed in the wavefunction, along with all the other dynamical parameters. These two descriptions are equivalent to one another in virtue of the group combination law. We shall mainly use the Schrödinger representation here since this is more in tune with the group theoretical ideology of symmetries and generators.

As explained in section 11.1, it is the operators themselves, for dimensional reasons, which are the positions and momenta, not the operators multiplying the fields. The observable values which correspond to the classical quantities are extracted from this function by considering the eigenvalues of the operators. Since the wavefunction $\psi(x)$ can always be written as a linear combination of the complete set of eigenvectors $E(x)$ belonging to any operator on Hilbert space, with constants λ_a,

$$\psi(x) = \sum_a \lambda_a E_a(x), \tag{14.94}$$

there is always a well defined eigenvalue problem which can convert a Hermitian operator into a real eigenvalue.

Commutation relations Since the field Poisson bracket is unhelpful, we look for a representation of position and momentum which distills the important property in eqn. (14.57) from the classical canonical theory and injects it into the quantum theory. One sees that, on choosing the following algebraic representation of the underlying Galilean symmetry group for the wavefunction[5]

$$\psi(x) = \sum_k a_k e^{i(\mathbf{k} \cdot \mathbf{x} - \omega t)}, \qquad (14.95)$$

a simple representation of the operators $\hat{\mathbf{x}}$ and $\hat{\mathbf{p}}$ may be constructed from

$$\hat{\mathbf{x}} = \mathbf{x}$$
$$\hat{\mathbf{p}} = -i\hbar\nabla. \qquad (14.96)$$

These operators live on the vector space of the Galilean group (i.e. real space), so it is natural to use their operator property directly in forming a canonical algebra. They are complete, as may be verified by computing the straightforward commutator

$$[\hat{\mathbf{x}}, \hat{\mathbf{p}}] = \hat{\mathbf{x}}\hat{\mathbf{p}} - \hat{\mathbf{p}}\hat{\mathbf{x}} = i\hbar. \qquad (14.97)$$

This clearly satisfies eqn. (14.57). Thus, with this representation of position and momentum, based directly on the underlying symmetry of spacetime, there is no need to introduce an abstract phase space in order to construct a set of vectors spanning the dynamics. The representations of the Galilean group suffice. The only contribution from empirical quantum theory is the expression of the wavenumber $\mathbf{k}$ and the frequency ω in terms of the de Broglie relation. In fact, this cancels from eqn. (14.97).

Dirac notation: bases In Dirac notation, Hilbert space vectors are usually written using angle brackets ($|x\rangle$, $|\psi\rangle$). To avoid confusing this notation with that for expectation values, used earlier, we shall use another fairly common notation, $|x)$, $|\psi)$, here. The components of such a vector are defined with respect to a particular basis. Adjoint vectors are written $(x|$, $(\psi|$, and so on.

The scalar product of two such vectors is independent of the basis, and is written

$$(\psi_1|\psi_2) = \int d\sigma \, \psi_1^\dagger(x)\psi_2(x)$$
$$= \int (dp)\psi_1^\dagger(p)\psi_2(p). \qquad (14.98)$$

In Dirac notation one considers the functional dependence of wavefunctions to be the basis in which they are defined. Thus, $\psi(x)$ is likened to the components

---

[5] Note that the representations of the Galilean group are simply the Fourier expansion.

Table 14.2. Matrix elements and operator bases.

$\hat{O}$	$(x'|\hat{O}|x)$	$(p'|\hat{O}|p)$
$\hat{\mathbf{p}}$	$-i\hbar\nabla\,\delta\,(\mathbf{x}, \mathbf{x}')$	$\mathbf{p}\,\delta(\mathbf{p}, \mathbf{p}')$
$\hat{\mathbf{x}}$	$\mathbf{x}\delta(\mathbf{x}, \mathbf{x}')$	$-i\hbar\frac{\partial}{\partial\mathbf{p}}\delta(\mathbf{p}, \mathbf{p}')$

of the general function ψ in an x basis. Similarly, the Fourier transform $\psi(p)$ is thought of as the components of ψ in a p basis. As in regular geometry, the components of a vector are obtained by taking the scalar product of the vector with a basis vector. In Dirac notation, the wavefunction and its Fourier transform are therefore written as

$$\psi(x) = (x|\psi)$$
$$\psi(p) = (p|\psi), \tag{14.99}$$

as a projection of the vector onto the basis vectors. The basis vectors $|x)$ and $|p)$ form a complete set of eigenstates of their respective operators, satisfying the completeness relation

$$(\mathbf{x}, \mathbf{x}') = \delta(\mathbf{x}, \mathbf{x}'). \tag{14.100}$$

Similarly, a matrix, or operator is also defined by an outer product according to what basis, or type of variable, it operates on. The identity operator in a basis x is

$$\hat{I} = \int d\sigma_x |x)(x|, \tag{14.101}$$

and similarly, in an arbitrary basis ξ, one has

$$\hat{I} = \int d\sigma_\xi |\xi)(\xi|. \tag{14.102}$$

See table 14.2. This makes the scalar product basis-independent:

$$(\psi_1|\psi_2) = \int d\sigma_x\,(\psi_1|x)(x|\psi_2), \tag{14.103}$$

as well as the expectation value of $\hat{O}$ with respect to the state $|\psi)$:

$$(\psi|\hat{O}|\psi) = \int d\sigma_x d\sigma_{x'}\,(\psi|x')(x'|\hat{O}|x)(x|\psi), \tag{14.104}$$

Transformation function The scalar product $(\psi_1|\psi_2)$ represents an overlap of one state of the system with another; thus, the implication is that transitions or transformations from one state to another can take place between these two states. $(\psi_2(x_2)|\psi_1(x_1))$ is often called the transformation function. It represents the probability amplitude of a transition from $\psi_1(x_1)$ to $\psi_2(x_2)$. The quantity

$$A = (\psi'|\hat{O}|\psi) \tag{14.105}$$

is not an expectation value, since it refers to two separate states; rather, it is to be interpreted as another transition amplitude, perturbed by the operation $\hat{O}$, since

$$\hat{O}|\psi) = |\psi''). \tag{14.106}$$

Thus

$$A = (\psi'|\psi''), \tag{14.107}$$

which is just another transition function. The transformation function plays a central role in Schwinger's action principle for quantum mechanics, and is closely related to the path integral formulation.

Operator variations and unitary transformations In order to define a variational theory of quantum mechanics, meaning must be assigned to the variation of an operator. An operator has no meaning without a set of vectors on which to operate, so the notion of an operator variation must be tied to changes in the states on which it operates. States change when they are multiplied by the elements of a transformation group U:

$$|\psi) \rightarrow U|\psi). \tag{14.108}$$

Similarly, the adjoint transforms by

$$(\psi| \rightarrow (\psi|U^\dagger. \tag{14.109}$$

The invariance of the scalar product $(\psi|\psi)$ implies that U must be a unitary transformation, satisfying

$$U^\dagger = U^{-1}. \tag{14.110}$$

Consider an infinitesimal unitary transformation with generator G such that $U = \exp(-iG/\hbar)$.

$$|\psi) \rightarrow e^{-iG/\hbar}|\psi) = (1 - iG/\hbar)|\psi). \tag{14.111}$$

The change in an expectation value due to an operator variation $\hat{X} \rightarrow \hat{X} + \delta\hat{X}$,

$$(\psi|\hat{X} + \delta\hat{X}|\psi) = (\psi|e^{iG/\hbar}\hat{X}e^{-i/\hbar G}|\psi)$$
$$= (\psi|(1 + iG/\hbar)\hat{X}(1 - iG/\hbar)|\psi), \tag{14.112}$$

or, equating $\delta \hat{X}$ to the first infinitesimal order on the right hand side,

$$\delta \hat{X} = \frac{1}{i\hbar}[\hat{X}, G]. \tag{14.113}$$

Eqn. (14.113) may be taken as the definition of operator variations, affected by unitary transformations. It can be compared with eqn. (14.35) for the canonical variations. It is this definition which permits an action principle to be constructed for quantum mechanics. From eqn. (14.113), one can define the expectation value

$$i\hbar \langle \delta X \rangle = \langle \xi | [X, G] | \xi \rangle, \tag{14.114}$$

and, by a now familiar argument for the time variation $G_t = -H\delta t$,

$$\frac{\mathrm{d}}{\mathrm{d}t}\langle X \rangle = \left\langle \frac{\partial X}{\partial t} + \frac{1}{i\hbar}[X, H] \right\rangle, \tag{14.115}$$

where the expectation value is interpreted with respect to a basis ξ in Hilbert space:

$$\langle \ldots \rangle = \int \mathrm{d}\xi \, (\xi | \ldots | \xi). \tag{14.116}$$

This relation can be compared with eqn. (14.90) from classical statistical mechanics.

It is straightforward to verify Hamilton's equations for the operators by taking

$$\hat{H} = -\frac{\hbar^2}{2m}\nabla^2 + V, \tag{14.117}$$

so that

$$\dot{\hat{p}} = \frac{1}{i\hbar}[\hat{p}, \hat{H}] = -\nabla \hat{H}$$

$$\dot{\hat{q}} = \frac{1}{i\hbar}[\hat{q}, \hat{H}] = i\hbar \left[-\frac{\nabla^2}{2m}\hat{x} \right] = -i\hbar \frac{\nabla}{m} = \frac{\hat{p}}{m} = \frac{\partial \hat{H}}{\partial \hat{p}}. \tag{14.118}$$

The last step here is formal, since the derivation with respect to an operator is really a distribution or Green function, which includes a specification of boundary conditions. In this case, the only possibility is for causal, retarded, boundary conditions, and thus the expression is unambiguous.

Statistical interpretation By comparing quantum expectation values, or scalar products, with statistical mechanics, it is possible to see that the states referred to in quantum mechanics must have a probabilistic interpretation. This follows directly from the canonical structure and from the analogy between the quantum state function and the density operator in eqn. (14.78).

If it were not already clear from phenomenology, it should now be clear from eqn. (14.37) that the quantum theory has the form of a statistical theory. Thus, the wavefunction can be regarded as a probabilistic object, and the waves which give rise to quantum interference are 'probability waves'.

The basis-independence of the quantum expectation value is analogous to the ergodicity property of classical mechanics: it implies that it is not important what variable one chooses to average over. A 'dynamically complete' average will lead to the same result.

The formalism of quantum theory makes no statements about wave–particle duality and no confusion arises with regard to this concept. Quantum mechanics must be regarded as a straightforward generalization of classical canonical mechanics, which admits a greater freedom of expression in terms of group theory and invariances.

Classical correspondence Although sometimes stated misleadingly, the correspondence between the Poisson bracket in classical mechanics and the commutator in quantum mechanics is not such that one recovers the Poisson bracket formulation from the classical limit of the commutator. They are completely independent, since they refer to different spaces. While the commutator function exists in the classical limit $\hbar \to 0$, the wavefunction does not, since $\mathbf{k} \to \infty$ and $\omega \to \infty$. Thus, the basis vectors cease to exist.

The true correspondence with classical physics is through expectation values of quantum operators, since these are independent of the operator basis. The classical theory is through the equations

$$\left(\psi \left| -\frac{\hbar^2}{2m} \nabla^2 + V = i\hbar \frac{\partial}{\partial t} \right| \psi \right) \to \frac{p^2}{2m} + V = E, \qquad (14.119)$$

and

$$\left\langle \frac{\mathrm{d}\hat{\mathbf{p}}}{\mathrm{d}t} \right\rangle = \frac{\mathrm{d}\langle \hat{\mathbf{p}} \rangle}{\mathrm{d}t}$$

$$= \frac{-i}{\hbar} \langle [\hat{\mathbf{p}}, H] \rangle$$

$$= \frac{-i}{\hbar} (\hat{\mathbf{p}} V(x) - V(x)\hat{\mathbf{p}})$$

$$= \frac{-i}{\hbar} (-i\hbar \nabla V(x))$$

$$= -\nabla V(x)), \qquad (14.120)$$

which is Newton's law.

14.4.2 Quantum mechanical action principle

Schwinger has shown that the complete unitary, dynamical structure of quantum mechanics can be derived from a quantum action principle, based on operator variations. The quantum action principle is directly analogous to its classical counterpart. We shall return to this quantum action principle in chapter 15 since it plays a central role in modern quantum field theory. For now, the action principle will not be proven; instead we summarize the main results. The algebraic similarities to the classical action principle are quite remarkable.

The central object in the quantum theory is the transformation function or transition amplitude $(\psi|\psi')$. The quantum action principle states that the action is a generating functional, which induces changes in the transformation function,

$$\delta(\psi(t_2)|\psi(t_1)) = \frac{1}{i\hbar}(\psi(t_2)|\delta\hat{S}_{12}|\psi(t_1)), \qquad (14.121)$$

where $\hat{S}$ is the action operator, which is constructed from the classical action by replacing each dynamical object by its operator counterpart:

$$\hat{S}_{12} = \int_{t_1}^{t_2} dt \left(\frac{1}{2}(\hat{\mathbf{p}}\dot{\hat{\mathbf{q}}} - \hat{\mathbf{q}}\dot{\hat{\mathbf{p}}}) - \hat{H}\right). \qquad (14.122)$$

In this simple case, the ordering of the operators is unambiguous. The variation in the action contains contributions only from within the time values corresponding to the end-points of the transformation function for causality.

If one now introduces the identity $I = |x) \times (x|$ into the transformation function and substitutes the real-space representations of the operators, eqn. (14.121) becomes

$$\delta(\psi(t_2)|\psi(t_1)) =$$
$$\frac{1}{i\hbar}\delta\int_{t_1}^{t_2}(dx)\psi^\dagger(x)\left[\frac{1}{2}(-i\hbar\nabla\dot{\mathbf{x}} + i\hbar\dot{\mathbf{x}}\nabla) + \frac{\hbar^2}{2m}\nabla^2 - V\right]\psi(x),$$
$$(14.123)$$

which is equal to

$$\delta(\psi(t_2)|\psi(t_1)) =$$

$$\frac{1}{i\hbar}\delta\int_{t_1}^{t_2}(dx)\,\psi^\dagger(x)\left[-\frac{i\hbar}{2}\left(\overrightarrow{\partial_t} - \overleftarrow{\partial_t}\right) + \frac{\hbar^2}{2m}\nabla^2 - V\right]\psi(x). \quad (14.124)$$

δ refers only to the contents of the square brackets. This expression may be compared with the action for the Schrödinger field in eqn. (17.4). For the purposes of computing the variation, the form in eqn. (14.122) is most convenient. In fact, it is easy to see that – given that the variation of an operator is well defined – the results are exactly analogous to the classical case.

If the end-points of the variation satisfy fixed boundary conditions, then

$$\delta(\psi(t_2)|\psi(t_1)) = 0, \tag{14.125}$$

since the field is constrained to admit no unitary transformations there, thus the right hand side of eqn. (14.121) is also zero. This, in turn, implies that the variation of the action operator vanishes, and this leads to the *operator equations of motion*, analogous to Hamilton's equations:

$$\delta_{\hat{\mathbf{x}}}\hat{S} = \int dt \left(-\dot{\hat{\mathbf{p}}} - \frac{\partial \hat{H}}{\partial \mathbf{x}}\right)\delta\mathbf{x}, \tag{14.126}$$

whence

$$\dot{\hat{\mathbf{p}}} = -\frac{\partial \hat{H}}{\partial \mathbf{x}}. \tag{14.127}$$

Similarly, the variation with respect to the momentum operator leads to

$$\dot{\hat{\mathbf{x}}} = \frac{\partial \hat{H}}{\partial \mathbf{p}}, \tag{14.128}$$

whose consistency was verified in eqns. (14.118). This tells us that quantum mechanics, with commutators in place of Poisson brackets and differential operators acting on a Hilbert space, forms a well defined Hamiltonian system. Eqn. (14.124) shows that this is compatible with Schrödinger field theory. The final piece of the puzzle is to generalize the variations of the action to include non-fixed end-points, in a way analogous to that in section 14.1.7. Then, using the equations of motion to set the bulk terms to zero, one has

$$\delta(\psi(t_2)|\psi(t_1)) = \frac{1}{i\hbar}(\psi(t_2)|G_2 - G_1|\psi(t_1)), \tag{14.129}$$

which shows that the extended variation merely induces an infinitesimal unitary transformation at the end-points of the variation. This variation is in accord with eqn. (14.113), and one may verify that

$$\delta\hat{\mathbf{x}} = \frac{1}{i\hbar}[\hat{\mathbf{x}}, G_{\mathbf{x}}]$$

$$= \frac{1}{i\hbar}[\hat{\mathbf{x}}, \hat{\mathbf{p}}\delta\hat{\mathbf{x}}], \tag{14.130}$$

which immediately gives the fundamental commutation relations

$$[\hat{\mathbf{x}}, \hat{\mathbf{p}}] = i\hbar. \tag{14.131}$$

This final piece of the puzzle verifies that the operator variational principle is self-consistent for quantum mechanics. In fact, it can be generalized to other operators too, as we shall see in chapter 15, when we consider the fully quantized theory of fields.

14.4.3 Relativistic quantum mechanics

A Lorentz-invariant theory of quantum mechanics may be obtained by repeating the previous construction for the non-relativistic theory, replacing the non-relativistic energy relation in eqn. (14.92) with

$$E^2 = \mathbf{p}^2 c^2 + m^2 c^4. \tag{14.132}$$

One writes

$$(-\hat{E}^2 + \hat{\mathbf{p}}^2 c^2 + m^2 c^4)\phi(x) = 0, \tag{14.133}$$

where $\hat{E} = i\hbar \partial_t$ and $\hat{\mathbf{p}} = -i\hbar \nabla$, and we call the field $\phi(x)$ to distinguish it from the non-relativistic field. This leads us directly to the Klein–Gordon equation

$$(-\hbar^2 c^2 \Box + m^2 c^4)\phi = 0. \tag{14.134}$$

However, all is not straightforward. The interpretation of this equation is full of subtleties, which leads inexorably to a full quantum field theory. To begin with its quadratic nature implies that it has solutions of both arbitrarily large positive and negative energy (see section 5.1.3). This further implies that the conserved quantities normally used to define probability measures can also be negative; this is difficult to interpret. Ultimately, the assumptions of quantum field theory save the relativistic formulation. Leaning on these, relativistic quantum mechanics survives as an approximation to the more complete quantum field theory under conditions of 'sufficient stability'.[6]

State vectors and wavefunctions In non-relativistic quantum mechanics it was easy to choose state vectors satisfying the Schrödinger equation because of the simple form of the conserved quantities arising from the linear time derivative (see eqn. (12.39)). The structural symmetry of the natural inner product:

$$(\psi_1, \psi_2) = \int d\sigma \ \psi_1^\dagger \psi_2, \tag{14.135}$$

means that the state vectors $|\psi_1\rangle$ and the adjoint $\langle\psi_1|$ were simply Hermitian conjugates of one another. In the case of the Klein–Gordon equation, matters are more complicated. The corresponding invariant inner product is

$$(\phi_1, \phi_2) = -i\hbar c^2 \int d\sigma \ \phi_1^* \overset{\leftrightarrow}{\partial_0} \phi_2 \tag{14.136}$$

---

[6] To make this woolly statement precise, one needs to address issues in the language of quantum field theory or renormalization group, which we shall only touch on in chapter 15.

the symmetry of which is made less obvious by the time derivative, and one is now faced with both positive and negative energy solutions. These two sets of solutions decouple, however. If one splits the field into its positive and negative energy parts,

$$\phi(x) = \phi^{(+)}(x) + \phi^{(-)}(x), \tag{14.137}$$

then one has, for a real scalar field,

$$
\begin{aligned}
(\phi(x), \phi(x)) &= (\phi^{(+)}(x), \phi^{(+)}(x)) + (\phi^{(-)}(x), \phi^{(-)}(x)) \\
&= 0;
\end{aligned} \tag{14.138}
$$

i.e.

$$
\begin{aligned}
(\phi^{(+)}(x), \phi^{(+)}(x)) &= -(\phi^{(-)}(x), \phi^{(-)}(x)) \\
(\phi^{(+)}(x), \phi^{(-)}(x)) &= 0.
\end{aligned} \tag{14.139}
$$

or, more generally,

$$(\phi_A, \phi_B) = -(\phi_B, \phi_A)^*. \tag{14.140}$$

By analogy with the non-relativistic case, we wish to view this scalar product as the definition of a vector space with vectors $|\phi)$ and adjoint vectors $(\phi|$, such that

$$(\phi_1|\phi_2) = (\phi_1, \phi_2), \tag{14.141}$$

i.e. the inner product on the vector space is identified with the conserved quantity for the field. The ϕ_A satisfy the Klein–Gordon equation:

$$
\begin{aligned}
\phi(x) &= \int \frac{d^{n+1}k}{(2\pi)^{n+1}} e^{ikx} \phi(k) \delta(p^2 c^2 + m^2 c^4) \\
&= \int \frac{d^n k}{(2\pi)^n} \frac{1}{2k_0 c^2 \hbar^2} e^{ikx} (\phi(p_0, \mathbf{p}) + \phi(-p_0, \mathbf{p})) \\
&= \int dV_k e^{ikx} (\phi^{(+)}(\mathbf{p}) + \phi^{(-)}(\mathbf{p})).
\end{aligned} \tag{14.142}
$$

What makes the relativistic situation different is the fact that the energy constraint surface is quadratic in k_0. The volume measure on momentum space is constrained by this energy relation. This is the so-called mass shell. On the manifold of only positive energy solutions, the volume measure is

$$
\begin{aligned}
V_k &= \int \frac{d^{n+1}k}{(2\pi)^{n+1}} \delta(p^2 c^2 + m^2 c^4) \theta(k_0) \\
&= \int \frac{d^n k}{(2\pi)^n} \frac{1}{2k_0 c^2 \hbar^2} \\
dV_k &= \frac{d^n k}{(2\pi)^n} \frac{1}{2k_0 c^2 \hbar^2}.
\end{aligned} \tag{14.143}
$$

Thus, if we examine complete sets of position and momentum eigenfunctions on this constraint manifold, we find that the normalization of momentum eigenfunctions is dictated by this constraint:

$$
(\mathbf{x}, \mathbf{x}') \equiv \delta(\mathbf{x}, \mathbf{x}')
$$

$$
= \int \frac{d^n k}{(2\pi)^n} e^{i\hat{\mathbf{k}} \cdot (\mathbf{x} - \mathbf{x}')}
$$

$$
= 2k_0 \hbar^2 c^2 \int dV_k e^{i\hat{\mathbf{k}} \cdot (\mathbf{x} - \mathbf{x}')}. \tag{14.144}
$$

From this expression, it follows that

$$
(\hat{\mathbf{k}} | \mathbf{x}) = \sqrt{2k_0 \hbar^2 c^2}\, e^{i\hat{\mathbf{k}} \cdot (\mathbf{x})} \tag{14.145}
$$

$$
(\hat{\mathbf{k}} | \hat{\mathbf{k}}') = \int d\sigma\, (\hat{\mathbf{k}} | \mathbf{x})(\mathbf{x} | \hat{\mathbf{k}}') = 2k_0 \hbar^2 c^2\, \delta(\hat{\mathbf{k}} - \hat{\mathbf{k}}'). \tag{14.146}
$$

Thus the one-particle positive energy wavefunction is

$$
\psi \equiv \phi_1(x) = (\mathbf{x}, \phi) = \int dV_k (\mathbf{x} | \hat{\mathbf{k}})(\hat{\mathbf{k}} | \phi)
$$

$$
= N \int dV_k \sqrt{2k_0 \hbar^2 c^2}\, e^{i\hat{\mathbf{k}} \cdot \mathbf{x}}. \tag{14.147}
$$

Compare this with the re-scaling in eqn. (13.7). It is normalized such that

$$
(\psi, \psi) = (\phi_1 | \phi_1) = 1
$$

$$
= N^2 \int dV_k dV_{k'} \phi_1^*(k)\, (2k_0 \hbar^2 c^2)\, \phi_1(k) \delta(\hat{\mathbf{k}} - \hat{\mathbf{k}}')
$$

$$
= N^2 \int \frac{d^{n+1} k}{(2\pi)^{n+1}} |\phi_1(k)|^2. \tag{14.148}
$$

The normalization factor, N, is fixed, at least in principle, by this relation, perhaps through box normalization. This inner product is unambiguously positive, owing to the restriction to only positive energies. An example is the one-particle wavefunction in $3 + 1$ dimensions:

$$
\psi = \phi_1(x) = N \int \frac{d^3 k}{(2\pi)^3} \frac{e^{i\hat{\mathbf{k}} \cdot \mathbf{x}}}{\sqrt{2k_0 \hbar^2 c^2}}
$$

$$
= \text{const.} \left(\frac{m}{\mathbf{x}}\right)^{\frac{5}{4}} H_{\frac{5}{4}}^{(1)}(im\mathbf{x}), \tag{14.149}
$$

where $H_{\frac{5}{4}}^{(1)}(x)$ is a Hankel (Bessel) function. What is significant here is that the one-particle wavefunction is not localized like a delta function. Indeed, it would be impossible to construct a delta function from purely positive energy functions. Rather, it extends in space, decaying over a distance of order $\hbar/mc^2$. See also section 11.2.

14.5 Canonically complete theories

The operational view of classical, statistical and quantum mechanics, which has been lain out above, could seem sterile from a physical perspective. In presenting it as a formal system of canonical equations, one eschews phenomenology entirely and uses only elementary notions based on symmetry. That such an approach is possible is surely an important insight. Mechanics should be regarded for what it is: a description of dynamics in terms of algebraic rules determined from necessary symmetries. Given the mathematical structure, more physical or philosophical discussions can follow of their own accord.

The Hamiltonian dynamical formulation can, for the most part, be circumvented completely by direct use of the action formalism in chapter 4. Again, we use a version of the action principle in which we allow infinitesimal canonical changes at the end-points of dynamical variations.

The quantum theory, being linear, is essentially a theory of small disturbances. The imprint left on the action by variation with respect to some variable is that variable's conjugate quantity. The conjugate quantity is said to be the generator of the variation of disturbance. If one varies the action with respect to a set of parameters ξ^i, and the action is invariant under changes of these parameters, the variation must be zero. Manipulating the symbols in the action and separating out the variation $\delta\xi$ to first order, one can write the infinitesimal variation in the form

$$\delta_\xi S = \int d\sigma^\mu \; G_i \delta\xi^i = 0, \tag{14.150}$$

where $d\sigma^\mu$ represents a spacelike hyper-surface. The quantity G_i is the *generator* of the symmetry in ξ^i. It is also called the variable conjugate to ξ^i. Notice that an external source J_{ext}, such that

$$S \rightarrow S + \int (dx) \; J_{\text{ext}}\phi \tag{14.151}$$

acts as a generator for the field, throughout the spacetime volume

$$\delta S \rightarrow 0 + \int (dx) \; J_{\text{ext}}\delta\phi, \tag{14.152}$$

since the dynamical variation of the regular action vanishes. This observation has prompted Schwinger to develop the quantum theory of fields almost entirely with the aid of 'sources' or generators for the different dynamical and symmetrical entities [119] (see table 14.3).

In this chapter, we have compared the way in which classical and quantum mechanics are derivable from action principles and obey canonical completeness relations, in the form of Poisson brackets or commutators. This is no accident. Since the action principle always generates a conjugate momentum, as exhibited

Table 14.3.   Some canonical transformations.

T_i	$\delta \xi^i$	Symmetry
$\int \mathrm{d}\sigma^\mu \theta_{\mu\nu}$	δx^ν	Lorentz invariance
$\int \mathrm{d}\sigma^\mu T_{\mu\nu}$	δx^ν	conformal invariance
$\mathbf{p}$	$\delta \mathbf{x}$	translational invariance
H	δt	time translation invariance
Π	$\delta q, \delta\phi, \delta\psi$	spacetime/canonical
J_{ext}	$\delta\phi, \delta\psi$	field canonical/unitary

by eqns. (4.62) and (4.66), one could *define* a canonical theory to be one which follows from an action principle. Thus, the action principle is always a good starting point for constructing a dynamical model of nature. To complete our picture of dynamics, it is necessary to resolve some of the problems which haunt the fringes of the classical relativistic theory. To do this, one must extend the dynamical content of the theory even more to allow the fields themselves to form a dynamical basis. As we saw in section 14.2.2, this was impractical using the Poisson bracket, so instead one looks for a commutator relation to generate infinitesimal variations. This completes the consistency of the formalism for relativistic fields.

15

Epilogue: quantum field theory

Where does this description of matter and radiation need to go next? The answer is that it needs to include interactions between different physical fields and between different excitations in the same field.

In order to pursue this course, one needs to extend the simple linear response analysis of classical field theory to a non-linear response analysis. In the presence of interactions, linear response is only a first-order approximation to the response of the field to a source. Interactions turn the fields themselves into sources (sources for the fields themselves and others to scatter from). Non-linear response theory requires quantum field theory, because the products of fields, which arise in interactions, bring up the issue of the ordering of fields, which only the second quantization can resolve. It means the possibility of creation and annihilation of particle pairs and negative probabilities, which the anti-particle concept and the vacuum concept repair the consistency.

An area which has not been touched upon in this book is that of Grassman variables [136], which describe fermionic interactions. These arose fleetingly in connection with TCP invariance (see section 10.5). In interacting theories, one needs to account for their anti-commuting properties.

A full discussion of quantum field theory, and all of its computational algorithms, is far beyond the scope of this book. The purpose of this chapter is only to indicate briefly how the quantum theory of fields leads to more of the same. As usual, the most pleasing way to derive corrections to the classical theory within a dynamically complete framework is, of course, through an action principle. Schwinger's generalization of the action principle, for quantum field theory, provides the most economical and elegant transition from the classical to the quantum universe. It leads, amongst other things, to the so-called *effective action* for quantum field theory. This effective action demonstrates, most forcibly, the way in which quantum field theory completes the cause–effect response theory we have used in previous chapters.

15.1 Classical loose ends

In classical physics, few problems may be described by single-particle equations. In many-particle theories, one normally invokes the continuum hypothesis and turns to effective equations in order to study bulk quantities. The same strategy is more difficult in the quantum theory, since an effective theory of the quantum nature is less obvious. Quantum field theory is a theory of multiple quanta, or discrete excitations within a field. More than that, it is a theory of multiple *identical* quanta. This is helpful in overcoming some of the problems with classical field theory.

Quantum mechanics deals primarily with one-particle systems. In quantum mechanics, many-particle systems must be handled as discrete N-body problems. Identical particles must be handled by cumbersome Slater determinants, or explicit symmetrization. In quantum field theory one replaces this with a continuum theory of field operators, subject to algebraic constraints.

It is this last point which leads one to modify quantum mechanics. Instead of trying to symmetrize over wavefunctions for identical particles, one uses the normalization properties of the wavefunction to generate the required multiplicity. The identical nature of the particles then follows, subject to certain restrictions on the spacetime symmetries of the fields. In particular, it is necessary to specify the topology of the field with respect to interchanges of particles. The Pauli principle, in particular, places a strong constraint on the spacetime properties of the field.

Finally, the existence of negative energy states requires additional assumptions to avoid the problem of decay. The anti-particle concept and the vacuum concept (the existence of a lowest possible state) have a formal expression in quantum field theory, but have to be put in by hand in a classical theory.

15.2 Quantum action principle

Schwinger has introduced an action principle for quantum mechanics [115, 120] which turns out to be equivalent to path integral formulations [47]. In quantum field theory one is interested in computing transition or scattering amplitudes of the form

$$\langle s_2 | s_1 \rangle, \tag{15.1}$$

where the states denoted by s_1 and s_2 are assumed to be complete at each arbitrary time, so that

$$\langle s_1 | s_2 \rangle = \int \langle s_1, t_1 | \alpha, t \rangle \, d\alpha \, \langle \alpha, t | s_2, t_2 \rangle. \tag{15.2}$$

Schwinger's quantum action principle states that

$$\delta \langle s_2 | s_1 \rangle = \frac{i}{\hbar} \langle s_2 | \delta S_{12} | s_1 \rangle \tag{15.3}$$

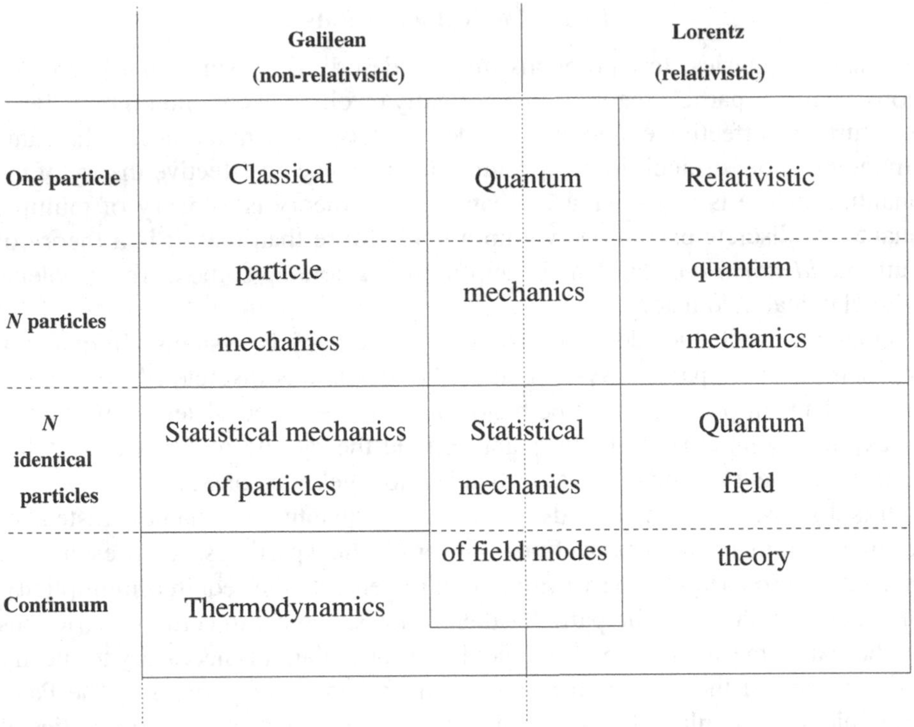

	Galilean (non-relativistic)		Lorentz (relativistic)
One particle	Classical	Quantum	Relativistic
N particles	particle mechanics	mechanics	quantum mechanics
N identical particles	Statistical mechanics of particles	Statistical mechanics	Quantum field
Continuum	Thermodynamics	of field modes	theory

Fig. 15.1. Overlap between classical and quantum and statistical theories.

where S is the action operator, obtained by replacing the classical fields ϕ by field operators ϕ. The form of the action is otherwise the same as in the classical theory (which is the great bonus of this formulation). Specifically,

$$S_{12} = \int_{\sigma_1}^{\sigma_2} (\mathrm{d}x)\, \mathcal{L}. \tag{15.4}$$

Since operators do not necessarily commute, one must adopt a specific operator-ordering prescription which makes the action self-adjoint

$$S^{\dagger} = S. \tag{15.5}$$

The action should also be symmetrical with respect to time reversals, as in the classical theory. Central to quantum field theory is the idea of 'unitarity'. This ensures the reversibility of physical laws and the conservation of energy. In view of the property expressed by eqn. (15.2), successive variations of the amplitude with respect to a source,

$$S \to S - \int (\mathrm{d}x)\, J\phi, \tag{15.6}$$

lead automatically to a time ordering of the operators:

$$\frac{\delta}{\delta J(x)} \langle s_2 | s_1 \rangle = \frac{i}{\hbar} \langle s_2 | \phi(x) | s_1 \rangle$$

$$= \int \frac{i}{\hbar} \langle s_2 | \alpha, x \rangle d\alpha \langle \alpha, x | \phi(x) | s_1 \rangle$$

$$\frac{\delta^2}{\delta J(x') \delta J(x)} \langle s_2 | s_1 \rangle = \frac{\delta}{\delta J(x')} \langle s_2 | x \rangle \frac{i}{\hbar} \langle x | \phi(x) | s_1 \rangle$$

$$= \int \left(\frac{i}{\hbar}\right)^2 \langle s_2 | \phi(x') | \alpha, x \rangle d\alpha \langle \alpha, x | \phi(x) | s_1 \rangle$$

$$= \left(\frac{i}{\hbar}\right)^2 \langle s_2 | T\phi(x')\phi(x) | s_1 \rangle, \tag{15.7}$$

where the T represents time ordering of the field operators. The classical limit of the action principle is taken by letting $\hbar \to 0$, from which we obtain $\delta S = 0$. Thus, only the operator equations of motion survive. The amplitude is suppressed, and thus so are the states. This makes the operator nature of the equations of motion unimportant.

15.2.1 Operator variations

The objects of variation, the fields, are now operators in this formulation, so we need to know what the variation of an operator means. As usual, this can be derived from the differential generating structure of the action principle.

It is useful to distinguish between two kinds of variation: variations which lead to unitary transformations of the field on a spacelike hyper-surface, and variations which are dynamical, or are orthogonal to, such a hyper-surface. Suppose we consider an infinitesimal change in the state $|s_1\rangle$, with generator G,

$$|s_2\rangle \to |s_2\rangle + \delta|s_2\rangle = (1 - iG)|s_2\rangle, \tag{15.8}$$

where G is a generator of infinitesimal unitary transformations $U = e^{iG}$, such that

$$U^\dagger U = 1. \tag{15.9}$$

Note that the transformation is the first term in an expansion of e^{-iG}. Then we have

$$\delta|a\rangle = -iG|a\rangle$$
$$\delta\langle a| = \langle a|iG. \tag{15.10}$$

So if F is any unitary operator, then the change in its value under this unitary variation can be thought of as being due to the change in the states, as a result of the unitary generator G:

$$\langle a|F'|b\rangle = \langle a'|(1 + iG)F(1 - iG)|b'\rangle + \cdots, \qquad (15.11)$$

which is the first terms in the infinitesimal expansion of

$$\langle a|F'|b\rangle = \langle a'|e^{iG} F e^{-iG}|b'\rangle. \qquad (15.12)$$

Eqn. (15.11) can be likened to what one would expect to be the definition of variation in the operator

$$\langle a|F'|b\rangle = \langle a'|F + \delta F|b'\rangle, \qquad (15.13)$$

in order to define the infinitesimal variation of an operator,

$$\delta F = -i[F, G]. \qquad (15.14)$$

15.2.2 Example: operator equations of motion

The same result can also be obtained from Hamilton's equations for dynamical changes. Consider variations in time. The generator of time translations is the Hamiltonian

$$\delta_t F = \left(\frac{\partial F}{\partial t} - \frac{dF}{dt}\right)\delta t, \qquad (15.15)$$

since

$$\delta F = F(t + \delta t) - F(t). \qquad (15.16)$$

(The numerical value of t is not affected by the operator transformation.) Hence, using our definition, we obtain

$$\frac{dF}{dt} = \frac{\partial F}{\partial t} - i[F, H], \qquad (15.17)$$

which is the time development equation for the operator F.

15.3 Path integral formulation

Feynman's famous path integral formulation of quantum field theory can be thought of as an integral solution to the differential Schwinger action principle in eqn. (15.3). To see this, consider supplementing the action S with a source term [22, 125, 128]

$$S \to S - \int (dx)\, J\phi. \qquad (15.18)$$

The operator equations of motion are now

$$\frac{\delta S}{\delta \phi} = J \equiv E[\phi], \tag{15.19}$$

and we define $E[\phi]$ as the operator one obtains from the first functional derivative of the action operator. From the Schwinger action principle, we have,

$$\frac{\delta^n}{\delta J^n} \langle 0|0 \rangle = \left(\frac{-i}{\hbar} \right)^n \langle 0|T\phi(x_1) \cdots \phi(x_n)|0 \rangle \bigg|_{J=0}, \tag{15.20}$$

where the T indicates time ordering implicit in the action integral. This may be summarized (including causal operator ordering) by

$$\langle 0|0 \rangle_J = \left\langle 0 \left| T \exp\left(-\frac{i}{\hbar} \int dV_t J\phi \right) \right| 0 \right\rangle_J. \tag{15.21}$$

We may now use this to write the operator $E[\phi]$ in terms of functional derivatives with respect to the source,

$$E[\delta_J]\langle 0|0 \rangle_J = J \langle 0|0 \rangle_J. \tag{15.22}$$

This is a functional differential equation for the amplitude $\langle 0|0 \rangle_J$. We can attempt to solve it by substituting a trial solution

$$\langle 0|0 \rangle_J = \int d\phi \, F[\phi] \exp\left(\frac{-i}{\hbar} \int dV_t J\phi \right). \tag{15.23}$$

Substituting in, and using $J = i\hbar \frac{\delta}{\delta \phi}$,

$$\begin{aligned}
0 &= \int d\phi \left\{ E[\delta_J] - J \right\} F[\phi] \exp\left(i \int dV_t J\phi \right) \\
&= \int d\phi \left\{ E[\delta_J] \exp\left(\frac{-i}{\hbar} \int dV_t J\phi \right) \right. \\
&\quad \left. - i\hbar F[\phi] \frac{\delta}{\delta \phi} \exp\left(-\frac{i}{\hbar} \int dV_t J\phi \right) \right\}.
\end{aligned} \tag{15.24}$$

Integrating by parts with respect, moving the derivative $\frac{\delta}{\delta \phi}$, yields

$$\begin{aligned}
0 &= \int d\phi \left\{ E[\phi]F[\phi] + i\hbar \frac{\delta F}{\delta \phi} \right\} T \exp\left(i \int dV_t J\phi \right) \\
&\quad - i \left[F[\phi] T \exp\left(i \int dV_t J\phi \right) \right]_{-\infty}^{+\infty}.
\end{aligned} \tag{15.25}$$

Assuming that the surface term vanishes independently gives

$$E[\phi]F[\phi] = -i\hbar\frac{\delta F}{\delta\phi}, \tag{15.26}$$

and thus

$$F[\phi] = C\exp\left(\frac{i}{\hbar}S[\phi]\right). \tag{15.27}$$

Thus, the transformation function for vacua, in the presence of a source, may be taken to be

$$\langle 0|0\rangle_J = \int d\phi\exp\left(\frac{i}{\hbar}(S[\phi] - \int dV_t\,J\phi)\right). \tag{15.28}$$

15.4 Postscript

For all of its limitations, classical covariant field theory is a remarkable stepping stone, both sturdy and refined, building on the core principles of symmetry and causality. Its second quantized extension has proven to be the most successful strategy devised for understanding fundamental physical law. These days, classical field theory tends to merit only an honourable mention, as a foundation for other more enticing topics, yet the theoretical toolbox of covariant classical field theory underpins fundamental physics with a purity, elegance and unity which are virtually unparalleled in science. By dwelling on the classical aspects of the subject, this book scratches the surface of this pivotal subject and celebrates more than a century of fascinating discovery.

Part 3
Reference: a compendium of fields

16
Gallery of definitions

16.1 Units

SI units are used throughout this book unless otherwise stated. Most books on modern field theory choose natural units in which $\hbar = c = \epsilon_0 = \mu_0 = 1$. With this choice of units, very many simplifications occur, and the full beauty of the covariant formulation is apparent. The disadvantage, however, is that it distances field theory from the day to day business of applying it to the measurable world. Many assumptions are invalidated when forced to bow to the standard SI system of measurements. The definitions in this guide are chosen, with some care, to make the dimensions of familiar objects appear as intuitive as possible.

Some old units, still encountered, include electrostatic units, rather than coulombs; ergs rather than joules; and gauss rather than tesla, or webers per square metre:

Old	SI
1 e.s.u.	$\frac{1}{3} \times 10^{-9}$ C
1 erg	10^{-7} J
1 eV	1.6×10^{-19} J
1 Å	10^{-10} m
1 G	10^{-4} Wb m^{-2}
1 γ	10^{-5} G

16.2 Constants

Planck's constant	$\hbar = 1.055 \times 10^{-34}$ J s
speed of light in a vacuum	$c = 2.998 \times 10^8$ m s^{-1}
electron rest mass	$m_{\mathrm{e}} = 9.1 \times 10^{-31}$ kg
proton rest mass	$m_{\mathrm{p}} = 1.67 \times 10^{-27}$ kg
Boltzmann's constant	$k_{\mathrm{B}} = 1.38 \times 10^{-23}$ J K^{-1}
Compton wavelength	$\hbar/mc$
structure constant	$\alpha = \frac{e^2}{4\pi\epsilon_0\hbar c} = \frac{1}{137.3}$
classical electron radius	$r_0 = \frac{e^2}{4\pi\epsilon_0 mc^2} = 2.2 \times 10^{-15}$ m
Bohr radius	$a_0 = \frac{4\pi\epsilon_0\hbar^2}{e^2 m_{\mathrm{e}}} = 0.5292$ Å
electron plasma frequency	$\omega_{\mathrm{p}} = \sqrt{\frac{Ne^2}{\epsilon_0 m_{\mathrm{e}}}}$ s^{-1}
	$\omega_{\mathrm{p}} \sim 56\sqrt{N}$ rad s^{-1}
cyclotron frequency	$\omega_{\mathrm{c}} = \omega_B = \frac{eB}{m}$ s^{-1}

16.3 Engineering dimensions

In n spatial dimensions plus one time dimension, we have the following engineering dimensions for key quantities (note that square brackets denote the engineering dimension of a quantity):

velocity (of light) $[c]$	LT^{-1}
Planck's constant $[\hbar]$	$\mathrm{ML}^2\mathrm{T}^{-1}$
electric charge $[e]$	$\mathrm{L}^{(n+1)/2}\mathrm{T}^{-2}$
gravitational constant G	$\mathrm{M}^{-1}\mathrm{L}^3\mathrm{T}^{-2}$
permittivity $[\epsilon_0]$	$\mathrm{M}^{-1}\mathrm{LT}^{-2}$
permeability $[\mu_0]$	$\mathrm{MT}^4\mathrm{L}^{-3}$
structure constant $[\alpha]$	L^{n-3}

The dynamical variables have dimensions:

Schrödinger field $[\psi]$	$\mathrm{L}^{-n/2}$
Dirac field $[\psi]$	$\mathrm{L}^{-n/2}$
Klein–Gordon $[\phi]$	$\mathrm{L}^{-n/2}\mathrm{T}^{\frac{1}{2}}[\hbar]^{-\frac{1}{2}}$
	$= \mathrm{L}^{-(n+2)/2}\mathrm{T}^{-\frac{1}{2}}\mathrm{M}^{-\frac{1}{2}}$
Maxwell $[A_\mu]$	$\mathrm{L}^{1-n}\mathrm{T}[\hbar] = \mathrm{MTL}^{-(n-1)/2}$
electric current density $J_\mu = \frac{\delta S}{\delta A^\mu}$	$[e]\mathrm{L}^{1-n}\mathrm{T}^{-1}$
particle number density N	L^{-n}

The plasma distributions are defined from the fact that their integral over a phase space variable gives the number density:

$$N(x) = \int d\mathbf{v} \, f_v(\mathbf{v}, x) \tag{16.1}$$

$$= \int d^n\mathbf{p} \, f_p(\mathbf{p}, x) \tag{16.2}$$

and so on. One is generally interested in the distribution as a function of velocity $\mathbf{v}$, the momentum $\mathbf{p}$ or the wavenumber $\mathbf{k}$. In common units, where $\hbar = c = 1$, the above may be simplified by setting $L = T = M^{-1}$. Notice that all coupling constants scale with spacetime dimension.

The constants ϵ_0 and μ_0 are redundant scales; it is not possible to identify the dimensions of the fields and couplings between matter and radiation uniquely. Dimensional analysis of the action, allows one to determine only two combinations:

$$[eA_\mu] = MLT^{-1}$$

$$\left[\frac{e^2}{\epsilon_0}\right] = L^n MT^{-2}. \tag{16.3}$$

These may be determined by identifying $J^\mu A_\mu$ as an energy density and from Maxwell's equations, respectively. If we assume that ϵ_0 and μ_0 do not change their engineering dimensions with the dimension of space n, then we can identify the scaling relations

$$[A_\mu] \sim L^{-\frac{n}{2}}$$

$$[e] \sim L^{\frac{n}{2}}, \tag{16.4}$$

where $\sim$ means 'scales like'. The former relation is demanded by the dimensions of the action; the latter is demanded by the dimensions of the coupling between matter and radiation, since the product eA_μ must be independent of the dimension of spacetime. Although the dimensions of e, A_μ, ϵ_0 and μ_0 are not absolutely defined, a solution is provided in the relations above.

From the above, one determines that the cyclotron frequency is independent of the spacetime dimension

$$\left[\frac{eB}{m}\right] = T^{-1}, \tag{16.5}$$

and that the structure constant α has dimensions

$$\left[\frac{e^2}{4\pi\epsilon_0\hbar c}\right] = L^{n-3}. \tag{16.6}$$

The so-called plasma frequency is defined only for a given plasma charge density ρ, since

$$\left[\frac{e^2}{\epsilon_0 m}\right] = L^n T^{-2}.$$ (16.7)

Thus, $\omega_p^2 = \frac{e\rho_N}{\epsilon_0 m} = \frac{eN}{\epsilon_0 m}$.

The Hall conductivity is a purely two-dimensional quantity. The dimensional equation $J = \sigma_H E$ can be verified for $n = 2$ and $\sigma_H = e^2/\hbar$, but it is noted that each of the quantities scales in a way which requires an additional scale for $n \neq 2$.

16.4 Orders of magnitude

16.4.1 Sizes

Planck length L_p	$\sqrt{G\hbar/c^3} = 1.6 \times 10^{-35}$ m
Planck time $T_p = L_p/c$	$\sqrt{G\hbar/c^5} = 5.3 \times 10^{-44}$ s
Planck mass M_p	$\sqrt{\hbar c/G} = 2.1 \times 10^{-8}$ kg
Planck energy $E_p = M_p c^2$	1.8×10^9 J $= 1.2 \times 10^{19}$ GeV
Hall conductance in $n = 2$	$\sigma_H = e^2/\hbar$
Landau length at $k_B T$	$l = e^2/(4\pi\epsilon_0 k_B T) = 1.67 \times 10^{-5}/T$ m
Debye length	$h = \sqrt{\epsilon KT/Ne^2} = 69 \times \sqrt{T/N}$ m

The Landau length is that length at which the electrostatic energy of electrons balances their thermal energy.

16.4.2 Densities and pressure

number density ('particles' per unit volume)	$N(x)$ or ρ_N
number current $N^0 = Nc$, $N^i = Nv^i$.	N_μ or $J_{N\mu}$
mass density current	$J_\mu^m = mN_\mu$
charge density current	$J_\mu^e = eN_\mu$
charge density or other sources	J_μ

interstellar gas	10^6 m^{-3}
ionosphere	10^8–10^{12} m^{-3}
solar corona	10^{13} m^{-3}
solar atmosphere	10^{18} m^{-3}
laboratory plasma	10^{18}–10^{24} m^{-3}
mean density of Earth	5520 kg m^{-3}
mean density of Jupiter/Saturn	1340/705 kg m^{-3}
solar wind particle number density	3–20 cm^{-3}
magneto-pause number density	10^5–10^6 m^{-3}

Pressure is denoted by P and has the dimensions of energy density or force per unit area.

16.4.3 Temperatures

interstellar gas	10^2 K
Earth ionosphere	10^4 K
solar corona	10^6 K
solar atmosphere	10^4 K
laboratory plasma	10^6 K
super-conducting transition	0–100 K
Bose–Einstein condensation	μK–nK

16.4.4 Energies

first ionization energies	~ 10 eV
Van der Waals binding energy	2 keV
covalent binding energy	20 keV
hydrogen bond binding energy	20 keV
plasma energies, solar wind	1–100 keV
Planck energy $E_\mathrm{p} = M_\mathrm{p} c^2$	1.8×10^9 J $= 1.2 \times 10^{19}$ GeV

Lorentz energy–momentum tensor	$\theta_{\mu\nu}$
conformal energy–momentum tensor	$T_{\mu\nu}$

16.4.5 Wavelengths

radio waves	$> 10^{-2}$ m
microwaves	10^{-2} m
infra-red (heat)	10^{-3}–10^{-6} m
visible light	10^{-6}–10^{-7} m
ultra-violet	10^{-7}–10^{-9} m
X-rays	10^{-9}–10^{-13} m
gamma rays	$< 10^{11}$ m
thermal de Broglie wavelength	$\lambda = \sqrt{\dfrac{\hbar^2}{2mk_\mathrm{B}T}}$
hydrogen atom at 273 K	2.9×10^{-11} m
hydrogen atom at 1 K	4.9×10^{-10} m
electron at 273 K	1.27×10^{-9} m

16.4.6 Velocities

speed of light in vacuum c	2.9×10^8 m s^{-1}
solar wind	300–800 km s^{-1}
phase velocity	$\frac{\omega}{k_i} = v_{\mathrm{ph}}^i(k)$
group velocity	$\frac{\partial \omega}{\partial k_i} = v_{\mathrm{g}}^i(k)$
energy transport velocity	$\frac{T_{0i}}{T_{00}} = v_{\mathrm{en}}^i$

16.4.7 Electric fields

geo-electric field at surface (fine weather)	100 V m^{-1}
geo-electric field at surface (stormy weather)	1000 V m^{-1}
auroral field	10^{-3}–10^{-2} V m^{-1}

16.4.8 Magnetic fields

intense laboratory field **H**	$H \sim 10^6$ A m^{-1} [102]
highest coercive field **H** in minerals	$H \sim 10^6$ A m^{-1} [102]
geo-magnetic field	$H \sim 10$ A m^{-1} [102]
geo-magnetic field	$B_0 = 1.88 \times 10^{-5}$ tesla
vertical geo-magnetic field	$B_v = B_0 \tan \delta$, $\delta =$ declination from north
Earth dipole moment	7.95×10^{22} A m^{-2}

16.4.9 Currents

atmospheric current from ionosphere to ground	10^{-12} A m^{-2} [50]
auroral current aligned with field	10^{-7} A m^{-2}
ionospheric dynamo	500 000 A (eastward)

16.5 Summation convention

Einstein's summation convention is used throughout this book. This means that repeated indices are summed over implicitly:

$$\sum_A \phi_A \phi_A \rightarrow \phi_A \phi_A, \tag{16.8}$$

and

$$\sum_\mu A^\mu A_\mu \rightarrow A^\mu A_\mu. \tag{16.9}$$

In other words, summation signs are omitted for brevity.

16.6 Symbols and signs

16.6.1 Basis notation

$g_{\mu\nu}$	the spacetime metric with signature $-+++\cdots$
$\eta_{\mu\nu}$	the constant Minkowski spacetime metric with value $\mathrm{diag}(-1, 1, 1, 1, \ldots)$
$g = -\det g_{\mu\nu}$	the unsigned determinant of the metric which appears in volume measures
$\mu, \nu, \lambda, \rho \ldots$	Greek indices are spacetime-covariant and run from $0, \ldots, n$ in $n+1$ dimensions
$i, j, k = 1, \ldots, n$	Latin indices refer to spatial dimensions
∂_t	shorthand for $\frac{\partial}{\partial t}$ etc.
$A, B = 1, \ldots, d_R$	upper case Latin indices are the components of a group multiplet for non-spacetime groups, e.g. charge, colour, in a general representation G_R
$a, b = 1, \ldots, d_G$	lower case Latin indices are group indices which belong to the adjoint representation G_{adj}
σ	signifies space
$\mathrm{d}\sigma = \mathrm{d}x^1 \ldots \mathrm{d}x^n$	the spatial volume element
$\mathrm{d}\sigma^\mu$	volume element for a spacelike hyper-surface
∂_σ	derivative normal to a spacelike hyper-surface has the canonical interpretation ∂_0
$U_\mu{}^\nu$ or $U_A{}^B$	matrix element of a transformation group

Some books make the abbreviation, $(\partial_\mu \phi)^2$, when – in fact – they mean $(\partial_\mu \phi)(\partial^\mu \phi)$. In this text $(\partial_\mu \phi)^2$ means only $(\partial_\mu \phi)(\partial_\mu \phi)$ which differs by a factor of the metric. Note that, because of the choice of metric above, $(\partial_i \phi)^2 = (\partial^i \phi)(\partial_i \phi) = (\partial_i \phi)(\partial_i \phi)$.

16.6.2 Volume elements

dV_x invariant volume element in $(n + 1)$ dimensional spacetime; $dV_x = dx^0 dx^1 dx^2 \ldots dx^n \sqrt{g}$

$dV_t = \frac{1}{c} dV_x$ the volume element which appears in in most dynamical contexts, such as the action

$(dx) = dV_t$ alternative notation for dV_t

$d\sigma^\mu$ the volume element on spacelike hyper-surface with a unit normal n^μ parallel to $d\sigma^\mu$

$d\sigma \equiv (dx)$ an abbreviation for $d\sigma^0$, the 'canonical' spacelike hyper-surface; $d\sigma = dx^1 \ldots dx^n \sqrt{-\det g_{ij}}$

The volume element appearing in the action, and in most transformations, is (dx), which differs from the spacetime volume element by a factor of $1/c$. This is because the action has dimensions of energy × time. Had the action been defined with an extra factor of c one could have avoided this blemish, but that is not traditionally the case. In natural units, $\hbar = c = 1$, this problem is concealed.

16.6.3 Symmetrical and anti-symmetrical combinations

A bar (like a mean value) is used for objects which are symmetrical, e.g.

$$\bar{x} = \frac{1}{2}(x_1 + x_2), \tag{16.10}$$

whereas a tilde is used to signify anti-symmetry:

$$\tilde{x} = (x_1 - x_2). \tag{16.11}$$

Similarly, tensor parts $\overline{T}_{ij}$ and $\tilde{T}_{ij}$ are, by assumption, symmetrical and anti-symmetrical parts.

16.6.4 Derivatives

Field theory abounds with derivatives. Since we often have use for more symbols than are available, some definitions depend on context.

$\frac{\mathrm{d}}{\mathrm{d}x^\mu}$ the total derivative (this object is seldom used)

$\frac{\partial}{\partial x^\mu} = \partial_\mu$ the partial derivative acting on x, e.g. $\overset{x}{\partial_\mu} G(x, x')$

D_μ a generic derivative; it commonly denotes the gauge-covariant derivative $D_\mu = \partial_\mu - ieA_\mu$

∇_μ the Lorentz-covariant derivative, which includes the 'affine connection' ∇_μ is the same as ∂_μ when acting on scalar fields, but for a vector field $\nabla_\mu A^\nu = \partial_\mu A^\nu + \Gamma^\lambda_{\mu\nu} A_\lambda$

$\nabla^2 = \nabla^i \nabla_i$ the spatial Laplacian

$\Box = \nabla^\mu \nabla_\mu$ the d'Alambertian operator; in Cartesian coordinates, $\Box = -\frac{1}{c^2}\frac{\partial^2}{\partial t^2} + \partial_i^2$, but generally $\Box = \frac{1}{\sqrt{g}}\partial_\mu \left(\sqrt{g}g^{\mu\nu}\partial_\nu\right)$

$\hat{\partial}_\mu$ a partial derivative in which the speed of light is replaced by an effective speed of light; also used for higher-dimensional indices in Kaluza–Klein theory.

16.6.5 Momenta

p_i the *kinetic* momentum also denoted **p**;
the generalization of mass $\times$ velocity in classical mechanics
Quantum theory replaces this by $-i\hbar\partial_i$

p_μ the $n + 1$ dimensional spacetime momentum,
a covariant representation of energy and momentum
The $\mu = 0$ component is E/c and the spatial
components are p_i

π_μ the covariant momentum. It is the analogue of
p_μ but includes any covariant connections,
e.g. the electromagnetic vector potential, or the
spacetime 'affine' connection
e.g. π_μ is $-i\hbar D_\mu$ or $-i\hbar\nabla_\mu$

Π_σ (or simply Π) is the *canonical* momentum, defined
by the surface term of the variation of the action (see eqn. (4.23))
The covariant definition of the momentum conjugate
to $q(x)$ is $\Pi_\sigma = \frac{\partial\mathcal{L}}{\partial(D^\sigma q(x))}$, where σ is a timelike direction;
e.g. $\Pi = \frac{\partial\mathcal{L}}{\partial(D^0 q(x))}$. This quantity does not have the dimensions of

momentum: it is referred to only as a momentum in the sense of being canonically conjugate to the field variable $q(x)$ (which does not have the dimensions of position)

$\hat{q}, \hat{p}_i$ coordinates and momenta which are re-scaled so as to have common engineering dimensions

(dk)   Schwinger notation for the integration measure $\frac{d^{n+1}k}{(2\pi)^{n+1}}$

(d**k**)   Schwinger notation for the integration measure $\frac{d^n k}{(2\pi)^n}$

16.6.6 Position, velocity and acceleration

$r_i = x_i - x_i'$ a ray between spatial position **x** and **x**′

$\hat{r}^i$ a hat can indicate a unit vector

v^i components of the velocity of an object in the laboratory frame

β^i same as above in units of the speed of light $\beta^i = v^i/c$

β^2 $\beta^i \beta_i$ where $\beta^i = v^i/c$

γ relativistic contraction factor $1/\sqrt{1-\beta^2}$

U^μ, β^μ velocity $(n+1)$-vector; $U^\mu = \gamma(c, v^i) = \gamma c \beta^\mu$
$U^\mu = \partial_\tau x^\mu$ is not directly measurable, but transforms as a vector under Lorentz transformations
$\beta^\mu = \partial_t x^\mu$ does not because ∂_t is not invariant

$a^\mu = \partial_0 \beta^\mu$ components of the acceleration in the laboratory frame $a^i = \dot{v}^i/c^2$.
This quantity does not transform as an $(n+1)$-vector

$A^\mu = \partial_\tau^2 x^\mu$ acceleration vector, transforms as a tensor of rank 1

$\frac{\omega_k}{k_i} = v_{\mathrm{ph}}^i(k)$ phase velocity

$\frac{\partial \omega_k}{\partial k_i} = v_{\mathrm{g}}^i(k)$ group velocity

$\frac{T_{0i}}{T_{00}}$ energy transport velocity

16.7 Limits

It is natural to check derived expressions in various limits to evaluate their reliability. Here we list a few special limits which might be taken in this context and caution against possibly singular limits. While it might be possible to set quantities such as the mass and magnetic field strength to zero without incurring any explicit singularities, one should not be surprised if these limits yield inconsistent answers compared with explicit calculations in their absence.

In many cases the singular nature of these limits is not immediately obvious, and one must be careful to set these to zero, only at the end of a calculation, or risk losing terms of importance:

$c \to \infty$ non-relativistic limit
$\hbar \to 0$ classical limit
$\mathbf{B} \to \mathbf{0}$ the limit of zero magnetic field is often singular
$m \to 0$ the limit of zero mass is often singular
$R \to 0$ the limit of zero curvature is often singular.

17

The Schrödinger field

The Schrödinger equation is the quantum mechanical representation of the non-relativistic energy equation

$$\frac{\mathbf{p}^2}{2m} + V = E \tag{17.1}$$

and is obtained by making the replacement $p_i \rightarrow -i\hbar\partial_i$ and $E = i\hbar\tilde{\partial}_t$, and allowing the equation to operate on a complex field $\psi(x)$. The result is the basic equation of quantum mechanics

$$\left(-\frac{\hbar^2}{2m}\nabla^2 + V\right)\psi = i\hbar\tilde{\partial}_t\psi. \tag{17.2}$$

which may also be written

$$H_{\mathrm{D}}\psi = i\hbar\tilde{\partial}_t\psi, \tag{17.3}$$

thereby defining the differential Hamiltonian operator. The free Hamiltonian operator H_0 is defined to be the above with $V = 0$.

17.1 The action

The action for the Schrödinger field is

$$S = \int \mathrm{d}\sigma\,\mathrm{d}t \left\{ -\frac{\hbar^2}{2m}(\partial^i\psi)^*(\partial_i\psi) - V\psi^*\psi \right.$$
$$\left. +\frac{i\hbar}{2}(\psi^*\tilde{\partial}_t\psi - \psi\tilde{\partial}_t\psi^*) - J^*\psi - \psi^*J \right\}. \tag{17.4}$$

Notice that this is not Lorentz-invariant, and cannot be expressed in terms of $n + 1$ spacetime dimensional vectors.

410

17.2 Field equations and continuity

The variation of the action can be performed with respect to both $\psi(x)$ and $\psi^*(x)$ since these are independent variables. The results are

$$\delta_{\psi^*}S = \int d\sigma\, dt\; \delta\psi^* \left(\frac{\hbar^2}{2m}\partial^i\partial_i\psi - V\psi + i\hbar\tilde{\partial}_t\psi\right)$$

$$+ \int d\sigma \left[\frac{i\hbar}{2}\delta\psi^*\psi\right] + \int d\sigma^i \left[\frac{\hbar^2}{2m}\delta\psi(\partial_i\psi)^\dagger\right] = 0$$

$$\delta_\psi S = \int d\sigma\, dt\; \delta\psi \left(\frac{\hbar^2}{2m}\partial^i\partial_i\psi^* - V\psi^* - i\hbar\tilde{\partial}_t\psi^*\right)$$

$$+ \int d\sigma \left[-\frac{i\hbar}{2}\delta\psi\psi^*\right] + \int d\sigma^i \left[\frac{\hbar^2}{2m}\delta\psi^*(\partial_i\psi)\right] = 0, \qquad (17.5)$$

where we have used integration by parts, and the two expressions are mutually conjugate. From the surface terms, we can now infer that the canonical momentum conjugate to $\psi(x)$ is

$$\Pi = i\hbar\psi, \qquad (17.6)$$

and that spatial continuity at an interface is guaranteed by the condition

$$\Delta\left(\frac{\hbar^2}{2m}(\partial_i\psi)\right) = 0, \qquad (17.7)$$

where Δ means the change in value across the interface.

17.3 Free-field solutions

The free-field solutions may be written in a compact form as a linear combination of plane waves satisfying the energy constraint:

$$\psi(x) = \int_0^\infty \frac{d\tilde{\omega}}{2\pi} \int_{-\infty}^{+\infty} \frac{d^n\mathbf{k}}{(2\pi)^n} e^{i(\mathbf{k}\cdot\Delta\mathbf{x} - \tilde{\omega}\Delta t)}\, \psi(\mathbf{k}, \tilde{\omega})$$

$$\times\, \theta(\tilde{\omega})\, \delta\left(\frac{\hbar^2\mathbf{k}^2}{2m} - \hbar\tilde{\omega}\right). \qquad (17.8)$$

The coefficients of the Fourier expansion $\psi(\mathbf{k}, \tilde{\omega})$ are arbitrary.

17.4 Expression for the Green function

The Schrödinger Green function contains purely retarded solutions. This is a consequence of its spectrum of purely positive energy solutions. If one views the Schrödinger field as the non-relativistic limit of a relativistic field, then the negative frequency Wightman function for the relativistic field vanishes in the non-relativistic limit as a result of choosing only positive energy solutions. The Fourier space expression for the free-field Green function is

$$G_{NR}(x, x') = \int_{-\infty}^{+\infty} d\tilde{\omega} \int_{-\infty}^{+\infty} \frac{d^n \mathbf{k}}{(2\pi)^n} \frac{e^{i(\mathbf{k}\cdot\Delta\mathbf{x} - \tilde{\omega}\Delta t)}}{(\frac{\hbar^2 \mathbf{k}^2}{2m} - \hbar\tilde{\omega}) - i\epsilon}. \tag{17.9}$$

This may be interpreted in the light of the more general expression:

$$G_{NR}(x, x') = \sum_n \theta(t - t') \, u_n(x) u_n^*(x')$$

$$= \int \frac{d\alpha}{2\pi} e^{-i\alpha(t-t')} \frac{u_n(x) u_n^*(x')}{\alpha - \omega_n + i\epsilon} \tag{17.10}$$

where $u_n(x)$ are a complete set of eigenfunctions of the free Hamiltonian, i.e.

$$(H_0 - E_n) u_n(x) = 0, \tag{17.11}$$

where $E_n = \hbar\omega_n$.

17.5 Formal solution by Green functions

The free Schrödinger Green function satisfies the equation

$$\left(-\frac{\hbar^2 \nabla^2}{2m} - i\hbar\tilde{\partial}_t\right) G_{NR}(x, x') = \delta(\mathbf{x}, \mathbf{x}')\delta(t, t'), \tag{17.12}$$

or

$$(H_0 - E) G_{NR}(x, x') = \delta(\mathbf{x}, \mathbf{x}')\delta(t, t'), \tag{17.13}$$

and provides the solution for the field perturbed by source $J(x)$,

$$\psi(x) = \int (dx') G_{NR}(x, x') J(x'). \tag{17.14}$$

The infinitesimal source J is not normally written as such, but rather in the framework of the potential V, so that $J = V\psi$:

$$(H_0 - E_n)\psi_n = -V\psi_n, \tag{17.15}$$

where $\psi(x) = \sum_n c_n \psi_n(x)$. Substitution of this into the above relation leads to an infinite regression:

$$\psi(x) = \int (dx') G_{NR}(x, x') J(x')$$

$$= \int (dx') G_{NR}(x, x') V(x') \psi(x')$$

$$= \int (dx')(dx'') G_{NR}(x, x') G_{NR}(x', x'') J(x''), \tag{17.16}$$

and so on. This multiplicative hierarchy is only useful if it converges. It is thus useful to make this into an additive series, which converges for sufficiently weak $V(x)$. To do this, one defines the free-field $\psi_0(x)$ as the solution of the free-field equation

$$(H_0 - E_n)\psi_{0n}(x) = 0, \tag{17.17}$$

and expands in the manner of a perturbation series. The solutions to the full-field equation are defined by

$$\psi_n(x) = \psi_{0n}(x) + \delta\psi_n \tag{17.18}$$

where the latter terms are assumed to be small in the sense that they lead to convergent results in calculations. Substituting this into eqn. (17.15) gives

$$(H_0 - E_n)\delta\psi_n = -V(x)\psi_n(x), \tag{17.19}$$

and thus

$$\delta\psi(x) = -\int (\mathrm{d}x')\, G_{\mathrm{NR}}(x, x')\, V(x')\psi(x'), \tag{17.20}$$

or

$$\psi(x) = \psi_0(x) - \int (\mathrm{d}x')\, G_{\mathrm{NR}}(x, x')\, V(x')\psi(x'). \tag{17.21}$$

This result is sometimes called the *Lippmann–Schwinger equation*. The equation can be solved iteratively by re-substitution, i.e.

$$\psi(x) = \psi_0(x) - \int (\mathrm{d}x')\, G_{\mathrm{NR}}(x, x')\, V(x')\psi_0(x'')$$
$$+ \int (\mathrm{d}x')(\mathrm{d}x'')\, G_{\mathrm{NR}}(x, x')G_{\mathrm{NR}}(x', x'')\, V(x')V(x'')\psi(x''), \tag{17.22}$$

and generates the usual quantum mechanical perturbation series, expressed in the form of Green functions.

17.6 Conserved norm and probability

The variation of the action with respect to constant δs under a phase transformation $\psi \to e^{is}\psi$ is given by

$$\delta S = \int (\mathrm{d}x)\left\{ -\frac{\hbar^2}{2m}\left[-i\delta s(\partial^i\psi^*)(\partial_i\psi) + (\partial^i\psi^*)i\delta s(\partial_i\psi)\right] \right.$$
$$\left. + i\left[-i\delta s\psi^*\tilde{\partial}_t\psi + i\delta s\psi^*\tilde{\partial}_t\psi\right] \right\}. \tag{17.23}$$

Integrating by parts and using the equation of motion, we obtain the expression for the continuity equation,

$$\delta S = \int (dx)\delta s \left(\tilde{\partial}_t J^t + \partial_i J^i \right) = 0, \qquad (17.24)$$

where

$$J^t = \psi^*\psi = \rho$$
$$J^i = \frac{i\hbar^2}{2m} \left[\psi^*(\partial^i \psi) - (\partial^i \psi^*)\psi \right], \qquad (17.25)$$

which can be compared to the current conservation equation eqn. (12.1). ρ is the probability density and J^i is the probability current. The conserved probability is therefore

$$P = \int d\sigma \psi^*(x)\psi(x), \qquad (17.26)$$

and this can be used to define the notion of an inner product between two wavefunctions, given by the overlap integral

$$(\psi_1, \psi_2) = \int d\sigma \psi_1^*(x)\psi_2(x). \qquad (17.27)$$

17.7 Energy–momentum tensor

Replacing $\eta_{\mu\nu}$ by $\delta_{\mu\nu}$ (the Euclidean metric), we have for the components of the energy–momentum tensor:

$$\theta_{tt} = \frac{\partial \mathcal{L}}{\partial(\tilde{\partial}_t \psi)}(\tilde{\partial}_t \psi) + (\tilde{\partial}_t \psi^*)\frac{\partial \mathcal{L}}{\partial(\tilde{\partial}_t \psi)^*} - \mathcal{L}$$
$$= \frac{\hbar^2}{2m}(\partial^i \psi)^\dagger (\partial_i \psi) + V\psi^*\psi, \qquad (17.28)$$
$$\equiv H. \qquad (17.29)$$

In the second-quantized theory, where $\psi(x)$ is a field operator, this quantity is often called the Hamiltonian density operator H. This is to be distinguished from H_D, the differential Hamiltonian operator. In the classical case, the spatial integral of θ_{tt} is the expectation value of the Hamiltonian, as may be seen by integration by parts:

$$H = \int d\sigma \theta_{tt} = \int d\sigma \ \psi(x)^* \left[-\frac{\hbar^2}{2m}\partial^2 + V \right] \psi(x)$$
$$= (\psi, H_D\psi)$$
$$\equiv \langle H_D \rangle. \qquad (17.30)$$

Thus, θ_{tt} represents the total energy of the fields in the action S. The off-diagonal spacetime components are related to the expectation value of the momentum operator

$$\theta_{ti} = \frac{\partial \mathcal{L}}{\partial \tilde{\partial}_t \psi}(\partial_i \psi) + (\partial_i \psi^*)\frac{\partial \mathcal{L}}{\partial \tilde{\partial}_t \psi^*} = \frac{i\hbar}{2}\psi^*(\partial_i \psi) - \frac{i\hbar}{2}(\partial_i \psi^*)\psi$$

$$\int d\sigma \theta_{ti} = (\psi, i\hbar \partial_i \psi)$$

$$= -\langle p_i \rangle, \qquad (17.31)$$

and

$$\theta_{it} = \frac{\partial \mathcal{L}}{\partial(\partial_i \psi)}(\tilde{\partial}_t \psi) + \frac{\partial \mathcal{L}}{\partial(\partial_i \psi^*)}(\tilde{\partial}_t \psi^*)$$

$$= -\frac{\hbar^2}{2m}\left\{(\partial_i \psi)^*(\tilde{\partial}_t \psi) + (\partial_i \psi)(\tilde{\partial}_t \psi)^*\right\}. \qquad (17.32)$$

Note that θ is not symmetrical in the spacetime components: $\theta_{it} \neq \theta_{ti}$. This is a result of the lack of Lorentz invariance. Moreover, the sign of the momentum component is reversed, as compared with the relativistic cases, owing to the difference in metric signature. Finally, the 'stress' in the field is given by the spatial components:

$$\theta_{ij} = \frac{\partial \mathcal{L}}{\partial(\partial_i \psi)}(\partial_j \psi) + \frac{\partial \mathcal{L}}{\partial(\partial_i \psi)^*}(\partial_j \psi)^* - \mathcal{L}\delta_{ij}$$

$$= -\frac{\hbar^2}{2m}\left\{(\partial_i \psi)^*(\partial_j \psi) + (\partial_i \psi)(\partial_j \psi)^* - (\partial^k \psi)^*(\partial_k \psi)\delta_{ij}\right\}$$

$$+ \left\{V\psi^*\psi - \frac{i\hbar}{2}(\psi^* \overset{\leftrightarrow}{\partial}_t \psi)\right\}\delta_{ij}. \qquad (17.33)$$

Using the field equation (17.2), the trace of the spatial part may be written

$$\mathrm{Tr}\,\theta_{ii} = (n-2)\frac{\hbar^2}{2m}(\partial^k \psi)^*(\partial_k \psi) + n\left(V\psi^*\psi - \frac{1}{2}\frac{\hbar^2}{2m}\psi^* \overset{\leftrightarrow}{\partial^2} \psi\right)$$

$$= (1-n)\,H + 2V\psi^*\psi, \qquad (17.34)$$

where the last line is obtained by partial integration over all space, and on identifying the first and last terms as being $H - V$, and is therefore true only up to a partial derivative, or under the integral sign. See also Jackiw and Pi for a discussion of a conformally improved energy–momentum tensor, coupled to electromagnetism [78].

18

The real Klein–Gordon field

The relativistic scalar field satisfies the Klein–Gordon equation. This equation can be interpreted as the quantum mechanical analogue of the relativistic energy relation

$$E^2 = p^2c^2 + m^2c^4,$$ (18.1)

and is found by making the usual replacement $p_\mu \to -i\hbar\partial_\mu$, and allowing the equation to operate on a real scalar field $\phi(x)$. The result is

$$\left(-\Box + \frac{m^2c^2}{\hbar^2}\right)\phi(x) = 0.$$ (18.2)

18.1 The action

If we generalize the single scalar field above to a set of N real scalar fields $\phi_A(x)$ for $A = 1, \ldots, N$, with a linear perturbation, J_A, then all of the physical information about this system can be derived from the following action:

$$S = \int (dx) \left\{ \frac{1}{2}\hbar^2c^2(\partial^\mu\phi_A)(\partial_\mu\phi_A) + \frac{1}{2}m^2c^4\phi_A\phi_A + V(\phi) - J_A\phi_A \right\}.$$ (18.3)

Note that the position of the A indices is immaterial here, since they only label the number of the field components. The repeated indices are summed using a Euclidean metric, for which there is no notion of 'up' or 'down' indices.

Looking at this action, it can be noted that it does not have the familiar form of an integral over $T - V$ (kinetic energy minus potential energy). Instead, it has the form of an integral over $-E^2 + p^2 + m^2 + V$. Although this looks dimensionally incorrect, this is not the case, since the dimensions of the field are simply chosen so that S has the dimensions of action.

416

In what follows, the position of an index A is chosen for clarity. The Lagrangian density $\mathcal{L}$ is defined by

$$S = \int (\mathrm{d}x)\mathcal{L}. \tag{18.4}$$

In the usual canonical tradition, we define the conjugate momentum to the field $\phi_a(x)$ by

$$\Pi_\sigma^A = \frac{\delta \mathcal{L}}{\delta(\partial^\sigma \phi_A)} = \hbar^2 c^2 \partial_\sigma \phi, \tag{18.5}$$

where σ is a specific direction, normal to a spacelike hyper-surface. Usually we do not need to be this general and we can just pick $\sigma = 0$ for the normal, which corresponds to the time direction (normal to space in an observer's rest frame). Then we have, more simply,

$$\Pi_A = \hbar^2 c^2 \partial_0 \phi_A. \tag{18.6}$$

The Hamiltonian density is then obtained straightforwardly from the Legendre transformation

$$\mathcal{H} = \Pi(\partial_0 \phi) - \mathcal{L} g_{00}. \tag{18.7}$$

Or, using the fully covariant form,

$$\mathcal{H} = \Pi_\sigma(\partial_\sigma \phi) - \mathcal{L} g_{\sigma\sigma}. \tag{18.8}$$

Note the positions of the indices here and the presence of the metric in the second term of the right hand side. The need for this factor will become apparent later when looking at transformations and the energy–momentum tensor. It makes the relativistic Legendre transformation more subtle than that in Euclidean space, because of the indefinite metric. Eqns. (18.7) and (18.8) evaluate to

$$\mathcal{H} = \frac{1}{2}\hbar^2 c^2 \left[(\partial_0 \phi)^2 + (\partial_i \phi)^2 \right] + \frac{1}{2} m^2 c^4 \phi^2 + V(\phi). \tag{18.9}$$

18.2 Field equations and continuity

The variation of the action (with $V = 0$) leads to

$$\delta S = \int (\mathrm{d}x) \left\{ \hbar^2 c^2 \delta\phi_A(-\Box)\phi_A + m^2 c^4 \phi_A \delta\phi_A - J_A \delta\phi_A \right\}$$
$$+ \frac{1}{c} \int \mathrm{d}\sigma^\mu \left\{ \frac{1}{2}\hbar^2 c^2 \delta\phi_A(\partial_\mu \phi_A) \right\}. \tag{18.10}$$

Appealing to the action principle (see chapter 4), we surmise that the field equations are

$$\left(-\Box + \frac{m^2c^2}{\hbar^2}\right)\phi_A = (\hbar^2c^2)^{-1}J_A, \tag{18.11}$$

and that the condition for continuity of the field through any n-dimensional surface is

$$\Delta\Pi_\sigma^A = 0. \tag{18.12}$$

If a delta-function source $\Delta J_A = \delta j_A \delta(x)$ is added to J_a exactly on the surface σ, then this continuity equation is modified, and the new condition is given by

$$\Delta\Pi_\sigma^A = \Delta j^A \, n_\sigma, \tag{18.13}$$

where n^μ is the unit normal vector to σ. This equation tells us that a sudden change in the momentum of the field can only be caused by an impulsive force (source) Δj.

18.3 Free-field solutions

The field $\phi(x)$ may be expanded as a linear combination of a complete set of plane wavefunctions satisfying the equation of motion,

$$\phi(x) = \int \frac{d^{n+1}k}{(2\pi)^{n+1}}\phi(k)e^{ikx}\delta\left(\hbar^2c^2k^2 + m^2c^4\right), \tag{18.14}$$

where $\phi(k)$ are arbitrary coefficients, independent of x. The reality of the field requires that

$$\Phi^*(k) = \Phi(-k). \tag{18.15}$$

The integral ranges over all energies, but one can separate the positive and negative energy solutions by writing

$$\phi(x) = \phi^{(+)}(x) + \phi^{(-)}(x), \tag{18.16}$$

where

$$\phi^{(+)}(x) = \int \frac{d^{n+1}k}{(2\pi)^{n+1}}\phi(k)e^{ikx}\theta(-k_0)\delta\left(\hbar^2c^2k^2 + m^2c^4\right)$$

$$\phi^{(-)}(x) = \int \frac{d^{n+1}k}{(2\pi)^{n+1}}\phi(k)e^{ikx}\theta(k_0)\delta\left(\hbar^2c^2k^2 + m^2c^4\right). \tag{18.17}$$

The symmetry of the energy relation then implies that

$$\phi^{(+)}(x) = \left(\phi^{(-)}(x)\right)^*. \tag{18.18}$$

The positive and negative energy solutions to the free relativistic field equations form independently complete sets, with respect to the scalar product,

$$(\phi^{(+)}(x), \phi^{(+)}(x)) = \text{const.}$$
$$(\phi^{(-)}(x), \phi^{(-)}(x)) = \text{const.}$$
$$(\phi^{(+)}(x), \phi^{(-)}(x)) = 0. \tag{18.19}$$

18.4 Reality of solutions

It should be noted that the uses of the real scalar field are somewhat limited. The boundary conditions one can apply to a real scalar field are only the retarded or advanced ones. The solution

$$\phi(x) = \int (dx') \, G(x, x') J(x') \tag{18.20}$$

is only real if the Green function itself is real. This excludes the use of the time-ordered (Feynman) Green function.

18.5 Conserved norm and probability

Since the real scalar field has no complex phase symmetry, Noether's theorem leads to no conserved quantities corresponding to a conserved inner product. It is possible to define an invariant inner product on the manifold of positive energy solutions, however. This is what introduces the complex symmetry in the non-relativistic limit;

$$\phi \partial_0 \phi \tag{18.21}$$

has no definite sign.

Since the relativistic energy equation $E^2 = p^2 c^2 + m^2 c^4$ admits both possibilities, we do this by writing the real field as a sum of two parts,

$$\phi = \phi^{(+)} + \phi^{(-)}, \tag{18.22}$$

where $\phi^{(+)*} = \phi^{(-)}$. $\phi^{(+)}$ is a complex quantity, but the sum $\phi^{(+)} + \phi^{(-)}$ is clearly real. What this means is that it is possible to define a conserved current and therefore an inner product on the manifold of positive energy solutions $\phi^{(+)}$,

$$(\phi_1^{(+)}, \phi_2^{(+)}) = i\hbar c \int d\sigma^\mu (\phi_1^{(+)*} \partial_\mu \phi_2^{(+)} - (\partial_\mu \phi_1^{(+)})^* \phi_2^{(+)}) \tag{18.23}$$

and another on the manifold of negative energy solutions $\phi^{(-)}$. Thus there is local conservation of probability (though charge still does not make any sense) of particles and anti-particles separately.

18.6 Normalization

The scalar product is only defined for normalizable wave-packet solutions, i.e. those for which $(\phi, \phi) < \infty$. A plane wave is a limiting case, which can only be defined by box normalization. It does not belong to the Hilbert space. However, adopting an invariant normalization in momentum space, one can express plane waves simply. Noting that the following construction is both invariant and 'on shell', i.e. satisfies the Klein–Gordon equation,

$$\phi = \int \frac{d^{n+1}k}{(2\pi)^{n+1}} e^{ikx} \theta(\pm k_0) \delta(p^2 c^2 + m^2 c^4)$$

$$= \int \frac{d^n k}{(2\pi)^n} \frac{e^{ikx}}{2p_0}. \tag{18.24}$$

Adopting the normalization

$$(\phi(p), \phi(p)) = 2p_0 \, \delta(\mathbf{p} - \mathbf{p}')(2\pi)^n, \tag{18.25}$$

a positive energy solution takes the form

$$\phi^+(p) = e^{ikx} \quad \left(p_0 = \sqrt{\mathbf{p}^2 + m^2} \right). \tag{18.26}$$

18.7 Formal solution by Green functions

The formal solution of the equations of motion can be written down in terms of Green functions. The essence of the procedure is to find the inverse of the differential operator on the left hand side of eqn. (18.11). Formally, we may write

$$\phi_A(x) = \left(-\Box + \frac{m^2 c^2}{\hbar^2} \right)^{-1} (\hbar^2 c^2)^{-1} J_A, \tag{18.27}$$

where this is given meaning by comparing it with the expression involving the Green function or 'kernel' $G_{AB}(x, x')$:

$$\phi_A(x) = (\hbar^2 c^2)^{-1} \int (dx') G_{AB}(x, x') J_B(x'). \tag{18.28}$$

Comparing eqns. (18.27) and (18.28), we see that $G(x, x')$ must satisfy the equation

$$\left(-\Box + \frac{m^2 c^2}{\hbar^2} \right) G_{AB}(x, x') = \delta_{AB} \delta(x, x'), \tag{18.29}$$

and thus we see that $G_{AB}(x, x')$ is the inverse of the differential operator, insofar as $\delta_{AB}\delta(x, x')$ can be regarded as the 'identity' operator.

In this case, the indices A, B on the Green function are superfluous, since

$$G_{AB}(x, x') = \delta_{AB} G(x, x'),\tag{18.30}$$

but non-diagonal terms in A, B might be important when the components of the field interact. This is the case in a gauge theory, for example.

The Green function $G(x, x')$ is not unique: there is still a freedom to choose the boundary conditions. By this we mean a specification of how the field is affected by changes in the source J_a both in the past and in the future. The 'causal' Green function, also referred to as the retarded Green function, is such that $\phi(x)$ is only affected by a change in $J(x')$ if $x > x'$.

18.8 All the Green functions

The symmetry of the Green functions is as follows:

$$\overline{G}_{AB}(x, x') = \overline{G}_{BA}(x', x)$$
$$\tilde{G}_{AB}(x, x') = -\tilde{G}_{BA}(x', x)$$
$$G_{FAB}(x, x') = G_{FBA}(x', x).\tag{18.31}$$

The symmetrical parts of the Wightman functions may be constructed explicitly. For example

$$\frac{1}{2}\left[G_{AB}^{(+)}(x, x') + G_{BA}^{(+)}(x', x)\right] = \frac{1}{2}\left[G_{AB}^{(+)}(x, x') - G_{AB}^{(-)}(x, x')\right]$$
$$= \frac{1}{2}\left[G_{AB}^{(+)}(x, x') - \left(G_{AB}^{(+)}(x, x')\right)^*\right]$$
$$= i\,\mathrm{Im}\,G_{AB}^{(+)}(x, x').\tag{18.32}$$

The retarded, advanced and Feynman Green functions are all constructed from causally selective combinations of the Wightman functions.

$$G_{AB}^{(+)}(x, x') = -G_{BA}^{(-)}(x', x)$$
$$\left(G_{AB}^{(+)}(x, x')\right)^* = G_{AB}^{(-)}(x, x').\tag{18.33}$$

The properties of the step function lead to a number of linear relations:

$$G_{\mathrm{r}}(x, x') = -\theta(t - t')\tilde{G}(x, x')$$
$$G_{\mathrm{a}}(x, x') = \theta(t' - t)\tilde{G}(x, x')$$
$$G_{\mathrm{r}}(x, x') = G_{\mathrm{F}}(x, x') - G^{(-)}(x, x')$$
$$G_{\mathrm{a}}(x, x') = G_{\mathrm{F}}(x, x') + G^{(+)}(x, x')$$
$$G_{\mathrm{F}}(x, x') = -\theta(t - t')G^{(+)}(x, x') + \theta(t' - t)G^{(-)}(x, x').\tag{18.34}$$

Some caution is needed in interpreting the latter two relations, which should be considered formal. The causal properties of the Green functions distinguish $G^{(\pm)}(x, x')$, which satisfy the homogeneous eqn. (5.65), from G_r, G_F, which pose as right-inverses for a differential operator and satisfy an equation such as eqn. (5.62). We can investigate this by calculating time derivatives. Starting with the definition in eqn. (18.34), we obtain the time derivatives using the relations in sections A.1 and A.2 of Appendix A:

$$\partial_t G_F(x, x') = -\delta(t, t')\tilde{G}(x, x') - \theta(t - t')\partial_t G^{(+)}(x, x')$$
$$+\theta(t' - t)\partial_t G^{(-)}(x, x'), \qquad (18.35)$$

where eqn. (5.71) was used. The second derivative is thus

$$\partial_t^2 G_F(x, x') = - \partial_t\delta(t - t')\tilde{G}(x, x') - \delta(t - t')\partial_t\tilde{G}(x, x')$$
$$- \delta(t - t')\partial_t\tilde{G}(x, x') - \theta(t - t')\partial_t^2 G^{(+)}(x, x')$$
$$+ \theta(t' - t)\partial_t^2 G^{(-)}(x, x'). \qquad (18.36)$$

The property in eqn. (A.14) was used here. Thus using eqn. (5.73) we may write

$$\partial_t^2 G_F(x, x') = \delta(t - t')\delta(\mathbf{x} - \mathbf{x}') - \theta(t - t')\partial_t^2\tilde{G}(x, x')$$
$$+\theta(t' - t)\partial_t^2 G^{(-)}(x, x'). \qquad (18.37)$$

From this it should be clear that

$$(-\Box + M^2)G_F(x, x') = \delta(t - t')\delta(\mathbf{x} - \mathbf{x}')$$
$$- \theta(t - t')(-\Box + M^2)\tilde{G}(x, x')$$
$$+ \theta(t' - t)(-\Box + M^2)G^{(-)}(x, x')$$
$$= c\delta(x, x'). \qquad (18.38)$$

The Green function for the scalar field is directly related to that for the electromagnetic field in the Lorentz–Feynman gauge, up to factors of $\hbar$ and μ_0.

$$D_{\mu\nu}(x, x')\Big|_{\alpha=1} = \mu_0\hbar^2\, G(x, x')\Big|_{m=0} g_{\mu\nu}. \qquad (18.39)$$

18.9 The energy–momentum tensor

The application of Noether's theorem for spacetime translations leads to a symmetrical energy–momentum tensor. Although the sign of the energy is ambiguous for the Klein–Gordon field, we can define a Hamiltonian with the interpretation of an energy density which is positive definite, from the zero–zero component of the energy–momentum tensor. Using the action and the formula

(11.44), we have

$$\theta_{00} = \frac{\partial \mathcal{L}}{\partial(\partial^0 \phi)_A}(\partial_0 \phi_A) - \mathcal{L}g_{00}$$

$$= \mathcal{H} = \frac{1}{2}\hbar^2 c^2 \left[(\partial_0 \phi_A)^2 + (\partial_i \phi_A)^2\right] + \frac{1}{2}m^2 c^4 \phi_A^2 + V(\phi). \quad (18.40)$$

This quantity has the interpretation of a Hamiltonian density. The integral over all space provides a definition of the Hamiltonian:

$$H = \int d\sigma \, \mathcal{H}. \qquad (18.41)$$

The explicit use of zero instead of a general timelike direction here makes this definition of the Hamiltonian explicitly non-covariant. Note that this is not a differential Hamiltonian operator analogous to that in eqn. (17.3), but more like an expectation value. In the quantum theory (in which the fields are operator-valued) this becomes the Hamiltonian operator.

The off-diagonal spacetime components give

$$\theta_{0i} = \theta_{i0} = \frac{\partial \mathcal{L}}{\partial(\partial^0 \phi)_A}(\partial_i \phi)_A$$

$$= \hbar^2 c^2 (\partial_0 \phi_A)(\partial_i \phi_A). \qquad (18.42)$$

Since there is no invariant inner product for the real scalar field, it is awkward to define this as a field momentum. However, on the manifold of positive energy solutions $\phi^{(+)}$, the integral over all space may be written

$$\int d\sigma \, \theta_{0i} = c \int d\sigma \, (\phi^{(-)} \overset{\leftrightarrow}{\partial_0} \partial_i \phi^{(+)})$$

$$\equiv -(\phi^{(+)}, p_i c \phi^{(+)}), \qquad (18.43)$$

where $p_i = -i\hbar \partial_i$. The diagonal spatial components are

$$\theta_{ii} = \frac{\partial \mathcal{L}}{\partial(\partial^i \phi_A)}(\partial_i \phi_A) - \mathcal{L}$$

$$= \hbar^2 c^2 (\partial_i \phi_A)^2 - \frac{1}{2}\hbar^2 c^2 (\partial^\mu \phi_A)(\partial_\mu \phi_A) - \frac{1}{2}m^2 c^4 \phi_A^2 - V(\phi),$$

$$(18.44)$$

where the repeated i index is not summed. The off-diagonal 'stress' tensor is

$$\theta_{ij} = \frac{\partial \mathcal{L}}{\partial(\partial^i \phi_A)}(\partial_j \phi_A)$$

$$= (\partial_i \phi_A)(\partial_j \phi_A), \qquad (18.45)$$

where $i \neq j$. Notice that the trace of the space parts in $n + 1$ dimensions gives

$$\sum_i \theta_{ii} = \mathcal{H} - m^2 c^4 \phi_A^2 - 2V(\phi) + (n - 1)\mathcal{L} \qquad (18.46)$$

so that the full trace is

$$\theta^\mu_{\ \mu} = g^{\mu\nu}\theta_{\nu\mu} = -m^2 c^4 \phi_A^2 - 2V(\phi) + (n - 1)\mathcal{L}, \qquad (18.47)$$

which vanishes in $1 + 1$ dimensions in the massless, potential-less theory.

19

The complex Klein–Gordon field

This chapter is a supplement to the material for a real scalar field. Much of the previous chapter applies here too.

19.1 The action

The free-field action is given by

$$S = \int (\mathrm{d}x)\left\{\hbar^2 c^2 (\partial^\mu \phi_A)^\dagger (\partial_\mu \phi_A) + m^2 c^4 \phi_A^* \phi_A \right.$$
$$\left. + V(\phi_A^\dagger \phi_A) - J_A^\dagger \phi_A - J_A \phi_A^\dagger \right\}. \quad (19.1)$$

The field now has effectively twice as many components as the real scalar field, coming from the real and imaginary parts.

19.2 Field equations and continuity

Since the complex field and its complex conjugate are independent variables, there are equations of motion for both of these. Varying S first with respect to ϕ_A^*, we obtain

$$\delta S = \int (\mathrm{d}x)\delta\phi_A^* \left(-\hbar^2 c^2 \Box\, \phi_A + m^2 c^4 \phi_A + V'(\phi_A^2, \phi_A) - J_A\right)$$
$$+ \frac{1}{c}\int \mathrm{d}\sigma^\mu \delta\phi_A^* (\partial_\mu \phi_A) = 0, \quad (19.2)$$

which gives rise to the field equation

$$\left(-\Box + \frac{m^2 c^2}{\hbar^2}\right) \phi_A(x) = (\hbar^2 c^2)^{-1} J_A(x), \quad (19.3)$$

and the continuity condition over a spacelike hyper-surface,

$$\Delta(\partial_\sigma \phi(x)) = \Delta \Pi_\sigma = 0, \tag{19.4}$$

identifies the conjugate momentum Π_σ. Conversely, the variation with respect to ϕ_A gives the conjugate field equations, which give rise to the field equation

$$\left(-\Box + \frac{m^2 c^2}{\hbar^2}\right) \phi_A^*(x) = (\hbar^2 c^2)^{-1} J_A^*(x), \tag{19.5}$$

and the corresponding continuity condition over a spacelike hyper-surface,

$$\Delta(\partial_\sigma \phi^*(x)) = \Delta \Pi_\sigma^* = 0. \tag{19.6}$$

19.3 Free-field solutions

The free-field solutions for the complex scalar field have the same form as those of the real scalar field,

$$\phi(x) = \int \frac{\mathrm{d}^{n+1} k}{(2\pi)^{n+1}} \phi(k) \mathrm{e}^{ikx} \, \delta\left(\hbar^2 c^2 k^2 + m^2 c^4\right); \tag{19.7}$$

however, the Fourier coefficients are no longer restricted by eqn. (18.15).

19.4 Formal solution by Green functions

The formal solution to the field equation may be expressed in terms of a Green function $G_{AB}(x, x')$ by

$$\phi_A(x) = \int (\mathrm{d}x') G_{AB}(x, x') J_B(x'), \tag{19.8}$$

where the Green function satisfies the equation

$$\left(-\overset{x}{\Box} + \frac{m^2 c^2}{\hbar^2}\right) G_{AB}(x, x') = \delta(x, x') \delta_{AB}. \tag{19.9}$$

Similarly,

$$\phi_A^\dagger(x) = \int (\mathrm{d}x') J_B^\dagger(x') G_{BA}(x', x). \tag{19.10}$$

Note that, although the fields are designated as conjugates ϕ and $\phi^\dagger$, this relationship is not necessarily preserved by the choice of boundary conditions. If the time-ordered, or Feynman Green function is used (which represents virtual processes), then the resulting fields do not remain conjugate to one another over time. The retarded Green function does preserve the conjugate relationship between the fields, since it is real.

19.5 Conserved norm and probability

Let s be independent of x, and consider the phase transformation of the kinetic part of the action:

$$S = \int (dx)\hbar^2 c^2 \left\{ (\partial^\mu e^{-is}\phi^*)(\partial_\mu e^{is}\phi) \right\}$$

$$\delta S = \int (dx) \left[(\partial^\mu \phi^*(-i\delta s)e^{-is})(\partial_\mu \phi e^{is}) + c.c. \right]$$

$$= \int (dx)\delta s (\partial_\mu J^\mu), \tag{19.11}$$

where

$$J^\mu = -i\hbar^2 c^2 (\phi^* \partial^\mu \phi - \phi \partial^\mu \phi^*). \tag{19.12}$$

The conserved 'charge' of this symmetry can now be used as the definition of the inner product between fields:

$$(\phi_1, \phi_2) = i\hbar c \int d\sigma^\mu (\phi_1^* \partial_\mu \phi_2 - (\partial_\mu \phi_1)^* \phi_2), \tag{19.13}$$

or, in non-covariant form,

$$(\phi_1, \phi_2) = i\hbar c \int d\sigma (\phi_1^* \partial_0 \phi_2 - (\partial_0 \phi_1)^* \phi_2). \tag{19.14}$$

This is now our notion of probability.

19.6 The energy–momentum tensor

The application of Noether's theorem for spacetime translations leads to a symmetrical energy–momentum tensor. Although the sign of the energy is ambiguous for the Klein–Gordon field, we can define a Hamiltonian with the interpretation of an energy density which is positive definite, from the zero–zero component of the energy–momentum tensor. Using the action and the formula (11.44), we have

$$\theta_{00} = \frac{\partial \mathcal{L}}{\partial(\partial^0 \phi_A)}(\partial_0 \phi_A) + \frac{\partial \mathcal{L}}{\partial(\partial^0 \phi_A^*)}(\partial_0 \phi_A^*) - \mathcal{L} g_{00}$$

$$= \hbar^2 c^2 \left[(\partial_0 \phi_A^*)(\partial_0 \phi_A) + (\partial_i \phi_A^*)(\partial_i \phi_A) \right] + m^2 c^4 + V(\phi). \tag{19.15}$$

Thus, the last line defines the Hamiltonian density $\mathcal{H}$, and the Hamiltonian is given by

$$H = \int d\sigma \mathcal{H}. \tag{19.16}$$

The off-diagonal spacetime components define a momentum:

$$
\theta_{0i} = \theta_{i0} = \frac{\partial \mathcal{L}}{\partial(\partial^0 \phi_A)}(\partial_i \phi_A) + \frac{\partial \mathcal{L}}{\partial(\partial^0 \phi_A^*)}(\partial_i \phi_A^*)
$$

$$
= \hbar^2 c^2 \left\{ (\partial_0 \phi_A^*)(\partial_i \phi_A) + (\partial_0 \phi_A)(\partial_i \phi_A^*) \right\}. \tag{19.17}
$$

Taking the integral over all space enables us to integrate by parts and show that this quantity is the expectation value (inner product) of the momentum:

$$
\int d\sigma \, \theta_{0i} = \hbar^2 c \int d\sigma \, \left(\phi^* \partial_i \partial_0 \phi - (\partial_0 \phi^*) \partial_i \phi \right)
$$

$$
= -(\phi, \, p_i c \phi), \tag{19.18}
$$

where $p = -i\hbar \partial_i$. The diagonal space components are given by

$$
\theta_{ii} = \frac{\partial \mathcal{L}}{\partial(\partial^i \phi_A)}(\partial_i \phi_A) + \frac{\partial \mathcal{L}}{\partial(\partial^i \phi_A^*)}(\partial_i \phi_A^*) - \mathcal{L}
$$

$$
= 2\hbar^2 c^2 (\partial_i \phi^*)(\partial_i \phi) - \mathcal{L}, \tag{19.19}
$$

where i is not summed. Similarly, the off-diagonal 'stress' components are given by

$$
\theta_{ij} = \frac{\partial \mathcal{L}}{\partial(\partial^i \phi_A)}(\partial_j \phi_A) + \frac{\partial \mathcal{L}}{\partial(\partial^i \phi_A^*)}(\partial_j \phi_A^*)
$$

$$
= \hbar^2 c^2 \left\{ (\partial_i \phi_A^*)(\partial_j \phi_A) + (\partial_j \phi_A^*)(\partial_i \phi_A) \right\}
$$

$$
= \hbar^{-1} c (\phi_A, \, p_i p_j \phi_A). \tag{19.20}
$$

We see that the trace over spatial components in $n + 1$ dimensions is

$$
\sum_i \theta_{ii} = \mathcal{H} - 2m^2 c^4 \phi_A^2 - 2V(\phi) + (n-1)\mathcal{L}, \tag{19.21}
$$

so that the full trace gives

$$
\theta^\mu{}_\mu = g^{\mu\nu} \theta_{\nu\mu} = -2m^2 c^4 \phi_A^2 - 2V(\phi) + (n-1)\mathcal{L}. \tag{19.22}
$$

This vanishes for $m = V = 0$ in $1 + 1$ dimensions.

19.7 Formulation as a two-component real field

The real and imaginary components of the complex scalar field can be parametrized as a two-component vector or real fields φ_A, where $A = 1, 2$. Define

$$
\Phi(x) = \frac{1}{\sqrt{2}} \left(\varphi_1(x) + i\varphi_2(x) \right). \tag{19.23}
$$

Substituting in, and comparing real and imaginary parts, one finds

$$D_\mu = \partial_\mu + ieA_\mu, \tag{19.24}$$

or

$$D_\mu\varphi_A = \partial_\mu\varphi_A - e\epsilon_{AB}A_\mu\varphi_B. \tag{19.25}$$

The action becomes

$$S = \int (\mathrm{d}x) \left\{ \frac{1}{2}(D^\mu\varphi_A)(D_\mu\varphi_A) + \frac{1}{2}m^2\varphi_A\varphi_A - J_A\varphi_A \right\}. \tag{19.26}$$

Notice that the operation charge conjugation is seen trivially here, due to the presence of ϵ_{AB}, as the swapping of field labels.

20

The Dirac field

The Klein–Gordon equation's negative energy solutions and corresponding negative probabilities prompted Dirac to look for a relativistically invariant equation of motion which was linear in time. His equation was formally the square-root of the Klein–Gordon equation.

The Dirac equation leads naturally to the existence of spin $\frac{1}{2}$. It is the basic starting point for the study of spin-$\frac{1}{2}$ particles such as the electron and quarks. It also appears in condensed matter physics as the relevant low-energy degrees of freedom in the strong-coupling limit of the Hubbard model [92], and has been used as an alternative formulation of gravity [84].

20.1 The action

The action for the Dirac field is given by

$$S_\mathrm{D} = \int (\mathrm{d}x) \left\{ -\frac{1}{2}\mathrm{i}\hbar c \overline{\psi} (\gamma^\mu \overrightarrow{\partial_\mu} - \gamma^\mu \overleftarrow{\partial_\mu})\psi \right.$$
$$\left. + (mc^2 + V)\overline{\psi}\psi - \overline{J}\psi - \overline{\psi}J \right\}, \tag{20.1}$$

where ψ and $\overline{\psi} = \psi^\dagger \gamma^0$ are d_R-component spinors. The γ^μ are $d_\mathrm{R} \times d_\mathrm{R}$ matrices, defined below. All quantities here are implicitly matrix-valued. They have hidden 'spinor' indices, which we shall write explicitly at times using Greek letters $\alpha, \beta, \ldots$.

The variation of the action with respect to a dynamical change in the field $\overline{\psi}$ gives the equation of motion for ψ is found by varying the action with respect to $\overline{\psi}$, and is given by

$$(-\mathrm{i}\hbar c \gamma^\mu \partial_\mu + mc^2 + V)\psi = J. \tag{20.2}$$

If we drop the source term J, this can also be written

$$\mathrm{i}\hbar c \partial_0 \psi = \gamma^0 (-\mathrm{i}\hbar c \gamma^i \partial_i + mc^2 + V)\psi = H_\mathrm{D}\psi, \tag{20.3}$$

where H_D is the differential Hamiltonian operator (to be distinguished from the field theoretical Hamiltonian below). The conjugate equation is found by varying the action with respect to ψ and may be written as

$$\overline{\psi}(i\hbar c\gamma^\mu \overleftarrow{\partial}_\mu +mc^2 + V) = 0, \qquad (20.4)$$

or in terms of the differential Hamiltonian operator H_D,

$$-i\hbar c(\partial_0\overline{\psi}) = \overline{\psi}\gamma^0 H_D\gamma^0. \qquad (20.5)$$

The free Dirac equation may be viewed as essentially the square-root of the Klein–Gordon equation. In the massless limit, the linear combination of derivatives is a Lorentz-scalar-representation of the square-root of $\Box$. This may be verified by squaring the Dirac operator and separating the product of γ-matrices into symmetric and anti-symmetric parts:

$$(\gamma^\mu\partial_\mu)^2 = \gamma^\mu\gamma^\nu \partial_\mu\partial_\nu \qquad (20.6)$$

$$= \frac{1}{2}\{\gamma^\mu, \gamma^\nu\}\partial_\mu\partial_\nu + \frac{1}{2}[\gamma^\mu, \gamma^\nu]\partial_\mu\partial_\nu \qquad (20.7)$$

$$= -g^{\mu\nu}\partial_\mu\partial_\nu + \frac{1}{2}\gamma^\mu\gamma^\nu[\partial_\mu, \partial_\nu] \qquad (20.8)$$

$$= -\Box . \qquad (20.9)$$

The commutator of two partial derivatives vanishes when the derivatives act on any non-singular function. Since the fields are non-singular, except in the presence of certain exceptional interactions which do not apply here, the Dirac operator can be identified as the square-root of the d'Alambertian.

20.2 The γ-matrices

In order to satisfy eqn. (20.9), the γ-matrices must satisfy the relation

$$\{\gamma^\mu, \gamma^\nu\} = -2g^{\mu\nu} \qquad (20.10)$$

$$(\gamma^0)^2 = -(\gamma^i)^2 = I. \qquad (20.11)$$

The matrices satisfy a *Clifford algebra*. The set of matrices which satisfies this constraint is of fundamental importance to the Dirac theory. They are not unique, but may have several representations. The form of the γ^μ matrices is dependent on the dimension of spacetime and, since they carry a spacetime index, on the Lorentz frame [61, 103].

Products of the γ^μ form a group of matrices Γ_a, where $a = 1, \ldots, d_G$, and the dimension of the group is $d_G = 2^{(n+1)}$. The elements Γ_a are proportional to the unique combinations:

$$I$$
$$\gamma^0, \quad \gamma^i, \quad \gamma^i\gamma^j, \quad \gamma^0\gamma^i \quad (i \neq j)$$
$$\gamma^0\gamma^i\gamma^j, \quad \gamma^1\gamma^2 \ldots \gamma^n$$
$$\gamma^0\gamma^1\gamma^2 \ldots \gamma^n.$$

The usual case is $n + 1 = 4$, where one has Γ_a,

$$I$$
$$\gamma^0, \quad i\gamma^1, \quad i\gamma^3, \quad i\gamma^3,$$
$$\gamma^0\gamma^1, \quad \gamma^0\gamma^2, \quad \gamma^0\gamma^3, \quad i\gamma^2\gamma^3, \quad i\gamma^3\gamma^1, \quad i\gamma^2\gamma^3$$
$$i\gamma^0\gamma^2\gamma^3, \quad i\gamma^0\gamma^1\gamma^3, \quad i\gamma^0\gamma^1\gamma^2, \quad \gamma^1\gamma^2\gamma^3$$
$$i\gamma^0\gamma^1\gamma^2\gamma^3$$

Factors of i have been introduced so that each matrix squares to the identity (see ref. [112]). These may also be grouped differently, in the more suggestive Lorentz-covariant form:

1	scalar
γ^μ	vector
$\sigma_{\mu\nu}$	anti-symmetric tensor
$\gamma^5\gamma^\mu$	pseudo-(axial) vector
γ^5	pseudo-scalar

where

$$\sigma^{\mu\nu} = \frac{1}{2i}[\gamma^\mu, \gamma^\nu]. \tag{20.12}$$

For each Γ_a, with the exception of the identity element, it is possible to find a suitably defined Γ_a, such that

$$\Gamma_a\Gamma_b\Gamma_a = -\Gamma_b \quad (b \neq 1). \tag{20.13}$$

By taking the trace of this relation, and noting that

$$\mathrm{Tr}(\Gamma_a\Gamma_b\Gamma_a) = \mathrm{Tr}(\Gamma_a\Gamma_a\Gamma_b) = \mathrm{Tr}(\Gamma_b) \tag{20.14}$$

one obtains

$$\mathrm{Tr}(\Gamma_b) = -\mathrm{Tr}(\Gamma_b) = 0. \tag{20.15}$$

From this, it follows that the 2^{n+1} elements are linearly independent, since, if one attempts to construct a linear combination which is zero,

$$\sum_a \lambda_a\Gamma_a = 0, \tag{20.16}$$

then taking the trace of this implies that each of the components $\lambda_a = 0$. This establishes that each component is linearly independent and that a matrix representation for the γ^μ must have at least this number of elements, in order to satisfy the algebra constraints. This divides the possibilities into two cases, depending on whether the dimension of spacetime is even or odd.

For even $n+1$, the γ^μ are most simply represented as $d_R \times d_R$ matrices, where

$$d_R = 2^{(n+1)/2}. \tag{20.17}$$

d_R^2 then contains exactly the right number of elements. Although the matrices are not unique (they can be transformed by similarity transformations), all such sets of matrices of this dimension are *equivalent* representations. Moreover, since there is no redundancy in the matrices, the $d_R \times d_R$ representations are also *irreducible*, or *fundamental*. In this case, the identity is the only element of the group which commutes with every other element (the group is said to have a trivial centre). Another common way of expressing this, in the literature, is to observe that other matrices, typically $\gamma^0\gamma^1\dots\gamma^n$, anti-commute with an arbitrary element γ^μ. There are thus more elements in the centre of the group than the identity. This is a sign of reducibility or multiple equivalent representations.

For odd $n+1$, it is not possible to construct a matrix with exactly the right number of elements. This is a symptom of the existence of several *inequivalent* representations of the algebra. In this case, one must either construct several sets of smaller matrices (which are inequivalent), or combine these into matrices of larger dimension, which are *reducible*. The reducible matrices reduce to block-diagonal representations, in which the blocks are the multiple, inequivalent, irreducible representations. In this case, the identity is not the only element of the group which commutes with every other element (the group is said to have a non-trivial centre), and the matrix $\gamma^0\gamma^1\dots\gamma^n$ anti-commutes with an arbitrary element γ^μ. Spinors in $n+1$ dimensions are discussed in ref. [10].

20.2.1 Example: $n+1 = 4$

In $3+1$ dimensions, the dimension of the algebra is $2^4 = 16$, and one has the standard representation,

$$\gamma^0 = \begin{pmatrix} -1 & 0 \\ 0 & 1 \end{pmatrix}, \quad \gamma^i = \begin{pmatrix} 0 & -\sigma^i \\ \sigma^i & 0 \end{pmatrix} \tag{20.18}$$

where σ^i are the Pauli matrices, defined by

$$\sigma_1 = \begin{pmatrix} 0 & 1 \\ 1 & 0 \end{pmatrix}, \quad \sigma_2 = \begin{pmatrix} 0 & -i \\ i & 0 \end{pmatrix}, \quad \sigma_3 = \begin{pmatrix} 1 & 0 \\ 0 & -1 \end{pmatrix}. \tag{20.19}$$

The product of two γ-matrices may be evaluated by separating into even and odd parts, in terms of the commutator and anti-commutator:

$$\gamma_\mu \gamma_\nu = \frac{1}{2}[\gamma_\mu, \gamma_\nu] + \frac{1}{2}\{\gamma_\nu, \gamma_\nu\}$$
$$= i\sigma_{\mu\nu} - g_{\mu\nu}, \tag{20.20}$$

where

$$\sigma_{\mu\nu} = \frac{1}{2i}[\gamma_\mu, \gamma_\nu]. \tag{20.21}$$

The product of all the γ's is usually referred to as γ^5, and is defined by

$$\gamma^5 = i\gamma^0\gamma^1\gamma^2\gamma^3$$
$$= \begin{pmatrix} 0 & -I \\ I & 0 \end{pmatrix}. \tag{20.22}$$

Clearly, this notation is poorly motivated in spacetime dimensions other than $3+1$. In $3+1$ dimensions, it is straightforward to show that

$$(\gamma^5)^2 = 1, \quad \{\gamma^\mu, \gamma^5\} = 0. \tag{20.23}$$

The cyclic nature of the trace can be used together with the anti-commutativity of γ^5 to prove that the trace of an odd number of γ-matrices vanishes in $3+1$ dimensions. To see this, one notes that

$$\mathrm{Tr}(\gamma_5 A \gamma_5^{-1}) = \mathrm{Tr}(A). \tag{20.24}$$

Thus, choosing a product of m such matrices $A = \gamma_\mu \gamma_\nu \ldots \gamma_\sigma$, such that

$$\gamma_5 A = (-1)^m A \gamma_5, \tag{20.25}$$

it follows immediately that

$$\mathrm{Tr}(A) = (-1)^m \mathrm{Tr}(A), \tag{20.26}$$

and hence the trace of an odd number m of the matrices must vanish. The hermiticity properties of the matrices are contained by the relation

$$\gamma^{\mu\dagger} = \gamma^0 \gamma^\mu \gamma^0, \tag{20.27}$$

which summarizes

$$\gamma^{0\dagger} = \gamma^0 \tag{20.28}$$
$$\gamma^{i\dagger} = -\gamma^i. \tag{20.29}$$

20.2.2 Example: $n + 1 = 3$

In $2 + 1$ dimensions, the dimension of the algebra is $2^3 = 8$, and thus the minimum representation is is terms of either two irreducible sets of 2 matrices, or a single set of reducible 4×4 matrices, with redundant elements.

The fundamental $\vec{2}$ representation is satisfied by

$$\gamma^0 = \sigma_3, \quad \gamma^i = -i\sigma_i \tag{20.30}$$

for $i = 1, 2$. This representation breaks parity invariance, thus there are two *inequivalent* representations which differ by a sign.

$$\gamma^\mu, -\gamma^\mu \tag{20.31}$$

The $\vec{4}$ representation is a symmetrized direct sum of these, padded with zeros:

$$\gamma^\mu(\vec{4}) = \begin{pmatrix} +\gamma^\mu(\vec{2}) & 0 \\ 0 & -\gamma^\mu(\vec{2}) \end{pmatrix}. \tag{20.32}$$

The matrices of the $\vec{2}$ representation satisfy

$$\gamma^\mu \gamma^\nu = -g^{\mu\nu} - i\epsilon^{\mu\nu\rho}\gamma^c \tag{20.33}$$

$$\mathrm{Tr}(\gamma^\mu \gamma^\mu \gamma^\rho) = 2i\epsilon^{\mu\nu\rho} \tag{20.34}$$

$$\gamma_5 = \gamma^0 \gamma^1 \gamma^2 \tag{20.35}$$

$$[\gamma_5, \gamma_\mu] = 0, \tag{20.36}$$

where the first of these relations is found by splitting into a commutator and anti-commutator and using the $su(2)$ Lie algebra relation for the Pauli matrices:

$$[\sigma_i, \sigma_j] = 2i\epsilon_{ijk}\sigma_k. \tag{20.37}$$

In the $\vec{4}$ representation in $2 + 1$ dimensions the results are as for $\vec{2}$ except that

$$\mathrm{Tr}(\gamma^\mu \gamma^\nu \gamma^\rho)_{\vec{4}} = 0. \tag{20.38}$$

Note that the product of all elements is usually referred to as γ^5 in the literature, rather than γ^4, by analogy with the $(3 + 1)$ dimensional case.

$$\gamma_5 = \gamma^5 = \gamma^4 = i\gamma^0 \gamma^1 \gamma^2$$

$$= \begin{pmatrix} -I & 0 \\ 0 & -I \end{pmatrix}. \tag{20.39}$$

Since this is a multiple of the identity matrix, it commutes with every element in the algebra. Thus there are two elements to the centre of the group: I and $-I$. The centre is the discrete group Z_2, and the complete fundamental representation of the algebra is

$$\gamma^\mu(\vec{2}) \otimes Z_2. \tag{20.40}$$

These two inequivalent representations correspond to the fact that, in a two-dimensional plane, spin up and spin down cannot be continuously rotated into one another (not even classically), and thus these two physical possibilities are disconnected regions of the rotation group. In much of the literature on two-dimensional physics, it is common to adopt either a spin up, or spin down 2×2 representation for the γ-matrices, not the complete $\bar{4}$ representation.

20.3 Transformation properties of the Dirac equation

Consider a Lorentz transformation of the Dirac spinor by a matrix representation of the Lorentz group:

$$\psi'(x') = S(L)\,\psi(x) = L_R\,\psi(x). \tag{20.41}$$

The matrix, usually denoted $S(L)$ in the literature, is just an example of a non-adjoint representation of the Lorentz group from section 9.4.3. This representation has to carry spinor indices α, β, which are suppressed above, in the usual way. These spinor indices correspond to the representation indices A, B of section 9.4.3.

A transformation of the free Dirac equation may be written as

$$(\gamma^\mu p'_\mu c + mc^2)\psi'(x') = 0 \tag{20.42}$$

$$(\gamma^\mu (L^{-1})^\mu{}_\nu p'_\mu c + mc^2)(L_R\psi(x)) = 0, \tag{20.43}$$

where one recalls that

$$p^\mu = L^\mu{}_\nu\, p^\nu \quad \rightarrow \quad p_\mu = (L^{-1})\mu{}^\nu p_\nu. \tag{20.44}$$

Multiplying on the left hand side by L_R^{-1}, and comparing with the untransformed equation, leads to a condition

$$L_R^{-1}\,\gamma^\mu\, L_R = L^\mu{}_\nu \gamma^\nu, \tag{20.45}$$

which is an identity, provided $L = L_{\text{adj}}$, the adjoint representation of the group.

The infinitesimal form of the spinor representation may be written in terms of the generators of this representation T_R:

$$L_R = S(L) = I + \theta^a T_R^a, \tag{20.46}$$

or, with spinor (representation) indices intact,

$$S(L)^\alpha{}_\beta = \delta^\alpha{}_\beta + \theta^a (T_R^a)^\alpha{}_\beta. \tag{20.47}$$

Consider an infinitesimal transformation

$$x'^\mu = x^\mu + \epsilon\omega^\mu{}_\nu x^\nu, \tag{20.48}$$

so that

$$L_R(I + \epsilon\omega) = I + \epsilon T_R. \tag{20.49}$$

The adjoint transformation can thus be expressed in two equivalent forms:

$$\begin{aligned} S^{-1}\gamma^\mu S &= (1 - \epsilon T)\gamma^\mu(1 - \epsilon T) \\ &= \gamma^\mu + \epsilon(\gamma^\mu T - T\gamma_\mu) \\ &= (L_{\mathrm{adj}})^\mu_\nu(1 + \epsilon_\omega)\gamma^\nu \\ &= \gamma^\mu + \epsilon\omega^\mu_\nu\gamma^\nu. \end{aligned} \tag{20.50}$$

Thus,

$$\gamma^\mu T - T\gamma^\mu = \omega^\mu_\nu\gamma^\nu, \tag{20.51}$$

which defines T up to a multiple of the identity matrix. Choosing unit determinant $\det(I + \epsilon T) = 1 + \epsilon\,\mathrm{Tr}\,T$, we have that $\mathrm{Tr}\,T = 0$, and one may write

$$(T_R)^\alpha_\beta = \frac{1}{8}\omega^{\mu\nu}(\gamma_\mu\gamma_\nu - \gamma_\nu\gamma_\mu)^\alpha_\beta, \tag{20.52}$$

or, compactly,

$$(T_R)^\alpha_\beta = \frac{i}{4}\omega^{\mu\nu}\sigma_{\mu\nu}. \tag{20.53}$$

20.3.1 Rotations

An infinitesimal rotation by angle ϵ about the x^1 axis has

$$\omega^{23} = -\omega^{32} = 1, \tag{20.54}$$

and all other components zero. Thus the generator in the spinor representation is

$$T^1 = \frac{1}{2}\gamma_2\gamma_3, \tag{20.55}$$

and the exponentiated finite element becomes

$$\begin{aligned} S(R_1) = e^{\theta_1 T_R} &= e^{-\frac{i}{2}\theta\Sigma_1}, \\ &= I\,\cos\frac{\theta}{2} + i\Sigma_1\sin\frac{\theta}{2}, \end{aligned} \tag{20.56}$$

where

$$\Sigma_1 = \begin{pmatrix} \sigma_1 & 0 \\ 0 & \sigma_1 \end{pmatrix}. \tag{20.57}$$

The half-angles are characteristic of the double-valued nature of spin:

$$S(\theta_1 + 2\pi) = -S(\theta_1). \tag{20.58}$$

20.3.2 Boosts

For a boost in the x^1 direction, $\omega^{10} = -\omega^{01} = 1$,

$$S(B_1) = \frac{1}{2}\gamma^0\gamma^1 = \frac{1}{2}\begin{pmatrix} 0 & \sigma_1 \\ \sigma_1 & 0 \end{pmatrix} \equiv \frac{1}{2}\alpha_1. \qquad (20.59)$$

The finite, exponentiated element is thus

$$S(B_1) = e^{\frac{\alpha}{2}\alpha_1} = I\cosh\frac{\alpha}{2} + \alpha_1\sinh\frac{\alpha}{2}, \qquad (20.60)$$

where $\tanh\alpha = v/c$. Notice that the half-valued arguments have no effect on translations.

20.3.3 Parity and time reversal

The meaning of parity invariance is intrinsically linked to the number of spacetime dimensions, since an even number of reflections about spatial axes is equivalent to a rotation, and is therefore simply connected to the infinitesimally generated group. In that case, spatial reflection is defined by a reflection in an odd number of axes. In odd numbers of spatial dimensions, reflections in all axes lead to a 'large' transformation which cannot be generated by exponentiated infinitesimal generators.

Consider the case in $3 + 1$ dimensions. For a space inversion, one has

$$\begin{aligned} L_R^{-1}\gamma^0 L_R &= \gamma^0 \\ L_R^{-1}\gamma^i L_R &= -\gamma^i. \end{aligned} \qquad (20.61)$$

Thus, the parity transformation can be represented by:

$$S(P) = e^{i\phi}\gamma^0 \qquad (20.62)$$

in Dirac space. This exchanges the upper and lower spinor contributions. Similarly, a time inversion

$$\begin{aligned} L_R^{-1}\gamma^0 L_R &= -\gamma^0 \\ L_R^{-1}\gamma^i L_R &= \gamma^i \end{aligned} \qquad (20.63)$$

can be given the form

$$S(T) = e^{i\phi}\gamma^5. \qquad (20.64)$$

20.3.4 Charge conjugation

Charge conjugation transforms a positive energy solution with charge q into a negative energy solution with charge $-q$. One searches for a unitary transformation C with the following properties:

$$C \psi C^{-1} = \eta C \overline{\psi}^{\mathrm{T}}$$
$$C A_\mu C^{-1} = -A_\mu. \qquad (20.65)$$

η is a possible intrinsic property of the field, $\eta^2 = 1$. The two features of this transformation are that it exchanges positive for negative energies and that it reflects the vector field like an axial vector. Since A_μ always multiplies the charge q, in the covariant derivative, this is equivalent to changing the sign of the charge. The action for the gauged Dirac equation is ($\hbar = c = 1$),

$$S = \int (\mathrm{d}x) \, \overline{\psi} (i\gamma^\mu D_\mu + m)\psi, \qquad (20.66)$$

where $D_\mu = \partial_\mu + iq A_\mu$. In order to find a transformation which exchanges ψ with $\overline{\psi}^{\mathrm{T}}$, one begins by integrating by parts:

$$S = \int (\mathrm{d}x) \, \overline{\psi} (-i\gamma^\mu (\overleftarrow{\partial_\mu} - iq A_\mu) + m)\psi, \qquad (20.67)$$

then, taking the transpose:

$$S = \int (\mathrm{d}x) \, \psi^{\mathrm{T}} (-i\gamma^{\mathrm{T}\mu} (\overrightarrow{\partial_\mu} - iq A_\mu)) \overline{\psi}^{\mathrm{T}}. \qquad (20.68)$$

This has almost the same form as the original, untransposed action, with opposite charge. In order to make it identical, we require a matrix which has the property

$$C \gamma^{\mathrm{T}\mu} C^{-1} = -\gamma^\mu. \qquad (20.69)$$

Introducing such a matrix, one has

$$S = \int (\mathrm{d}x) \, (\gamma^{\mathrm{T}} C^{-1})(i\gamma^\mu D_\mu^* + m)(C \overline{\psi}^{\mathrm{T}})$$
$$= \int (\mathrm{d}x) \, \overline{\psi}^{\mathrm{c}} (i\gamma^\mu D_\mu^* + m)\psi^{\mathrm{c}}, \qquad (20.70)$$

where the charge conjugated field is $\psi^{\mathrm{c}} = C \overline{\psi}^{\mathrm{T}}$.

The existence of a matrix C, in $3 + 1$ dimensions, possessing the above properties can be determined as follows [112]. Taking the transpose of the Clifford algebra relation,

$$\{\gamma^{\mathrm{T}\mu}, \gamma^{\mathrm{T}\nu}\} = -2g^{\mu\nu}, \qquad (20.71)$$

one sees that the transposed γ-matrices also satisfy the algebra, and must therefore be related to the untransposed ones by a similarity transformation

$$\gamma^{\mathrm{T}\mu} = B^{-1} \gamma^\mu B. \qquad (20.72)$$

In $3+1$ dimensions, the 4×4 γ-matrices are irreducible, and thus the existence of a non-singular, unitary B is guaranteed. Taking the transpose of eqn. (20.72) and re-using the relation to replace for γ^μ, one obtains

$$\gamma^{T\mu} = (B^{-1}B^{T})\, \gamma^{T\mu}\, (B^{-1T}B), \tag{20.73}$$

thus establishing that $B^{-1}B^{T}$ commutes with all the γ-matrices. From Schur's lemma, it follows that this must be a multiple of the identity:

$$B^{-1}B^{T} = cI. \tag{20.74}$$

Taking the inverse and then the complex conjugate of this relation, one finds

$$\frac{1}{c^*} = B^*(B^{T})^{-1*}$$
$$= B^* B^{**} \tag{20.75}$$
$$= B^* B. \tag{20.76}$$

where we have used the unitarity $B^\dagger B = I$. This means that c is real, and furthermore that $c = \pm 1$, i.e.

$$B = \pm B^{T}, \tag{20.77}$$

so, from this, the matrix is either symmetrical or anti-symmetrical. An additional constraint comes from the number of symmetrical and anti-symmetrical degrees of freedom in the 4×4 γ-matrices. If B is anti-symmetric, then the six matrices $\gamma^\mu B, \gamma^5 B, B$ are also anti-symmetric, whereas the ten matrices $B\gamma^5\gamma^\mu, B\sigma^{\mu\nu}$ are symmetrical. This matches the number of anti-symmetrical degrees of freedom in a 4×4 matrix representation. Conversely, if one takes B to be symmetrical, then the numbers are reversed and it does not match. One concludes, then, that B is an anti-symmetric, unitary matrix. This result was shown by Pauli in 1935. It has now been shown that it is possible to construct a similarity transformation which turns γ-matrices into their transposes. The matrix we require is now

$$C = -i\gamma^5 B. \tag{20.78}$$

With this definition, we have

$$C^{-1}\gamma^\mu C = -B^{-1}i\gamma_5 \gamma^\mu i\gamma_5 B = -B^{-1}\gamma^\mu B$$
$$= -\gamma^{T\mu}. \tag{20.79}$$

C is thus a charge conjugation matrix for Dirac spinors.

20.4 Chirality in 3 + 1 dimensions

The field equations of massless spinors are, from eqn. (20.101),

$$\begin{pmatrix} -p_0 + & -\sigma^i p_i \\ \sigma^i p_i & p_0 \end{pmatrix} \begin{pmatrix} \chi_1 \\ \chi_2 \end{pmatrix} = 0 \tag{20.80}$$

or

$$\sigma^i \hat{p}_i \, \chi_L = 2\lambda \chi_L = -\chi_L$$
$$\sigma^i \hat{p}_i \, \chi_R = 2\lambda = +\chi_R, \tag{20.81}$$

where $\lambda = \frac{1}{2}\sigma^i \hat{p}_i$, and

$$\chi_L = \chi_1 + \chi_2$$
$$\chi_R = \chi_1 - \chi_2. \tag{20.82}$$

These equations are referred to as the Weyl equations, and χ_L and χ_R are Weyl-spinors. For such massless particles, the eigenvalue of γ^5 is referred to as the *chirality* of the solution:

$$\gamma^5 u(p, \lambda) = 2\lambda u(p, \lambda)$$
$$\gamma^5 v(p, \lambda) = -2\lambda v(p, \lambda). \tag{20.83}$$

A projection operator for the chirality states is thus

$$\mathcal{P}_\pm = \frac{1}{2}(1 \pm \gamma^5). \tag{20.84}$$

Particles with helicity $+\frac{1}{2}$ are referred to as right handed, while particles with helicity $-\frac{1}{2}$ are referred to as left handed. Only left handed neutrinos interact by the weak interaction and appear in the Standard Model. Symmetry under the continuous transformation

$$\psi(x) \to e^{i\lambda\gamma_5}\psi \tag{20.85}$$

is known as chiral symmetry.

20.5 Field continuity

The variation of the action leads to surface terms,

$$\hbar \int d\sigma^\mu \left(\delta\overline{\psi}\gamma_\mu\psi \right), \tag{20.86}$$

for ψ variations, and

$$\hbar \int d\sigma^\mu \left(\overline{\psi}\gamma_\mu\delta\psi \right), \tag{20.87}$$

for $\overline{\psi}$ variations. This provides us with definitions for the conjugate momenta across spacelike hyper-surfaces σ:

$$\hbar\overline{\Pi}_\sigma^A = \gamma_\sigma \psi^A, \tag{20.88}$$

for the variable conjugate to $\overline{\psi}$, and

$$\Pi_\sigma^A = \hbar\overline{\psi}^A \gamma_\sigma, \tag{20.89}$$

for the variable conjugate to ψ. The canonical values for these momenta are

$$\overline{\Pi} = \hbar\gamma_0 \psi$$
$$\Pi = \hbar\overline{\psi}\gamma_0 = \psi^\dagger. \tag{20.90}$$

20.6 Conserved norm and probability

The linear nature of the Dirac action implies that the conserved current is independent of derivatives. This means that the sign of the energy $i\hbar c\partial_0$ cannot change the sign of the conserved probability, thus the Dirac equation does not suffer the problem of negative norms or probabilities as does the Klein–Gordon equation.

To determine the conserved current, one considers the effect of an infinitesimal x-independent phase transformation δ:

$$\begin{aligned}
\delta S = \int (\mathrm{d}x)\Big\{&(\overline{\psi}e^{-is}(-i\delta s)\gamma^\mu\partial_\mu(e^{is}\psi)) \\
+&(\overline{\psi}e^{-is}(-i\delta s)\gamma^\mu\partial_\mu((-i\delta s)e^{is}\psi)) \\
-&(\partial_\mu(\overline{\psi}e^{-is}(-i\delta s))\gamma^\mu(e^{is}\psi)) \\
-&(\partial_\mu(\overline{\psi}e^{-is}(-i\delta s))\gamma^\mu((-i\delta s)e^{is}\psi))\Big\}
\end{aligned} \tag{20.91}$$

Integrating by parts to remove derivatives from δs, and using the equations of motion, one arrives at the simple expression

$$\delta S = \hbar \int \mathrm{d}\sigma^\mu (\overline{\psi}\gamma_\mu\psi)\delta s, \tag{20.92}$$

which defines a conserved current $\delta S = \int \mathrm{d}\sigma^\mu J_\mu \delta s$. This motivates the definition of an inner product given by

$$(\psi_1, \psi_2) = -\frac{1}{2}\int \mathrm{d}\sigma^\mu (\overline{\psi}_1\gamma_\mu\psi_2 + \overline{\psi}_2\gamma_\mu\psi_1), \tag{20.93}$$

giving the norm of the field as

$$(\psi, \psi) = -\int d\sigma^\mu \overline{\psi} \gamma_\mu \psi. \tag{20.94}$$

The canonical interpretation of this is

$$(\psi_1, \psi_2) = -\frac{1}{2} \int d\sigma (\overline{\psi}_1 \gamma_0 \psi_2 + \overline{\psi}_2 \gamma_0 \psi_1), \tag{20.95}$$

which means that the norm may also be written

$$(\psi, \psi) = -\int d\sigma \, \psi^\dagger \gamma^0 \gamma_0 \psi = \int d\sigma \, \psi^\dagger \psi. \tag{20.96}$$

The norm of the field is defined separately on the manifold of positive and negative energy solutions.

20.7 Free-field solutions in $n = 3$

The free-field equation is

$$(-i\hbar c \gamma^\mu \partial_\mu + mc^2)_{\alpha\beta} \psi_\beta(x) = 0, \tag{20.97}$$

where $\psi_\alpha(x)$ is a $2l$-component vector for some $l \geq n/2$, which lives on spinor space (usually these indices are suppressed). In a given number of dimensions, we may express this equation in terms of a representation of the γ-matrices. In three dimensions we may use eqn. (20.18) to write

$$\begin{pmatrix} i\hbar\partial_t + mc^2 & i\hbar c\sigma^i\partial_i \\ -i\hbar c\sigma^i\partial_i & -i\hbar\partial_t + mc^2 \end{pmatrix} \psi = 0, \tag{20.98}$$

where we suppress the α, β spinor indices. The blocks are now 2×2 matrices, and the spinor may also be written in terms of two two-component spinors u:

$$\psi = \begin{pmatrix} u_1 \\ u_2 \end{pmatrix}. \tag{20.99}$$

If we transform the spinors to momentum space,

$$\psi(x) = \int \frac{d^{n+1}k}{(2\pi)^{n+1}} e^{ikx} \psi(k), \tag{20.100}$$

then the field equations may be written as

$$\begin{pmatrix} -p_0 c + mc^2 & -c\sigma^i p_i \\ c\sigma^i p_i & p_0 c + mc^2 \end{pmatrix} \begin{pmatrix} u_1 \\ u_2 \end{pmatrix} = 0, \tag{20.101}$$

where $p_\mu = \hbar k_\mu$. This matrix equation has non-zero solutions for ψ only if the determinant of the operator vanishes. Thus,

$$\det = p^2 c^2 + m^2 c^4 = 0. \tag{20.102}$$

Here one makes use of the fact that

$$
\begin{aligned}
(\sigma^i p_i)^2 &= \sigma^i \sigma^j p_i p_j \\
&= \left(\frac{1}{2}[\sigma^i, \sigma^j] + \frac{1}{2}\{\sigma^i, \sigma^j\} \right) p_i p_j \\
&= (i\epsilon^{ijk}\sigma_k + \delta^{ij}) p_i p_j \\
&= p^i p_i.
\end{aligned}
\tag{20.103}
$$

Eqn. (20.102) indicates that the solutions of the Dirac equation must satisfy the relativistic energy relation. Thus the Dirac field also satisfies a Klein–Gordon equation, which may be seen by operating on eqn. (20.97) with the conjugate of the Dirac operator:

$$
(i\hbar c\gamma^\mu \partial_\mu + mc^2)(-i\hbar c\gamma^\mu \partial_\mu + mc^2)\psi(x) = 0
$$
$$
(-\hbar^2 c^2 \Box + m^2)\psi = 0. \tag{20.104}
$$

The last line follows from eqn. (20.20). The vanishing of the determinant also gives us a relation which will be useful later, namely

$$\frac{(p_0 c + mc^2)}{c\sigma^i p_i} = \frac{-c\sigma^i p_i}{(-p_0 c + mc^2)}. \tag{20.105}$$

The 2×2 components of eqn. (20.101) are now

$$
\begin{aligned}
(-p_0 c + mc^2)u_1 - c(\sigma^i p_i)u_2 &= 0 \\
c(\sigma^i p_i)u_1 + (p_0 c + mc^2)u_2 &= 0,
\end{aligned}
\tag{20.106}
$$

which implies that the two-component spinors u are linearly dependent:

$$
\begin{aligned}
u_1 &= \frac{c(\sigma^i p_i)}{(-p_0 c + mc^2)} u_2, \\
u_1 &= -\frac{(p_0 c + mc^2)}{c(\sigma^i p_i)} u_2.
\end{aligned}
\tag{20.107}
$$

The consistency of these apparently contradictory relations is secured by the determinant constraint in eqn. (20.105).

In spite of the linear (first-order) derivative in the Dirac action, the determinant condition for non-trivial solutions leads us straight back to a quadratic constraint on the allowed spectrum of energy and momenta. This means that both positive and negative energies are allowed in the Dirac equation, exactly as in the Klein–Gordon case. The linear derivative does cure the negative probabilities, however, as we show below.

The solutions of the Dirac equation may be written in various forms. A direct attempt to apply the field equation constraint in a delta function, by analogy with the scalar field, cannot work directly, since the delta function cannot have a matrix argument. However, by introducing a projection operator $(-\gamma^\mu p_\mu c + mc^2)/2|p_0|$, it is possible to write

$$\psi(x) = \int \frac{d^{n+1}k}{(2\pi)^{n+1}} e^{ikx} (-\gamma^\mu p_\mu c + mc^2) \delta(p^2 c^2 + m^2 c^4) u(k),$$

(20.108)

where $p = \hbar k$ and $u(k)$ is a mass shell spinor. The projection term ensures that application of the Dirac operator leads to the squared mass shell constraint. By inserting $\theta(\pm k_0)$ alongside the delta function, one can also restrict this to the manifold of positive or negative energy solutions, i.e.

$$\psi(x)^{(\pm)} = \int \frac{d^{n+1}k}{(2\pi)^{n+1}} e^{ikx} \theta(\mp k_0)(-\gamma^\mu p_\mu c + mc^2) \delta(p^2 c^2 + m^2 c^4) u_\pm(k),$$

$$= \int \frac{d^n k}{(2\pi)^n} e^{ikx} \frac{mc^2}{|k_0|} \gamma^0 u_\pm(k),$$

(20.109)

since

$$(-\gamma^\mu p_\mu c + mc^2) = 2mc^2,$$

(20.110)

when p_μ is on the mass shell $\gamma^\mu p_\mu c + mc^2 = 0$. In the literature it is customary to proceed by examining the positive and negative energy cases separately. As we shall see below, solutions of the Dirac equation can be normalized on either the positive or negative energy solution spaces.

It is more usual to consider positive and negative energy solutions to the Dirac equation separately. To this end, there is sufficient freedom in the expression

$$\psi^{(\pm)}(x) = \int \frac{d^{n+1}k}{(2\pi)^{n+1}} e^{ikx} \delta(p^2 c^2 + m^2 c^4) \theta(\mp k_0) \begin{pmatrix} u_1 \\ u_2 \end{pmatrix} N_\pm(k)$$

$$= \int \frac{d^n k}{(2\pi)^n} \frac{e^{i(\mathbf{k}\mathbf{x} - \omega t)}}{2|E|c\hbar} \begin{pmatrix} \frac{c\sigma^i p_i}{(\pm E + mc^2)} \\ 1 \end{pmatrix} N_\pm(k) u.$$

(20.111)

The two-component spinors u are taken to be a linear combination of the spin eigenfunctions for spin up and spin down, as measured conventionally along the z axis

$$u_i = c_1 \begin{pmatrix} 1 \\ 0 \end{pmatrix} + c_2 \begin{pmatrix} 0 \\ 1 \end{pmatrix},$$

(20.112)

where $i = 1, 2$ and $c_1^2 + c_2^2 = 1$. Unlike, the case of the Klein–Gordon equation, both positive and negative energy solutions can be normalized to unity, although this is not necessarily an interesting choice of normalization. An example: consider the normalization of the positive energy solutions $(-p_0 c = E)$,

$$1 = (\psi^{(+)}(x), \psi^{(+)}(x))$$

$$= \int d\sigma \int \frac{d^{n+1}k}{(2\pi)^{n+1}} \frac{d^{n+1}k'}{(2\pi)^{n+1}} \frac{e^{i(k-k')x}}{4E^2 c^2 \hbar^2} N_+^2 \left(\frac{c^2 (\sigma^i p_i)^2}{(E+mc^2)} + 1 \right) |u|^2$$

$$= \int \frac{d^n k}{(2\pi)^n} \frac{N_+^2}{4E^2 c^2 \hbar^2} \left(\frac{2E}{E+mc^2} \right). \tag{20.113}$$

Assuming a box normalization, where $d^n k / (2\pi)^n \sim L^{-n} \sum_k$, we have

$$N_+ = \sqrt{2L^n c^2 \hbar^2 (E + mc^2)}, \tag{20.114}$$

and hence

$$\psi^{(+)}(k) = L^{n/2} e^{i(\mathbf{kx} - \omega t)} \sqrt{\frac{(E+mc^2)}{2E}} \left(\begin{array}{c} \frac{c\sigma^i p_i}{(E+mc^2)} \\ 1 \end{array} \right) \chi^{(s)}, \tag{20.115}$$

where

$$\chi^{(\frac{1}{2})} = \left(\begin{array}{c} 1 \\ 0 \end{array} \right) , \quad \chi^{(-\frac{1}{2})} = \left(\begin{array}{c} 0 \\ 1 \end{array} \right). \tag{20.116}$$

20.8 Invariant normalization in p-space

The normalization of Dirac fields is a matter of some subtlety. Different invariant normalizations are used for different purposes. The usual case is to consider plane wave solutions, or wave-packets. Consider the probability on a spacelike hyper-surface, transforming as the zeroth component of a vector:

$$\overline{\psi}(x) \gamma_0 \psi(x) = \int \frac{d^{n+1}k}{(2\pi)^{n+1}} u(k)^\dagger u(k)\, \theta(\mp k_0)$$

$$\times (-\gamma^\mu p_\mu c + mc^2)\, \delta(p^2 c^2 + m^2 c^4)$$

$$= \int \frac{d^n k}{(2\pi)^n} \frac{2mc^2}{2|p_0|} u(k)^\dagger u(k) = 1. \tag{20.117}$$

The factor of $2mc^2 / 2|p_0|$ is required to ensure that the spinors satisfy the equations of motion for the Dirac field. This indicates that the invariant normalization for the spinors should be

$$u^\dagger(k) u(k) = \frac{|p_0|}{mc^2}. \tag{20.118}$$

Consider now what this means for the product $\bar{u}(k)u(k)$. The field equations for these, in momentum space, are:

$$(\gamma^0 p_0 c + \gamma^i p_i c + mc^2)u(k) = 0$$
$$\bar{u}(k)(\gamma^0 p_0 c + \gamma^i p_i c + mc^2) = 0. \tag{20.119}$$

Multiplying the first of these by $u^\dagger$, and using the fact that $\bar{u} = u^\dagger \gamma^0$, gives:

$$\bar{u}(\gamma^0 p_0 c + \gamma^0 \gamma^i p_i c + mc^2 \gamma^0)u(k) = 0. \tag{20.120}$$

Now, multiplying the second (adjoint) equation on the right hand side by $\gamma^0 u$ and commuting γ^0 through the left hand side, one has:

$$\bar{u}(p_0 c - \gamma^0 \gamma^i p_i c + mc^2 \gamma^0)u(k) = 0. \tag{20.121}$$

Thus, adding eqns. (20.121) and (20.120), leaves

$$2p_0 c\, \bar{u}u + 2mc^2\, u^\dagger u = 0. \tag{20.122}$$

Taking the normalization for $u^\dagger u$ in eqn. (20.118), we find that

$$\bar{u}u = \frac{-p_0}{|p_0|} = \frac{E}{|E|}. \tag{20.123}$$

Thus, a positive energy spinor is normalized with positive norm, whilst a negative energy spinor has a negative norm, in momentum space. It is custom to refer to the positive and negative energy spinors as $u(k)$ and $v(k)$, respectively. Accordingly, one takes the invariant normalization to be

$$\bar{u}_r u_s = \delta_{rs}$$
$$\bar{v}_r v_s = -\delta_{rs}, \tag{20.124}$$

with spinor indices shown.

20.9 Formal solution by Green functions

The formal solution to the free equations of motion ($V = 0$) may be written

$$\psi(x) = \int (dx')S(x, x')J(x'), \tag{20.125}$$

and the conjugate form

$$\psi^\dagger(x) = \int (dx')J^\dagger(x')S(x', x). \tag{20.126}$$

20.10 Expressions for the Green functions

The Green functions can be obtained from the corresponding Green functions for the scalar field; see section 5.6:

$$(-i\hbar c\gamma^\mu\partial_\mu + mc^2)(i\hbar c\gamma^\mu\partial_\mu + mc^2) =$$
$$-\hbar^2 c^2\Box + m^2 c^4 + \frac{1}{2}[\gamma^\mu, \gamma^\nu]\partial_\mu\partial_\nu, \qquad (20.127)$$

and the latter term vanishes when operating on non-singular objects. It follows for the free field that

$$(i\hbar c\gamma^\mu\partial_\mu + mc^2)G^{(\pm)}(x, x') = S^{(\pm)}(x, x') \qquad (20.128)$$
$$(i\hbar c\gamma^\mu\partial_\mu + mc^2)G_F(x, x') = S_F(x, x') \qquad (20.129)$$
$$(-i\hbar c\gamma^\mu\partial_\mu + mc^2)S^{(\pm)}(x, x') = 0 \qquad (20.130)$$
$$(-i\hbar c\gamma^\mu\partial_\mu + mc^2)S_F(x, x') = \delta(x, x'). \qquad (20.131)$$

20.11 The energy–momentum tensor

The application of Noether's theorem for spacetime translations leads to a symmetrical energy–momentum tensor. In accordance with the other fields, the zero–zero component of the energy–momentum tensor has the interpretation of an energy density or Hamiltonian density. This is to be distinguished from the differential Hamiltonian operator, which generates the time evolution of the field. We have,

$$\theta_{00} = \frac{\partial\mathcal{L}}{\partial(\partial^0\psi)}(\partial_0\psi) + (\partial_0\overline{\psi})\frac{\partial\mathcal{L}}{\partial(\partial^0\overline{\psi})} - \mathcal{L}g_{00}$$
$$= -\frac{i\hbar c}{2}\overline{\psi}\gamma_0(\partial_0\psi) + \frac{i\hbar c}{2}(\partial_0\overline{\psi})\gamma_0\psi + \mathcal{L}. \qquad (20.132)$$

Using the equation of motion (20.2), the integral of this quantity over all space may be written as

$$\int d\sigma\,\theta_{00} = \int d\sigma\,\overline{\psi}(-i\hbar c\gamma^i\partial_i + mc^2 + V)\psi$$
$$= (\psi, H_D\psi), \qquad (20.133)$$

where we have used $(\gamma^0)^2 = 1$. This expression is formally the expectation value of the differential Hamiltonian operator, but it is also used as the definition of a 'field theoretical' Hamiltonian. In the second quantization, where the fields are operator-valued, this expression is referred to as the Hamiltonian operator and may be thought of as generating the time evolution of the fully quantized field.

The spacetime components of the energy–momentum tensor are not explicitly symmetrical. This is a consequence of the linear derivative in the field equation (20.2). However, the off-diagonal components can be shown to be equal provided the field satisfies the equations of motion. We have

$$\theta_{0i} = \frac{\partial \mathcal{L}}{\partial(\partial^0 \psi)}(\partial_i \psi) + (\partial_i \overline{\psi})\frac{\partial \mathcal{L}}{\partial(\partial^0 \overline{\psi})}$$

$$= -\frac{i\hbar c}{2}\overline{\psi}\gamma_0(\partial_i \psi) + \frac{i\hbar c}{2}(\partial_i \overline{\psi})\gamma_0 \psi. \tag{20.134}$$

Taking the integral over all space allows us to integrate by parts, giving

$$\int d\sigma \theta_{0i} = -i\hbar c \int d\sigma \overline{\psi}\gamma_0 \partial_i \psi$$

$$= -(\psi, p_i c \psi), \tag{20.135}$$

where $p_i = -i\hbar \partial_i$. Thus, this component is identified with the momentum in the field. Switching the order of the indices, we have

$$\theta_{0i} = \frac{\partial \mathcal{L}}{\partial(\partial^i \psi)}(\partial_0 \psi) + (\partial_0 \overline{\psi})\frac{\partial \mathcal{L}}{\partial(\partial^i \overline{\psi})}$$

$$= -\frac{i\hbar c}{2}\overline{\psi}\gamma_i(\partial_0 \psi) + \frac{i\hbar c}{2}(\partial_0 \overline{\psi})\gamma_i \psi. \tag{20.136}$$

This is clearly not the same as eqn. (20.134). However on using the field equation and its conjugate in eqns. (20.2) and (20.5), it may be shown that

$$\theta_{i0} = \frac{i\hbar c}{2}\overline{\psi}(\gamma_i \gamma^0 \gamma^j \overset{\rightarrow}{\partial_j} - \gamma^j \overset{\leftarrow}{\partial_j} \gamma^0 \gamma_i)\psi - \frac{1}{2}\{\gamma_i, \gamma^0\}(mc^2 + V)\overline{\psi}\psi, \tag{20.137}$$

so that the integral over all space can be integrated by parts to give

$$\int d\sigma \theta_{i0} = \int d\sigma \left\{ \frac{i\hbar c}{2}\overline{\psi}\gamma^0\{\gamma_i, \gamma^j\}\partial_j \psi - \frac{1}{2}\{\gamma_i, \gamma^0\}(mc^2 + V)\overline{\psi}\psi \right\}. \tag{20.138}$$

On using the anti-commutation relations for the γ-matrices, we find

$$\int d\sigma \theta_{i0} = -i\hbar c \int d\sigma \overline{\psi}\gamma_0 \partial_i \psi = \int d\sigma \theta_{0i}. \tag{20.139}$$

The diagonal space components are given by

$$\theta_{ii} = \frac{\partial \mathcal{L}}{\partial(\partial^i \psi)}(\partial_i \psi) + (\partial_i \overline{\psi})\frac{\partial \mathcal{L}}{\partial(\partial^i \overline{\psi})}$$

$$= -\frac{i\hbar c}{2}\overline{\psi}\gamma_i(\partial_i \psi) + \mathcal{L}, \tag{20.140}$$

where i is not summed. The off-diagonal space components are

$$\theta_{ij} = \frac{\partial \mathcal{L}}{\partial(\partial^i \psi)}(\partial_j \psi) + (\partial_j \overline{\psi})\frac{\partial \mathcal{L}}{\partial(\partial^i \overline{\psi})}$$

$$= -\frac{i\hbar c}{2}\left(\overline{\psi}\gamma_i \partial_j \psi - (\partial_i \overline{\psi})\gamma_j \psi\right), \qquad (20.141)$$

where $i \neq j$. Although not explicitly symmetrical in this form, the integral over all space of this quantity is symmetrical by partial integration. Note that the trace of the space components is given in $n + 1$ dimensions by

$$\sum_i \theta_{ii} = \mathcal{H} + (mc^2 + V)\overline{\psi}\psi + (n - 1)\mathcal{L}, \qquad (20.142)$$

so that the total trace of the energy–momentum tensor is

$$\theta^\mu{}_\mu = g^{\mu\nu}\theta_{\mu\nu} = (mc^2 + V)\overline{\psi}\psi + (n - 1)\mathcal{L}. \qquad (20.143)$$

This vanishes for $m = V = 0$ in $1 + 1$ dimensions.

20.12 Spinor electrodynamics

The action for spinor electrodynamics is

$$S_{\mathrm{QED}} = \int (\mathrm{d}x)\left\{\overline{\psi}\left(-\frac{1}{2}i\hbar c(\gamma^\mu \overrightarrow{D}_\mu - \gamma^\mu \overleftarrow{D}_\mu^\dagger) + mc^2\right)\psi\right.$$

$$\left. +\frac{1}{4\mu_0}F^{\mu\nu}F_{\mu\nu}\right\}. \qquad (20.144)$$

This is the basis of the quantum theory of electrodynamics for electrons (QED).

Pauli [104] has shown that the Dirac action may be modified by a term of the form

$$S \to S + \int (\mathrm{d}x)\,\overline{\psi}\frac{1}{2}\frac{\mu c^2}{\hbar}\sigma^{\rho\lambda}F_{\rho\lambda}\psi, \qquad (20.145)$$

whereupon the field behaves as though it has an additional (anomalous) magnetic moment $e\hbar/2m$. Later, Foldy investigated generalizations of the Dirac action which preserve Lorentz invariance and gauge invariance [51]. One makes two restrictions: linearity in A_μ (weak field) and finiteness in the zero momentum limit (independent of $\partial_\mu \psi$). The result is

$$S \to S + \int (\mathrm{d}x)\overline{\psi}\left[c\sum_{i=0}^\infty \left(\alpha_i \Box^i \gamma^\mu A_\mu + \frac{1}{2}\beta_n \sigma^{\mu\nu}\Box^i F_{\mu\nu}\right)\right]\psi, \qquad (20.146)$$

where α_i, β_i are constants representing anomalous charge and magnetic moments respectively.

There is a number of problems for which spinor electrodynamics can be solved exactly. These include:

- the spherically symmetrical Coulomb potential [31, 38, 62, 75, 98];

- the homogeneous magnetic field [73, 81, 106, 109];

- the field of an electromagnetic plane wave [131].

A review of these is given in many books. See, for example, ref. [8].

21

The Maxwell radiation field

There are two ways of describing the interaction between matter and the electromagnetic field: the first and most fundamental way is to consider the electromagnetic field to be coupled to every individual microscopic charge in a physical system explicitly. Apart from these charges, the field lives on a background vacuum. The charges are represented by an $(n + 1)$ dimensional current vector J_μ.

In systems with very complex distributions of charge, this approach is too cumbersome, and an alternative view is useful: that of an electromagnetic field in dielectric media. This approach is an effective-field-theory approach in which the average effect of a very complex, on average neutral, distribution of charges is taken into account by introducing an effective speed of light, or equivalently effective permittivities and permeabilities. Any remainder charges which make the system non-neutral can then be handled explicitly by an $(n + 1)$ dimensional current vector. Although the second of these two approaches is a popular simplification in many cases, it has only a limited range of validity, whereas the first approach is fundamental. We shall consider these two cases separately.

21.1 Charges in a vacuum

21.1.1 The action

The action for the electromagnetic field in a vacuum is given by

$$S = \int (\mathrm{d}x) \left\{ \frac{1}{4\mu_0} F^{\mu\nu} F_{\mu\nu} \right\}, \tag{21.1}$$

where the anti-symmetric field strength tensor is defined by

$$F_{\mu\nu} = \partial_\mu A_\nu - \partial_\nu A_\mu, \tag{21.2}$$

and

$$A^\mu = \begin{pmatrix} c^{-1}\phi \\ \mathbf{A} \end{pmatrix}. \tag{21.3}$$

In $3 + 1$ dimensions, the field components are given by

$$E_i = -\partial_i\phi - \partial_t A_i = c\, F_{i0}$$
$$\epsilon_{ijk}B_k = F_{ij}, \tag{21.4}$$

where $i = 1, 2, 3$. The latter equation may be inverted to give

$$B_i = \frac{1}{2}\epsilon_{ijk}F_{jk}. \tag{21.5}$$

Note that the indices on the electric and magnetic field *vectors* in $3 + 1$ dimensions are always written as subscripts, never as superscripts. In $2 + 1$ dimensions, the magnetic field is a pseudo-scalar, and one has

$$E_i = -\partial_i\phi - \partial_t A_i = c\, F_{i0}$$
$$B = F_{12}, \tag{21.6}$$

where $i = 1, 2$. In higher dimensions, the tensor character of $F_{\mu\nu}$ is unavoidable, and E and B cease to lose their separate identities. A further important point is that the derivatives in the action are purely classical – there are no factors of $\hbar$ present here.

A phenomenological source can be added to the Maxwell action, in an ambient vacuum:

$$S = \int (\mathrm{d}x) \left\{ \frac{1}{4\mu_0}F^{\mu\nu}F_{\mu\nu} - J^\mu A_\mu \right\}. \tag{21.7}$$

This describes an electromagnetic field, extended in a vacuum around source charges. What we really mean here is that there is no ambient background matter present: we allow positive and negative charges to exist freely in a vacuum, but there is no overall neutral, polarizable matter present. The case of polarization in the ambient medium is dealt with later.

21.1.2 Field equations and continuity

The variation of the action leads to

$$\begin{aligned}
\delta S &= \int (\mathrm{d}x) \left\{ (\partial^\mu \delta A^\nu)F_{\mu\nu} - J^\mu \delta A_\mu \right\} \\
&= \int (\mathrm{d}x) \left\{ \delta A^\nu(-\partial^\mu F_{\mu\nu}) - J^\mu \delta A_\mu \right\} \\
&\quad + \int \mathrm{d}\sigma^\mu \left\{ \delta A^\nu F_{\mu\nu} \right\}.
\end{aligned} \tag{21.8}$$

Thus, the field equations $\delta S = 0$ are given by

$$\partial_\mu F^{\mu\nu} = -\mu_0 J^\nu, \tag{21.9}$$

and the continuity condition tells us that the conjugate momentum $(\mathrm{d}\sigma^\mu = \mathrm{d}\sigma^0)$ is

$$\Pi_i = F_{0i}. \tag{21.10}$$

If the surface σ is taken to separate two regions of space rather than time, one has the standard continuity conditions for the electromagnetic field in a vacuum:

$$\Delta F_{i0} = 0$$
$$\Delta F_{ij} = 0, \tag{21.11}$$

and we have assumed that δA_μ is a continuous function. The momentum conjugate to the field A_μ is

$$\Pi_\mu^\sigma = \frac{\delta \mathcal{L}}{\delta(\partial^\sigma A^\mu)} = F_{\sigma\mu}, \tag{21.12}$$

where σ points outward from a spacelike hyper-surface. The canonical choice for this momentum is $\sigma = 0$, where one has

$$\Pi_\mu = \frac{\delta \mathcal{L}}{\delta(\partial^0 A^\mu)} = F_{0i}, \tag{21.13}$$

which means that μ can only take values $i = 1, \ldots, n$ in n spatial dimensions, owing to the anti-symmetry of $F_{\mu\nu}$.

The velocity analogous to $\dot{q}$ is given by the derivative of the field $\partial_\sigma A^\mu = \partial_0 A^\mu$. Thus, the canonical definition of the Hamiltonian is

$$\mathcal{H} = F_{0\mu}(\partial_0 A^\mu) - g_{\mu\nu}\frac{1}{4\mu_0}F^{\mu\nu}F_{\mu\nu}. \tag{21.14}$$

However, this expression is not gauge-invariant, whereas the Hamiltonian must be. The problem lies in the naive interpretation of the Legendre transform. The problem may be cured by defining the Hamiltonian in terms of the variation of the action:

$$H = -\frac{\delta S}{\delta t}. \tag{21.15}$$

This is a special case (the zero–zero component θ_{00}) of the energy–momentum tensor, which is discussed below in more general terms. The result for the Hamiltonian density is

$$\mathcal{H} = \frac{1}{2}\left(\epsilon_0 E_i E_i + \mu_0^{-1} B_i B_i\right) \tag{21.16}$$

where $i = 1, \ldots, n$.

21.1.3 The Jacobi–Bianchi identity

The Bianchi identity in $n + 1$ dimensions provides two of Maxwell's equations. The equations implied by this identity are different in each new number of dimensions. In $3 + 1$ dimensions, we have

$$\epsilon^{\mu\nu\lambda\rho} \partial_\nu F_{\lambda\rho} = 0. \tag{21.17}$$

Separating out the space and time components of μ, we obtain, for $\mu = 0$,

$$\epsilon^{ijk} \partial_i F_{jk} = \partial^i B_i = \text{div}\mathbf{B} = 0. \tag{21.18}$$

For $\mu = i$, i.e. the spatial components, we have

$$\epsilon^{i0jk} \partial_0 F_{jk} + \epsilon^{ik0j} \partial_k F_{0j} + \epsilon^{ijk0} \partial_j F_{k0} = 0, \tag{21.19}$$

which may be re-written as

$$2\frac{1}{c} \partial_t B_i - \epsilon^{ijk} \partial_k F_{0j} + \epsilon^{ijk} \partial_j F_{k0} = 0. \tag{21.20}$$

Thus, using the definition of the electric field in eqn. (21.12), together with the anti-symmetry of $F_{\mu\nu}$, we obtain

$$(\text{curl }\mathbf{E})_i = -\frac{\partial B_i}{\partial t}, \tag{21.21}$$

which completes the proof.

In $2 + 1$ dimensions, eqn. (21.18) is absent, since the Bianchi identity now has the form

$$\epsilon^{\mu\nu\lambda} \partial_\mu F_{\nu\lambda} = 0. \tag{21.22}$$

The full expansion of this equation is

$$\epsilon^{0jk} \partial_0 F_{jk} + \epsilon^{k0j} \partial_k F_{0j} + \epsilon^{jk0} \partial_j F_{k0} = 0, \tag{21.23}$$

which can be written as

$$\epsilon^{jk} \partial_j E_k = -\frac{\partial B}{\partial t}. \tag{21.24}$$

Note that, in $2 + 1$ dimensions, the B field is a pseudo-scalar.

21.1.4 Formal solution by Green functions

The formal solution to the equations of motion is most conveniently expressed in terms of the vector potential. Re-writing the field equation (21.9) in terms of the vector potential, we have

$$-\Box A_\nu + \partial^\mu \partial_\nu A_\mu = \mu_0 J_\nu \tag{21.25}$$

or

$$(-\Box \, \delta^\mu_{\ \nu} + \partial^\mu \partial_\nu) A_\mu = \mu_0 J_\nu. \tag{21.26}$$

The formal solution therefore requires the inverse of the operator on the left hand side of this equation. This presents problem, though: the determinant of this matrix-valued operator vanishes! This is easily seen by separating space and time components as a 2×2 matrix,

$$\begin{vmatrix} -\Box + \partial_0 \partial^0 & \partial_0 \partial^i \\ \partial_i \partial^0 & -\Box + \partial_i \partial^i \end{vmatrix} = 0. \tag{21.27}$$

The problem here is related to the gauge symmetry, or non-uniqueness, of A_μ and can be fixed by choosing a gauge for the potential. The choice of gauge is arbitrary, but *two* conditions are required in general to fix the gauge freedom fully (see chapter 9), and ensure a one-to-one correspondence between the potentials and the physical fields.

21.1.5 Lorentz gauge

To solve the inverse problem in the 'Lorentz gauge', it is sufficient to take

$$\partial^\mu A_\mu = 0. \tag{21.28}$$

This is only a single condition, so it does not fix the gauge completely, but it is sufficient for our purposes. Using this condition directly in eqn. (21.26) we get the modified field equation,

$$-\Box \, A_\mu = \mu_0 J_\mu. \tag{21.29}$$

This equation now presents the appearance of a massless Klein–Gordon field. The formal inverse of the differential operator is therefore the scalar Green function $G(x, x')$, for $m = 0$, giving the solution

$$A_\mu = \mu_0 \int (\mathrm{d}x') G(x, x') J_\mu(x'). \tag{21.30}$$

Another way of imposing this condition, which is frequently used in the literature, is to add a Lagrange multiplier term to the action:

$$S' = \int (\mathrm{d}x) \left\{ \frac{1}{4\mu_0} F^{\mu\nu} F_{\mu\nu} - J^\mu A_\mu + \frac{1}{2\alpha} \mu_0^{-1} (\partial^\mu A_\mu)^2 \right\}, \tag{21.31}$$

where α^{-1} is the Lagrange multiplier. The field equations and continuity conditions resulting from the variation of the action are now

$$\left[-\Box \, \delta^\mu_{\ \nu} + \left(1 - \frac{1}{\alpha} \right) \partial^\mu \partial_\nu \right] A_\mu = \mu_0 J_\nu \tag{21.32}$$

and

$$\Delta(F_{\sigma\nu} + \partial_\sigma A_\nu) = 0. \tag{21.33}$$

The inverse of the differential operator in eqn. (21.32) is found by solving the equation

$$\left[-\Box \, \delta^\mu_{\ \nu} + \left(1 - \frac{1}{\alpha}\right) \partial^\mu \partial_\nu\right] D_\mu^{\ \lambda}(x, x') = \delta_\nu^{\ \lambda} \delta(x, x'), \tag{21.34}$$

which gives a formal solution for the potential

$$A_\mu(x) = \mu_0 \int (dx') D_{\mu\nu}(x, x') J^\nu(x'). \tag{21.35}$$

21.1.6 Coulomb/radiation gauge

The Coulomb gauge is based on the condition

$$\partial^i A_i = 0. \tag{21.36}$$

Again, this is only a single condition, and it is usually supplemented by the condition $A_0 = 0$, or by the use of the zeroth-component field equation to eliminate A_0 entirely. Using eqn. (21.36) in eqn. (21.26), we separate the space and time parts of the field equations (a step backwards from covariance):

$$(-\Box \, A_0 + \partial_0(\partial^0 A_0)) = \mu_0 \, J_0 \tag{21.37}$$
$$(-(\partial_0\partial^0 + \nabla^2)A_i + \partial_i\partial^0 A_0 + \partial_i(\partial^0 A_0)) = \mu_0 J_i, \tag{21.38}$$

or, simplifying,

$$-\nabla^2 A_0 = \mu_0 J_0 \tag{21.39}$$
$$-\Box \, A_i + \partial_i(\partial^0 A_0) = \mu_0 J_i. \tag{21.40}$$

At this point, it is usual to use the first of these equations to eliminate A_0 from the second, thereby fixing the gauge completely. Formally, we may write

$$-\Box \, A_i + \partial_i\partial^0 \left(\frac{J_0}{-\nabla^2}\right) = \mu_0 J_i, \tag{21.41}$$

where $(-\nabla^2)^{-1}$ really implies the inverse (or Green function) for the differential operator $-\nabla^2$, which we denote $g(x, x')$ and which satisfies the equation

$$-\nabla^2 g(x, x') = \delta(x, x'). \tag{21.42}$$

Thus, eqn. (21.41) is given (still formally, but more explicitly) by

$$-\Box \, A_i + \mu_0\partial_i\partial^0 \int d\sigma_{x'} g(x, x') J_0(x') = \mu_0 J_i. \tag{21.43}$$

21.1.7 Retarded Green function in $n = 3$

In the Lorentz gauge, with $\alpha = 1$, we have

$$D_{\mu\nu}(x, x') = g_{\mu\nu}G(x, x'). \tag{21.44}$$

From Cauchy's residue theorem in eqn. (5.76), we have

$$G_{\rm r}(x, x') = -2\pi{\rm i}(\hbar^2 c)^{-1} \int \frac{d^3k}{(2\pi)^4} \left[\frac{e^{{\rm i}(\mathbf{k}\cdot\Delta\mathbf{x}-\omega_k\Delta t)}}{2\omega_k} - \frac{e^{{\rm i}(\mathbf{k}\cdot\Delta\mathbf{x}+\omega_k\Delta t)}}{2\omega_k} \right]. \tag{21.45}$$

Using the derivation in section 5.4.1, this evaluates to

$$G_{\rm r}(x, x') = \frac{1}{4\pi\hbar^2 c\Delta X}\delta(ct - \Delta X)$$

$$= \frac{1}{4\pi\hbar^2 c|\mathbf{x} - \mathbf{x}'|}\, \delta\left(c(t' - t_{\rm ret})\right), \tag{21.46}$$

where the retarded time is defined by

$$t_{\rm ret} = t - |\mathbf{x} - \mathbf{x}'|. \tag{21.47}$$

Note that the retarded time is a function of the position.

21.1.8 The energy–momentum tensor

The gauge-invariant definition of the energy–momentum tensor is (see section 11.5),

$$\theta'_{\mu\nu} = \frac{\partial\mathcal{L}}{\partial(\partial^\mu A^\alpha)}F_\nu{}^\alpha - \mathcal{L}g_{\mu\nu}$$

$$= 2\frac{\partial\mathcal{L}}{\partial F^{\mu\alpha}}F_\nu{}^\alpha - \mathcal{L}g_{\mu\nu}$$

$$= \mu_0^{-1}F_{\mu\alpha}F_\nu{}^\alpha - \frac{1}{4\mu_0}F^{\lambda\rho}F_{\lambda\rho}g_{\mu\nu}. \tag{21.48}$$

This result is manifestly gauge-invariant and can be checked against the traditional expressions obtained from Maxwell's equations for the energy density and the momentum flux.

The zero–zero component, in $3 + 1$ dimensions, evaluates to:

$$\theta_{00} = \mu_0^{-1}\left(F_{0i}F_0{}^i - \mathcal{L}g_{00}\right)$$

$$= \frac{E_i E_i}{c^2\mu_0} + \frac{1}{2\mu_0}\left(B_i B_i - \frac{E_i E_i}{c^2}\right)$$

$$= \frac{1}{2\mu_0}(\mathbf{E}^2/c^2 + \mathbf{B}^2)$$

$$= \frac{1}{2}(\epsilon_0\mathbf{E}\cdot\mathbf{E} + \mu_0^{-1}\mathbf{B}\cdot\mathbf{B}), \tag{21.49}$$

which has the interpretation as an energy or Hamiltonian density. The spacetime off-diagonal components are given by

$$
\begin{aligned}
\theta_{0j} = \theta_{j0} &= \mu_0^{-1} F_{0i} F_j{}^i \\
&= \mu_0^{-1} \epsilon_{ijk} E_i B_k / c \\
&= -\frac{(\mathbf{E} \times \mathbf{H})_k}{c},
\end{aligned} \tag{21.50}
$$

which have the interpretation of a momentum density for the field. This vector is also known as Poynting's vector. The off-diagonal space parts are

$$
\begin{aligned}
\mu_0 \theta_{ij} = F_{i\mu} F_j{}^\mu &= F_{i0} F_j{}^i + F_{ik} F_j{}^k \\
&= -\frac{E_i}{c} \frac{E_j}{c} - 2 B_i B_j,
\end{aligned} \tag{21.51}
$$

with $i \neq j$. The diagonal terms with i not summed are

$$
\begin{aligned}
\mu_0 \theta_{ii} = F_{i0} F_i{}^0 + F_{ij} F_i{}^j - \frac{1}{4} F^{\mu\nu} F_{\mu\nu} \\
= -\frac{E_i^2}{c^2} + 2 B_i^2 - \frac{1}{2} (B_j B_j - E_j E_j / c^2).
\end{aligned} \tag{21.52}
$$

The invariant trace of this tensor in $n + 1$ dimensions is

$$
\theta^\mu{}_\mu = \mu_0^{-1} F_{\mu\alpha} F^{\mu\alpha} - \frac{(n+1)}{4\mu_0} F^{\mu\nu} F_{\mu\nu}, \tag{21.53}
$$

which vanishes when $n = 3$, indicating that Maxwell's theory is conformally invariant in $3 + 1$ dimensions.

21.2 Effective theory of dielectric and magnetic media

This section contains a brief summarial discussion of the effective fields for the radiation field in the presence of a passive medium. The dielectric approach to electromagnetism in near-neutral media is often used since it offers an enormous simplification of very many systems. Its main weaknesses are that it makes two assumptions: namely, that the response of background matter is dipole-like and linear in the applied fields, and that the background matter is smoothly homogeneous throughout a given region. The first of these assumptions breaks down for strong fields, and the latter breaks down on very small length scales; thus, the theory provided in this section must be treated as a long-wavelength approximation to electromagnetism for weak fields.

Maxwell's equations in a dielectric/magnetic medium are most conveniently written in terms of the dielectric displacement vector **D**, defined by any one of the equivalent relations

$$\mathbf{D} = \epsilon_0\epsilon_r\mathbf{E} + \mathbf{P} \tag{21.54}$$

$$= \epsilon_0(1 + \chi_e)\mathbf{E} \tag{21.55}$$

$$= \epsilon_0\epsilon_r\mathbf{E}, \tag{21.56}$$

where $\mathbf{P}$ is the dielectric polarization and χ_e is the electric susceptibility, described below. The magnetic field intensity $\mathbf{H}$, defined by the equivalent forms

$$\mathbf{H} = \frac{1}{\mu_0\mu_r}\mathbf{B} - \mathbf{M} \tag{21.57}$$

$$= \frac{\mathbf{B}}{\mu_0(1 + \chi_m)} \tag{21.58}$$

$$= \frac{\mathbf{B}}{\mu_0\mu_r}. \tag{21.59}$$

$\mathbf{M}$ is called the magnetization and χ_m is the magnetic susceptibility, also defined below. In terms of these quantities, Maxwell's equations take on the form

$$\vec{\nabla} \cdot \mathbf{D} = \rho_e$$

$$\vec{\nabla} \times \mathbf{E} = -\frac{\partial \mathbf{B}}{\partial t}$$

$$\vec{\nabla} \cdot \mathbf{B} = 0$$

$$\vec{\nabla} \times \mathbf{H} = \mathbf{j} + \frac{\partial \mathbf{D}}{\partial t}. \tag{21.60}$$

This form of Maxwell's equations is valid inside any linear medium. As a further point, the energy density of the electromagnetic field is given by

$$\mathcal{E} = \frac{1}{2}(\mathbf{E} \cdot \mathbf{D} + \mathbf{B} \cdot \mathbf{H}). \tag{21.61}$$

This must agree with the Hamiltonian density.

21.2.1 The Maxwell action and Hamiltonian in a medium

To express Maxwell's equations in *covariant* form, we had to introduce the fields $\mathbf{D}$ and $\mathbf{H}$. We should therefore expect that, in the covariant description, we need to introduce a new tensor. We shall call this tensor $G_{\mu\nu}$, and define it by

$$G_{\mu\nu} = \begin{pmatrix} 0 & -cD_1 & -cD_2 & -cD_3 \\ cD_1 & 0 & H_3 & -H_2 \\ cD_2 & -H_3 & 0 & H_1 \\ cD_3 & H_2 & -H_1 & 0 \end{pmatrix}. \tag{21.62}$$

We see that this tensor has the same structure as $F_{\mu\nu}$, but with $c\mathbf{D}$ replacing $\mathbf{E}/c$ and $\mathbf{H}$ replacing $\mathbf{B}$.

In terms of this tensor, we can write the action in the form

$$S = \int (\mathrm{d}x) \left\{ \frac{1}{4} F^{\mu\nu} G_{\mu\nu} - J^\mu A_\mu \right\}. \tag{21.63}$$

The canonical momentum can be written

$$\Pi_\mu = \frac{\partial \mathcal{L}}{\partial(\partial^0 A^\mu)} = G_{0i}. \tag{21.64}$$

Again, one has the same problem of gauge invariance with the generalized 'velocity' as in the vacuum case. The Hamiltonian is best computed from the energy–momentum tensor. It is given by

$$\mathcal{H} = \frac{1}{2}(B_i H_i + E_i D_i) + J^\mu A_\mu. \tag{21.65}$$

21.2.2 Field equations and continuity

Using the property of the trace that

$$\delta F^{\mu\nu} G_{\mu\nu} = \delta G^{\mu\nu} F_{\mu\nu}, \tag{21.66}$$

for linear equations of motion (i.e. $G_{\mu\nu}$ does not depend on $F_{\mu\nu}$), we find the variation of the action is given by

$$\delta S = \int (\mathrm{d}x) \left\{ -\delta A^\nu \, \partial^\mu G_{\mu\nu} - J^\mu \delta A_\mu \right\} + \int \mathrm{d}\sigma^\mu \left\{ \delta A_\nu G_{\mu\nu} \right\} = 0. \tag{21.67}$$

We see immediately that the field equations are given by

$$\partial^\mu G_{\mu\nu} = -J_\nu, \tag{21.68}$$

which may be compared with eqn. (21.9), and that the continuity condition implies that the canonical momentum is ($\mu = 0$)

$$\Pi_\mu = D_{0\mu}, \tag{21.69}$$

and that the condition for continuity across a surface dividing two regions of space is $\mu = i$ divides into two cases,

$$\Delta D_{i0} = \Delta \mathbf{D} = 0$$
$$\Delta D_{ij} = \Delta \mathbf{H} = 0. \tag{21.70}$$

These are the well known continuity conditions for the field at a dielectric boundary.

In terms of the vector potential, we can write the field equations

$$c^2 \epsilon \left\{ \partial_0 (\partial_i A^i) + \nabla^2 A^0 \right\} + J^0 = 0$$

$$c^2 \epsilon \left\{ \partial_0 \partial^0 A^i - \partial^i (\partial^0 A_0) \right\} + \frac{1}{\mu} (\partial^j \partial_j A^i - \partial^i (\partial_j A^j)) + J^i = 0.$$

$$(21.71)$$

This ugly mess can be compared with eqn. (21.26) for the vacuum case. At first sight, it appears that covariance is irretrievably lost in these expressions, but this is only an illusion caused by the spurious factors of ϵ and μ as explained below.

21.2.3 Reinstating covariance with $c \to c/n$

To reinstate covariance, we note that the introduction of a modified gauge condition helps to unravel the equations:

$$c^2 \epsilon \mu (\partial^0 A_0) + \partial_j A^j = 0. \tag{21.72}$$

This can also be written as

$$n^2 (\partial^0 A_0) + \partial_j A^j = 0, \tag{21.73}$$

which suggests that we re-define the derivative as

$$\hat{\partial}_\mu = \left(\frac{n}{c} \partial_t, \vec{\nabla} \right). \tag{21.74}$$

In terms of this equation, the field equations now combine to give

$$-\hat{\Box} A^\mu = \mu J^\mu. \tag{21.75}$$

To re-write the gauge condition in terms of this new derivative, we must also define $\hat{A}_\mu$ and $\hat{J}_\mu$, replacing c by c/n in each case:

$$\hat{A}^\mu = \begin{pmatrix} \frac{n}{c} \phi \\ \mathbf{A} \end{pmatrix} \qquad \hat{J}^\mu = \begin{pmatrix} \rho_e \frac{c}{n} \\ \mathbf{J} \end{pmatrix}. \tag{21.76}$$

Then we have the complete, covariant $(n+1)$-vector form of the field equations in a medium:

$$-\hat{\Box} \hat{A}^\mu = \mu \hat{J}^\mu$$

$$\hat{\partial}^\mu \hat{A}_\mu = 0. \tag{21.77}$$

21.2.4 Green function

The Green function is easily obtained by direct analogy to the vacuum case. The most elegant form is obtained using the careted notation. The photon Green function in a dielectric satisfies the equation

$$\left[\hat{\Box} \, g_{\mu\nu} + \left(1 - \frac{1}{\alpha} \right) \hat{\partial}_\mu \hat{\partial}_\nu \right] \hat{D}^{\nu\rho}(x, x') \delta_\mu^{\ \rho} \delta(x, x'). \tag{21.78}$$

Thus, defining the careted momentum by

$$i\hat{k}_\mu \equiv \hat{\partial}_\mu e^{ikx}, \tag{21.79}$$

i.e. such that $\hat{k}_\mu = (-n\omega/c, \mathbf{k})$, one has straightforwardly that

$$\hat{D}_{\mu\nu}(x, x') = \int \frac{d^{n+1}k}{(2\pi)^{n+1}} e^{ik(x-x')} \left[\frac{g_{\mu\nu}}{\hat{k}^2} + (\alpha - 1) \frac{\hat{k}_\mu \hat{k}_\nu}{\hat{k}^4} \right]. \tag{21.80}$$

Note carefully which of the quantities are careted and which are not. This Green function relates the careted field to the careted source,

$$\hat{A}_\mu(x) = \int (dx') \hat{D}_{\mu\nu}(x, x') \hat{J}^\nu(x'). \tag{21.81}$$

22

The massive Proca field

The massive vector field is the model which describes massive vector bosons, such as the W and Z particles of the electro-weak theory.

22.1 Action and field equations

The action for the Proca field is

$$S = \int (\mathrm{d}x) \left\{ \frac{1}{4} F^{\mu\nu} F_\mu + \frac{1}{2} m^2 A^\mu A_\mu - J^\mu A_\nu \right\}, \tag{22.1}$$

where

$$F_{\mu\nu} = \partial_\mu A_\nu - \partial_\nu A_\mu. \tag{22.2}$$

The variation of the action gives

$$\delta A = \int (\mathrm{d}x) \left[-\partial^\nu F_{\mu\nu} + m^2 A_\mu - J_\mu \right] \delta A_\mu + \int \mathrm{d}\sigma^\mu \ F_{\mu\nu} \delta A^\nu. \tag{22.3}$$

This yields the field equation

$$-\partial^\nu F_{\mu\nu} + m^2 A_\mu = J_\mu, \tag{22.4}$$

also writable as

$$-\Box A_\mu - \partial_\mu (\partial^\nu A_\nu) + m^2 A_\mu = J_\mu, \tag{22.5}$$

and associated continuity conditions identical to those of the Maxwell field. The conjugate momentum ($\mathrm{d}\sigma^\mu = \mathrm{d}\sigma^0$) is

$$\Pi_i = F_{0i}. \tag{22.6}$$

464

If the surface σ is taken to separate two regions of space rather than time, one has the continuity conditions for the Proca field in a vacuum:

$$\Delta F_{i0} = 0$$
$$\Delta F_{ij} = 0, \tag{22.7}$$

and we have assumed that δA_μ is a continuous function. Taking the $n + 1$ divergence of eqn. (22.4), we obtain

$$\partial^\mu A_\mu = 0. \tag{22.8}$$

Here we have used the anti-symmetry of $F_{\mu\nu}$ and the assumption that the source is conserved, $\partial^\mu J_\nu$. Thus the field equations, in the form of eqn. (22.5), become

$$(-\Box + m^2)A_\mu = J_\mu \tag{22.9}$$
$$\partial^\mu A_\mu = 0. \tag{22.10}$$

In contrast to the electromagnetic field, this has both transverse and longitudinal components.

23

Non-Abelian fields

Positive and negative electrical charges label the different kinds of matter that respond to the electromagnetic field; there is the gravitational mass, for example, which only seems to have positive sign and labels matter which responds to the gravitational field. More kinds of charge are required to label particles which respond to the nuclear forces. With more kinds of charge, there are many more possibilities for conservation than simply that the sum of all positive and negative charges is constant. Non-Abelian gauge theories are physical models analogous to electromagnetism, but with more general ideas of charge. Some have three kinds of charge: red, green and blue (named whimsically after the primary colours); other theories have more kinds with very complicated rules about how the different charges are conserved. This chapter is about such theories.

23.1 Lie groups and algebras

In chapter 9 it was noted that the gauge invariance of matter and electromagnetic radiation could be thought of as a symmetry group called $U(1)$: the group of phase transformations on matter fields:

$$\Phi \rightarrow e^{i\theta(x)}\Phi, \tag{23.1}$$

for some scalar function $\theta(x)$. Since phase factors of this type commute with one another,

$$[e^{i\theta(x)}, e^{i\theta'(x)}] = 0 \tag{23.2}$$

such a symmetry group is called commutative or *Abelian*. The symmetry group was identified from the anti-symmetry properties of the curls in Maxwell's equations, but the full beauty of the symmetry only became apparent in the covariant formulation of field theory.

466

In the study of angular momentum in chapter 9, it was noted that the symmetry group of rotations in two spatial dimensions was $U(1)$, but that in three spatial dimensions it was $O(3)$. The latter is a *non-Abelian* group, i.e. its generators do not commute; instead they have a commutator which satisfies a relation called a Lie algebra.

Non-Abelian gauge theories have, for the most part, been the domain of particle physicists trying to explain the elementary nature of the nuclear forces in collision experiments. In recent times, some non-Abelian field theories have also been used by condensed matter physicists. In the latter case, it is not fundamental fields which satisfy the exotic symmetry properties, but composite excitations in matter referred to as quasi-particles.

The motivation for non-Abelian field theory is the existence of families of excitations which are related to one another by the fact that they share and respond to a common form of charge. Each so-called flavour of excitation is represented by an individual field which satisfies an equation of motion. The fields are grouped together so that they form the components of a column vector, and *matrices*, which multiply these vectors exact symmetry transformations on them – precisely analogous to the phase transformations of electromagnetism, but now with more components. The local form of the symmetry requires the existence of a non-Abelian gauge field, A_μ, which is matrix-valued.

Thus one asks the question: what happens if fields are grouped into multiplets (analogous to the components of angular momentum) by postulating hidden symmetries, based on non-Abelian groups.

This idea was first used by Yang and Mills in 1954 to develop the isospin $SU(2)$ model for the nuclear force [141]. The unfolding of the experimental evidence surrounding nucleons led to a series of deductions about conservation from observed particle lifetimes. Charge labels such as baryon number, isospin and strangeness were invented to give a name to these, and the supposition that conserved charges are associated with symmetries led to the development of non-Abelian symmetry models. For a summary of the particle physics, see, for example, refs. [34, 108].

Non-Abelian models have been used in condensed matter physics, where quasi-fields for mean-field spin systems have been formulated as field theories with $SU(N)$ symmetry [1, 54].

23.2 Construction

We can now extend the formalism in the remainder of this book to encompass non-Abelian fields. To do this, we have to treat the fields as multi-component vectors on the abstract internal space of the symmetry group, since the transformations which act on the fields are now matrices. The dimension of the matrices which act on matter fields (Klein–Gordon or Dirac) does not have to be the same as those which attach to the gauge field A_μ – the only requirement is that both sets of matrices satisfy the same algebra. This will become clearer when we examine the nature of gauge transformations for non-Abelian groups.

We begin with some notation. Let $\{T_R^a\}$, where $a = 1, \ldots, d_G$, denote a set of matrices which acts as the generators of the simple Lie algebra for the group G. These matrices satisfy the Lie algebra

$$[T^a, T^b] = -\mathrm{i} f^{ab}{}_c T^c \,. \tag{23.3}$$

T^a are chosen here to be Hermitian. This makes the structure constants real numbers. It is also possible to find an anti-Hermitian representation by multiplying all of the T^a by a factor of $\mathrm{i} = \sqrt{-1}$, but we shall not use this convention here. With anti-Hermitian conventions, the Abelian limit leads to an imaginary electric charge which does not agree with the conventions used in other chapters.

The T^a are $d_R \times d_R$ matrices. In component form, one may therefore write them explicitly $(T^a)_{AB}$, where $A, B = 1, \ldots, d_R$, but normally the explicit components of T^a are suppressed and a matrix multiplication is understood. We denote the group which is obtained from these by G_R, which means the representation R of the group G. The normalization of the generators is fixed by defining

$$\mathrm{Tr}\big(T_R^a T_R^b\big) = I_2(G_R)\delta^{ab}, \tag{23.4}$$

where I_2 is called the Dynkin index for the representation G_R. The Dynkin index may also be written

$$I_2(G_R) = \frac{d_R}{d_G} C_2(G_R), \tag{23.5}$$

where d_R is the dimension (number of rows/columns) of the generators in the representation G_R and d_G is the dimension of the group. $C_2(G_R)$ is the quadratic Casimir invariant for the group in the representation G_R: $C_2(G_R)$ and $I_2(G_R)$ are constants which are listed in tables for various representations of Lie groups. d_G is the same as the dimension of the adjoint representation of the algebra G_{adj}, by definition of the adjoint representation. Note therefore that $I_2(G_{\mathrm{adj}}) = C_2(G_{\mathrm{adj}})$.

In many texts, authors make the arbitrary choice of replacing the right hand side of eqn. (23.4) with $\frac{1}{2}\delta^{ab}$. This practice simplifies formulae in a small number of special cases, but can lead to confusion later. Also, it makes the identification of group constants (for arbitrary groups) impossible and leads therefore to expressions which are not covariant with respect to changes of symmetry group.

To construct a physical theory with such an internal symmetry group we must look to the behaviour of the fields under a symmetry transformation. We require the analogue of a gauge transformation in the Abelian case. We begin by assuming that the form of a symmetry transformation on matter fields is

$$\Phi \to U\Phi, \tag{23.6}$$

for some matter field Φ and some matrix

$$U = \exp\left(i\theta^a(x)T^a\right), \tag{23.7}$$

which is the element of some Lie group, with an algebra generated by T^a, $(a = 1, \ldots, d_G)$. Eqn. (23.6) contains an implicit matrix multiplication: the components are normally suppressed; if we write them explicitly, eqn. (23.6) has the appearance:

$$\Phi^A \to U^A_{\ B}\Phi^B. \tag{23.8}$$

Since the generators do not commute with one another, and since U is a combination of these generators, T^a and U cannot commute; moreover, consecutive gauge transformations do not commute,

$$[U, U'] \neq 0, \tag{23.9}$$

in general. The exception to this statement is if the group element U lies in the centre of the group (i.e. the group's Abelian sub-group) which is generated purely by the Cartan sub-algebra:

$$U_c = \exp\left(i\theta^i(x)H^i\right), \qquad (i = 1, \ldots, \text{rank } G) \tag{23.10}$$
$$0 = [U_c, U]. \tag{23.11}$$

Under such a transformation, the spacetime-covariant derivative is not gauge-covariant:

$$\partial_\mu(U\Phi) \neq U(\partial_\mu\Phi). \tag{23.12}$$

We must therefore follow the analogue of the procedure in chapter 10 to define a covariant derivative for the non-Abelian symmetry. We do this in the usual way, by introducing a gauge connection, or vector potential

$$A_\mu = A_\mu^a(x)T^a, \tag{23.13}$$

which is a linear combination of all the generators. The basis components $A_\mu^a(x)$ are now the physical fields, which are to be varied in the action. There is one such field for each generator, i.e. the total number of fields is equal to the dimension of the group d_G. In terms of this new field, we write the covariant derivative

$$D_\mu = \partial_\mu + i\frac{g}{\hbar}A_\mu, \tag{23.14}$$

where g is a new charge for the non-Abelian symmetry. As in the Abelian case, D_μ will only satisfy

$$D_\mu(U\Phi) = U(D_\mu\Phi), \tag{23.15}$$

if Φ and A_μ both transform together. We can determine the way in which A_μ must transform by writing

$$D_\mu(U\Phi) = (\partial_\mu U)\Phi + U(\partial_\mu\Phi) + i\frac{g}{\hbar}A_\mu U\Phi$$

$$= U\left(\partial_\mu\Phi + U^{-1}(\partial_\mu U)\Phi + i\frac{g}{\hbar}U^{-1}A_\mu U\Phi\right). \tag{23.16}$$

From this, we deduce that

$$i\frac{g}{\hbar}A'_\mu\Phi = i\frac{g}{\hbar}U^{-1}A_\mu U\Phi + U^{-1}(\partial_\mu U)\Phi, \tag{23.17}$$

so that the complete non-Abelian gauge transformation has the form

$$\Phi' = U\Phi$$

$$A'_\mu = U^{-1}A_\mu U - \frac{i\hbar}{g}U^{-1}(\partial_\mu U). \tag{23.18}$$

The transformation of the field strength tensor in a non-Abelian field theory can be derived from its definition:

$$F_{\mu\nu} = \partial_\mu A_\nu - \partial_\nu A_\mu + i\frac{g}{\hbar}[A_\mu, A_\nu], \tag{23.19}$$

and has the form

$$F_{\mu\nu} \to U^{-1}F_{\mu\nu}U. \tag{23.20}$$

Note that the field strength is not gauge-invariant: it transforms in a non-trivial way. This means that $F_{\mu\nu}$ is not an observable in non-Abelian field theory. The field strength tensor can also be expressed directly in terms of the covariant derivative by the formula

$$[D_\mu, D_\nu] = i\frac{g}{\hbar}F_{\mu\nu}, \tag{23.21}$$

or

$$F_{\mu\nu} = D_\mu A_\nu - D_\nu A_\mu - i\frac{g}{\hbar}[A_\mu, A_\nu]. \tag{23.22}$$

The field strength can also be expressed as a linear combination of the generators of the Lie algebra, and we define the physical components relative to a given basis set T^a by

$$F_{\mu\nu} = F^a_{\mu\nu}T^a. \tag{23.23}$$

Using the algebra relation (23.3), these components can be expressed in the form

$$F^a_{\mu\nu} = \partial_\mu A^a_\nu - \partial_\nu A^a_\mu + gf^a_{bc}A^b_\mu A^c_\nu. \tag{23.24}$$

23.3 The action

We are now in a position to postulate a form for the action of a non-Abelian gauge theory. We have no way of knowing what the 'correct' action for such a theory is (nor any way of knowing if such a theory is relevant to nature), so we allow ourselves to be guided by the invariant quantities which can be formed from the non-Abelian fields. For free scalar matter fields, it is natural to write

$$S_M = \int (\mathrm{d}x) \left\{ \hbar^2 c^2 (D^\mu \Phi)^\dagger (D_\mu \Phi) + m^2 c^4 \Phi^\dagger \Phi \right\}, \tag{23.25}$$

where

$$D_\mu \Phi = \partial_\mu \Phi + ig A_\mu \Phi \tag{23.26}$$

which has the form of a matrix acting on a vector. Clearly, the number of components in the vector Φ must be the same as the number of rows and columns in the matrix A_μ in order for this to make sense. The dagger symbol implies complex conjugation and transposition.

For the non-Abelian Yang–Mills field the action analogous to the Maxwell action is $S_{YM}[\overline{A} + A]$, where

$$S_{YM}[A] = \frac{1}{4\mu_{NA} I_2(G_{adj})} \int (\mathrm{d}x) \mathrm{Tr} \left(F^{\mu\nu} F_{\mu\nu} \right), \tag{23.27}$$

where μ_{NA} is analogous to the permeability in electromagnetism. The trace in eqn. (23.27) refers to the trace over implicit matrix components of the generators. The cyclic property of the trace ensures that this quantity is gauge-invariant. Under a gauge transformation, one has

$$\mathrm{Tr} \left(F^{\mu\nu} F_{\mu\nu} \right) \to \mathrm{Tr} \left(U^{-1} F^{\mu\nu} F_{\mu\nu} U \right) = \mathrm{Tr} \left(F^{\mu\nu} F_{\mu\nu} \right). \tag{23.28}$$

23.4 Equations of motion and continuity

The Wong equations describe classical point particles coupled to a non-Abelian gauge field [140]:

$$m \frac{\mathrm{d}x^\mu}{\mathrm{d}\tau} = p^\mu$$

$$m \frac{\mathrm{d}p^\mu}{\mathrm{d}\tau} = g Q^a F^{a\mu\nu} p_\nu$$

$$m \frac{\mathrm{d}Q}{\mathrm{d}\tau} = -g f^{abc} p^\mu A_\mu^b Q^c. \tag{23.29}$$

23.5 Multiple representations

The gauge field A_μ appears several times in the action: both in connection with the covariant derivative acting on matter fields, and in connection with the Yang–Mills term. The dimension d_R of the matrix representation used in the different parts of the action does not have to be equal throughout. Indeed, the number of components in the matter vector is chosen on 'phenomenological' grounds to match the number of particles known to exist in a multiplet. A *common* choice is:

- the fundamental representation for matter fields, i.e. $A_\mu = A_\mu^a T_f^a$ in D_μ;

- the adjoint representation for the Yang–Mills terms, i.e. $A_\mu = A_\mu^a T_{\text{adj}}^a$ in $\text{Tr} F^2$. *common* situation is to choose

Although this is a common situation, it is not a necessity. The choice of representation for the matter fields should be motivated by phenomenology. In the classical theory, there seems to be no good reason for choosing the adjoint matrices for gauge fields. It is always true that the components of the field transform in the adjoint representation regardless of the matrices used to define the action.

23.6 The adjoint representation

One commonly held belief is that the gauge field, A_μ, must be constructed from the generators of the adjoint representation. The components of the gauge field A_μ^a transform like a vector in the adjoint representation, regardless of the matrix representations used to define the gauge fields in the action. This follows simply from the fact that A_μ is a linear combination of all the generators of the algebra [21]. To show this, we begin by noting that, in a given representation, the structure constants which are identical for any matrix representation form the components of a matrix representation for the adjoint representation, by virtue of the Jacobi identity (see section 8.5.2).

Consider an arbitrary field Λ, with components θ^a relative to a set of basis generators T^a in an arbitrary representation, defined by

$$\Lambda = T^a \lambda^a. \tag{23.30}$$

The generator matrices may be in a representation with arbitrary dimension d_R. Under a gauge transformation, we shall assume that the field transforms like

$$\Lambda' = U^{-1} \Lambda U, \tag{23.31}$$

where U is in the same matrix representation as T^a and may be written

$$U = \exp(i\theta^a T^a). \tag{23.32}$$

Using the matrix identity,

$$\exp(A)B\exp(-A) = B + [A, B] + \frac{1}{2!}[A, [A, B]]$$

$$+ \frac{1}{3!}[A, [A, [A, B]]] + \cdots, \qquad (23.33)$$

it is straightforward to show that

$$\Lambda' = \lambda^a \left\{ \delta^a{}_r - \theta_b f^{ab}{}_r + \frac{1}{2} \theta_b \theta_c f^{ca}{}_s f^{bs}{}_r \right.$$

$$\left. - \frac{1}{3!} \theta_b \theta_c \theta_d f^{da}{}_q f^{cq}{}_p f^{bp}{}_r + \cdots \right\} T^r, \qquad (23.34)$$

where the algebra commutation relation has been used. In our notation, the generators of the adjoint representation may be written

$$(T^a_{\text{adj}})^b{}_c = i f^{ab}{}_c, \qquad (23.35)$$

and the structure constants are real. Eqn. (23.34) may therefore be identified as

$$\Lambda' = \lambda^a (U_{\text{adj}})^a{}_b T^b, \qquad (23.36)$$

where

$$U_{\text{adj}} = \exp(i\theta^a T^a_{\text{adj}}). \qquad (23.37)$$

If we now define the components of the transformed field by

$$\Lambda' = \lambda'^a T^a, \qquad (23.38)$$

in terms of the original generators, then it follows that

$$\lambda'^a = (U_{\text{adj}})^a{}_b \lambda^b. \qquad (23.39)$$

We can now think of the set of components λ^a and λ'^a as being grouped into d_G component column *vectors* λ and λ', so that

$$\lambda' = U_{\text{adj}}\lambda. \qquad (23.40)$$

In matrix notation, the covariant derivative of the matrix-valued field Λ is

$$D_\mu \Lambda = \partial_\mu \Lambda + ig[A_\mu, \Lambda], \qquad (23.41)$$

for any representation. Using the algebra commutation relation this becomes

$$D_\mu \Lambda = \partial_\mu \Lambda + ig A^{\text{adj}}_\mu \Lambda, \qquad (23.42)$$

where $A^{\text{adj}}_\mu = A^a_\mu T^a_{\text{adj}}$. We have therefore shown that the vectorial components of the gauge field transform according to the adjoint representation, regardless of the matrices which are used in the matrix form.

23.7 Field equations and continuity

$$S = \int (\mathrm{d}x) \left\{ (D^\mu \Phi)^\dagger (D_\mu \Phi) + m^2 \Phi^\dagger \Phi + \frac{1}{4I_2(G_{\mathrm{adj}})} \mathrm{Tr}\left(F^{\mu\nu} F_{\mu\nu}\right) \right\}.$$

$$(23.43)$$

The variation of the action with respect to $\Phi^\dagger$ yields the equation of motion for Φ:

$$\delta S = \int (\mathrm{d}x) \left\{ (D^\mu \delta \Phi)^\dagger (D_\mu \Phi) + m^2 \delta \Phi^\dagger \Phi \right\}$$

$$= \int (\mathrm{d}x) \delta \Phi^\dagger \left\{ -D^2 \Phi + m^2 \Phi \right\}$$

$$+ \int \mathrm{d}\sigma^\mu \left\{ \delta \Phi^\dagger (D_\mu \Phi) \right\}.$$

$$(23.44)$$

The gauge-fixing term is

$$S_{\mathrm{GF}} = \frac{1}{2\alpha \mu_{\mathrm{NA}} I_2(G_{\mathrm{adj}})} \int \mathrm{d}v_x \mathrm{Tr}\left(D_\mu A^\mu\right)^2.$$

$$(23.45)$$

23.8 Commonly used generators

It is useful to have explicit forms for the generators in the fundamental and adjoint representations for the two most commonly discussed groups. For $SU(N)$, the matrices of the fundamental representation have dimension N.

23.8.1 *SU*(2) *Hermitian fundamental representation*

Here, the generators are simply one-half the Pauli matrices in the usual basis:

$$T^1 = \frac{1}{2} \begin{pmatrix} 0 & 1 \\ 1 & 0 \end{pmatrix}$$

$$T^2 = \frac{1}{2} \begin{pmatrix} 0 & -i \\ i & 0 \end{pmatrix}$$

$$T^3 = \frac{1}{2} \begin{pmatrix} 1 & 0 \\ 0 & -1 \end{pmatrix}$$

$$(23.46)$$

In the Cartan–Weyl basis, we construct

$$H = T^3 = \frac{1}{2} \begin{pmatrix} 1 & 0 \\ 0 & -1 \end{pmatrix}$$

$$E_\alpha = \frac{1}{\sqrt{2}} (T^1 + iT^2) = \frac{1}{\sqrt{2}} \begin{pmatrix} 0 & 1 \\ 0 & 0 \end{pmatrix}$$

$$E_{-\alpha} = \frac{1}{\sqrt{2}}(T^1 - iT^2) = \frac{1}{\sqrt{2}}\begin{pmatrix} 0 & 0 \\ 1 & 0 \end{pmatrix}, \qquad (23.47)$$

where the eigenvalue $\alpha = 1$ and

$$[H, E_\alpha] = \alpha E_\alpha$$
$$[E_\alpha, E_{-\alpha}] = \alpha H. \qquad (23.48)$$

The diagonal components of H are the weights of the representation.

23.8.2 *SU(2) Hermitian adjoint representation*

In the adjoint representation, the generators are simply the components of the structure constants in the regular basis:

$$T^1 = \begin{pmatrix} 0 & 0 & 0 \\ 0 & 0 & -i \\ 0 & i & 0 \end{pmatrix}$$

$$T^2 = \begin{pmatrix} 0 & 0 & i \\ 0 & 0 & 0 \\ -i & 0 & 0 \end{pmatrix}$$

$$T^3 = \begin{pmatrix} 0 & -i & 0 \\ i & 0 & 0 \\ 0 & 0 & 0 \end{pmatrix}. \qquad (23.49)$$

To find a Cartan–Weyl basis, in which the Cartan sub-algebra matrices are diagonal, we explicitly look for a transformation which diagonalizes one of the matrices. The same transformation will diagonalize the entire Cartan sub-algebra. Pick arbitrarily T^1 to diagonalize. The self-inverse matrix of eigenvectors for T^1 is easily found. It is given by

$$\Lambda = \begin{pmatrix} -1 & 0 & 0 \\ 0 & \frac{1}{\sqrt{2}} & \frac{-i}{\sqrt{2}} \\ 0 & \frac{i}{\sqrt{2}} & \frac{-1}{\sqrt{2}} \end{pmatrix}. \qquad (23.50)$$

Constructing the matrices $\Lambda^{-1}T^a\Lambda$, one finds a new set of generators,

$$T^1 = \begin{pmatrix} 0 & 0 & 0 \\ 0 & 1 & 0 \\ 0 & 0 & -1 \end{pmatrix}$$

$$T^2 = \begin{pmatrix} 0 & 1 & i \\ 1 & 0 & 0 \\ -i & 0 & 0 \end{pmatrix}$$

$$T^3 = \begin{pmatrix} 0 & i & 1 \\ -i & 0 & 0 \\ 1 & 0 & 0 \end{pmatrix}.$$ (23.51)

The Cartan–Weyl basis is obtained from these by constructing the combinations

$$E_{\pm 1} = \frac{1}{\sqrt{2}}(T^3 \mp iT^2)$$

$$H = T^1.$$ (23.52)

Explicitly,

$$E_1 = \begin{pmatrix} 0 & i & 0 \\ 0 & 0 & 0 \\ 1 & 0 & 0 \end{pmatrix}$$

$$E_{-1} = \begin{pmatrix} 0 & 0 & 1 \\ -i & 0 & 0 \\ 0 & 0 & 0 \end{pmatrix}.$$ (23.53)

It may be verified that

$$[H, E_\alpha] = \alpha E_\alpha$$ (23.54)

for $\alpha = \pm 1$. The diagonal values of H are the roots of the Lie algebra. It is interesting to note that the footprint of $SU(2)$ crops up often in the generators of other groups. This is because $SU(2)$ sub-groups are a basic entity where the roots show the simplest reflection symmetry. Since roots occur in signed pairs, $SU(2)$ is associated with root pairs.

23.8.3 SU(3) Hermitian fundamental representation

The generators of $SU(3)$'s fundamental representation are the Gell-Mann matrices:

$$T_1 = \frac{1}{2} \begin{pmatrix} 0 & -1 & 0 \\ -1 & 0 & 0 \\ 0 & 0 & 0 \end{pmatrix}$$

$$T_2 = \frac{1}{2} \begin{pmatrix} 0 & i & 0 \\ -i & 0 & 0 \\ 0 & 0 & 0 \end{pmatrix}$$

$$T_3 = \frac{1}{2} \begin{pmatrix} -1 & 0 & 0 \\ 0 & 1 & 0 \\ 0 & 0 & 0 \end{pmatrix}$$

$$T_4 = \frac{1}{2} \begin{pmatrix} 0 & 0 & -1 \\ 0 & 0 & 0 \\ -1 & 0 & 0 \end{pmatrix}$$

$$T_5 = \frac{1}{2} \begin{pmatrix} 0 & 0 & i \\ 0 & 0 & 0 \\ -i & 0 & 0 \end{pmatrix}$$

$$T_6 = \frac{1}{2} \begin{pmatrix} 0 & 0 & 0 \\ 0 & 0 & -1 \\ 0 & -1 & 0 \end{pmatrix}$$

$$T_7 = \frac{1}{2} \begin{pmatrix} 0 & 0 & 0 \\ 0 & 0 & i \\ 0 & -i & 0 \end{pmatrix}$$

$$T_8 = \frac{1}{2\sqrt{3}} \begin{pmatrix} -1 & 0 & 0 \\ 0 & -1 & 0 \\ 0 & 0 & 2 \end{pmatrix}. \tag{23.55}$$

The generators of the Cartan sub-algebra T^3 and T^8 are already diagonal in this representation. Forming a matrix which is an explicit linear combination of these generators $\theta^a T^a$, the following linear combinations are seen to parametrize the algebra naturally:

$$E_{\mp 1} = \frac{i}{\sqrt{2}} (T^1 \pm iT^2)$$

$$E_{\mp 2} = \frac{i}{\sqrt{2}} (T^4 \pm iT^5)$$

$$E_{\mp 3} = \frac{i}{\sqrt{2}} (T^6 \pm iT^7)$$

$$H^1 = T^3$$

$$H^2 = T^8. \tag{23.56}$$

These matrices satisfy the Cartan–Weyl relations

$$\left[H^i, E_\alpha \right] = \alpha^i E_\alpha$$

$$\left[E_\alpha, E_{-\alpha} \right] = \alpha^i H_i, \tag{23.57}$$

where i is summed over the elements of the Cartan sub-algebra. This last relation tells us that the commutator of the generators for equal and opposite roots always generates an element of the centre of the group. The coefficients α^i are the components of the root vectors on the sub-space spanned by the Cartan sub-algebra. Explicitly,

$$E_1 = \frac{1}{\sqrt{2}} \begin{pmatrix} 0 & 0 & 0 \\ -1 & 0 & 0 \\ 0 & 0 & 0 \end{pmatrix}$$

$$E_{-1} = \frac{1}{\sqrt{2}} \begin{pmatrix} 0 & -1 & 0 \\ 0 & 0 & 0 \\ 0 & 0 & 0 \end{pmatrix}$$

$$E_2 = \frac{1}{\sqrt{2}} \begin{pmatrix} 0 & 0 & 0 \\ 0 & 0 & 0 \\ -1 & 0 & 0 \end{pmatrix}$$

$$E_{-2} = \frac{1}{\sqrt{2}} \begin{pmatrix} 0 & 0 & -1 \\ 0 & 0 & 0 \\ 0 & 0 & 0 \end{pmatrix}$$

$$E_3 = \frac{1}{\sqrt{2}} \begin{pmatrix} 0 & 0 & 0 \\ 0 & 0 & 0 \\ 0 & -1 & 0 \end{pmatrix}$$

$$E_{-3} = \frac{1}{\sqrt{2}} \begin{pmatrix} 0 & 0 & 0 \\ 0 & 0 & -1 \\ 0 & 0 & 0 \end{pmatrix}. \tag{23.58}$$

Constructing all of the opposite combinations in the second relation of eqn. (23.57), one finds the root vectors in the Cartan–Weyl basis,

$$\alpha_{\pm 1} = \pm(1, 0)$$

$$\alpha_{\pm 2} = \pm \left(\frac{1}{2}, \frac{\sqrt{3}}{2} \right)$$

$$\alpha_{\pm 3} = \pm \left(-\frac{1}{2}, \frac{\sqrt{3}}{2} \right). \tag{23.59}$$

23.8.4 SU(3) Hermitian adjoint representation

The generators in the adjoint representation are obtained from the observation in eqn. (23.35) that the structure constants form a representation of the Lie algebra with the same dimension as the group:

$$(T^a)^b{}_c = i f^{ab}{}_c, \tag{23.60}$$

where $a, b, c = 1, \ldots, 8$. The structure constants are

$$f_{123} = 1$$

$$f_{147} = -f_{156} = f_{246} = f_{257} = f_{345} = -f_{367} = \frac{1}{2}$$

$$f_{458} = f_{678} = \frac{\sqrt{3}}{2}, \tag{23.61}$$

together with anti-symmetric permutations. In explicit form, we have

$$T^1 = i \begin{pmatrix} 0 & 0 & 0 & 0 & 0 & 0 & 0 & 0 \\ 0 & 0 & -1 & 0 & 0 & 0 & 0 & 0 \\ 0 & 1 & 0 & 0 & 0 & 0 & 0 & 0 \\ 0 & 0 & 0 & 0 & 0 & 0 & -\frac{1}{2} & 0 \\ 0 & 0 & 0 & 0 & 0 & \frac{1}{2} & 0 & 0 \\ 0 & 0 & 0 & 0 & -\frac{1}{2} & 0 & 0 & 0 \\ 0 & 0 & 0 & \frac{1}{2} & 0 & 0 & 0 & 0 \\ 0 & 0 & 0 & 0 & 0 & 0 & 0 & 0 \end{pmatrix}$$

$$T^2 = i \begin{pmatrix} 0 & 0 & 1 & 0 & 0 & 0 & 0 & 0 \\ 0 & 0 & 0 & 0 & 0 & 0 & 0 & 0 \\ -1 & 0 & 0 & 0 & 0 & 0 & 0 & 0 \\ 0 & 0 & 0 & 0 & 0 & -\frac{1}{2} & 0 & 0 \\ 0 & 0 & 0 & 0 & 0 & 0 & -\frac{1}{2} & 0 \\ 0 & 0 & 0 & \frac{1}{2} & 0 & 0 & 0 & 0 \\ 0 & 0 & 0 & 0 & \frac{1}{2} & 0 & 0 & 0 \\ 0 & 0 & 0 & 0 & 0 & 0 & 0 & 0 \end{pmatrix}$$

$$T^3 = i \begin{pmatrix} 0 & -1 & 0 & 0 & 0 & 0 & 0 & 0 \\ 1 & 0 & 0 & 0 & 0 & 0 & 0 & 0 \\ 0 & 0 & 0 & 0 & 0 & 0 & 0 & 0 \\ 0 & 0 & 0 & 0 & -\frac{1}{2} & 0 & 0 & 0 \\ 0 & 0 & 0 & \frac{1}{2} & 0 & 0 & 0 & 0 \\ 0 & 0 & 0 & 0 & 0 & 0 & \frac{1}{2} & 0 \\ 0 & 0 & 0 & 0 & 0 & -\frac{1}{2} & 0 & 0 \\ 0 & 0 & 0 & 0 & 0 & 0 & 0 & 0 \end{pmatrix}$$

$$
T^4 = i\begin{pmatrix}
0 & 0 & 0 & 0 & 0 & 0 & \tfrac{1}{2} & 0 \\
0 & 0 & 0 & 0 & 0 & \tfrac{1}{2} & 0 & 0 \\
0 & 0 & 0 & 0 & \tfrac{1}{2} & 0 & 0 & 0 \\
0 & 0 & 0 & 0 & 0 & 0 & 0 & 0 \\
0 & 0 & -\tfrac{1}{2} & 0 & 0 & 0 & 0 & -\tfrac{\sqrt{3}}{2} \\
0 & -\tfrac{1}{2} & 0 & 0 & 0 & 0 & 0 & 0 \\
-\tfrac{1}{2} & 0 & 0 & 0 & 0 & 0 & 0 & 0 \\
0 & 0 & 0 & 0 & \tfrac{\sqrt{3}}{2} & 0 & 0 & 0
\end{pmatrix}
$$

$$
T^5 = i\begin{pmatrix}
0 & 0 & 0 & 0 & 0 & -\tfrac{1}{2} & 0 & 0 \\
0 & 0 & 0 & 0 & 0 & 0 & \tfrac{1}{2} & 0 \\
0 & 0 & 0 & -\tfrac{1}{2} & 0 & 0 & 0 & 0 \\
0 & 0 & \tfrac{1}{2} & 0 & 0 & 0 & 0 & \tfrac{\sqrt{3}}{2} \\
0 & 0 & 0 & 0 & 0 & 0 & 0 & 0 \\
\tfrac{1}{2} & 0 & 0 & 0 & 0 & 0 & 0 & 0 \\
0 & -\tfrac{1}{2} & 0 & 0 & 0 & 0 & 0 & 0 \\
0 & 0 & 0 & -\tfrac{\sqrt{3}}{2} & 0 & 0 & 0 & 0
\end{pmatrix}
$$

$$
T^6 = i\begin{pmatrix}
0 & 0 & 0 & 0 & \tfrac{1}{2} & 0 & 0 & 0 \\
0 & 0 & 0 & -\tfrac{1}{2} & 0 & 0 & 0 & 0 \\
0 & 0 & 0 & 0 & 0 & 0 & -\tfrac{1}{2} & 0 \\
0 & \tfrac{1}{2} & 0 & 0 & 0 & 0 & 0 & 0 \\
-\tfrac{1}{2} & 0 & 0 & 0 & 0 & 0 & 0 & 0 \\
0 & 0 & 0 & 0 & 0 & 0 & 0 & 0 \\
0 & 0 & \tfrac{1}{2} & 0 & 0 & 0 & 0 & -\tfrac{\sqrt{3}}{2} \\
0 & 0 & 0 & 0 & 0 & 0 & \tfrac{\sqrt{3}}{2} & 0
\end{pmatrix}
$$

$$
T^7 = i\begin{pmatrix}
0 & 0 & 0 & -\tfrac{1}{2} & 0 & 0 & 0 & 0 \\
0 & 0 & 0 & 0 & -\tfrac{1}{2} & 0 & 0 & 0 \\
0 & 0 & 0 & 0 & 0 & \tfrac{1}{2} & 0 & 0 \\
\tfrac{1}{2} & 0 & 0 & 0 & 0 & 0 & 0 & 0 \\
0 & \tfrac{1}{2} & 0 & 0 & 0 & 0 & 0 & 0 \\
0 & 0 & -\tfrac{1}{2} & 0 & 0 & 0 & 0 & \tfrac{\sqrt{3}}{2} \\
0 & 0 & 0 & 0 & 0 & 0 & 0 & 0 \\
0 & 0 & 0 & 0 & 0 & -\tfrac{\sqrt{3}}{2} & 0 & 0
\end{pmatrix}
$$

$$
T^8 = i \begin{pmatrix}
0 & 0 & 0 & 0 & 0 & 0 & 0 & 0 \\
0 & 0 & 0 & 0 & 0 & 0 & 0 & 0 \\
0 & 0 & 0 & 0 & 0 & 0 & 0 & 0 \\
0 & 0 & 0 & 0 & -\frac{\sqrt{3}}{2} & 0 & 0 & 0 \\
0 & 0 & 0 & \frac{\sqrt{3}}{2} & 0 & 0 & 0 & 0 \\
0 & 0 & 0 & 0 & 0 & 0 & -\frac{\sqrt{3}}{2} & 0 \\
0 & 0 & 0 & 0 & 0 & \frac{\sqrt{3}}{2} & 0 & 0 \\
0 & 0 & 0 & 0 & 0 & 0 & 0 & 0
\end{pmatrix}.
\tag{23.62}
$$

The anti-Hermitian form of these matrices is obtained by dropping the leading factor of i. The Cartan–Weyl basis for the adjoint representation is obtained by diagonalizing one (and thereby several) of the generators. We choose to diagonalize T^8 because of its simple form. This matrix has four zero eigenvalues representing an invariant sub-space, so eigenvectors must be constructed for these manually. A set of normalized eigenvectors can be formed into a matrix which will diagonalize the generators of the Cartan sub-algebra:

$$
\Lambda = \begin{pmatrix}
\frac{i}{\sqrt{2}} & \frac{1}{\sqrt{2}} & 0 & 0 & 0 & 0 & 0 & 0 \\
\frac{1}{\sqrt{2}} & \frac{i}{\sqrt{2}} & 0 & 0 & 0 & 0 & 0 & 0 \\
0 & 0 & -1 & 0 & 0 & 0 & 0 & 0 \\
0 & 0 & 0 & \frac{i}{\sqrt{2}} & \frac{1}{\sqrt{2}} & 0 & 0 & 0 \\
0 & 0 & 0 & \frac{1}{\sqrt{2}} & \frac{i}{\sqrt{2}} & 0 & 0 & 0 \\
0 & 0 & 0 & 0 & 0 & \frac{i}{\sqrt{2}} & \frac{1}{\sqrt{2}} & 0 \\
0 & 0 & 0 & 0 & 0 & \frac{1}{\sqrt{2}} & \frac{i}{\sqrt{2}} & 0 \\
0 & 0 & 0 & 0 & 0 & 0 & 0 & 1
\end{pmatrix}.
\tag{23.63}
$$

The inverse of this is simply the complex conjugate. The new basis is now constructed by forming $\Lambda^{-1} T^a \Lambda$:

$$
T^1 = \begin{pmatrix}
0 & 0 & \frac{i}{\sqrt{2}} & 0 & 0 & 0 & 0 & 0 \\
0 & 0 & -\frac{1}{\sqrt{2}} & 0 & 0 & 0 & 0 & 0 \\
-\frac{i}{\sqrt{2}} & -\frac{1}{\sqrt{2}} & 0 & 0 & 0 & 0 & 0 & 0 \\
0 & 0 & 0 & 0 & 0 & \frac{1}{2} & 0 & 0 \\
0 & 0 & 0 & 0 & 0 & 0 & -\frac{1}{2} & 0 \\
0 & 0 & 0 & \frac{1}{2} & 0 & 0 & 0 & 0 \\
0 & 0 & 0 & 0 & -\frac{1}{2} & 0 & 0 & 0 \\
0 & 0 & 0 & 0 & 0 & 0 & 0 & 0
\end{pmatrix}
$$

$$
T^2 = \begin{pmatrix}
0 & 0 & \frac{1}{\sqrt{2}} & 0 & 0 & 0 & 0 & 0 \\
0 & 0 & -\frac{i}{\sqrt{2}} & 0 & 0 & 0 & 0 & 0 \\
\frac{1}{\sqrt{2}} & \frac{i}{\sqrt{2}} & 0 & 0 & 0 & 0 & 0 & 0 \\
0 & 0 & 0 & 0 & 0 & -\frac{i}{\sqrt{2}} & 0 & 0 \\
0 & 0 & 0 & 0 & 0 & 0 & -\frac{i}{\sqrt{2}} & 0 \\
0 & 0 & 0 & \frac{i}{\sqrt{2}} & 0 & 0 & 0 & 0 \\
0 & 0 & 0 & 0 & \frac{i}{\sqrt{2}} & 0 & 0 & 0 \\
0 & 0 & 0 & 0 & 0 & 0 & 0 & 0
\end{pmatrix}
$$

$$
T^3 = \begin{pmatrix}
1 & 0 & 0 & 0 & 0 & 0 & 0 & 0 \\
0 & -1 & 0 & 0 & 0 & 0 & 0 & 0 \\
0 & 0 & 0 & 0 & 0 & 0 & 0 & 0 \\
0 & 0 & 0 & \frac{1}{2} & 0 & 0 & 0 & 0 \\
0 & 0 & 0 & 0 & -\frac{1}{2} & 0 & 0 & 0 \\
0 & 0 & 0 & 0 & 0 & -\frac{1}{2} & 0 & 0 \\
0 & 0 & 0 & 0 & 0 & 0 & \frac{1}{2} & 0 \\
0 & 0 & 0 & 0 & 0 & 0 & 0 & 0
\end{pmatrix}
$$

$$
T^4 = \begin{pmatrix}
0 & 0 & 0 & 0 & 0 & 0 & \frac{i}{\sqrt{2}} & 0 \\
0 & 0 & 0 & 0 & 0 & \frac{i}{\sqrt{2}} & 0 & 0 \\
0 & 0 & 0 & -\frac{i}{2\sqrt{2}} & -\frac{1}{2\sqrt{2}} & 0 & 0 & 0 \\
0 & 0 & \frac{i}{2\sqrt{2}} & 0 & 0 & 0 & 0 & -\frac{i\sqrt{3}}{2\sqrt{2}} \\
0 & 0 & -\frac{1}{2\sqrt{2}} & 0 & 0 & 0 & 0 & \frac{\sqrt{3}}{2\sqrt{2}} \\
0 & -\frac{i}{2} & 0 & 0 & 0 & 0 & 0 & 0 \\
-\frac{i}{2} & 0 & 0 & 0 & 0 & 0 & 0 & 0 \\
0 & 0 & 0 & -\frac{i\sqrt{3}}{2\sqrt{2}} & \frac{\sqrt{3}}{2\sqrt{2}} & 0 & 0 & 0
\end{pmatrix}
$$

$$
T^5 = \begin{pmatrix}
0 & 0 & 0 & 0 & 0 & 0 & \frac{1}{2} & 0 \\
0 & 0 & 0 & 0 & 0 & -\frac{1}{2} & 0 & 0 \\
0 & 0 & 0 & \frac{1}{2\sqrt{2}} & \frac{i}{2\sqrt{2}} & 0 & 0 & 0 \\
0 & 0 & \frac{1}{2\sqrt{2}} & 0 & 0 & 0 & 0 & -\frac{\sqrt{3}}{2\sqrt{2}} \\
0 & 0 & -\frac{i}{2\sqrt{2}} & 0 & 0 & 0 & 0 & \frac{i\sqrt{3}}{2\sqrt{2}} \\
0 & -\frac{1}{2} & 0 & 0 & 0 & 0 & 0 & 0 \\
\frac{1}{2} & 0 & 0 & 0 & 0 & 0 & 0 & 0 \\
0 & 0 & 0 & -\frac{\sqrt{3}}{2\sqrt{2}} & -\frac{i\sqrt{3}}{2\sqrt{2}} & 0 & 0 & 0
\end{pmatrix}
$$

$$T^6 = \begin{pmatrix} 0 & 0 & 0 & -\frac{1}{2} & 0 & 0 & 0 & 0 \\ 0 & 0 & 0 & 0 & \frac{1}{2} & 0 & 0 & 0 \\ 0 & 0 & 0 & 0 & 0 & \frac{i}{2\sqrt{2}} & \frac{1}{2\sqrt{2}} & 0 \\ -\frac{1}{2} & 0 & 0 & 0 & 0 & 0 & 0 & 0 \\ 0 & \frac{1}{2} & 0 & 0 & 0 & 0 & 0 & 0 \\ 0 & 0 & i\frac{i}{2\sqrt{2}} & 0 & 0 & 0 & 0 & -\frac{i\sqrt{3}}{2\sqrt{2}} \\ 0 & 0 & \frac{1}{2\sqrt{2}} & 0 & 0 & 0 & 0 & \frac{\sqrt{3}}{2\sqrt{2}} \\ 0 & 0 & 0 & 0 & 0 & \frac{i\sqrt{3}}{2\sqrt{2}} & \frac{\sqrt{3}}{2\sqrt{2}} & 0 \end{pmatrix}$$

$$T^7 = \begin{pmatrix} 0 & 0 & 0 & -\frac{i}{2} & 0 & 0 & 0 & 0 \\ 0 & 0 & 0 & 0 & -\frac{i}{2} & 0 & 0 & 0 \\ 0 & 0 & 0 & 0 & 0 & -\frac{1}{2\sqrt{2}} & -\frac{i}{2\sqrt{2}} & 0 \\ \frac{i}{2} & 0 & 0 & 0 & 0 & 0 & 0 & 0 \\ 0 & \frac{i}{2} & 0 & 0 & 0 & 0 & 0 & 0 \\ 0 & 0 & -\frac{1}{2\sqrt{2}} & 0 & 0 & 0 & 0 & -\frac{\sqrt{3}}{2\sqrt{2}} \\ 0 & 0 & \frac{i}{2\sqrt{2}} & 0 & 0 & 0 & 0 & \frac{i\sqrt{3}}{2\sqrt{2}} \\ 0 & 0 & 0 & 0 & 0 & -\frac{\sqrt{3}}{2\sqrt{2}} & -\frac{i\sqrt{3}}{2\sqrt{2}} & 0 \end{pmatrix}$$

$$T^8 = \begin{pmatrix} 0 & 0 & 0 & 0 & 0 & 0 & 0 & 0 \\ 0 & 0 & 0 & 0 & 0 & 0 & 0 & 0 \\ 0 & 0 & 0 & 0 & 0 & 0 & 0 & 0 \\ 0 & 0 & 0 & \frac{\sqrt{3}}{2} & 0 & 0 & 0 & 0 \\ 0 & 0 & 0 & 0 & -\frac{\sqrt{3}}{2} & 0 & 0 & 0 \\ 0 & 0 & 0 & 0 & 0 & \frac{\sqrt{3}}{2} & 0 & 0 \\ 0 & 0 & 0 & 0 & 0 & 0 & -\frac{\sqrt{3}}{2} & 0 \\ 0 & 0 & 0 & 0 & 0 & 0 & 0 & 0 \end{pmatrix}. \tag{23.64}$$

The Cartan–Weyl basis is now obtained by constructing the linear combinations

$$E_{\mp 1} = \frac{1}{\sqrt{2}}(T^1 \pm iT^2)$$

$$E_{\mp 2} = \frac{1}{\sqrt{2}}(T^4 \pm iT^5)$$

$$E_{\mp 3} = \frac{1}{\sqrt{2}}(T^6 \pm iT^7). \tag{23.65}$$

Explicitly,

$$E_1 = \begin{pmatrix} 0 & 0 & 0 & 0 & 0 & 0 & 0 & 0 \\ 0 & 0 & -1 & 0 & 0 & 0 & 0 & 0 \\ -i & 0 & 0 & 0 & 0 & 0 & 0 & 0 \\ 0 & 0 & 0 & 0 & 0 & 0 & 0 & 0 \\ 0 & 0 & 0 & 0 & 0 & 0 & -\frac{1}{\sqrt{2}} & 0 \\ 0 & 0 & 0 & \frac{1}{\sqrt{2}} & 0 & 0 & 0 & 0 \\ 0 & 0 & 0 & 0 & 0 & 0 & 0 & 0 \\ 0 & 0 & 0 & 0 & 0 & 0 & 0 & 0 \end{pmatrix}$$

$$E_{-1} = \begin{pmatrix} 0 & 0 & i & 0 & 0 & 0 & 0 & 0 \\ 0 & 0 & 0 & 0 & 0 & 0 & 0 & 0 \\ 0 & -1 & 0 & 0 & 0 & 0 & 0 & 0 \\ 0 & 0 & 0 & 0 & 0 & \frac{1}{\sqrt{2}} & 0 & 0 \\ 0 & 0 & 0 & 0 & 0 & 0 & 0 & 0 \\ 0 & 0 & 0 & 0 & 0 & 0 & 0 & 0 \\ 0 & 0 & 0 & 0 & -\frac{1}{\sqrt{2}} & 0 & 0 & 0 \\ 0 & 0 & 0 & 0 & 0 & 0 & 0 & 0 \end{pmatrix}$$

$$E_2 = \begin{pmatrix} 0 & 0 & 0 & 0 & 0 & 0 & 0 & 0 \\ 0 & 0 & 0 & 0 & 0 & \frac{1}{\sqrt{2}} & 0 & 0 \\ 0 & 0 & 0 & -\frac{i}{2} & 0 & 0 & 0 & 0 \\ 0 & 0 & 0 & 0 & 0 & 0 & 0 & 0 \\ 0 & 0 & -\frac{1}{2} & 0 & 0 & 0 & 0 & -i\frac{\sqrt{3}}{2} \\ 0 & 0 & 0 & 0 & 0 & 0 & 0 & 0 \\ -\frac{i}{\sqrt{2}} & 0 & 0 & 0 & 0 & 0 & 0 & 0 \\ 0 & 0 & 0 & 0 & i\frac{\sqrt{3}}{2} & 0 & 0 & 0 \end{pmatrix}$$

$$E_{-2} = \begin{pmatrix} 0 & 0 & 0 & 0 & 0 & 0 & \frac{i}{\sqrt{2}} & 0 \\ 0 & 0 & 0 & 0 & 0 & 0 & 0 & 0 \\ 0 & 0 & 0 & 0 & -\frac{1}{2} & 0 & 0 & 0 \\ 0 & 0 & \frac{i}{2} & 0 & 0 & 0 & 0 & -i\frac{\sqrt{3}}{2} \\ 0 & 0 & 0 & 0 & 0 & 0 & 0 & 0 \\ 0 & -\frac{i}{\sqrt{2}} & 0 & 0 & 0 & 0 & 0 & 0 \\ 0 & 0 & 0 & 0 & 0 & 0 & 0 & 0 \\ 0 & 0 & 0 & 0 & 0 & \frac{\sqrt{3}}{2} & 0 & 0 \end{pmatrix}$$

$$
E_3 = \begin{pmatrix}
0 & 0 & 0 & -\frac{1}{\sqrt{2}} & 0 & 0 & 0 & 0 \\
0 & 0 & 0 & 0 & 0 & 0 & 0 & 0 \\
0 & 0 & 0 & 0 & 0 & \frac{i}{2} & 0 & 0 \\
0 & 0 & 0 & 0 & 0 & 0 & 0 & 0 \\
0 & \frac{1}{\sqrt{2}} & 0 & 0 & 0 & 0 & 0 & 0 \\
0 & 0 & 0 & 0 & 0 & 0 & 0 & 0 \\
0 & 0 & \frac{1}{2} & 0 & 0 & 0 & 0 & \frac{\sqrt{3}}{2} \\
0 & 0 & 0 & 0 & 0 & i\frac{\sqrt{3}}{2} & 0 & 0
\end{pmatrix}
$$

$$
E_{-3} = \begin{pmatrix}
0 & 0 & 0 & 0 & 0 & 0 & 0 & 0 \\
0 & 0 & 0 & 0 & \frac{1}{\sqrt{2}} & 0 & 0 & 0 \\
0 & 0 & 0 & 0 & 0 & 0 & \frac{1}{2} & 0 \\
-\frac{1}{\sqrt{2}} & 0 & 0 & 0 & 0 & 0 & 0 & 0 \\
0 & 0 & 0 & 0 & 0 & 0 & 0 & 0 \\
0 & 0 & -\frac{i}{2} & 0 & 0 & 0 & 0 & 0 \\
0 & 0 & 0 & 0 & 0 & 0 & 0 & 0 \\
0 & 0 & 0 & 0 & 0 & 0 & \frac{\sqrt{3}}{2} & 0
\end{pmatrix} . \tag{23.66}
$$

These generators satisfy the relations in eqns. (23.57) and define the components of the root vectors in two ways. The diagonal components of the generators spanning the Cartan sub-algebra are the components of the root vectors. We define

$$
H_1 = T_3
$$
$$
H_2 = T_8. \tag{23.67}
$$

The commutators in eqns. (23.57) may now be calculated, and one identifies

$$
\alpha_{\pm 1} = \mp(1, 0)
$$
$$
\alpha_{\pm 2} = \mp \left(\frac{1}{2}, \frac{\sqrt{3}}{2} \right)
$$
$$
\alpha_{\pm 3} = \mp \left(-\frac{1}{2}, \frac{\sqrt{3}}{2} \right) . \tag{23.68}
$$

24

Chern–Simons theories

In $2+1$ dimensions it is possible to construct actions which include a 'topological' interaction called the Chern–Simons term. The Chern–Simons term takes the form

$$
S_{\text{CS}} = \int \frac{(\text{d}x)}{\sqrt{g}} \frac{1}{2} \mu \epsilon^{\mu\nu\lambda} A_\mu \partial_\nu A_\lambda
$$
$$
= \int \text{d}^{n+1}x \, \frac{1}{2} \mu \epsilon^{\mu\nu\lambda} A_\mu \partial_\nu A_\lambda, \tag{24.1}
$$

for Abelian theories, and the extended form,

$$
S_{\text{CS--NA}} = \int \frac{\text{d}V_t}{\sqrt{g}} \frac{kg^2}{2\hbar^2 C_2(G_{\text{adj}})} \epsilon^{\mu\nu\lambda} \text{Tr} \left(A_\mu \partial_\nu A_\lambda - \mathrm{i} \frac{2}{3} A_\mu A_\nu A_\lambda \right),
$$

$$\tag{24.2}$$

in the non-Abelian case, where Hermitian generators are used. This action is real, as may be seen by applying the Lie algebra relation in eqn. (23.3). The effect of the Chern–Simons term on the dynamics of a field theory depends on whether the Maxwell or Yang–Mills term is also present. Since the Chern–Simons term is purely linear in all derivatives, and there are no additional constraints, as in the Dirac equation, it does not carry any independent dynamics of its own.

In the absence of dynamics from a Maxwell or Yang–Mills-like contribution to the action, the effect of this term is to induce a duality of variables, i.e. an equivalence relation between F_μ and J^μ. Coupled together with a Maxwell or Yang–Mills term, the Chern–Simons term endows the vector potential with a gauge-invariant mass [35, 36, 64, 110].

An unusual but important feature of the Chern–Simons action is that it is independent of the spacetime metric. Since the Levi-Cevita tensor transforms like a tensor density, a factor of $\sqrt{g}$ is therefore required to cancel the one already in the volume element. This has obvious implications for the usefulness of the variational definition of energy and momentum, by $T_{\mu\nu}$.

24.1 Parity- and time reversal invariance

The Chern–Simons terms violates parity- and time reversal invariance, both of which are defined for $2 + 1$ dimensions as the reflection in the axis of a single coordinate [64]. This is clearly not a fundamental property of nature. It is therefore only expected to play a role in physical systems where such a breakdown of parity invariance is present by virtue of special physical conditions. There are several such situations. Ferromagnetic states of spin fields, in the Hall effect, strong magnetic fields and vortices are examples [54].

The presence of a Chern–Simons term in the action of a field theory would lead to a rotation of the plane of polarization of radiation passing through a two-dimensional system, as in the Faraday effect (see sections A.6.1 and 7.3.3 and refs. [14, 16]). In ref. [24], the authors use the formalism of parity-violating terms to set limits on parity-violation from astronomical observations of distant galaxies. Spin polarized systems can be made into junctions, where Chern–Simons coefficients can appear with variable strength and sign [11, 14].

24.2 Gauge invariance

The transformation of the Chern–Simons action under gauge transformations, with its independence of the metric tensor, is what leads to its being referred to as a topological term. Consider the transformation of the non-Abelian action under a gauge transformation

$$A_\mu \to U A_\mu U^{-1} - i\frac{\hbar}{g}(\partial_\mu U)U^{-1}; \tag{24.3}$$

it transforms to

$$S \to S + \int (\mathrm{d}x)\,(\partial_\mu V^\mu)$$
$$+ \frac{k}{6C_2(G_{\mathrm{adj}})}\varepsilon^{\mu\nu\lambda}\,\mathrm{Tr}\int (\mathrm{d}x)\,\left[U(\partial_\mu U^{-1})U(\partial_\nu U^{-1})U(\partial_\lambda U^{-1})\right]. \tag{24.4}$$

The second term in the transformed expression is a total derivative and therefore vanishes, provided $U(\infty) = U(0)$: for instance, if $U \to 1$ in both cases (this effectively compactifies the spacetime to a sphere). The remaining term is:

$$\delta S = \frac{\hbar k}{6C_2(G_{\mathrm{adj}})}\varepsilon^{\mu\nu\lambda}\,\mathrm{Tr}\int (\mathrm{d}x)\,\left[U(\partial_\mu U^{-1})U(\partial_\nu U^{-1})U(\partial_\lambda U^{-1})\right]$$
$$= 8\hbar\pi^2 k\,W(U), \tag{24.5}$$

where $W(U)$ is the winding number of the mapping of the spacetime into the group, which is determined by the cohomology of the spacetime manifold. It takes integer values n. Clearly the action is not invariant under large gauge transformations. However, if k is quantized, such that $4\pi k$ is an integer, the action only changes by an integral multiple of 2π, leaving the phase $\exp(iS/\hbar + 2\pi i n)$ invariant. This quantization condition has been discussed in detail by a number of authors [35, 36, 39, 40, 43].

24.3 Abelian pure Chern–Simons theory

The Abelian Chern–Simons theory is relatively simple and has been used mainly in connection with studies of fractional statistics and the quantum Hall effect, where it gives rise to 'anyons' [4].

24.3.1 Field equations and continuity

Pure Chern–Simons theory is described by the Chern–Simons action together with a gauged matter action. In the literature, Chern–Simons theory is usually analysed by coupling it only to some unspecified gauge-invariant source:

$$S = \int \left(-\frac{1}{2}\mu\epsilon^{\mu\nu\lambda}A_\mu\partial_\nu A_\lambda + J^\mu A_\mu \right). \qquad (24.6)$$

The variation of the action is given by

$$\delta S = \int (dx)\{-\mu\epsilon^{\mu\nu\lambda}\delta A_\mu\partial_\nu A_\lambda + J^\mu\delta A_\mu\} - \int d\sigma_\nu \frac{1}{2}\mu A_\mu\delta A_\lambda, \qquad (24.7)$$

implying that the field equations are

$$\frac{1}{2}\mu\epsilon^{\mu\nu\lambda}F_{\nu\lambda} = J^\mu, \qquad (24.8)$$

with associated boundary (continuity) conditions

$$\Delta\left(\frac{1}{2}\mu\epsilon_{\mu\sigma\lambda}A_\mu\right) = 0, \qquad (24.9)$$

where the boundary of interest points in the direction of x^σ. Notice that, whereas the field equations are gauge-invariant, the boundary conditions are not. The physical interpretation of this result requires a specific context.

24.4 Maxwell–Chern–Simons theory

24.4.1 Field equations and continuity

In the literature, Chern–Simons theory is usually analysed by coupling it only to some unspecified gauge-invariant current:

$$S = \int \left(\frac{1}{4\mu_0} F^{\mu\nu} F_{\mu\nu} - \frac{1}{2} \mu\epsilon^{\mu\nu\lambda} A_\mu \partial_\nu A_\lambda + J^\mu A_\mu \right). \tag{24.10}$$

The same warnings about the generality of this notation apply as for pure Chern–Simons theory. The field equations are now given by

$$\frac{1}{\mu_0} \partial^\mu F_{\mu\nu} + \frac{1}{2} \mu\epsilon^{\mu\nu\lambda} F_{\nu\lambda} = J^\mu, \tag{24.11}$$

with associated boundary conditions

$$\Delta \left(-\mu_0^{-1} F_{\sigma\lambda} + \frac{1}{2} \mu\epsilon_{\mu\sigma\lambda} A^\mu \right) = 0. \tag{24.12}$$

24.4.2 Topological mass

To see that the derivative terms of the Chern–Simons action lead to a gauge-invariant massive mode, one may perform a diagonalization to the eigenbasis of the action operator:

$$S = \frac{1}{2} \int (dx) \, A^\mu (-\Box \, g^\mu{}_\nu + \mu\epsilon^\mu{}_\nu{}^\lambda \partial_\lambda)$$

$$= \int (dx) A_\mu \mathcal{O}^\mu{}_\nu A^\nu. \tag{24.13}$$

In a flat space, Cartesian coordinate basis, where all derivatives commute, this is seen most easily by writing the components in matrix form:

$$\mathcal{O}^\mu{}_\nu = \begin{pmatrix} -\Box & \mu\partial_2 & -\mu\partial_1 \\ -\mu\partial^2 & -\Box & \mu\partial_0 \\ \mu\partial^1 & \mu\partial^0 & -\Box \end{pmatrix}. \tag{24.14}$$

The determinant of this basis-independent operator is the product of its eigenvalues, which is the product of dispersion constraints. Noting that $-\Box = -\partial_0^2 + \partial_1^2 + \partial_2^2$, it is straightforward to show that

$$\det \mathcal{O} = (-\Box)^2 (-\Box + \mu^2), \tag{24.15}$$

showing that the dispersion of the Maxwell–Chern–Simons field contains two massless modes and one mode of mass μ^2. This massive theory has been studied in refs. [35, 36, 64, 110].

24.4.3 Energy–momentum tensors

In Chern–Simons theory, the energy–momentum tensors $\theta_{\mu\nu}$ and $T_{\mu\nu}$ do not agree. The reason for this is that the Chern–Simons action is independent of the metric tensor (it involves only anti-symmetric symbols), thus the variational definition of $T_{\mu\nu}$ inevitably leads to a zero value. If we use eqn. (11.68) and assume the gauge-invariant variation in eqn. (4.81), we obtain the following contributions to $\theta_{\mu\nu}$ from the action in eqn. (24.1),

$$
\theta_{\mu\nu} = \frac{1}{4}\mu g_{\mu\nu}\epsilon^{\rho\sigma\lambda}A_\rho F_{\sigma\lambda} - \frac{1}{2}\mu\epsilon^{\rho}{}_{\mu}{}^{\sigma}A_\rho F^{\nu}{}_{\sigma}. \tag{24.16}
$$

The fact that these two tensors do not agree can be attributed to the failure of the variational definition of $T_{\mu\nu}$ in eqn. (11.79). Since the Chern–Simons term is independent of the spacetime metric, it cannot be used as a generator for the conformal symmetry.

The contribution in eqn. (24.16) is not symmetrical but, in using the Bianchi identity $\epsilon^{\mu\nu\rho}\partial_\mu F_{\nu\rho}$, it is seen to be gauge-invariant, provided the Chern–Simons coefficient is a constant [11, 12, 14].

24.5 Euclidean formulation

In its Wick-rotated, Hermitian form, the Chern–Simons action acquires a factor of $i = \sqrt{-1}$, unlike most other action terms, since the Levi-Cevita tensor does not transform under Wick rotation. It has the Abelian form

$$
S_{\text{CS–E}} = i\int \frac{(\mathrm{d}x)}{\sqrt{g}}\frac{1}{2}\mu\epsilon^{\mu\nu\lambda}A_\mu \partial_\nu A_\lambda, \tag{24.17}
$$

and the non-Abelian form, for Hermitian generators

$$
S_{\text{CS–NA–E}} = i\int \frac{(\mathrm{d}x)}{\sqrt{g}}\frac{kg^2}{2C_2(G_{\text{adj}})}\epsilon^{\mu\nu\lambda}\mathrm{Tr}\left(A_\mu \partial_\nu A_\lambda - i\frac{2}{3}A_\mu A_\nu A_\lambda\right). \tag{24.18}
$$

25

Gravity as a field theory

This chapter provides the briefest, tangential encounter with the Einsteinian gravity viewed as a field theory. Gravity is a huge topic, full of subtleties, and it deserves to be introduced as a systematic tower of thought, rather than as a gallery of sketchy assertions. The purpose of this chapter is therefore no more than to indicate, to those who already know the general theory of relativity, how gravity fits into the foregoing discussions, i.e. why the foregoing ideas are still valid in the presence of gravity, and how we generalize our notion of covariance to include the gravitational force.

25.1 Newtonian gravity

Newtonian gravity plays virtually no role in field theory, for the simple reason that gravity barely couples to any of the fields. Gravity is such a weak force at the scale of elementary particles that it is almost completely negligible. There are occasions, however, when we use field theory outside of the realm of the elementary physics. For instance, fluid dynamics is a field theory where gravity plays an often significant role.

In order to include gravity in terrestrial systems, we do not need to think about Einstein or relativity. Gravity is simply an effective potential

$$V = mgx + \text{const.,} \tag{25.1}$$

where x is the height above the centre of gravity. In this effective theory of gravity, planets and large objects are considered to be point particles, located at the centre of gravity of the system. Eqn. (25.1) expresses a linear, flat-Earth geometry, in which the potential is usually measured from the ground up (for small distances of a few hundred metres). The arbitrary constant in the potential is analogous to the arbitrariness in the electromagnetic potential A_μ. Instead of gauge invariance, we have a corresponding arbitrariness in the origin of the gravitational potential.

25.2 Curvature

On astrophysical scales, gravity is the dominant force, and we need to consider the subtleties of general relativity. There are two motivations for wanting to do this

- Einsteinian gravity can be formulated as a field theory, in which the metric tensor (spacetime itself) is also a dynamical field. This enables us to understand gravity and spacetime as a dynamical system, leaning on all of the lessons we have learned from electromagnetism etc.

- In the early universe, there was an important coupling between gravity and other fundamental fields. Thus, relativistic, covariant formulations of fields which include gravity are important models to consider.

Gravity therefore means Einsteinian gravity here, and this, we know, has a natural expression in terms of the intrinsic curvature of spacetime. For the reasons discussed in the previous section, it makes no sense to look at non-relativistic theories in the presence of a relativistically generalized gravitational potential; such combinations would not be consistently compatible. We therefore dispense with the non-relativistic theories for the remainder of this short chapter.

25.3 Particles in a gravitational field

The essence of general relativity is that gravitational effects can be considered as physics in non-inertial frames. A non-inertial frame is a coordinate basis which is either accelerating or which contains a gravitational field. These two situations are indistinguishable, according to the equivalence principle, and so this is a kind of tautology. Indeed, we could go on to refer to the gravitational field as an acceleration field.

How shall we describe physics in such frames? Non-linear coordinate transformations can always map us from a locally inertial frame,[1] so covariance will help us to formulate theories optimally. The discussion which follows is based on the conventions and notations of Weinberg [133]. Readers who are unfamiliar with gravity could do worse than to consult his book, since there is no room for more than a cursory sketch here.

Let us denote the coordinates and derivatives and metric in a locally inertial Cartesian frame by ξ^μ, $\overset{\xi}{\partial_\mu}$, $\eta_{\mu\nu}$, and the corresponding quantities in any other coordinate system (flat, curvilinear, curved, accelerating etc.) by x^μ, ∂_μ, $g_{\mu\nu}$. The transformation which relates the two metrics is written according to the

---

[1] Suppose you are in a fighter plane and are suffering from the effects of strong acceleration G forces: to transform to a locally inertial frame, simply press the ejector seat button and you will soon be in a freely falling coordinate system.

usual tensor rules,

$$\eta^{\alpha\beta} = g^{\mu\nu} L^{\alpha}_{\mu} L^{\beta}_{\nu}$$
$$= g^{\mu\nu} (\partial_\mu \xi^\alpha)(\partial_\nu \xi^\beta). \tag{25.2}$$

In its locally inertial, or freely falling, coordinate frame, a moving particle seems to be following a straight-line path (although, since the frame is only inertial locally, we should not extrapolate too far from our position of observation). The equation of motion of such a particle would then be

$$m\frac{d^2\xi^\alpha}{d\tau^2} = 0, \tag{25.3}$$

where the proper time τ is defined in the usual way by

$$-c^2 d\tau^2 = \eta_{\alpha\beta} d\xi^\alpha d\xi^\beta. \tag{25.4}$$

Suppose now we transform into a general set of coordinates, using the Lorentz transformation L^{ν}_{μ}. We then have to transform ξ^α, so that eqn. (25.3) becomes

$$\frac{d}{d\tau}\left(\frac{d\xi^\alpha(x)}{d\tau}\right) = \frac{d}{d\tau}\left(\frac{d\xi^\alpha(x)}{dx^\mu}\frac{dx^\mu}{d\tau}\right). \tag{25.5}$$

Thus, the equation of motion becomes

$$(\partial_\mu \xi^\alpha)\frac{d^2 x^\mu}{d\tau^2} + (\partial_\mu \partial_\nu \xi^\alpha)\frac{dx^\mu}{d\tau}\frac{dx^\nu}{d\tau} = 0. \tag{25.6}$$

This can be simplified by multiplying through by $\overset{\xi}{\partial_\alpha} x^\lambda$ and using the chain-rule $(\overset{\xi}{\partial_\alpha} x^\lambda)(\partial_\lambda \xi^\beta) = \delta^\beta_\alpha$ to give

$$\frac{d^2 x^\lambda}{d\tau^2} + \Gamma^\lambda_{\mu\nu}\frac{dx^\mu}{d\tau}\frac{dx^\nu}{d\tau} = 0, \tag{25.7}$$

which is the geodesic equation, where

$$\Gamma^\lambda_{\mu\nu} = (\partial_\mu \partial_\nu \xi^\alpha)(\overset{\xi}{\partial_\alpha} x^\lambda). \tag{25.8}$$

The presence of the affine connection $\Gamma^\lambda_{\mu\nu}$ signals the non-linear nature of the coordinates. The connection may also be expressed in terms of the metric tensor as

$$\Gamma^\sigma_{\lambda\mu} = \frac{1}{2}g^{\nu\sigma}\left\{\partial_\lambda g_{\mu\nu} + \partial_\mu g_{\lambda\nu} - \partial_\nu g_{\mu\lambda}\right\}. \tag{25.9}$$

25.4 Geodesics

The geodesic equation can also be understood in a different way, from the action principle. The geodesic equation is, in a sense, the structure of empty space, so what if we take an empty action in a locally inertial rest frame of a general curved spacetime and vary it with respect to different paths, as follows:

$$x^\mu \to x^\mu(\lambda) + \delta x^\mu(\lambda)? \tag{25.10}$$

The action would then be

$$S = a \int d\tau, \tag{25.11}$$

where τ is the proper time, defined in eqn. (3.38) and a is a constant with the dimensions of energy. Writing this in general coordinates, we have

$$S = a \int \sqrt{g_{\mu\nu}(x)dx^\mu dx^\nu}, \tag{25.12}$$

or – introducing a parameter λ,

$$S = a \int d\lambda \frac{d\tau}{d\lambda} = \int d\lambda \sqrt{g_{\mu\nu}(x)\frac{dx^\mu}{d\lambda}\frac{dx^\nu}{d\lambda}}. \tag{25.13}$$

This equation can now be varied with respect to x^μ to obtain the path of 'least action' in the coordinate system x. We already know that, in a locally inertial frame, the path of an object would be a straight line, and in a rest frame there is no motion. So the question is: how does this look to a different observer in possibly accelerating coordinates? The variation of the action is

$$\delta S = a \int d\lambda \frac{1}{2}\frac{d\lambda}{d\tau}\left\{\delta g_{\mu\nu}\frac{dx^\mu}{d\lambda}\frac{dx^\nu}{d\lambda} + 2g_{\mu\nu}\frac{d\delta x^\mu}{d\lambda}\frac{dx^\nu}{d\lambda}\right\} = 0. \tag{25.14}$$

Since we are looking at a coordinate variation, we have

$$\delta g_{\mu\nu} = (\partial_\lambda g_{\mu\nu})\,\delta x^\lambda; \tag{25.15}$$

see eqn. (4.85). Thus, integrating by parts and writing $d\lambda\frac{d\lambda}{d\tau}$ as $d\lambda\frac{d\lambda d\tau}{d\tau d\tau}$,

$$\delta S = \frac{a}{2}\int d\tau \left\{(\partial_\lambda g_{\mu\nu})\frac{dx^\mu}{d\tau}\frac{dx^\nu}{d\tau} - 2(\partial_\rho g_{\mu\nu})\frac{dx^\rho}{d\tau}\frac{dx^\nu}{d\tau}g_{\mu\lambda}\right.$$
$$\left. -2g_{\mu\nu}\frac{d^2x^\nu}{d\tau^2}g_{\mu\lambda}\right\}\delta x^\lambda = 0. \tag{25.16}$$

Here we have assumed that the surface term

$$\Delta\left(\frac{dx^\mu}{d\tau}\delta x_\mu\right) = 0 \tag{25.17}$$

vanishes for continuity. From eqn. (25.9), the result may be identified as

$$\delta S = a \int \left\{ -\Gamma^\lambda_{\mu\nu} \frac{dx^\mu}{d\tau} \frac{dx^\nu}{d\tau} - \frac{dx^\lambda}{d\tau^2} \right\} g_{\lambda\sigma} \delta x^\sigma d\tau = 0. \qquad (25.18)$$

We have used the symmetry on the lower indices of $\Gamma^\lambda_{\mu\nu}$. Thus we end up with the geodesic equation once again:

$$\frac{d^2 x^\lambda}{d\tau^2} + \Gamma^\lambda_{\mu\nu} \frac{dx^\mu}{d\tau} \frac{x^\nu}{d\tau} = 0. \qquad (25.19)$$

25.5 Curvature

The curvature of a vector field ξ^σ may be defined by the commutator of covariant derivatives, just as in the case of the electromagnetic field (see eqn. (10.45)). This defines a process of parallel transport of vectors and a tensor known as the Riemann curvature tensor:

$$[\nabla_\mu, \nabla_\nu] \xi^\sigma = -R^\lambda_{\sigma\mu\nu} \xi_\lambda. \qquad (25.20)$$

Also analogous to electromagnetism is the expression of the curvature as a covariant curl:

$$R^\lambda_{\mu\nu\kappa} = \nabla_\kappa \Gamma^\lambda_{\mu\nu} - \nabla_\nu \Gamma^\lambda_{\mu\kappa}. \qquad (25.21)$$

This may be compared with eqn. (2.24). The Riemann tensor has the following symmetry properties:

$$R_{\lambda\mu\nu\kappa} = R_{\nu\kappa\lambda\mu} \qquad (25.22)$$

$$R_{\lambda\mu\nu\kappa} = -R_{\mu\lambda\nu\kappa} = R_{\lambda\mu\kappa\nu} = R_{\mu\lambda\kappa\nu} \qquad (25.23)$$

$$R_{\lambda\mu\nu\kappa} + R_{\lambda\kappa\mu\nu} + R_{\lambda\nu\kappa\mu} = 0. \qquad (25.24)$$

The Ricci tensor is defined as the contraction

$$R_{\mu\kappa} = R_{\lambda\mu\nu\kappa} g^{\lambda\nu} = R^\nu_{\mu\nu\kappa}, \qquad (25.25)$$

and satisfies

$$R_{\mu\nu} = R_{\nu\mu}. \qquad (25.26)$$

The scalar curvature is the total contraction

$$R = R^{\mu\nu}_{\mu\nu}. \qquad (25.27)$$

The curvature satisfies Bianchi identities, just like the electromagnetic field:

$$\nabla_\rho R_{\lambda\mu\nu\kappa} + \nabla_\kappa R_{\lambda\mu\rho\nu} + \nabla_\nu R_{\lambda\mu\kappa\rho} = 0. \qquad (25.28)$$

Contracting with $g^{\lambda\nu}$ gives

$$\nabla_\mu \left[R^{\mu\nu} - \frac{1}{2} g^{\mu\nu} R \right] = 0. \qquad (25.29)$$

25.6 The action

The action for matter coupled to gravity is written

$$S = S_M + S_G, \tag{25.30}$$

where

$$S_G = -\frac{c^4}{16\pi G} \int (\mathrm{d}x) \, [R - 2\Lambda]; \tag{25.31}$$

$(\mathrm{d}x) = \mathrm{d}t \, \mathrm{d}^n \mathbf{x} \sqrt{g}$ and $g = -\det g_\mu$. S_M is the action for matter fields. These act as the source of the gravitational field, i.e. they carry gravitational charge (mass/energy).

Λ is the cosmological constant, which is usually set to zero. The variation of the action with respect to the metric is

$$\delta\sqrt{g} = -\frac{1}{2}\sqrt{g} g^{\mu\nu} \, \delta^{\mu\nu}$$
$$\delta R = \delta(g^{\mu\nu} R_{\mu\nu})$$
$$= \delta g^{\mu\nu} \, R_{\mu\nu}. \tag{25.32}$$

Thus,

$$\delta S = -\frac{c^4}{16\pi G} \int (\mathrm{d}x) \left[-\frac{1}{2} g_{\mu\nu} [R - 2\Lambda/c^2] + R_{\mu\nu} \right] \delta g^{\mu\nu}$$
$$+ \frac{\delta S_M}{\delta g^{\mu\nu}} \delta g^{\mu\nu} = 0. \tag{25.33}$$

The last term is the conformal energy–momentum tensor

$$R_{\mu\nu} - \frac{1}{2} R g_{\mu\nu} + \frac{\Lambda}{c^2} g_{\mu\nu} = \frac{8\pi G}{c^4} \, T_{\mu\nu}. \tag{25.34}$$

This is Einstein's field equation for gravity. It is, of course, supplemented by the field equations for matter to complete the dynamical system. Notice that matter and energy (the energy–momentum tensor) is the source of gravitation. Matter, in other words, carries the gravitational charge: mass/energy.

The solution of these field equations is non-trivial and beyond the scope of this book.

25.7 Kaluza–Klein theory

Following Maxwell's treatise on the electromagnetic field, Theodore Kaluza was amongst the first to propose a scheme for unifying the forces of nature using a classical field theory, based in Einstein's equations. Kaluza's paper, communicated to Einstein, endured a long delay before its publication in 1921. His main idea, later refined by Oskar Klein, made the bold assertion that, if one postulated the existence of extra dimensions, then both of the known forces of nature (electromagnetism and gravity) could be unified, using Einstein's idea of spacetime curvature. In Kaluza–Klein theory, the line element is assumed to have the usual form

$$ds^2 = \hat{g}_{\hat{\mu}\hat{\nu}} \, dx^{\hat{\mu}} dx^{\hat{\nu}} \tag{25.35}$$

where the careted indices run from $0, \ldots, 5$ and $x^\mu = (ct, x^1, x^2, x^3, y) = (x^\mu, y)$. Uncareted indices represent the usual $3 + 1$ dimensional vectors of general relativity. In order to account for the $U(1)$ symmetry, Klein proposed that the extra dimension should have the topology of a circle, with length L. The electromagnetic field plays the role of a vector field on the $3 + 1$ dimensional spacetime, seen as the projection of the curvature of the extra dimension:

$$ds^2 = \hat{g}_{\hat{\mu}\hat{\nu}} \, dx^{\hat{\mu}} dx^{\hat{\nu}}$$
$$= g_{\mu\nu} \, dx^\mu dx^\nu + (dy + \kappa A_\mu(x)dx^\mu)^2, \tag{25.36}$$

where κ is a constant. Covariance in the extra dimension determines the transformation rule for A_μ under coordinate transformations $y' = \theta(y, x^\mu)$:

$$dy' = \frac{\partial \theta}{\partial y} \, dy + \partial_\mu \theta \, dx^\mu. \tag{25.37}$$

For consistency with eqn. (25.36), one requires $\partial\theta/\partial y = 1$, so that under a change of y only,

$$dy + \kappa A_\mu dx^\mu \rightarrow dy' + \kappa A'_\mu dx^\mu$$
$$= \left(dy + \partial_\mu \theta dx^\mu\right) + \kappa A' dx^\mu$$
$$= dy + \kappa \left(A'_\mu(x) + \kappa^{-1}\partial_\mu\theta\right) dx^\mu. \tag{25.38}$$

Invariance of ds^2 therefore requires

$$A'_\mu(x) = A_\mu(x) - \kappa^{-1}\partial_\mu\theta, \tag{25.39}$$

which is the electromagnetic gauge transformation. From the line element, the metric is

$$\hat{g}_{\hat{\mu}\hat{\nu}} = \begin{pmatrix} g_{\mu\nu} + \kappa A_\mu A_\nu & \kappa A_\mu \\ \kappa A_\nu & 1 \end{pmatrix}; \tag{25.40}$$

however, by changing coordinates to the so-called horizontal lift basis, with 1-forms:

$$\tilde{\omega}^\mu = dx^\mu$$
$$\tilde{\omega}^5 = dy + \kappa A_\mu(x)dx^\mu, \tag{25.41}$$

the metric may be diagonalized, at the expense of non-Cartesian coordinates:

$$\tilde{g}'_{\hat{\mu}\hat{\nu}} = \begin{pmatrix} g_{\mu\nu} & 0 \\ 0 & 1 \end{pmatrix}, \tag{25.42}$$

The basis vectors conjugate to the 1-forms are $\tilde{\omega}^{\hat{\mu}} \hat{e}_{\hat{\nu}} = \delta^{\hat{\mu}}_{\hat{\nu}}$, i.e.

$$\hat{e}_{\mu} = \partial_{\mu} - \kappa A_{\mu}(x)\partial_{y}$$
$$\hat{E}_{5} = \partial_{y}. \tag{25.43}$$

In this anholonomic basis, there is one non-zero commutator:

$$[\hat{e}_{\mu}, \hat{e}_{\nu}] = -\kappa F_{\mu\nu}(x)\, \partial_{y}, \tag{25.44}$$

where $A_{\mu\nu} = \partial_{\mu}A_{\nu} - \partial_{\nu}A_{\mu}$, which gives the Lie algebra relation

$$[\hat{e}_{\hat{\mu}}, \hat{e}_{\hat{\mu}}] = C_{\hat{\mu}\hat{\nu}}{}^{\hat{\rho}}\, \hat{e}_{\hat{\rho}}. \tag{25.45}$$

The affine connection, in a non-holonomic basis, is

$$\Gamma_{\mu\nu\lambda} = \frac{1}{2}\left[\hat{e}_{\lambda}\, g_{\mu\nu} + \hat{e}_{\nu}\, g_{\mu\lambda} - \hat{e}_{\mu}\, g_{\lambda\nu} + C_{\mu\nu\lambda} + C_{\mu\lambda\nu} + C_{\lambda\nu\mu}\right], \tag{25.46}$$

so that we have non-zero components

$$\hat{\Gamma}_{\mu\nu 5} = \hat{\Gamma}_{\mu 5\nu} = -\hat{\Gamma}_{5\mu\nu} = -\frac{1}{2}\kappa F_{\mu\nu}$$
$$\hat{\Gamma}_{555} = 0 \quad , \quad \hat{\Gamma}_{\mu\nu\lambda} = \Gamma_{\mu\nu\lambda}. \tag{25.47}$$

From these, one may calculate the scalar curvature for the Einstein action,

$$\hat{R} = \hat{R}^{\mu\nu}{}_{\mu\nu} + 2\hat{R}^{\mu 5}{}_{\mu 5}$$
$$= R + \frac{\kappa^{2}}{4}F^{\mu\nu}F_{\mu\nu}. \tag{25.48}$$

Thus, the Einstein action, in five dimensions, automatically incorporates and extrapolates the Maxwell action:

$$S = -\frac{c^{4}}{16\pi GL}\int d^{4}x\, dy\, \sqrt{\hat{g}}\left[\hat{R} - 2\Lambda\right]. \tag{25.49}$$

Kaluza–Klein theory came into trouble when it attempted to incorporate the newly discovered nuclear forces in a common framework, and was eventually abandoned in its original form. However, the essence of Kaluza–Klein theory lives on, in a more sophisticated guise, in super-string theory.

Part 4
Appendices

Appendix A
Useful formulae

A.1 The delta function

The Dirac delta function is a bi-local distribution defined by the relations

$$
\delta(t, t') =
\begin{cases}
0 & t - t' \neq 0 \\
\infty & t - t' = 0
\end{cases}
$$

$$(A.1)$$
$$(A.2)$$

$$\delta(x, x') = \delta(x - x') \tag{A.3}$$

$$\int_{-a}^{+a} dx' \, \delta(x, x') f(x') = f(x) \tag{A.4}$$

$$\int_{-a}^{+a} dx' \, \delta(x - x') = 1. \tag{A.5}$$

If $f(x)$ is a function which is symmetrical about x_0, then

$$\int_{-\infty}^{x_0} \delta(x_0 - x') f(x') dx' + \int_{x_0}^{\infty} \delta(x_0 - x') f(x') dx' = f(x_0); \tag{A.6}$$

thus, by symmetry,

$$\int_{-\infty}^{x_0} \delta(x_0 - x') f(x') dx' = \frac{1}{2} f(x_0). \tag{A.7}$$

A useful, integral representation of the delta function is given by the Fourier integral

$$\delta(x_1 - x_1') = \int \frac{dk}{2\pi} e^{ik(x_1 - x_1')}. \tag{A.8}$$

501

Various integral representations of the delta function are useful. For instance

$$\delta(x) = \lim_{\alpha \to 0} \frac{1}{2\pi\alpha} e^{-\frac{1}{\alpha}x^2}.$$ (A.9)

The Fourier representation on the $(n + 1)$ dimensional delta function

$$\delta(x, x') \equiv \delta(x - x') = \int \frac{d^{n+1}k}{(2\pi)^{n+1}} e^{ik(x-x')}$$

$$= \delta(x^0 - x^{0'})\delta(x^1 - x^{1'}) \ldots \delta(x^n - x^{n'}) \quad (A.10)$$

in particular is used in solving for Green functions. Here the shorthand notation $k(x - x')$ in the exponential stands for $k_\mu(x - x')^\mu$.

Derivatives of the delta function normally refer to derivatives of the test functions which they multiply. Meaning may be assigned to these as follows. Consider the boundary value of a function $f(x)$. From the property of the delta function,

$$\int \delta(x - a)f(x - a)dx = f(0).$$ (A.11)

Now, differentiating with respect to a,

$$\frac{d}{da}f(0) = \int \left[\frac{d}{da}\delta(x - a)f(x - a) + \delta(x - a)\frac{d}{da}f(x - a) \right] dx$$

$$= 0.$$ (A.12)

From this, we discover that

$$f(x - a)\frac{d}{da}\delta(x - a) = -\delta(x - a)\frac{d}{da}f(x - a),$$ (A.13)

or

$$f(t)\, \partial_t\delta(t) = -\delta(t)\, \partial_t f(t),$$ (A.14)

which effectively defines the derivative of the delta function.

A useful relation for the one-dimensional delta function of a function $g(x)$ with several roots satisfying $g(x_i) = 0$ is:

$$\delta(g(x)) = \sum_i \frac{1}{g'(x_i)}\delta(g(x_i)),$$ (A.15)

where x_i are the roots of the function $g(x)$ and the prime denotes the derivative with respect to x. This is easily proven by change of variable. As with all delta-function relations, this is only strictly valid under the integral sign. Given

$$I = \int dx f(x)\delta(g(x)),$$ (A.16)

change variables to $x' = g(x)$. This incurs a Jacobian in the measure $|J| = \frac{\partial x}{\partial x'} = \frac{\partial x}{\partial g(x)} = \left(\frac{\partial g(x)}{\partial x}\right)^{-1}$. So,

$$I = \int dx' \frac{1}{g'(x')} f(g^{-1}(x))\delta(x'). \tag{A.17}$$

In replacing x by g^{-1}, we satisfy the rules of the change of variable, but the inverse function $g^{-1}(x')$ is not usually known. Fortunately, the singular nature of the delta function simplifies the calculation, since it implies that contributions can only come from the roots of $g(x')$, thus, the expression becomes,

$$I = \int dx' \frac{1}{g'(x_i)} f(x_i)\delta(x'). \tag{A.18}$$

In summary, one may use this eqn. (A.15) under the integral sign generally, thanks to the extremely singular nature of the delta function, provided all multiplying functions in the integrand are evaluated at the roots of the original function $g(x)$.

A.2 The step function

$$\theta(t, t') = \begin{cases} 1 & t - t' > 0 \\ \frac{1}{2} & t = t' \\ 0 & t - t' < 0. \end{cases} \tag{A.19}$$

An integral representation of these may be expressed in two equivalent forms:

$$\theta(t - t') = i \int_{\infty}^{\infty} \frac{d\alpha}{2\pi} \frac{e^{-i\alpha(t-t')}}{\alpha + i\epsilon}$$

$$\theta(t' - t) = -i \int_{\infty}^{\infty} \frac{d\alpha}{2\pi} \frac{e^{-i\alpha(t-t')}}{\alpha - i\epsilon}, \tag{A.20}$$

where the limit $\epsilon \to 0$ is understood. The derivative of the step function is a delta function,

$$\partial_t \theta(t - t') = \delta(t - t'). \tag{A.21}$$

A.3 Anti-symmetry and the Jacobi identity

The commutator (or indeed any anti-symmetrical quantity) has the purely algebraic property that:

$$[A, [B, C]] + [B, [C, A]] + [C, [A, B]] = 0. \tag{A.22}$$

A.4 Anti-symmetric tensors in Euclidean space

Anti-symmetric tensors arise in many situations in field theory. In most cases, we shall only be interested in the two-, three- and four-dimensional tensors, defined respectively by

$$\epsilon_{ij} = \begin{cases} +1 & ij = 12 \\ -1 & ij = 21 \\ 0 & \text{otherwise,} \end{cases} \tag{A.23}$$

$$\epsilon_{ijk} = \begin{cases} +1 & ijk = 123 \text{ and even permutations} \\ -1 & ijk = 321 \text{ and other odd permutations} \\ 0 & \text{otherwise,} \end{cases} \tag{A.24}$$

$$\epsilon_{ijkl} = \begin{cases} +1 & ijkl = 1234 \text{ and even permutations} \\ -1 & ijkl = 1243 \text{ and other odd permutations} \\ 0 & \text{otherwise.} \end{cases} \tag{A.25}$$

There are as many values for the indices as there are indices on the tensors in the above relations. Because of the anti-symmetric properties, the following relations are also true.

$$\epsilon_{ij} = = -\epsilon_{ji}$$
$$\epsilon_{ijk} = \epsilon_{kij} = \epsilon_{jki} = -\epsilon_{kji} = \epsilon_{ikj} = -\epsilon_{jik}$$
$$\epsilon_{ii} = 0$$
$$\epsilon_{iij} = 0$$
$$\epsilon_{iijk} = 0. \tag{A.26}$$

The number of different permutations increases as the factorial of the number of indices on the tensor. The different permutations can easily be generated by computing the determinant

$$\begin{vmatrix} i & j & k & l \\ i & j & k & l \\ i & j & k & l \\ i & j & k & l \end{vmatrix} \tag{A.27}$$

as a mnemonic, but the signs will not automatically distinguish even and odd permutations, so this is not a practical procedure.

Contractions of indices on anti-symmetric objects are straightforward to work out. The simplest of these are trivial to verify:

$$\epsilon^{ij}\epsilon_{kl} = \delta^i{}_k\delta^j{}_l - \delta^i{}_l\delta^j{}_k$$
$$\epsilon^{ij}\epsilon_{jk} = -\delta^i{}_k$$
$$\epsilon^{ij}\epsilon_{kj} = \delta^i{}_k$$
$$\epsilon^{ij}\epsilon_{ij} = 2. \tag{A.28}$$

More general contractions can be calculated by expressing the anti-symmetric tensor products as combinations of delta functions with varying signs and permutations of indices. These are most easily expressed using a notational shorthand for anti-symmetrization. Embedded square brackets are used to denote the anti-symmetrization over a set of indices. For example,

$$X_{[a}Y_{b]} \equiv \frac{1}{2!}(X_aY_b - X_bY_a),$$ (A.29)

$$X_{[a}Y_bZ_{c]} \equiv \frac{1}{3!}(X_aY_bZ_c + X_cY_aZ_b + X_bY_cZ_a$$
$$-X_cY_bZ_a - X_aY_cZ_b - X_bY_aZ_c),$$ (A.30)

and higher generalizations.

Consider then the product of two three-dimensional Levi-Cevita symbols. It may be proven on the grounds of symmetry alone that

$$\epsilon^{ijk}\epsilon_{lmn} = 3!\delta^i_{[l}\delta^j_m\delta^k_{n]},$$ (A.31)

where

$$\delta^i_{[l}\delta^j_m\delta^k_{n]} = \frac{1}{3!}\left(\delta^i_l\delta^j_m\delta^k_n + \delta^i_n\delta^j_l\delta^k_m + \delta^i_m\delta^j_n\delta^k_l\right.$$
$$\left. -\delta^i_n\delta^j_m\delta^k_l - \delta^i_l\delta^j_n\delta^k_m - \delta^i_m\delta^j_l\delta^k_n \right).$$ (A.32)

Contracting this on one index (setting $i = l$), and writing the outermost permutation explicitly, we have

$$\epsilon^{ijk}\epsilon_{imn} = 2!\left(\delta^i_i\delta^j_{[m}\delta^k_{n]} - \delta^i_m\delta^j_{[i}\delta^k_{n]} - \delta^i_n\delta^j_{[m}\delta^k_{i]}\right).$$ (A.33)

Summing over i gives

$$\epsilon^{ijk}\epsilon_{imn} = 2!(3-1-1)\delta^j_{[m}\delta^k_{n]}$$
$$= \delta^j_m\delta^k_n - \delta^j_n\delta^k_m.$$ (A.34)

It is not difficult to see that this procedure may be repeated for n-dimensional products,

$$\epsilon^{ij...k}\epsilon_{lm...n} = n!\delta^i_{[l}\delta^j_m\ldots\delta^k_{n]}.$$ (A.35)

Again, setting $i = l$ and expanding the outermost permutation gives,

$$\epsilon^{ij...k}\epsilon_{lm...n} = (n-1)!(\delta^i_i\delta^j_{[m}\ldots\delta^k_{n]} - \cdots)$$
$$= (n-1)!(\delta^i_i - (n-1))\delta^j_{[m}\ldots\delta^k_{n]}.$$ (A.36)

Since $\delta^i{}_i = n$, the first bracket in the result above always reduces to unity. We may also write this result in a more general way, for the contraction of two p-index anti-symmetric products in n dimensions:

$$p!\delta^i{}_{[i}\delta^j{}_m \ldots \delta^k{}_{n]} = (p-1)!(n-p+1)\delta^j{}_{[m} \ldots \delta^k{}_{n]}. \tag{A.37}$$

This formula leads to a number of frequently used results:

$$\epsilon^{ijk}\epsilon_{imn} = \delta^j{}_m\delta^k{}_n - \delta^j{}_n\delta^k{}_m$$

$$\epsilon^{ijkl}\epsilon_{imnp} = 3!\delta^j{}_{[m}\delta^k{}_n\delta^l{}_{p]}$$

$$\epsilon^{ijkl}\epsilon_{ijnp} = 2!(n-2)\delta^k{}_{[n}\delta^l{}_{p]}$$

$$= 2(\delta^k{}_n\delta^l{}_p - \delta^k{}_p\delta^l{}_n)$$

$$\epsilon^{ijkl}\epsilon_{ijkl} = 2(4^2 - 4) = 24. \tag{A.38}$$

A.5 Anti-symmetric tensors in Minkowski spacetime

In Minkowski spacetime, we have to distinguish between up and down indices. It is normal to define

$$\epsilon^{\mu\nu} = \begin{cases} +1 & \mu\nu = 01 \\ -1 & \mu\nu = 10 \\ 0 & \text{otherwise,} \end{cases} \tag{A.39}$$

$$\epsilon^{\mu\nu\lambda} = \begin{cases} +1 & \mu\nu\lambda = 012 \text{ and even permutations} \\ -1 & \mu\nu\lambda = 210 \text{ and other odd permutations} \\ 0 & \text{otherwise,} \end{cases} \tag{A.40}$$

$$\epsilon^{\mu\nu\lambda\rho} = \begin{cases} +1 & \mu\nu\lambda\rho = 0123 \text{ and even permutations} \\ -1 & \mu\nu\lambda\rho = 0132 \text{ and other odd permutations} \\ 0 & \text{otherwise.} \end{cases} \tag{A.41}$$

Indices are raised and lowered using the metric for the appropriate dimensional spacetime. Since the zeroth component always incurs a minus sign,

$$\epsilon^{\mu\nu\lambda\rho} = g^{\mu\sigma}g^{\nu\kappa}g^{\lambda\tau}g^{\rho\delta}\epsilon_{\sigma\kappa\tau\delta}$$

$$\epsilon^{0123} = -1.1.1.1.\epsilon_{0123}, \tag{A.42}$$

one has all of the above definitions with indices lowered on the left hand side and minus signs changed on the right hand side. This also means that all of the contraction formulae incur an additional minus sign. This formula leads to a number of frequently used results:

$$\epsilon^{\mu\nu\lambda}\epsilon_{\mu\rho\sigma} = -\delta^\nu{}_\rho\delta^\lambda{}_\sigma + \delta^\nu{}_\sigma\delta^\lambda{}_\rho$$

$$\epsilon^{\mu\nu\lambda\rho}\epsilon_{\mu\sigma\tau\zeta} = -3!\delta^\nu{}_{[\sigma}\delta^\lambda{}_\tau\delta^\rho{}_{\zeta]}$$

$$\epsilon^{\mu\nu\lambda\rho}\epsilon_{\mu\nu\sigma\tau} = -2(\delta^\lambda{}_\sigma\delta^\rho{}_\tau - \delta^\lambda{}_\tau\delta^\rho{}_\sigma)$$

$$\epsilon^{\mu\nu\lambda\rho}\epsilon_{\mu\nu\lambda\rho} = -24. \tag{A.43}$$

A.6 Doubly complex numbers

Complex numbers $z = x + iy$ and the conjugates $z^* = x - iy$ are vectors in the Argand plane. They form a complete covering of the two-dimensional space and, because of de Moivre's theorem,

$$e^{i\theta} = \cos\theta + i\sin\theta, \tag{A.44}$$

they are particularly suited to problems where rotation or circular symmetry is expected. But what of problems where rotation occurs in two separate, orthogonal planes? It seems logical to suppose that a complex representation of such rotation could be applied to each orthogonal plane individually. But such a description would require two separate kinds of vectors $x + iy$ for one plane and $x + jz$ for the orthogonal plane, where $i = \sqrt{-1}$ and $j = \sqrt{-1}$. We must treat these two imaginary numbers as independent vectors, such that

$$i^2 = -1$$
$$j^2 = -1$$
$$ij \neq -1. \tag{A.45}$$

Using these quantities, we can formulate doubly complex numbers

$$w = x + iy - jz$$
$$W = X + iY - jZ \tag{A.46}$$

as an alternative representation to the three-dimensional vectors $\mathbf{w} = x\hat{\mathbf{i}} + y\hat{\mathbf{j}} + z\hat{\mathbf{k}}$. The final line in eqn. (A.45) above leads to an interesting question. What commutation properties should we assign to these objects? There are two possibilities:

$$ij = \pm ji. \tag{A.47}$$

Interestingly, these two signs correspond to the representations of two different groups. When i and j commute, the w form a representation of the group $U(1) \times U(1)$ which corresponds to independent rotations about two orthogonal axes z and y, but no rotation about the third axis x, in a three-dimensional space. This result is, in fact, trivial from de Moivre's theorem.

When i and j anti-commute, the w form a representation of $SU(2)$, the group of three-dimensional rotations. To show this we must introduce some notation for complex conjugation with respect to the i and j parts. Let us denote

$$\overset{i}{w} = x - iy - jz$$
$$\overset{j}{w} = x + iy + jz$$
$$\overset{ij}{w} = x - iy + jz, \tag{A.48}$$

which have the following algebraic products:

$$w \overset{ij}{w} = x^2 + y^2 z^2 - 2\mathrm{ij} yz$$
$$\overset{i}{w}\overset{j}{w} = x^2 + y^2 + z^2 + 2\mathrm{ij} yz. \tag{A.49}$$

Thus, the length of a vector is

$$\frac{1}{2}(w \overset{ij}{w} + \overset{i}{w}\overset{j}{w}) = x^2 + y^2 + z^2 = \mathbf{w} \cdot \mathbf{w}, \tag{A.50}$$

and the scalar product is

$$\mathbf{w} \cdot \mathbf{W} = \frac{1}{4} \left(w \overset{ij}{W} + \overset{i}{w}\overset{j}{W} + \overset{j}{w}\overset{i}{W} + \overset{ij}{w} W \right). \tag{A.51}$$

It is interesting, and significant, that – concealed within these products are the vector and scalar products for Euclidean space. If we assume that i and j anti-commute, we have

$$(w, W) = \overset{i}{w}\overset{j}{W} = (xX + yY + zZ) + \mathrm{i}(xY - Yx)$$
$$-\mathrm{j}(zX - xZ) - \mathrm{ij}(yZ - zY)$$
$$= (\mathbf{w} \cdot \mathbf{W})\mathbf{1} + (\mathbf{w} \times \mathbf{W}), \tag{A.52}$$

where we have identified the complex numbers with Euclidean unit vectors as follows:

$$1 \leftrightarrow \text{scalars}$$
$$\mathrm{i} \leftrightarrow \hat{\mathbf{k}}$$
$$\mathrm{j} \leftrightarrow -\hat{\mathbf{j}}$$
$$\mathrm{ij} \leftrightarrow -\hat{\mathbf{i}}. \tag{A.53}$$

When the coupling between planes is unimportant, i and j commute and the power of this algebraic tool is maximal. An application of this method is given in section A.6.1.

A.6.1 Refraction in a magnetized medium

The addition of a magnetic field leads to the interesting phenomenon of plane wave rotation, studied in section 7.3.3. Neglecting attenuation, $\gamma = 0$, the forcing term can be written in the form of a general Lorentz force

$$m\frac{d^2\mathbf{s}}{dt^2} + k\mathbf{s} = -e\left(\mathbf{E} + \frac{d\mathbf{s}}{dt} \times \mathbf{B}\right); \tag{A.54}$$

or writing out the components and defining $\omega_0 = k/m$, $B = B_z$,

$$\frac{d^2 s_x}{dt^2} + \frac{e}{m} B \frac{ds_y}{dt} + \omega_0^2 s_x = -\frac{e}{m} E_x, \tag{A.55}$$

$$\frac{d^2 s_y}{dt^2} - \frac{e}{m} B \frac{ds_x}{dt} + \omega_0^2 s_y = -\frac{e}{m} E_y. \tag{A.56}$$

These two equations may be combined into a single equation by defining complex coordinates $s = s_x + i s_y$ and $E = E_x + i E_y$, provided ω_0 is an isotropic spring constant, i.e. $\omega_{0x} = \omega_{0y}$

$$\frac{d^2 s}{dt^2} - i \frac{e}{m} B \frac{ds}{dt} + \omega_0^2 s = -\frac{e}{m} E. \tag{A.57}$$

Plane polarized waves enter the medium, initially with their E vector parallel to the x axis. These waves impinge upon the quasi-elastically bound electrons, forcing the motion

$$s = \text{Re}\left[s_0 e^{j(kz - \omega t - \phi)}\right], \tag{A.58}$$

$$E = \text{Re}\left[E_0 e^{j(kz - \omega t)}\right], \tag{A.59}$$

where $j^2 = -1$, but $ij \neq -1$. We use j as a vector, orthogonal to i and to the real line. Re is the real part with respect to the j complex part of a complex number. Substituting for E and s, we obtain

$$\left[-\omega^2 + ij\omega \frac{eB}{m} + \omega_0^2\right] s_0 = -\frac{e}{m} E_0 e^{j\phi}. \tag{A.60}$$

The amplitudes s_0 and E_0 are purely real both in i and j. Comparing real and imaginary parts in j, we obtain two equations:

$$(\omega_0^2 - \omega^2)s_0 = -\frac{e}{m} E_0 \cos\phi, \tag{A.61}$$

$$i\frac{eB\omega}{m} s_0 = -\frac{e}{m} E_0 \sin\phi. \tag{A.62}$$

The phase ϕ is i complex, and this leads to rotation of the polarization plane – but this is not the best way to proceed. We shall show below that it is enough that the wavevector k be an i complex number to have rotation of the polarization plane vector E. To find k, we must find the dispersion relation for waves in a magnetized dielectric. It is assumed that the resonant frequency of the system is greater than the frequency of the electromagnetic waves ($\omega_0 > \omega$). For low-energy radiation, this is reasonable. Defining the usual relations

$$\mathbf{P} = -\rho_N e \mathbf{s}$$
$$\mathbf{D} = \mathbf{P} + \epsilon_0 \mathbf{E}$$
$$\mathbf{B} = \mu \mathbf{H}$$
$$c^2 = \frac{1}{\epsilon_0 \mu_0}, \tag{A.63}$$

from Maxwell's equations, one has that

$$\nabla^2 \mathbf{E} = \mu_0 \frac{\partial^2 \mathbf{D}}{\partial t^2} = \mu_0 \frac{\partial^2 \mathbf{P}}{\partial t^2} + \frac{1}{c^2} \frac{\partial^2 \mathbf{E}}{\partial t^2}. \tag{A.64}$$

Differentiating twice with respect to t allows one to substitute for s in terms of P and therefore E, so that eliminate s altogether to obtain a dispersion relation for the waves.

$$\left[\frac{\partial^2}{\partial t^2} - i \frac{eB}{m} \frac{\partial}{\partial t} + \omega_0^2 \right] \left(\nabla^2 \mathbf{E} - \frac{1}{c^2} \frac{\partial^2 \mathbf{E}}{\partial t^2} \right) = \frac{\mu_0 \rho_N e^2}{m} \frac{\partial^2 \mathbf{E}}{\partial t^2}. \tag{A.65}$$

For linearly polarized plane waves, it then follows that

$$\left(-\omega^2 + \mathrm{ij} \frac{eB}{m} \omega + \omega_0^2 \right) \left(-k^2 + \frac{\omega^2}{c^2} \right) E = -\frac{\mu_0 \rho_N e^2 E \omega^2}{m}. \tag{A.66}$$

This is the dispersion relation. The ij complex nature is a direct result of the coupling to the magnetic field. Re-arranging:

$$k^2 = \frac{\omega^2}{c^2} \left[1 + \frac{\mu_0 \rho_N e^2 c^2 / m}{(-\omega^2 + \mathrm{ij} \frac{eB}{m} \omega + \omega_0^2)} \right] \quad (\omega_0 > \omega). \tag{A.67}$$

The wavevector is therefore a complex number. Writing the wavenumber with real and imaginary parts separated:

$$k = k_\mathrm{r} - \mathrm{ij} k_\mathrm{i} \quad (k_\mathrm{i} > 0), \tag{A.68}$$

one can substitute back into the plane wave:

$$E = E_0 \cos(kz - \omega t) = \mathrm{Re}\, E_0\, e^{\mathrm{j}(kz - \omega t)}$$
$$= \mathrm{Re}\, E_0 \exp \mathrm{j}(k_\mathrm{r} z - \omega t + \mathrm{ij} k_\mathrm{i} z)$$
$$= E_0 \exp(\mathrm{i} k_\mathrm{i} z) \mathrm{Re} \exp(\mathrm{j}(k_\mathrm{r} z - \omega t))$$
$$E = E_0 \cdot \underbrace{[\cos(k_\mathrm{i} z) + \mathrm{i} \sin(k_\mathrm{i} z)]}_{\text{rotation} \, \propto \, z} \cdot \underbrace{\cos(k_\mathrm{r} z - \omega t)}_{\text{travelling wave}}.$$
$$\tag{A.69}$$

Thus, we have a clockwise rotation of the polarization plane.

A.7 Vector identities in $n = 3$ dimensions

For general vectors **A** and **B**, and scalar ϕ,

$$\nabla \cdot (\phi \mathbf{A}) = \phi (\nabla \cdot \mathbf{A}) + \mathbf{A} \cdot (\nabla \phi) \tag{A.70}$$

$$\nabla \cdot (\mathbf{A} \times \mathbf{B}) = \mathbf{B} \cdot (\nabla \times \mathbf{A}) - \mathbf{A} \cdot (\nabla \times \mathbf{B}) \tag{A.71}$$

$$\nabla \times \phi \mathbf{A} = \phi (\nabla \times \mathbf{A}) + (\nabla \phi) \times \mathbf{A} \tag{A.72}$$

$$\nabla \times \nabla \phi = 0 \tag{A.73}$$

$$\nabla \cdot (\nabla \times \mathbf{A}) = 0 \tag{A.74}$$

$$(\nabla \times (\nabla \times \mathbf{A})) = \nabla (\nabla \cdot \mathbf{A}) - \nabla^2 \mathbf{A}. \tag{A.75}$$

A.8 The Stokes and Gauss theorems ·

Stokes' theorem in three spatial dimensions states that

$$\int_R (\nabla \times \mathbf{A}) \cdot d\mathbf{S} = \oint_C \mathbf{A} \cdot d\mathbf{l}, \tag{A.76}$$

i.e. the integral over a surface region R of the curl of a vector, also called the flux of the curl of that vector, is equal to the value of the vector integrated along a loop which encloses the region.

The Gauss divergence theorem in three-dimensional vector language states that

$$\int_\sigma (\nabla \cdot \mathbf{A}) d\sigma_x = \int_S \mathbf{A} \cdot d\mathbf{S}; \tag{A.77}$$

i.e. the integral over a spatial volume, σ, of the divergence of a vector is equal to the integral over the surface enclosing the volume of the vector itself. In index notation this takes on the trivial form:

$$\int d\sigma \, \partial^i A_i = \int dS^i A_i, \tag{A.78}$$

and the spacetime generalization to $n + 1$ dimensions (which we use frequently) is

$$\int dV_x \, \partial^\mu A_\mu = \int d\sigma^\mu A_\mu. \tag{A.79}$$

Notice that Gauss' law is really just the generalization of integration by parts in a multi-dimensional context. In action expressions we frequently use the quantity $(dx) = dV_t = \frac{1}{c} dV_x$, whence

$$\int (dx) \, \partial^\mu A_\mu = \frac{1}{c} \int d\sigma^\mu A_\mu. \tag{A.80}$$

A.9 Integrating factors

Differential equations of the form

$$\frac{dy}{dx} + f(x)\, y = g(x) \tag{A.81}$$

can often be solved by multiplying through by a factor $I(x)$

$$I(x)\, \frac{dy}{dx} + f(x)\, I(x)\, y = g(x)\, I(x), \tag{A.82}$$

which makes the left hand side a perfect differential:

$$\frac{d(uv)}{dx} = u\, \frac{dV}{dx} + v\, \frac{du}{dx}. \tag{A.83}$$

Comparing these equations and identifying $u = I$ and $v = y$, one finds

$$\frac{dI}{dx} = I(x) f(x), \tag{A.84}$$

which solves to give

$$I(x) = \exp\left(\int_0^x f(x')\, dx' \right). \tag{A.85}$$

Thus the differential equation (A.81) may be written

$$\frac{d}{dx}\, (I(x)y) = g(x)\, I(x). \tag{A.86}$$

A.10 Matrix formulae

The so-called Baker–Campbell–Hausdorf identity for non-singular matrices A and B states that

$$e^{-A} B e^A = B + \frac{1}{1!}[B, A] + \frac{1}{2!}[[B, A], A] + \cdots. \tag{A.87}$$

A.11 Matrix factorization

A formula which is useful in diagonalizing systems is:

$$\begin{pmatrix} \Delta_1 & A \\ B & \Delta_2 \end{pmatrix} = \begin{pmatrix} \Delta_1 - A\Delta_2^{-1}B & A\Delta_2^{-1} \\ 0 & 1 \end{pmatrix} \begin{pmatrix} 1 & 0 \\ B & \Delta_2 \end{pmatrix}. \tag{A.88}$$

$$\det \begin{pmatrix} \Delta_1 & A \\ B & \Delta_2 \end{pmatrix} = \det(\Delta_1 - A\Delta_2^{-1}B)\det\Delta_2. \tag{A.89}$$

Appendix B
Recommended reading

- J.M. Cassels, *Basic Quantum Mechanics* (2nd edition). Macmillan Press, London (1970). An excellent summary of basic quantum mechanics.

- B. DeWitt, *Dynamical Theory of Groups and Fields*. Gordon and Breach, New York (1965). This demanding book contains deep insights into basic field theory, prior to the understanding of non-Abelian gauge theories. There is no other book like it. Metric conventions are the same as in this book.

- K. Huang, *Statistical Mechanics*. John Wiley and Sons, New York (1963). A classic book on statistical mechanics, which details the foundations of the subject, in a scholarly fashion, prior to the renormalization group era.

- H.F. Jones, *Groups, Representations and Physics* (2nd edition). Institute of Physics IoP Press, Bristol (1998). A very nice introduction to group theory for physicists, with much more attention to relevant detail than most group theory texts. A very nice summary of Dirac notation.

- S. Schweber, *Relativistic Quantum Field Theory*, Harper & Row, New York (1961). Although a little dated, this is still one of the most scholarly books on quantum field theory. It is one of the few books which answers more probing questions than it raises about the formulation of field theory. This book cannot be praised highly enough. The opposite metric signature is used.

- J. Schwinger, *Particles, Sources and Fields, Volume I*. Addison Wesley, Redwood, CA (1970). This book is Schwinger's motivation for, and treatise on, *source theory*, which is a formulation of effective quantum field theory. This is a classic work, which is full of important insights for the dedicated reader. The conventions are largely the same as those used here.

- J. Schwinger, L.L. DeRaad, K.A. Milton and W. Tsai, *Classical Electrodynamics*, Perseus, Reading MA (1998). A long awaited book on the Green function approach to classical electrodynamics. Alas, it uses old gaussian units, which can be confusing with regard to dimensions and factors of c. Notations otherwise resemble those used here.

- B. Schutz, *Geometrical Methods in Mathematical Physics*. Cambridge University Press (1980). A uniquely readable, and unpretentious, introduction to geometrical methods with carefully crafted examples.

- S. Weinberg, *Gravitation and Cosmology*. J. Wiley and Sons, New York (1972). An excellent introduction to the general theory of relativity and its influence on physics. The conventions used are the same as those used in this book.

- S. Weinberg, *Quantum Theory of Fields, Volume I*, Cambridge University Press (1995). A new book, which takes over where Schweber leaves off and one of the few books on quantum field theory which tries to explain what field theory is really about. A must for any field theorist. Conventions are similar to this book, but the Lagrangian functions differ by an overall sign.

References

[1] I.K. Affleck and J.B. Marston. *Phys. Rev.*, B**37**:3774, 1988.

[2] S.C. Anco and G. Bluman. *Phys. Rev. Lett*, **78**:2869, 1997.

[3] P.W. Anderson. *Phys. Rev.*, **130**:439, 1963.

[4] D. Arovas, J.R. Schrieffer, F. Wilczek, and A. Zee. *Nucl. Phys.*, B**251**:117, 1985.

[5] J. Bardeen, L.N. Cooper, and J.R. Schrieffer. *Phys. Rev.*, **108**:1975, 1957.

[6] V Bargmann and E.P. Wigner. *Proc. Natl. Acad. Sci.*, **34**:211, 1948.

[7] G. Benivegna and A. Messina. *J. Phys.*, A**27**:L625, 1994.

[8] H.A. Bethe and E.E. Salpeter. *Handbuch der Physik*, **XXXV/1**:(Springer Verlag, Berlin), 1957.

[9] B. Binegar. *J. Math Phys.*, **23**:1511, 1982.

[10] R. Brauer and H. Weyl. *Am. J. Math*, page 425, 1935.

[11] M. Burgess. Gauge invariance and disequilibrium in Chern–Simons theory. Proceedings of the third workshop on thermal fields and their applications (World Scientific, Singapore), 1993.

[12] M. Burgess. *Phys. Rev. Lett*, **72**:2823, 1994.

[13] M. Burgess. *Phys. Rev.*, D**52**:7103, 1995.

[14] M. Burgess and M. Carrington. *Phys. Rev.*, B**52**:5052, 1995.

[15] M. Burgess and B. Jensen. *Phys. Rev.*, A**48**:1861, 1993.

[16] M. Burgess, J.M. Leinaas, and O.M. Løvvik. *Phys. Rev.*, B**48**:12912, 1993.

[17] M. Burgess, A. McLachlan, and D.J. Toms. *Phys. Rev.*, D**43**:1956, 1991.

[18] M. Burgess, A. McLachlan, and D.J. Toms. *Int. J. Mod. Phys.*, A**8**:2623, 1993.

[19] M. Burgess and D.J. Toms. *Phys. Lett.*, B**252**:596, 1990.

[20] M. Burgess and D.J. Toms. *Phys. Rev. Lett.*, **64**:1639, 1990.

[21] M. Burgess and D.J. Toms. *Ann. Phys. (New York)*, **210**:438–463, 1991.

[22] W.K. Burton. *Nuovo Cimento*, **1**:355, 1955.

[23] G.G. Callan, S. Soleman, and R. Jackiw. *Ann. Phys. (N.Y.)*, **59**:42, 1970.

[24] S.M. Carroll, G.B. Field, and R. Jackiw. *Phys. Rev.*, D**41**:1231, 1990.

[25] P.A. Cherenkov. *C.R. Acad. Sci. (USSR)*, **3**:413, 1936.

[26] N.A. Chernikov and E.A. Tagirov. *Ann. Inst. Henri Poincaré*, **9**A:109, 1968.

[27] S. Coleman. *Commun. Math. Phys.*, **31**:264, 1975.

[28] S. Coleman and E. Weinberg. *Phys. Rev.*, **7**:1888, 1973.

[29] R.C. da Costa. *Phys. Rev.*, A**23**:1982, 1981.

[30] R.C. da Costa. *Phys. Rev.*, A**25**:2893, 1982.

[31] C.G. Darwin. *Proc. Roy. Soc. (London)*, A**118**:654, 1928.

[32] A.T. Davies and A. McLachlan. *Phys. Lett.*, B**200**:305, 1988.

[33] A.T. Davies and A. McLachlan. *Nucl. Phys.*, B**317**:237, 1989.

[34] P.C.W. Davies. *The Forces of Nature* (Cambridge University Press, Cambridge), 1979.

[35] S. Deser, R. Jackiw, and S. Templeton. *Ann. Phys.*, **140**:372, 1982.

[36] S. Deser, R. Jackiw, and S. Templeton. *Phys. Rev. Lett.*, **48**:975, 1982.

[37] R.H. Dicke. *Phys. Rev.*, **93**:99, 1954.

[38] P.A.M. Dirac. *Proc. Roy. Soc. (London)*, A**117**:610, 1928.

[39] G.V. Dunne, R. Jackiw, and C.A. Trugenberger. *Ann. Phys.*, **194**:197, 1989.

[40] G.V. Dunne, R. Jackiw, and C.A. Trugenberger. *Mod. Phys. Lett.*, A**4**:1635, 1989.

[41] F. Dyson. *Phys. Rev.*, A**75**:486, 1949.

[42] F. Dyson. *Am. J. Phys.*, **58**:209, 1990.

[43] S. Elitzer, G. Moore, A. Schwimmer, and N. Seiberg. *Nucl. Phys.*, B**326**:108, 1989.

[44] E. Eriksen and J.M. Leinaas. *Physica Scripta*, **22**:199–202, 1980.

[45] J. Fang and C. Fronsdal. *Phys. Rev.*, D**18**:3630, 1978.

[46] R.P. Feynman. Spacetime approach to non-relativistic quantum mechanics. *Rev. Mod. Phys.*, A**20**:267, 1948.

[47] R.P. Feynman. *Phys. Rev.*, A**76**:769, 1949.

[48] R.P. Feynman. *Phys. Rev.*, A**80**:440, 1950.

[49] R.P. Feynman. *Statistical Mechanics* (Addison Wesley, Redwood), 1972.

[50] R.P. Feynman, R. Leighton, and M. Sands. *The Feynman Lectures on Physics* (Addison Wesley, California), 1964.

[51] L.L. Foldy. *Phys. Rev.*, **87**:688, 1952.

[52] L. Ford. *Phys. Rev.*, D**21**:933, 1980.

[53] D. Forster. *Hydrodynamic Fluctuations, Broken Symmetry and Correlation Functions* (Addison Wesley, New York), 1975.

[54] E. Fradkin. *Field Theories of Condensed Matter Systems. Frontiers in Physics* (Addison Wesley, New York), 1991.

[55] C. Fronsdal. *Phys. Rev.*, D**18**:3624, 1978.

[56] J. Fuchs and C. Schwigert. *Symmetries, Lie Algebras and Representations* (Cambridge University Press, Cambridge), 1997.

[57] R. Gilmore. *Catastrophe Theory for Scientists and Engineers* (Dover, New York), 1981.

[58] D.M. Gitman and A.L. Shelpin. *J. Math Phys.*, Gen. **30**:6093, 1997.

[59] J. Goldstone. *Nuovo Cimento*, **19**:154, 1961.

[60] J. Goldstone, A. Salam, and S. Weinberg. *Phys. Rev.*, **127**:965, 1962.

[61] R.H. Good. *Rev. Mod. Phys.*, **27**:187, 1955.

[62] W. Gordon. *Z. Phys.*, **48**:11, 1928.

[63] I.S.Gradshteyn and I.M.Ryzhik. *Table of Integrals, Series and Products (5th edn)* (Academic Press, New York), 1994.

[64] C.R. Hagen. *Ann. Phys.*, **157**:342, 1984.

[65] F.D.M. Haldane and E.H. Rezeyi. *Phys. Rev.*, B**31**:2529, 1985.

[66] B.I. Halperin, T.C. Lubensky, and S. Ma. *Phys. Rev. Lett.*, **32**:292, 1974.

[67] P. Havas. *Nuovo Cimento (Suppl.)*, **5**:363, 1957.

[68] P.W. Higgs. *Phys. Lett.*, **12**:132, 1964.

[69] P.W. Higgs. *Phys. Rev. Lett.*, **13**:508, 1964.

[70] P.W. Higgs. *Phys. Rev.*, **145**:1156, 1966.

[71] V. Hnizdo. *Am. J. Phys.*, **65**:55, 1997.

[72] Y. Hosotani. *Phys. Lett.*, B**126**:309, 1983.

[73] L.D. Huff. *Phys. Rev.*, **38**:501, 1931.

[74] R. Hughes. *Am. J. Phys.*, **60**:301, 1992.

[75] E.A. Hylleraas. *Z. Phys.*, **140**:626, 1955.

[76] R. Jackiw. *Phys. Rev.*, D**41**:1635, 1978.

[77] R. Jackiw and V.P. Nair. *Phys. Rev. Lett.*, **43**:1933, 1991.

[78] R. Jackiw and S.Y. Pi. *Phys. Rev.*, D**42**:3500–3513, 1990.

[79] E.T. Jaynes and F.W. Cummings. *Proc. IEEE*, **51**:89, 1963.

[80] H. Jensen and H. Koppe. *Ann. Phys.*, **63**:586, 1971.

[81] M.H. Johnson and B.A. Lippmann. *Phys. Rev.*, **76**:828, 1949.

[82] R. Kubo. *J. Phys. Soc. Jap.*, **12**:570, 1957.

[83] M. G. G. Laidlaw and C. M. DeWitt. *Phys. Rev*, D**3**:1375, 1971.

[84] G. Landi and C. Rovelli. *Phys. Rev. Lett.*, **78**:3051, 1997.

[85] R.B. Laughlin. *Ann. Phys.*, **191**:163, 1989.

[86] I.D. Lawrie. *Nucl. Phys.*, B**200**[FS4]:1, 1982.

[87] G. Lüders. *Kong. Dansk. Vid. Selskab, Mat-Fys. Medd.*, **28**:5, 1954.

[88] G. Lüders. *Ann. Phys.*, **2**:1, 1957.

[89] J.M. Leinaas and J. Myrheim. *Nuovo Cimento*, B**37**:1, 1977.

[90] E.M. Lifshitz and L.P. Pitaevskii. *Physical Kinetics* (Pergamon Press, Oxford), 1981.

[91] P. Mandel and E Wolf. *Optical Coherence and Quantum Optics* (Cambridge University Press, Cambridge), 1995.

[92] J.B. Marston and I. Affleck. *Phys. Rev.*, B**39**:11538, 1989.

[93] P.C. Martin and J. Schwinger. *Phys. Rev.*, **115**:1342, 1959.

[94] S. Matsutani. *J. Phys. Soc. Jap.*, **61**:55, 1992.

[95] S. Matsutani and H. Tsuru. *Phys. Rev.*, A**46**:1144, 1992.

[96] W.G. McKay and J. Patera. *Tables of Dimensions, Indices and Branching Rules for Represemtations of Simple Lie Algebras* (Dekker, New York), 1981.

[97] N.D. Mermim and H. Wagner. *Phys. Rev. Lett.*, **17**:1133, 1966.

[98] N.F. Mott. *Proc. Roy. Soc. (London)*, A**124**:425, 1929.

[99] Y. Nambu. *Phys. Rev. Lett.*, **4**:380, 1960.

[100] Y. Nambu and G. Jona-Lasinio. *Phys. Rev.*, **122**:345, 1961.

[101] T.D. Newton and E.P. Wigner. *Rev. Mod. Phys.*, **21**:400, 1949.

[102] W. O'Reilly. *Rock and Mineral Magnetism* (Blackie, London), 1884.

[103] W. Pauli. *Ann. Inst. Henri Poincaré*, **6**:137, 1936.

[104] W. Pauli. *Rev. Mod. Phys.*, **13**:203, 1941.

[105] W. Pauli. *Nuovo Cimento*, **6**:204, 1957.

[106] I.I. Rabi. *Z. Phys.*, **49**:7, 1928.

[107] F. Reif. *Fundamentals of Statistical Mechanics* (McGraw-Hill, Singapore), 1965.

[108] L.H. Ryder. *Quantum Field Theory* (Cambridge University Press, Cambridge), 1985.

[109] F. Sauter. *Z. Phys.*, **69**:742, 1931.

[110] J. Schonfeld. *Nucl. Phys.*, B**185**:157, 1981.

[111] B. Schutz. *Geometrical Methods of Mathematical Physics* (Cambridge University Press, Cambridge), 1980.

[112] S. Schweber. *Relativistic Quantum Field Theory* (Harper and Row, New York), 1961.

[113] J. Schwinger. *Proc. Natl. Acad. Sci.*, A**37**:452, 1951.

[114] J. Schwinger. *Phys. Rev.*, **82**:664, 1951.

[115] J. Schwinger. *Phys. Rev.*, **82**:914, 1951.

[116] J. Schwinger. *J. Math. Phys.*, **2**:407, 1961.

[117] J. Schwinger. *Phys. Rev.*, **125**:397, 1962.

[118] J. Schwinger. *Phys. Rev.*, **152**:1219, 1966.

[119] J. Schwinger. *Particles, Sources and Fields (Volume 1)* (Addison Wesley, California), 1970.

[120] J. Schwinger. *Quantum Kinematics and Dynamics* (Addison Wesley, New York), 1991.

[121] J. Schwinger, L.L. DeRaad, K.A. Milton, and W. Tsai. *Classical Electrodynamics* (Perseus Books, Reading), 1998.

[122] J. Schwinger, Wu-Yang Tsai, and Thomas Erber. *Ann. Phys.*, **96**:303, 1976.

[123] C.E. Shannon and W. Weaver. *The Mathematical Theory of Communication* (University of Illinois Press, Urbana), 1949.

[124] R. Slansky. *Phys. Rep.*, **79**(1):1–128, 1981.

[125] K. Symanzik. *Zeit. f. Naturfor.*, **9a**:809, 1954.

[126] G. 't Hooft. *Nucl. Phys.*, B**153**:141, 1979.

[127] I.E. Tamm and I.M. Frank. *Dokl. Akad. Nauk. USSR*, **14**:109, 1937.

[128] D.J. Toms. Private communication.

[129] D.J. Toms. *Phys. Lett.*, B**126**:445, 1983.

[130] N.J. Vilenkin. *Trans. Math. Mono., Am. Math. Soc.*, **22**:447, 1968.

[131] D.M. Volkow. *Z. Phys.*, **94**:25, 1935.

[132] R.M. Wald. *Phys. Rev.*, D**33**:3613, 1986.

[133] S. Weinberg. *Gravitation and Cosmology* (J. Wiley and Sons, New York), 1972.

[134] S. Weinberg. *Physica*, **96**A:327, 1979.

[135] S. Weinberg. *Prog. Theor. Phys.*, **86**:43, 1986.

[136] S. Weinberg. *The Quantum Theory of Fields I,II,III* (Cambridge University Press, Cambridge), 1995–1999.

[137] E. Wigner. *Ann. Math.*, **40**:149, 1939.

[138] F. Wilczek. *Phys. Rev. Lett.*, **48**:3174, 1982.

[139] K.G. Wilson. *Phys. Rev.*, B **4**:1144, 1971.

[140] S. Wong. *Nuovo Cimento*, **65**A:689, 1970.

[141] C.N. Yang and R.L. Mills. *Phys. Rev.*, **96**:191, 1954.

Index

Printed in the United States
by Baker & Taylor Publisher Services